Ecophysiology of Tropical Crops

CONTRIBUTORS

PAULO de T. ALVIM
RAIMUNDO SANTOS BARROS
D. P. BARTHOLOMEW
JAKOB BENNEMA
JAN-DIRK FERWERDA
R. FORDHAM
S. K. HAHN
S. B. KADZIMIN
MOACYR MAESTRI
J. L. MONTEITH
VICENTE H. F. MORAES
G. O. MOTT
D. B. MURRAY
M. C. NAMBIAR
LOUIS G. NICKELL
H. L. POPENOE
WALTER REUTHER
L. B. SINGH
EGBERT A. TAI
L. A. WILSON
SHOUICHI YOSHIDA

Ecophysiology of Tropical Crops

Edited by

PAULO de T. ALVIM

Centro de Pesquisas do Cacau
Comissão Executiva do Plano da Lavoura Cacaueira
Itabuna, Bahia, Brazil

T. T. KOZLOWSKI

Russell Laboratories
Department of Forestry
University of Wisconsin
Madison, Wisconsin

ACADEMIC PRESS New York San Francisco London 1977

A Subsidiary of Harcourt Brace Jovanovich, Publishers

ACADEMIC PRESS, INC.
111 Fifth Avenue, New York, New York 10003

United Kingdom Edition published by
ACADEMIC PRESS, INC. (LONDON) LTD.
24/28 Oval Road, London NW1

Library of Congress Cataloging in Publication Data

Main entry under title:

Ecophysiology of tropical crops.

"Based to a large extent on material presented at an international symposium . . . held in Manaus, Brazil in May, 1975."
Includes bibliographies and index.
1. Tropical crops–Ecology–Congresses.
2. Tropical crops–Physiology–Congresses.
I. Alvim, Paulo de T. II. Kozlowski, Theodore Thomas, Date
SB111.A2E26 631′.0913 76-55967
ISBN 0-12-055650-2

PRINTED IN THE UNITED STATES OF AMERICA

Contents

Chapter 3 Rice

Shouichi Yoshida

Chapter 4 Sugarcane

Louis G. Nickell

Chapter 5 Pineapple

D. P. Bartholomew and S. B. Kadzimin

Chapter 6 Grasslands

G. O. Mott and H. L. Popenoe

Chapter 7 Root Crops

L. A. Wilson

Chapter 8 Sweet Potato

S. K. Hahn

Chapter 9 Coffee

Moacyr Maestri and Raimundo Santos Barros

Chapter 10 Cacao

Paulo de T. Alvim

List of Contributors

Numbers in parentheses indicate the pages on which the authors' contributions begin.

PAULO de T. ALVIM (279), Centro de Pesquisas do Cacau, Comissão Executiva do Plano da Lavoura Cacaueira, Itabuna, Bahia, Brazil

RAIMUNDO SANTOS BARROS (249), Divasão de Fisiologia Vegetal, Centro de Pesquisas do Cacau, Itabuna, Bahia, Brazil

D. P. BARTHOLOMEW (113), Department of Agronomy and Soil Science, University of Hawaii, Honolulu, Hawaii

JAKOB BENNEMA (29), Department of Soil Science and Geology, Agricultural University, Wageningen, The Netherlands

JAN-DIRK FERWERDA (351), Department of Tropical Crop Science, Agricultural University, Wageningen, The Netherlands

R. FORDHAM (333), Commodity Development Unit, Mars Ltd., Slough, England

S. K. HAHN (237), International Institute of Tropical Agriculture, Ibadan, Nigeria

S. B. KADZIMIN (113), Department of Agronomy and Horticulture, University of Agriculture Malaysia, Serdang, Selangor, Malaysia

MOACYR MAESTRI (249), Departmento de Biologia, Universidade Federal de Viçosa, Minas Gerais, Brazil

J. L. MONTEITH (1), Department of Physiology and Environmental Studies, University of Nottingham School of Agriculture, Sutton Bonington, Loughborough, Leics, England

VICENTE H. F. MORAES (315), Centro Nacional de Pesquisas da Seringueira, EMBRAPA, Manaus, Brazil

G. O. MOTT (157), Institute of Food and Agricultural Sciences, University of Florida, Gainesville, Florida

D. B. MURRAY (383), Faculty of Agriculture, University of the West Indies, St. Augustine, Trinidad, West Indies

M. C. NAMBIAR (461), Central Plantation Crops Research Institute, Kasaragod, Kerala, India

LOUIS G. NICKELL (89), Research Division, W. R. Grace & Co., Washington Research Center, Columbia, Maryland

H. L. POPENOE (157), Institute of Food and Agricultural Sciences, University of Florida, Gainesville, Florida

WALTER REUTHER (409), Department of Plant Science, University of California, Riverside, California

L. B. SINGH (479), Horticultural Research Institute, Saharanpur, Uttar Pradesh, New Delhi, India

EGBERT A. TAI (441), Ministry of Agriculture and Supply, St. John's, Antigua, West Indies

L. A. WILSON (187), Faculty of Agriculture, Department of Crop Science, University of the West Indies, St. Augustine, Trinidad, West Indies

SHOUICHI YOSHIDA (57), International Rice Research Institute, Manila, Philippines

Preface

The threat of impending food shortages at a time when population is increasing at an alarming rate emphasizes the importance of tropical crops to the health and economy of people within the developing tropical countries and elsewhere. Improving the productive potential of tropical crops on a sustained basis will depend on increased understanding of tropical climates and soils and the way in which environmental and cultural factors affect plant growth through the intermediation of physiological processes. These considerations, as well as a rapidly expanding body of research results, prompted our bringing together in a single volume the current state of knowledge and opinion on ecophysiology of the major tropical crop plants.

The opening chapter reviews fundamental ideas about numerical description of plant development and considers effects of climatic factors (e.g., temperature, light, and water) on physiological processes in plants. The second chapter presents an overview of physical and chemical characteristics of tropical soils. Separate chapters then follow on ecophysiology of the major crop plants, particularly those suitable for the wet tropics, including rice, sugarcane, pineapple, grasslands, root crops, sweet potato, coffee, cacao, rubber, banana, tea, oil palm, coconut palm, citrus, cashew, and mango.

The book is both authoritative and interdisciplinary. It should be useful to students and researchers, as well as those interested in production of tropical crops. The subject matter will be of interest to plant ecologists, plant physiologists, biochemists, horticulturists, agronomists, meteorologists, soil scientists, food technologists, and plant breeders.

This volume is based to a large extent on material presented at an International Symposium on Ecophysiology of Tropical Crops held in Manaus, Brazil in May, 1975. Invitations were extended to leading investigators concerned with growth and ecological requirements of tropical crops. Their contributions reflect many years of experience with tropical crops in Africa, Southeast Asia, Latin America, and the Caribbean Islands.

We thank the following organizations in Brazil for generous financial support and facilities for the Symposium: National Council for Scientific and Technological Development (CNPq.), Superintendency of the Free Zone of Manaus (SUFRAMA), the Executive Committee of the Plan for Cacao Production (CEPLAC), and the Inter-American Institute of Agricultural Sciences (IAIAS).

We also thank each author for his contribution, patience, and cooperation during the production phases. The senior editor expresses his deep gratitude to the following who assisted in organizing the Manaus Symposium: Luiz Carlos Cruz (CEPLAC), W. Martin Aitken (CEPLAC), Antonio Dantes Machado (CNPq.), and Marcilio Junqueira (SUFRAMA). Mr. T. L. Noland and Mr. R. J. Norby assisted with the preparation of the subject index.

Paulo de T. Alvim
T. T. Kozlowski

CHAPTER

1

Climate

J. L. MONTEITH

I. INTRODUCTION

How does the climate of a region determine what crop species will thrive? To what extent are yields determined by sequences of weather from month to month and from year to year? These are formidable questions, but they are central to agricultural meteorology; they have been attacked from many different directions; and they are still largely unanswered.

As we are concerned with environmental physiology of tropical crops, the effects of climate must be discussed with reference to specific physiological processes. The first section of this chapter is therefore concerned with some fundamental but simple points about the numerical description of plant development; subsequent sections will consider the effects of temperature, light, and water supply on growth and yield. By adopting this treatment, I have tried to avoid three major pitfalls of conventional crop ecology: (1) statistical treatments of crop–weather relations, which lack generality and usually conceal an ignorance of mechanisms; (2) conventional growth analysis in terms of net assimilation rates and relative growth rates, valid for single, isolated plants, but not for plant communities in which leaves are competing for light; and (3) complex dynamic models of crop growth, which are becoming so bulky and unwieldy that they may suffer the fate of the dinosaurs.

II. SPECIFICATION OF DEVELOPMENT

In mathematical terms, plants consist of an assembly of discrete units each of which has measurable attributes such as number, size, and weight. If the development of a plant is defined as the outcome of processes of differentiation and growth (Wareing and Phillips, 1970), the rate of differentiation can be expressed by an increase in primordial numbers, whereas the growth of an organ is expressed and measured by an increase in its size or weight. Cell division and expansion provide a basic example of the distinction between numbers and size, but it is applicable to any higher level of organization – to numbers of leaf primordia or florets, to lengths of stems, areas of leaves, or weights of tubers.

The physical dimensions of differentiation are number per unit time (T^{-1}), whereas the rate of growth of a single organ or a single plant will have dimensions of $L^n\ T^{-1}$ (where n = 1, 2, or 3 for length, area, or volume, respectively) or MT^{-1} (for weight). It is often more logical to express the growth rate of a crop on a field-area basis rather than on a plant basis in which case the area of foliage becomes a dimensionless leaf area index ($LL^{-1} = L^0$) and the dimensions of leaf expansion become $L^0\ T^{-1}$.

Weather—indeed, environmental factors generally—can determine the time at which a specific developmental process begins, the subsequent rate of development, and the time when the process stops. The study of developmental timing in relation to the calendar, otherwise known as phenology, is returning to fashion after many years of neglect (Lieth, 1974).

For the past 30 years, crop ecologists have been more concerned with measuring rates, particularly changes of weight per unit time related to transpiration and photosynthesis (see the volumes edited by Monteith, 1975a, 1976).

Changes in size (e.g., leaf areas) per unit time have usually been measured over periods which are too long to reveal the sensitivity of rates of expansion to temperature changes or water stress. If auxanometers had been developed as fast as infrared gas analyzers, crop ecology might now be in a much more balanced state.

Progress in the study of crop–weather relationships needs a much closer integration of phenological and physiological studies. We need to ask "when?" as well as "how fast?" Answers to both questions will emerge from studies of crop development in which numbers, sizes, and weights are measured as functions of time and environmental variables.

III. TEMPERATURE

A. Physical Aspects

Close to sea level, the mean monthly air temperature at most stations in the tropics is between 20° and 30°C in every month of the year; the decrease of temperature attributable to altitude is about 0.6°C per 100 m in the lowest 1.5 km (Lockwood, 1974). In some parts of the hot, wet region close to the equator, the annual range of monthly air temperature is only ±1°C and even in savanna regions, the difference in mean temperature between "cool" and "hot" months is rarely more than 7°C.

Mean temperatures recorded from day to day fluctuate around the monthly mean in response to changes in the synoptic pattern of weather, but these changes are usually less frequent and less abrupt in the tropics than in temperate latitudes. In terms of plant growth, the diurnal temperature cycle at many tropical stations is more important than either the regular seasonal cycle or the random effects of weather. For example, the mean *daily* temperature range at Manaus is ±3.2°C in April and ±4.7°C in September but the annual range of *monthly* mean temperatures is only ±1.2°C (Lamb, 1972). Whatever the cause, extremes of temperature occurring for a period of several days or even a few hours may severely reduce the growth of annual crops or even destroy perennial species such as coffee if grown in regions subject to occasional frost.

Even more important for plant processes are the effects of microclimate. Lal (1974) reported that growth of young maize plants is inhibited when the soil surface temperature exceeds 35°C, a condition not uncommon in the tropics, particularly when the soil surface is dry. Alvim found that the temperature of coffee leaves sometimes exceeded air temperature by 20°C (Chapter 9) and Franco reported that in Brazil, young coffee trees were exposed to damaging soil temperatures above 50°C (Chapter 9). Fruit and leaf temperature of pineapple may also exceed 50°C (Chapter 5). When tissue temperatures are likely to exceed the optimum value for any metabolic process, it is essential to take account of the microclimate of vegetation as well as the temperature regime at screen height.

In principle, the temperature of an organ exposed to sunlight can be estimated from its complete heat balance (see, e.g., Monteith, 1975b; Gates and Papian, 1971), but in practice, the rate at which heat is lost by evaporation of water is seldom known *a priori.* At best it is possible to calculate how the temperature of a leaf in a defined environment depends on the diffusion resistance of stomata; but even in this simple case, there is considerable uncertainty about the choice of an appropriate heat transfer coefficient (Clark and Wigley, 1975).

B. Physiological Aspects

Between the extremes of heat and cold which can damage tissue reversibly or irreversibly, most plant processes are strongly dependent on temperature over a range of 20° to 25°C. According to most plant physiology textbooks, rates of differentiation and growth are expected to obey the Arrhenius relation, i.e., to increase logarithmically with temperature. In a form of analysis which is consistent with this expectation, the dependence of a rate on temperature is often expressed as a Q_{10}—the increase in rate when temperature rises by 10°C. For many plant processes Q_{10} appears to be between 2 and 3 when the mean temperature of the experimental range is between 15° and 25°C. Close inspection of a large number of results, however, reveals that Q_{10} nearly always decreases with increasing temperature (James, 1953) and that the relation between development rate and temperature is often *linear rather than logarithmic.* A significant departure from linearity is sometimes observed when the development rate is less than 20% of the maximum rate observed at the optimum temperature; above this optimum the rate decreases rapidly with increasing temperature. Blacklow's measurements on maize (Fig. 1) are characteristic of the response to temperature shown by many other species and processes. The relation between germination rate and soil temperature appears to be particularly linear (Hegarty, 1973; Bierhuizen and Wagenvoort, 1974); several other examples of linear temperature relations can be found in this volume, e.g., growth of rice seedlings (Chapter 3) and the extension of oil palm leaves (Chapter 13).

The difference between the lowest temperature at which the rate of development is perceptible and the temperature at which it reaches a well-defined maximum is between 20° and 25°C for many processes both in temperate and tropical species. But, whereas most temperate plants grow between 5° and 30°C,

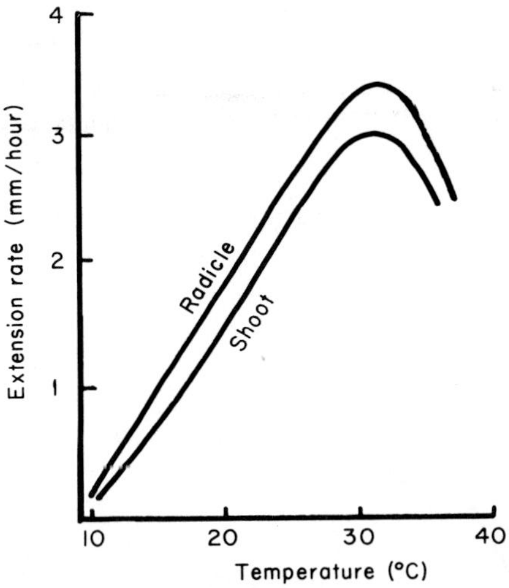

Fig. 1. Rate of extension of radicle and shoot of young maize plants as a function of growth at room temperature (after Blacklow, 1972).

the range for tropical species is often from 10° to 35° or 15° to 40°C. Examples will be found in later chapters on grasses and legumes, rice, sugarcane, pineapple, sweet potatoes, etc.

Although a simple linear relationship often fits rates of differentiation or growth which depend on cell division or expansion, the dependence of metabolic rates on temperature is usually more complex. In particular, the rate of photosynthesis of many C_3 plants reaches a maximum between 20° and 30°C and is almost independent of temperature over a range of 15° to 20°C. In contrast, photosynthesis in C_4 plants increases rapidly with leaf temperature up to a maximum rate between 30° and 40°C, and then at higher temperatures decreases rapidly (Fig. 2). The behavior of carboxylating enzymes provides one clue to this difference (Treharne and Cooper, 1969), but it is likely that the effect of temperature on stomatal opening is also involved. The temperature dependence of stomatal aperture is referred to in several other contributions (e.g., Chapters 5, 9, and 12). Björkman (1975) has emphasized the complexity of the relation between the rate of photosynthesis and temperature, and has drawn attention to the important role which membrane function may play in this context.

The dependence of respiration rate on temperature is usually assumed to have a Q_{10} of between 2 and 3 (James, 1953), but recent measurements by Ryle *et al.* (1976) suggest that the relation between respired CO_2 and temperature must depend on (1) the relative contribution from different metabolic processes and (2) the period over which respiration is measured.

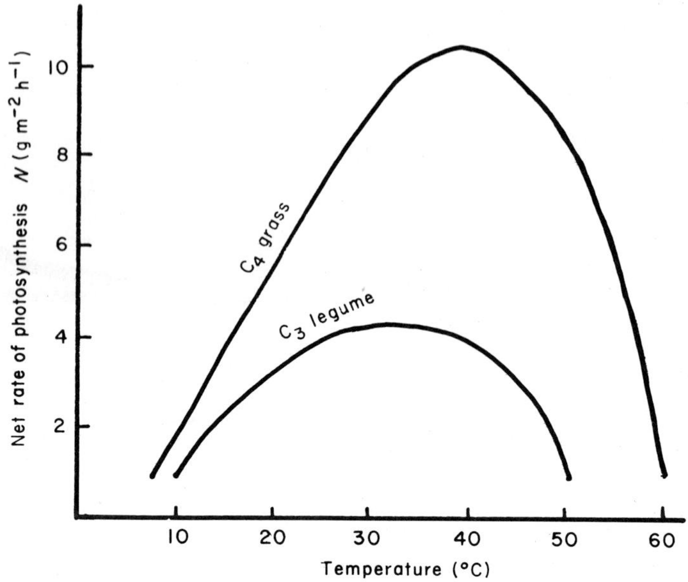

Fig. 2. Net rate of photosynthesis as a function of leaf temperature for two tropical pasture species: C_4 grass (Hamil grass, *Panicum maximum*) and C_3 legume (Calopo, *Calopogonium mucunoides*) (after Ludlow and Wilson, 1971).

Many other aspects of growth and differentiation proceed at rates which are strongly temperature dependent; reviews by Bierhuizen (1973) and Landsberg (1975) provide access to a large number of recent papers on this topic.

C. Day-Degrees

The existence of a linear relation between rate processes and temperature is an essential empirical basis for the use of day-degrees. Many crop ecologists seem to overlook the awkward incompatibility of day-degrees and the expression of temperature dependence by a Q_{10} or an activation energy. Lowry (1969) provides a good example of the confusion in the literature.

Suppose that the stage of development of a plant can be represented by a quantity X, which may be either a number related to a process of differentiation or a quantity, such as size or weight, related to growth. Then, the rate of development can be expressed as $\dot{X}$ and within the range where the rate is a linear function of temperature above a base of T_0

$$\dot{X} = A(T - T_0)$$

where A is a constant. Integration over a time interval τ then gives

$$X = A \int_0^\tau (T - T_0)\, dt = A(\bar{T} - T_0)\tau$$

or

$$X/\tau = A(\bar{T} - T_0)$$

where $\bar{T}$ is the mean temperature over τ.

If the mean rate of development X/τ is plotted against temperature as in Fig. 3, T_0 can be found from the intercept with the temperature axis and A from the slope. The quantity X/A is the number of day-degrees required to reach the stage of development specified by X.

As a special case of this analysis one specific phase of development, such as germination, can be represented by $X = 1$ so that the day-degree requirement for the phase is simply $1/A$. (When $X = 1$, the quantity τ, e.g., a time to germination, is often plotted against T to give a hyperbola from which it is impossible to extract either the base temperature or the day-degree requirement.)

As a relevant example of the utility of the day-degree concept, Alvim (Chapter 10) gives the mean number of days between the first setting of cacao pods and harvesting as

$$1/N = 1/\tau = (T - 9)/2500$$

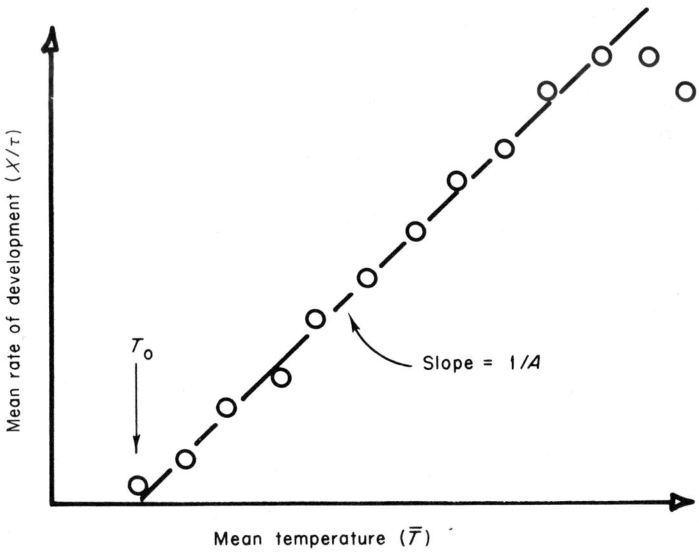

Fig. 3. Hypothetical relation between mean rate of development X/τ and temperature.

In this case $X = 1$ and $1/A = 2500$ day-degrees measured above a base temperature of 9°C.

In the preceding analysis, it was assumed implicitly that τ, a duration of development, was independent of temperature. Below an optimum, however, increasing temperature may shorten the duration of a specific phase of development as well as increasing the rate of growth or differentiation during that phase. It follows that the maximum number of organs differentiated or the maximum size achieved by a set of organs may either increase or decrease with increasing temperature, depending on whether it is the duration or the rate of growth which is most sensitive to temperature.

If the dependence of τ on temperature is assumed to obey the relation

$$1/\tau = B(\overline{T} - T_1)$$

where T_1 is a second base temperature and B is a constant, then

$$X = (A/B)(\overline{T} - T_0)/(\overline{T} - T_1)$$

It can be shown by differentiating this expression with respect to $\overline{T}$ that X will increase with temperature when $T_0 > T_1$ and vice versa (irrespective of the day-degree requirements defined by A and B). When $T_0 = T_1$, X will be independent of temperature. For example, when maximum leaf size is found to be independent of temperature, the base temperature for rate of expansion and for duration of expansion should be identical. On the other hand, Yoshida shows that the number of spikelets initiated on rice decreases with increasing temperature (Chapter 3) suggesting that the base temperature for the duration of initiation may be higher than for the rate of initiation.

The relation between development and day-degrees breaks down when temperature exceeds the optimum value for the process for a substantial part of the period over which it is integrated.

IV. LIGHT

A. Physical Aspects

1. Quantity

The annual income of solar radiation in the tropics ranges from about 8 GJ/m²* on the margins of the subtropical deserts to about 5 GJ/m² in wet equatorial regions. At most stations, the annual range in daily insolation, like the daily temperature amplitude, is much smaller than in temperate latitudes. For example, at Samaru, North Nigeria (11°N), the average income per day ranges

*1 GJ/m² = 10^9 J/m² = 23.9 kcal/cm².

from a minimum of about 17 MJ/m^2 in September to a maximum of about 24 MJ/m^2 in March. The mean diurnal variation of solar energy in the tropics, as in temperature regions, can be represented by a sine wave (Gloyne, 1972) and the maximum irradiance recorded under cloudless skies at noon is about 1.1 kW/m^2.*

2. *Quality*

In clean air, the fraction of solar energy in the photosynthetically useful waveband from 0.4 to 0.7 μm is usually close to 0.50 of the energy in the total solar spectrum (Unsworth and Monteith, 1972; Szeicz, 1974); it is probably safe to adopt this convenient round number throughout the tropics. The figure of 0.50 represents a weighted mean between the fraction for the direct solar beam (ca. 0.40 to 0.50) and diffuse blue skylight (0.60 to 0.70). Spectral balance is little affected by cloud (Szeicz, 1974).

Within the visible spectrum, the rate of photosynthesis of crop plants depends on the quantum flux density rather than the energy flux density (McCree, 1972), but the mean number of quanta per unit of solar energy is approximately constant. McCartney (1975) obtained measurements close to 4.60 microeinsteins (μE)† per joule of PAR and showed that almost the same figure was valid within crop canopies even though the spectral distribution of radiant energy was modified by scattering and absorption. It follows that quantum flux density in μE m^{-2} s^{-1} can be found by multiplying the irradiance of PAR in W/m^2 by 4.6.

3. *Duration*

Daylength, defined as the time from sunrise to sunset, is just over 12 hours throughout the year at the equator and ranges from 10.6 to 13.7 hours at latitude 25°. Although the range is much smaller than at higher latitudes, the timing of development in many tropical crops is determined by changes in daylength and several examples can be found in this volume, e.g., the flowering of grasses (Chapter 6) and the formation of tubers (Chapter 7). In some species, changes in daylength operate as on–off switches and determine when or whether a new phase of development will start. In other species, the photoperiodic mechanism acts as a proportional controller which can induce a graduated or "quantitative" response.

The phenomenon of photoperiodism has been reviewed extensively by Vince-Prue (1975) and succinctly, with special reference to tropical crops, by Williams and Joseph (1970).

*1 kW/m^2 = 1.43 cal cm^{-2} min^{-1}.

†1 Einstein (E) = energy content of 6.02 × 10^{23} quanta (i.e., Avogadro's number); 1 μE = energy content of 6.02 × 10^{17} quanta. PAR is "photosynthetically active radiation" i.e., in the waveband 0.4–0.7 μm.

B. Physiological Aspects

Apart from the photoperiodic control of development, the main function of light in plant growth is to provide energy for the photosynthetic system. Figure 4 illustrates the relation between CO_2 exchange and quantum flux density for a typical C_3 and C_4 plant (see also Chapter 6). In weak light, the relation for both species is almost linear implying that the rate of photosynthesis is limited exclusively by the absorption of light quanta. In strong light and for C_3 species, the rate of photosynthesis of young fully expanded leaves tends to a maximum value, but this maximum declines with age and the leaves of many cereals and grasses make a negligible contribution to carbon assimilation six to eight weeks after they emerge. C_4 species show less evidence of light saturation and age less rapidly.

These general features of photosynthesis in relation to irradiance can be expressed by two parameters to be considered separately: (1) the maximum *efficiency* of the photosynthetic system in weak light; (2) the maximum *rate* of photosynthesis in strong light.

1. Maximum Photosynthetic Efficiency

In principle, the efficiency of photosynthesis can be defined as the ratio of energy stored by the assimilation of carbon dioxide to the radiant energy

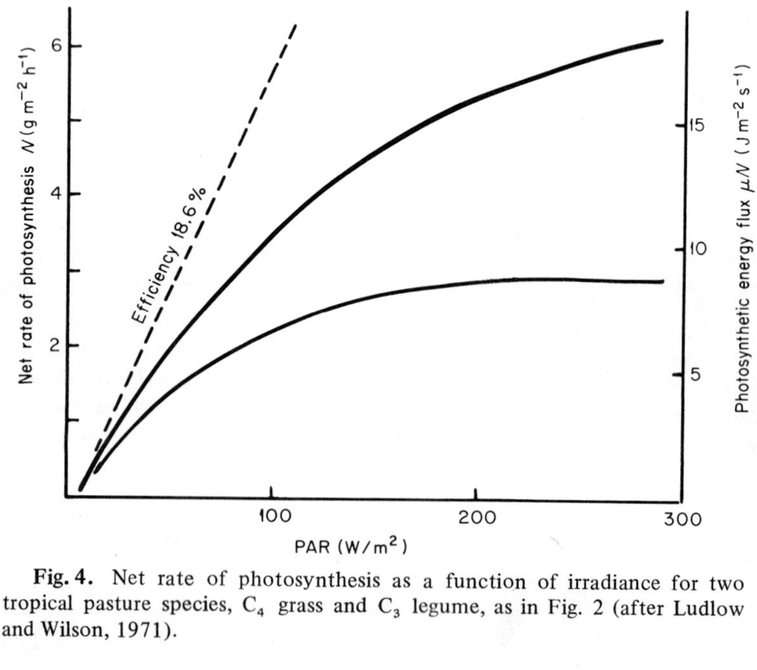

Fig. 4. Net rate of photosynthesis as a function of irradiance for two tropical pasture species, C_4 grass and C_3 legume, as in Fig. 2 (after Ludlow and Wilson, 1971).

absorbed by the photosynthetic system. In practice, neither part of this ratio can be measured directly. In the first place, the rate of photosynthesis, as measured, is usually the balance between a notional "true" or "gross" rate of photosynthesis and the contemporary respiration rate. When a component of respiration is coupled to photosynthesis, as in C_3 species, the associated gains and losses of CO_2 cannot be differentiated. In the second place, it is not possible to distinguish between the major fraction of light absorbed by the chloroplasts in a leaf and the minor fraction absorbed by other structures and organelles.

It is convenient, therefore, to redefine the maximum efficiency of photosynthesis in operational terms as the maximum of the differential quantity $\mu(dN/dI)$, where μ is the energy stored photochemically per gram of CO_2 reduced (10.8 kJ/g), N is the *net* assimilation rate of CO_2 (g m^{-2} s^{-1}), and I is the incident light flux (W/m^2). In Fig. 4, for example, the maximum efficiency achieved in weak light is given by the slope of the straight line portion of the light response curves (dN/dI) multiplied by the energy equivalent μ. The right-hand axis of the graph shows the value of μN directly.

For a system in which the respiration rate is independent of the rate of photosynthesis (but not necessarily zero in the light), the maximum value of dN/dI can be calculated theoretically as the product of four quantities: (a) the gram molecular weight of CO_2, 44 g/mole; (b) the number of Einsteins per unit energy of radiation (PAR), 4.6×10^{-6} E/J (see Section IV,A,2); (c) the quantum yield of the photochemical process, i.e., the number of moles of CO_2 assimilated per Einstein of absorbed radiation, often assumed to be 0.1 mole/E; (d) the fraction of incident energy absorbed by chloroplasts, assumed identical to the total absorbed fraction; 0.85 is an appropriate average figure for crop plants.

The product of these terms is

$$dN/dI = 44 \times (4.6 \times 10^{-6}) \times 0.1 \times 0.85 = 17.2\ \mu\text{g/J}$$

In more convenient units, an increase in PAR flux of 1 W/m^2 would be responsible for an increase in the rate of photosynthesis of 0.062 g CO_2 m^{-2} h^{-1} and the corresponding maximum efficiency of photosynthesis is

$$\mu(dN/dI) = (17.2 \times 10^{-6}) \times (10.8 \times 10^{3}) = 18.6\%$$

For C_3 plants, which are known to respire in light, the component of respiration coupled with photosynthesis must reduce the value of dN/dI below the maximum value achieved if the respiration rate were constant. For example, if photorespiration were 48% of the net rate of photosynthesis (a mean value for four species derived from a table given by Zelitch, 1975), the maximum efficiency would be reduced to 12.6% and if the radiant flux increased by 1 W/m^2, the rate of photosynthesis would increase by 0.048 g m^{-2} h^{-1}. These figures are consistent with measurements reported for a range of C_3 species and summarized by Monteith (1965).

2. *Maximum Rates of Photosynthesis*

In full sunlight, the leaves of C_3 plants usually behave as if the photosynthetic system was "light saturated," i.e., as if the rate of CO_2 assimilation was independent of the irradiance at a maximum value which is often between 3 and 5 g m^{-2} h^{-1} before the onset of senescence. The phenomenon of light saturation may be more aptly described as "CO_2 starvation" because the rate of operation of the photosynthetic system is limited by a shortage of CO_2 molecules. In the illuminated leaves of C_4 plants, the minimum intercellular concentration of CO_2 is close to zero, apparently because respiration is negligible, but in C_3 leaves the corresponding minimum value is often between 50 and 100 μl/liter. Using these minimum values and other information from gas exchange measurements, it is possible to estimate the CO_2 diffusion resistance for a leaf and distinguish components associated with the external boundary layer, stomatal pores, and mesophyll tissue. For C_3 leaves with a maximum rate of photosynthesis in bright light of 3 to 5 g m^{-2} h^{-1}, the corresponding total diffusion resistance is between 6 and 3 sec/cm. Provided stomata are wide open, the combined resistances of boundary layer and stomata usually account for about one-half of the total value. For C_4 plants, the minimum total resistance is usually less, say 2 to 3 sec/cm, attributable mainly to the boundary layer and stomata. The photosynthetic advantage of C_4 species can, therefore, be ascribed to a smaller mesophyll resistance, as well as to the absence of respiration in the light; but Gifford (1975) has argued that these advantages tend to disappear when the productivity of the two groups of plants is compared.

Attempts to distinguish a photochemical component of resistance (Monteith, 1963) or a carboxylation component (Chartier, 1970) have been unrewarding and may be misdirected. Prioul *et al.* (1975) found that when *Lolium* was grown at different levels of irradiance, the stomatal, mesophyll, and carboxylation resistances were *all* inversely related to the maximum photosynthetic rate and were correlated with several structural and metabolic variables.

3. *Photosynthesis and Production*

Although the analysis of photosynthetic rates in terms of resistances has not led to a deeper understanding of physiological processes, it has proved a useful way of summarizing and comparing results from different species; simple resistance models have been successfully used to estimate the rates of photosynthesis of whole canopies. Figure 5 shows an example in which the maximum rate of photosynthesis for three types of crops was expressed as a fraction of the rate expected if light were the only limiting factor. This fraction may be regarded as a measure of the limit to photosynthesis imposed by the finite diffusion resistance of leaves. As a further step in the same analysis, rates of dry matter production for a number of tropical crops were estimated on the assumption that 50% of the assimilated carbon dioxide was respired (Fig. 6). Estimates based

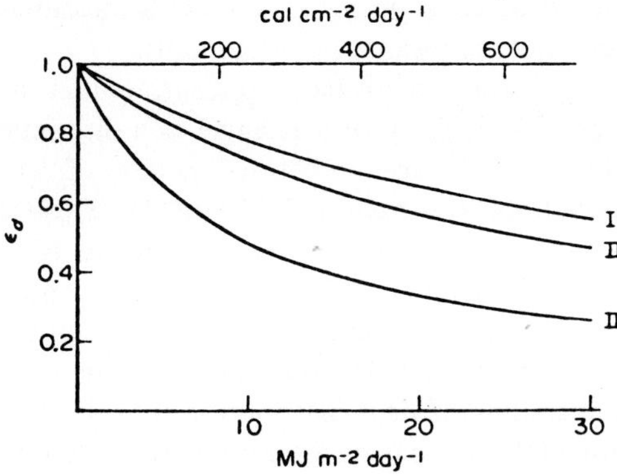

Fig. 5. Ratio of actual rate of photosynthesis for three types of crop canopy expressed as a fraction of the rate expected in the absence of light saturation (details given by Monteith, 1972).

Type		Total diffusion resistance (sec/cm)	Species
I	C_4	1.75	maize, sugarcane, sorghum, etc.
III	C_3	3.5	rice, wheat, barley, etc.
III	C_3	7.0	groundnuts, tobacco, cotton, etc.

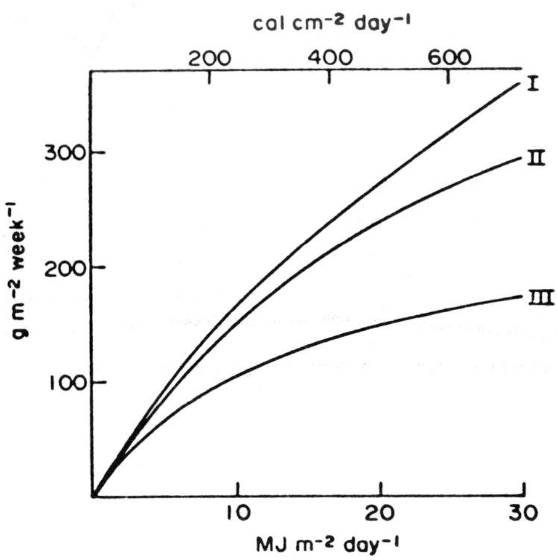

Fig. 6. Rate of dry matter production for three types of crop as specified in legend of Fig. 5. Complete ground cover was assumed and respiration was set at 50% of gross photosynthesis (Monteith, 1972).

on Fig. 6 were found to be in good agreement with measurements on maize, bulrush millet, sorghum, rice, soybeans, and groundnuts.

Many model studies of crop growth have concentrated on the significance of canopy architecture, and particularly of leaf angle as a discriminant of the rate of photosynthesis. In a recent comprehensive review of leaf inclination in relation to crop production, Trenbath and Angus (1975) concluded that the photosynthetic advantages of erect leaves are most likely to be expressed by densely sown stands of C_3 plants growing in the tropics and subtropics. The success of dwarf rices such as IR 8 supports this conclusion (Chandler, 1969); the significance of erectness in grasses is discussed in Chapter 6.

Many of the complications of light distribution models can be bypassed by exploiting the experimental fact that in the early stages of crop growth, rates of photosynthesis and dry matter production are very nearly proportional to the fraction of intercepted light (Monteith, 1972; Biscoe and Gallagher, 1976; Loomis and Gerakis, 1975). It follows that in the period before the canopy of a crop is complete, factors such as temperature and water stress, which govern the rate of leaf expansion, play a very important part in determining the rate of dry matter production of a stand and its photosynthetic efficiency.

Accounting for respiration is one of the major problems in estimating the dry matter production of crops from CO_2 exchange. McCree (1974) suggested the respiration rate of crop plants could be expressed as a linear function of the rate of photosynthesis (growth respiration) and of dry weight (maintenance respiration). Field measurements of respiration from a stand of barley were consistent with McRee's analysis (Biscoe *et al.*, 1975), but Ryle *et al.* (1976) have described a more detailed study in which maize and grass plants were kept in the dark after a period of illumination. Both components of respiration decreased with time but at different rates.

A second problem in estimating dry matter production is to make an appropriate allowance for the decrease in the rate of leaf photosynthesis with age, which depends on factors such as nutrition, water supply, and temperature. In most crop stands, it is this aspect of senescence, rather than the increasing respiratory load of nonphotosynthetic tissue, which slows and eventually stops the production of dry matter at the end of the growing season.

V. WATER

A. Physical Aspects

1. Heat Balance

The amount of water available to the roots of crop plants depends both on meteorological factors (the balance between rainfall and evaporation) and on soil factors (the relation between soil water content, water potential, and conductivity). As the physical properties of soil are dealt with in Chapter 2, this section is concerned with climatic aspects of water input and loss.

In most parts of the tropics, the distribution of rainfall during the year is much more variable than evaporation rate so that seasonal changes in soil water content are determined mainly by the rainfall pattern. This generalization is the basis of a simple classification of tropical climates into (a) wet equatorial regions where rainfall exceeds evaporation in every month of the year; (b) monsoon regions with one marked wet season in which rainfall exceeds evaporation and a dry season in which it is less; (c) savanna regions with one (or two) very dry seasons separated by wet seasons in which rainfall substantially exceeds evaporation.

To a good first approximation, the *maximum* rate of evaporation from crops or from soil over a period of say, one month, can be estimated from the net income of radiant energy over the same period. The latent heat of evaporation (λE) must be equal to the net input of energy from radiation (R_n) less sensible heat stored in the atmosphere and in the soil. During most of the day, soil and air temperatures both increase and this heating represents a positive storage of solar energy. During the night, soil and air cool because a large fraction of the heat stored during the day returns to the surface to compensate for the loss of long-wave radiation. Provided there is adequate water for evaporation from a surface of soil or vegetation, the net storage of heat over 24 hours is usually only a small fraction of the net radiation so that the loss of latent heat is nearly equal to the net gain of radiant energy, i.e., $\lambda E \triangleq R_n$. Figure 7 shows how closely this relation is observed throughout the year at Manaus. [The approximation $\lambda E \triangleq R_n$ is the basis of the "radiational index of dryness" used by some climatologists, i.e., the ratio of $\lambda P/R_n$ where P is precipitation in appropriate units (Budyko, 1974).]

There are two major exceptions to the general rule that the energy used for evaporation from vegetation should be close to the net heat available from radiation. First, when an area of rain-fed or irrigated vegetation is surrounded by dry territory, foliage may be cooled by evaporation below the temperature of the air passing over it. The atmosphere then becomes an additional source of heat for evaporation so that the daily mean value of λE may exceed the corresponding income of net radiation R_n. In a series of measurements reported by Rosenberg (1974), $\lambda E/R_n$ for an irrigated field was seldom more than 1.5, but even higher values are possible close to the upwind edge of an irrigated area. Second, when the evaporation rate is restricted by the supply of water to the root system rather than the supply of radiant energy to the tops, stomata tend to close and the rates $\lambda E/R_n$ may then assume any value between 0 and 1. In a few species such as pineapple, stomatal closure during the day is a normal state

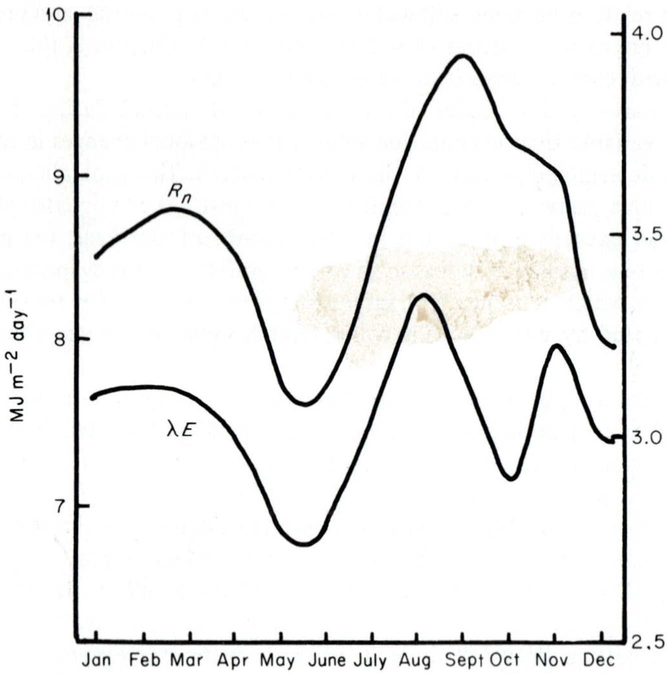

Fig. 7. Annual variation of net radiation (R_n) and latent heat of evaporation (λE) at Manaus, Brazil (after Lockwood, 1974). Note suppressed zero.

and the amount of water transpired per unit is only a small fraction of the potential evaporation rate (see Chapter 5).

2. *Water Balance*

In the tropics, the daily net income of radiation, expressed as a monthly mean value, ranges from a minimum of about 7 MJ/m^2 during the cloudiest part of the year near the equator to a maximum of about 12 MJ/m^2 during the dry season on the margin of the subtropical deserts. Over a large portion of the tropics, the annual mean income of daily net radiation is within ±20% of 9 MJ/m^2. Dividing by the latent heat of vaporization (2450 J/g) to obtain equivalent amounts of water, the corresponding range is approximately 3 to 5 mm per day with a mean of 4 mm/day similar to the summer maximum of temperate regions. Much higher rates of the order of 10 to 15 mm/day reported for irrigated crops in the semiarid tropics and subtropics are a consequence of the advection of hot, dry air.

By combining measurements of monthly rainfall with estimates of evaporation at the same station, it is possible to construct diagrams such as Fig. 8 or

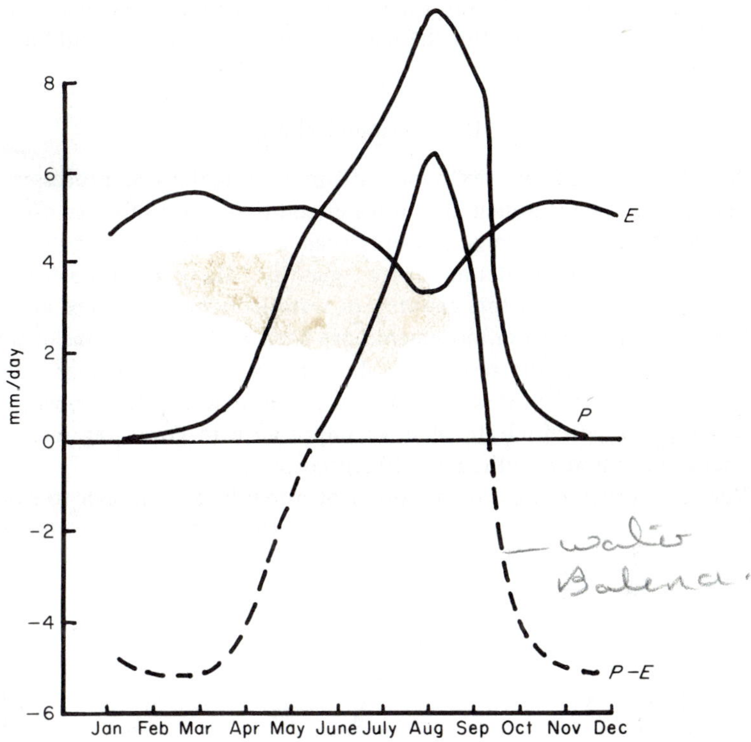

Fig. 8. Annual variation of rainfall (*P*), potential evaporation (*E*), and excess of rainfall over precipitation (*P–E*) at Samaru, North Nigeria (after Bunting, 1975).

regional maps which reveal the distribution of water surpluses and deficits through the year. Davies and Robinson (1969) presented maps for the whole of Africa and showed how well the water balance in Nigeria was correlated with distribution of natural vegetation.

In general, the length of the growing season in the tropics is determined by the period in which (1) the mean monthly rainfall exceeds about 150 mm/month (approximately the maximum evaporation rate) and (2) the distribution of rainfall is reliable enough from year to year and uniform enough within each season to avoid damaging droughts. Once a crop is established, roots have access to water stored in the soil which provides a buffer against periods of restricted rainfall and also extends the growing period beyond the date at which the rains stop. The size of the store depends on the physical composition of the soil and the extent of the root system. Roots in the top meter of a light sandy soil may be able to extract only 5 cm of stored water equivalent to 10 days of evaporation at 5 mm/day. But a meter of clay soil could sustain evaporation for a month at the same rate and the addition of organic matter could increase this duration.

B. Physiological Aspects

All plant physiology textbooks affirm that metabolic processes in green plants depend on the availability of water and that water shortage can restrict or stop growth. Unfortunately, concepts such as "availability" or "shortage" are extremely difficult to define or to quantify. Many physiologically minded plumbers have explored the anatomy of water-conducting tissues in crop plants and the dynamics of water movement have been analyzed in terms of free-energy potentials and hydraulic resistances. But it is still far from clear how the free energy of water in leaves is related to the rate of water flow or which components of water potential are of most importance in governing rates of photosynthesis, extension growth, and differentiation.

Very slight water stress of the order of a few bars is enough to slow or stop cell division and expansion (Fig. 9) (Slatyer, 1969; Hsaio *et al.,* 1976). When a

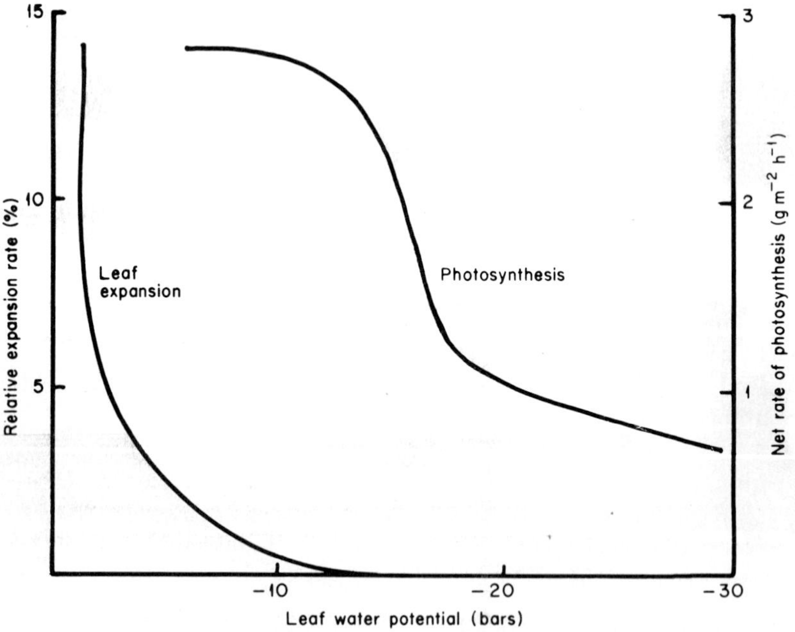

Fig. 9. Rates of leaf expansion and net rate of photosynthesis of soybean as a function of leaf water potential. Expansion is expressed as a fraction of the maximum rate in the absence of stress (after Boyer, 1970).

plant is exposed to increasing water stress as a result of a lack of rain or irrigation, the differentiation of new organs and the expansion of existing organs are affected first. With further stress, rates of photosynthesis are reduced by a combination of stomatal closure, increasing mesophyll resistance, and a decrease in the efficiency of the photosynthetic system (Slatyer, 1969). The evidence reviewed by Slatyer suggests that although a rate of primordial initiation may be slowed by moderate stress, it may later be accelerated when the stress is removed so that the final primordial number is almost unaffected by the episode. Processes such as leaf extension which depend on cell enlargement are more likely to be permanently affected by a period of stress. However, the number of metabolic processes which are sensitive to water stress is so large and their integration in plant growth is so complex that generalizations are dangerous. Modeling the effects of stress is inappropriate because there are so many unknowns; but a common framework of ideas is needed to integrate and reconcile results from the large volume of work reviewed by Salter and Goode (1967). This volume contains many examples of the effects of stress, e.g., on the survival of cherelles in cacao (Chapter 10), on floral abortion in the oil palm (Chapter 13), on expansion of coffee leaves (Chapter 9), and on the determination of grain number in upland rice (Chapter 3).

C. Ecological Aspects

In the last two sections, the energetics of evaporation and the response of plants to water stress were discussed independently and, at least until recently, field studies of plant–water relations have been inhibited by a similar dichotomy of concepts and techniques. Much closer integration of physical and physiological studies has now become possible, largely through the development of convenient instruments for measuring water potential and stomatal diffusion resistance in the field. Related papers by Biscoe *et al.* (1976), Elston *et al.* (1976), and Jarvis (1976) show how changes in the water relations of vegetation can be followed through the growing season and how they can be analyzed to quantify the response of plants to water stress. Although work of this type is still at a relatively primitive stage and knowledge is fragmentary (Elston, 1975), two general points are worth summarizing: (1) a moderate degree of stress is inescapable even when plants are growing in wet soil; and (2) field-grown plants have the ability to adapt to an increasing shortage of water.

During darkness, the water loss from crop plants is usually very slow because stomata are closed and the saturation deficit of the air is small. In consequence there is little difference in water potential between roots and leaves. When transpiration begins after sunrise, a significant amount of water is withdrawn from storage in mesophyll cells and in other parts of the plant, and by this means gradients of water potential are established which maintain a flow of water from the soil through the plant to the atmosphere. Oertli (1969) and others have emphasized that a decrease in leaf water content is a prerequisite of water movement rather than an indication that roots cannot absorb enough water to satisfy the demand for transpiration. Figures for the substantial amounts of water stored in plant tissues were recently reviewed by Jarvis (1975); in general, a loss of stored water can maintain transpiration from a field crop for several hours.

As a result of the contribution of cellular water to transpiration, wet soil is not an insurance against development of water stress in crop plants. On a sunny day when the mean transpiration rate is 0.3 mm/hour or more, the leaf water potential may fall below −5 bars or even below −10 bars. The imprecision of these figures is unimportant. If (negative) water potentials of a few bars are enough to affect cellular growth and metabolic processes generally, then the retardation of development by stress must be a normal daytime phenomenon in most crop plants grown in the tropics where high temperatures and strong insolation maintain rapid transpiration throughout the growing season. Stress is even more severe when an irrigated crop is surrounded by dry land, with the result that advected heat supplies energy for additional evaporation. As rainfall over a whole region removes this advected energy, the stress developed in a rainfed crop should be significantly less than in an irrigated crop growing in soil at an identical water content. By the same argument, the beneficial effects of irrigation are maximal when water is applied over a large area such as the Sudan Gezira.

Adaptation to a persistent water stress can take many different forms—the preferential growth of roots at the expense of tops, partial closure of stomata which may become complete under severe stress, and an increase in the osmotic potential of leaf cells which appears to play a very important part in maintaining turgidity. For example, Hsiao *et al.* (1976) reported how the solute potential of leaves from an unirrigated stand of sorghum decreased during the season in parallel with the decrease in total potential so that the midafternoon turgor pressure was maintained at about 5 bars (Fig. 10). In unirrigated wheat, turgor potential actually increased during the afternoon. Less marked but significant diurnal changes of osmotic potential have been reported for other species, e.g., wheat by Biscoe *et al.* (1976) and sugar beet by Biscoe (1972). The main cause appears to be an accumulation of sugars as a result of photosynthesis, with a minor contribution from diurnal changes in cell water content. The need for bright light to promote rapid photosynthesis may be one of several reasons why room-grown plants often appear much more sensitive to water stress than field-grown plants of the same species and age.

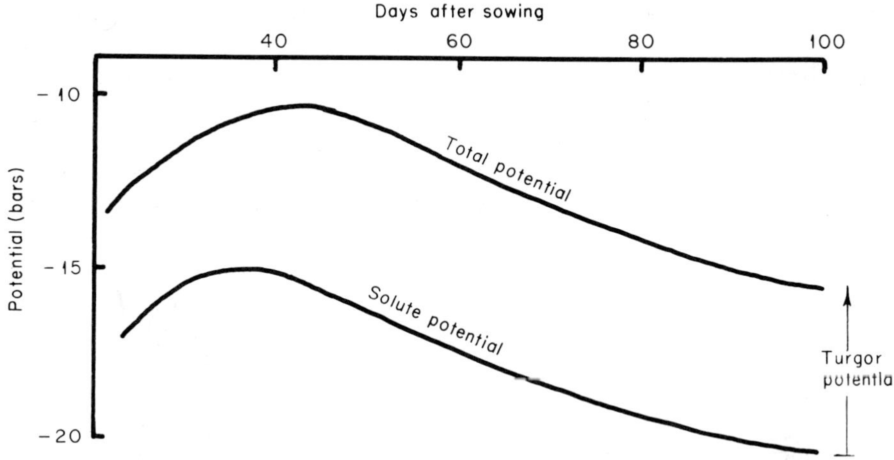

Fig. 10. Seasonal change of total, solute, and turgor potentials in unirrigated sorghum grown in California. Measurements on leaves between 1200 and 1500 hours local time (after Hsiao *et al.*, 1976).

VI. CONCLUSIONS

A. Integration

The progress of a crop plant from germination to maturity depends on the interplay of genetic and environmental factors which determine the timing and rate of developmental process. In Section II, some elementary quantitative aspects of development were briefly considered; the following three sections dealt with the effects of temperature, light, and water on development as if they acted separately. The scheme presented in this final section is an attempt to integrate quantitative aspects of development and related interacting factors.

In Fig. 11, the ordinate is time and the abscissa is a measure of development represented by X, which may be a number or a size or a weight. The points on the figure are the measurements from a hypothetical experiment in which X was measured at regular time intervals. It would be possible to fit a sigmoid curve through these points and the appropriate mathematical functions were discussed in great detail by Richards (1969) with specific reference to leaf growth. A more straightforward approach is to note that, provided the environment does not change, a large fraction of the total increase in X usually occurs at an almost

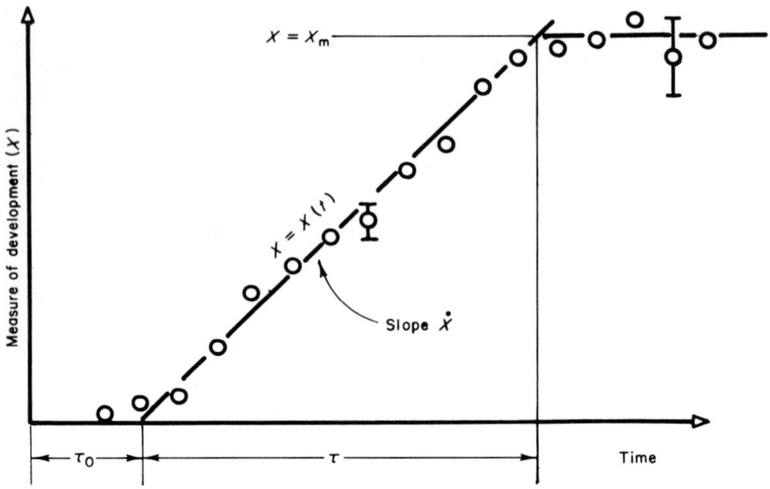

Fig. 11. Hypothetical relation between stage and development measured in terms of a quantity X and time. Error bars are indicated for two points only.

constant rate, i.e., X is a linear function of time (see Gallagher *et al.,* 1976). Two straight lines can now be drawn, one representing the increase in development with time [$X = X(t)$] and the second representing maximum development ($X = X_m$). The two lines define three quantities: (a) a mean rate of development $\bar{\dot{X}}$ [from the slope of $X = X(t)$]; (b) the apparent time τ_0 that elapses before development starts [from the intercept of $X = X(t)$ with $X = 0$]; (c) the apparent duration of development τ [from the intercept of $X = X(t)$ with $X = X_m$ such that $X_m = \tau\bar{\dot{X}}$].

The advantages of this scheme are that the values of the variables $\bar{X}$, τ_0, and τ can be readily determined by inspection and can then be related to the state of the environment. When one environmental factor changes systematically during development, it is sometimes possible to identify related changes in the value of $\dot{X}$.

When X represents a number of organs of a particular class, an extension of the scheme is sometimes needed to account for death occurring after the maximum number of organs has been initiated. The phenomenon of cherelle wilt in cacao is an extreme example (Chapter 10). If the number of surviving organs is represented as X_f, the number of deaths $X_m - X_f$ is often determined by the state of the environment.

Figure 11 can be applied either to differentiation (X–number) or to growth (X–size or weight). For a process of differentiation, the main environmental controls which determine the variables of the development curve are

τ_0 Daylength, providing the stimulus for a new phase of development; also environmental factors determining the rate at which previous phases of development were completed, particularly temperature and water supply.

$\bar{\dot{X}}$ Temperature of and water supply to meristematic tissue determining the rate of differentiation of primordial cells; daylength when the photoperiodic response is quantiative.

τ Temperature and water status of meristematic tissue; daylength.

X_m In some circumstances, the value of X_m $(= \tau\bar{\dot{X}})$ will depend on environmental factors through the individual values of τ and $\bar{\dot{X}}$. However, the phenomenon of plasticity implies the existence of a range of environmental states in which the value of X_m for a particular genotype is an expression of its genetic potential. The environmental control of $\bar{\dot{X}}$ and τ must then be complementary. For example, if an *increase* of temperature increases $\bar{\dot{X}}$ by 10%, τ must *decrease* by 10% to keep $\tau\bar{\dot{X}}$ constant. As shown in Section III,C, this type of complementarity implies the same base temperature for $\bar{\dot{X}}$ and $1/\tau$.

X_f Environmental factors such as high temperature, water stress, or weak light may be responsible for the death of organs which have started to grow. The death of tillers, florets, immature fruits, etc., has often been recorded in field experiments and has been ascribed to environmental factors. Quantitative analyses are rare.

For a process of growth measured as an increase of size or weight, it is necessary to distinguish between the activity of sources and sinks of assimilate (see Chapter 8). When the growth of an individual organ is limited by its demand for assimilates ("sink strength") the main environmental controls will be

τ_0 Previous temperature and water supply; daylength in cases of dormancy.

$\bar{\dot{X}}$ Temperature of and water supply to tissue determining the rate of division and expansion of cells.

τ Temperature and water status of tissue.

X_m See comments for differentiation. Within the limits of plasticity X_m may be independent of environmental factors because the response of $\bar{\dot{X}}$ and τ is complementary. Outside the limits, X_m may depend on temperature and water stress.

X_f At the end of the growth phase, the weight of a fruit or other organ may decrease slowly as a result of respiration at a rate determined by temperature. It has also been pointed out that the total weight of a crop plant would decrease if it produced more respiring tissue than it could support by photosynthesis, e.g., if leaf area index exceeded some hypothetical "optimum" value. Transitory negative growth rates may be associated with periods of adverse weather but substantial losses of dry matter recorded before harvest are likely to be the result of abscission or predation rather than respiration.

When the growth of an organ is limited by the rate of assimilate supply ("source strength"), the rate of growth will depend mainly on the factors which govern the photosynthesis rate. In principle, irradiance is the main determinant of the photosynthetic rate, but in tropical climates there is often little variation in insolation from day to day. Mean temperature is also relatively constant. In practice, therefore, water stress will often be the main climatic discriminant either for the increase in weight of an organ or for the rate of dry matter production of a whole plant or a crop stand. The duration of the period for which assimilate is available will depend on the rate of senescence of the organ or organs responsible for photosynthesis and this rate may be determined by temperature and/or water status. For perennial tree crops, daylength may also be implicated in senescence.

B. Omissions

Although Fig. 11 provides a framework for discussing the environmental control of development in very general terms, several important topics have been omitted. In particular, the effects of wind referred to in several later chapters have not been incorporated in the scheme, mainly because the mechanisms of wind damage are not well understood. Some effects of wind are visible by inspection, e.g., serious damage to trees or bending of sugarcane leaves (Chapter 4) or the dissection of banana leaves. Damage to the cuticle, invisible to the naked eye, may increase the loss of water from leaves and so increase water stress (Grace and Thompson, 1973). There are also several reports of growth being slowed by vibration but the mechanism is obscure. Studies of the damage which wind causes to horticultural crops were reviewed by Waister (1972) and a more recent review by Sturrock (1975) deals with the theory and practice of providing shelter for crops.

A second omission is the effect of temperature on the *quality* of the harvested product referred to in the chapters dealing with pineapple, tea, and citrus fruits. For these and other species, the economic significance of temperature in relation to quality is well recognized but systematic field studies are rare, and the associated biochemistry has not been explored. By analyzing measurements over a series of years, Smith (1973) was able to correlate the α-acid content of hops growing in northern Europe with the mean temperature over the final stages of growth and he identified an optimum temperature at which acid content reached a maximum. Day-to-day variations in acid content could also be correlated with short-term changes in temperature so that the best quality (most acid) was obtained when the hops were picked after a spell of warm days. Although hops are grown in an environment more thermally variable than the tropics, the success of this analysis underlines the need for more detailed studies on the chemical composition of material harvested.

REFERENCES

Bierhuizen, J. F. (1973). The effect of temperature on plant growth. *In* "Plant Response to Climatic Factors" (R. O. Slatyer, ed.), pp. 89–98. UNESCO, Paris.

Bierhuizen, J. F., and Wagenvoort, J. (1974). Some aspects of seed germination in vegetables. 1. The determination and application of heat sums and minimum temperature for germination. *Sci. Hort.* **2,** 213–219.

Biscoe, P. V. (1972). The diffusion resistance and water status of leaves of *Beta vulgaris. J. Exp. Bot.* **23,** 930–940.

Biscoe, P. V., and Gallagher, J. N. (1976). Weather, dry matter production and yield. *In* "Environmental Effects on Crop Physiology" (J. J. Landsberg and C. V. Cutting, eds.). Academic Press, London and New York, in press.

Biscoe, P. V., Scott, R. K., and Monteith, J. L. (1975). Barley and its environment. III. Carbon budget of the stand. *J. Appl. Ecol.* **12,** 269–293.

Biscoe, P. V., Cohen, Y., and Wallace, J. S. (1976). Daily and seasonal changes of water potential in cereals. *Philos. Trans. R. Soc. London, Ser. B* **273,** 565–580.

Björkman, O. (1975). Environmental and biological control of photosynthesis. *In* "Environmental and Biological Control of Photosynthesis" (R. Marcelle, ed.), pp. 1–16. Junk, The Hague.

Blacklow, W. M. (1972). Influence of temperature on germination and elongation of the radicle and shoot of corn. *Crop Sci.* **12,** 647–650.

Boyer, J. S. (1970). Leaf enlargement and metabolic rates in corn, soybean and sunflower at various leaf water potentials. *Plant Physiol.* **46,** 233–235.

Budyko, M. I. (1974). "Climate and Life." Academic Press, New York.

Bunting, A. H. (1975). Time, phenology and the yields of crops. *Weather* **30,** 312–325.

Chandler, R. F. (1969). Plant morphology and stand geometry. *In* "Physiological Aspects of Crop Yields" (J. D. Eastin *et al.,* eds.), pp. 265–285. Am. Soc. Agron., Madison, Wisconsin.

Chartier, P. (1970). A model of CO_2 assimilation in the leaf. *In* "Prediction and Measurement of Photosynthetic Activity" (I. Setlik, ed.), pp. 307–315. Pudoc, Wageningen.

Clark, J. A., and Wigley, G. (1975). Heat and mass transfer from real and model leaves. *In* "Heat and Mass Transfer in the Biosphere" (D. A. de Vries and N. H. Afgan, eds.), pp. 413–422. Wiley, New York.

Davies, J. A., and Robinson, P. 1969. A simple energy balance approach to the moisture balance climatology of Africa. *In* "Environment and Land Use in Africa" (M. F. Thomas and G. W. Whittington, eds.), pp. 23–56. Methuen, London.

Elston, J. (1975). The role of water in plants. *In* "Progress in Biometeorology: Division C, Progress in Plant Biometeorology" (L. P. Smith, ed.), pp. 51–57. Swets & Zeitlinger, Amsterdam.

Elston, J., Karamanos, A. J., Kassam, A. H., and Wadsworth, R. M. (1976). The water relations of the field bean crop. *Philos. Trans. R. Soc. London, Ser. B* **273,** 580–592.

Gallagher, J. N., Biscoe, P. V., and Scott, R. K. (1976). Barley and its environment. VI. Growth and development in relation to yield. *J. Appl. Ecol.* **13** (in press).

Gates, D. M., and Papian, V. N. (1971). "Atlas of Energy Budgets of Plant Leaves." Academic Press, New York.
Gifford, R. (1975). A comparison of potential photosynthesis, productivity and yield of plant species with differing photosynthetic metabolism. *Aust. J. Plant Physiol.* **1**, 107–117.
Gloyne, R. (1972). The diurnal variation of global radiation. *Meteorol. Mag.* **101**, 44–51.
Grace, J., and Thompson, J. R. (1973). The after effect of wind on the photosynthesis and transpiration of *Festuca arundinacea. Physiol. Plant.* **28**, 541–547.
Hegarty, T. W. (1973). Temperature coefficient (Q_{10}) of seed germination and other biological processes. *Nature (London)* **243**, 305–306.
Hsiao, T. C., Acevedo, E., Fereres, E., and Henderson, D. W. (1976). Water stress, growth, and osmotic adjustment. *Philos. Trans. R. Soc. London, Ser. B* **273**, 479–500.
James, W. O. (1953). "Plant Respiration." Oxford Univ. Press, London and New York.
Jarvis, P. J. (1975). Water transfer in plants. *In* "Heat and Mass Transfer in the Biosphere" (D. A. de Vries and N. H. Afgan, eds.), pp. 369–394. Wiley, New York.
Jarvis, P. G. (1976). The interpretation of variations in leaf water potential and stomatal conductance in the field. *Philos. Trans. R. Soc. London, Ser. B.* **273**, 593–600.
Lal, R. (1974). Effects of constant and fluctuating soil temperature on growth, development, and nutrient uptake of maize seedlings. *Plant Soil* **40**, 589–606.
Lamb, H. H. (1972). "Climate: Past, Present, and Future." Methuen, London.
Landsberg, J. J. (1975). Temperature effects and plant response. *In* "Progress in Biometeorology: Division C, Progress in Plant Biometeorology" (L. P. Smith, ed.), pp. 86–107. Swets & Zeitlinger, Amsterdam.
Lieth, H., ed. (1974). "Phenology and Seasonality Modeling." Springer-Verlag, Berlin and New York.
Lockwood, J. G. (1974). "World Climatology." Arnold, London.
Loomis, R. S., and Gerakis, P. A. (1975). Productivity of agricultural ecosystems. *In* "Photosynthesis and Productivity in Different Environments" (J. P. Cooper, ed.), pp. 145–172. Cambridge Univ. Press, London and New York.
Lowry, W. P. (1969). "Weather and Life." Academic Press, New York and London.
Ludlow, M. M., and Wilson, G. L. (1971). Photosynthesis of tropical pasture plants. *Aust. J. Biol. Sci.* **24**, 449–470.
McCartney, A. (1975). Spectral distribution of solar radiation within and above crops. Ph.D. Thesis, University of Nottingham.
McCree, K. J. (1972). The action spectrum, absorption and quantum yield of photosynthesis in crop plants. *Agric. Meterol.* **9**, 191–216.
McCree, K. J. (1974). Equations for the rate of dark respiration of white clover and grain sorghum, as functions of dry weight, photosynthetic rate, and temperature. *Crop Sci.* **14**, 409–514.
Monteith, J. L. (1963). Gas exchange in plant communities. *In* "Environmental Control of Plant Growth" (L. T. Evans, ed.), pp. 95–110. Academic Press, New York and London.
Monteith, J. L. (1965). Light and crop production. *Field Crop Abstr.* **18**, 213–219.
Monteith, J. L. (1972). Solar radiation and productivity in tropical ecosystems. *J. Appl. Ecol.* **9**, 747–766.
Monteith, J. L., ed. (1975a). "Vegetation and the Atmosphere," Vol. 1. Academic Press, London and New York.
Monteith, J. L. (1975b). "Principles of Environmental Physics." Arnold, London.
Monteith, J. L., ed. (1976). "Vegetation and the Atmosphere," Vol. 2. Academic Press. London and New York.
Oertli, J. J. (1969). Internal water relations (discussion). *In* "Physiological Aspects of Crop Yield" (J. D. Eastin *et al.*, eds.), pp. 85–88. Am. Soc. Agron., Madison, Wisconsin.

Prioul, J. L., Reyss, A., and Chartier, P. (1975). Relationships between carbon dioxide transfer resistances and some physiological and anatomical features. *In* "Environmental and Biological Control of Photosynthesis" (R. Marcelle, ed.), pp. 17–28. Junk, The Hague.

Richards, F. J. (1969). The quantitative analysis of growth. *In* "Plant Physiology" (F. C. Steward, ed.), Vol. 5A, pp. 3–76. Academic Press, New York.

Rosenberg, N. (1974). "Microclimate: The Biological Environment." Wiley, New York.

Ryle, G. J. A., Cobby, J. M., and Powell, C. E. (1976). Synthetic and maintenance respiratory losses of $^{14}CO_2$. *Ann. Bot (London)* **40,** 571–586.

Salter, P. J., and Goode, J. E. (1967). "Crop Responses to Water at Different Stages of Growth," Res. Rev. No. 2. Commonwealth Agriculture Bureau, Farnham Royal, England.

Slatyer, R. O. (1969). Physiological significance of internal water relations to crop yield. *In* "Physiological Aspects of Crop Yield" (J. D. Eastin *et al.,* eds.), pp. 53–79. Am. Soc. Agron., Madison, Wisconsin.

Slatyer, R. O. (1973). The effect of internal water status on plant growth. *In* "Plant Response to Climatic Factors" (R. O. Slatyer, ed.), pp. 177–188. UNESCO, Paris.

Smith, L. P. (1973). The effect of temperature on the quality of hops. *In* "Plant Response to Climatic Factors" (R. O. Slatyer, ed.), pp. 99–106. UNESCO, Paris.

Sturrock, J. W. (1975). Wind effects and their amelioration in crop production. *In* "Physiological Aspects of Dryland Farming" (U. S. Gupta, ed.), Oxford Univ. Press, London and New York.

Szeicz, G. (1974). Solar radiation for plant growth. *J. Appl. Ecol.* **11,** 617–636.

Treharne, K. J., and Cooper, J. P. (1969). Effect of temperature on the activity of carboxylases in tropical and temperate gramineae. *J. Exp. Bot.* **20,** 170–175.

Trenbath, B. P., and Angus, J. F. (1975). Leaf inclination and crop production. *Field Crop Abstr.* **28,** 231–244.

Unsworth, M. H., and Monteith, J. L. (1972). Aerosol and solar radiation in Britain. *Q. J. R. Meteorol. Soc.* **98,** 778–797.

Vince-Prue, D. (1975). "Photoperiodism in Plants." McGraw Hill, London and New York.

Waister, P. D. (1972). Wind damage in horticultural crops. *Hortic. Abstr.* **42,** 609–615.

Wareing, P. F., and Phillips, I. D. J. (1970). "The Control of Growth and Differentiation in Plants." Pergamon, Oxford.

Williams, C. N., and Joseph, K. T. (1970). "Climate, Soil and Crop Production in the Humid Tropics." Oxford Univ. Press, Singapore.

Zelitch, I. (1975). Environmental and biological control of photosynthesis: general assessment. *In* "Environmental and Biological Control of Photosynthesis" (R. Marcelle, ed.), pp. 251–262. Junk, The Hague.

CHAPTER

2

Soils

JAKOB BENNEMA

I. INTRODUCTION

This chapter deals with the ecological soil conditions in the humid tropics. The main soils are described first. Attention is then focused on the major ecological soil qualities, namely, root volume, availability of water for plant growth, and availability of nutrients (see Beek and Bennema, 1972).

Only those aspects of nutrient availability are discussed with respect to which main soils of the humid tropics differ largely from most other soils: low cation exchange capacity of the clay fraction, often low amounts of exchangeable bases, relatively high aluminum saturation, high phosphorus fixation, and, in general, the easily upset balance among nutrients (see also van Wambeke, 1974).

Because tropical agriculture often has to deal with low inputs, the use of fertilizers is often restricted. The nutrients present in the natural ecosystem, both in soil and vegetation, are often an important source for agricultural plant production. Nutrient storage and cycling in natural ecosystems are therefore dealt with in Section V.

Tropical soils are subject to deterioration with agricultural use mainly because of a decrease in organic matter content. Management practices that control the amounts of organic matter are therefore an important aspect of tropical agriculture. The last section discusses the factors that determine the amounts and distribution of organic matter in tropical soils.

II. RED AND YELLOW SOILS OF THE HUMID TROPICS

Attention is given in this section to the characteristics of different red and yellow soils, especially those occurring in the humid part of the Tropical Lowlands; emphasis is also given to characteristics of the most weathered soils, the Oxisols, not because they are the most important ones in humid tropics, but because they are most extreme in their characteristics and behavior. Many other soils have characteristics in common with the Oxisols, although these often are not obvious.

The red and yellow soils are mostly old (mature) soils, but some are older than others. The very oldest soils are found on ancient erosion surfaces with subdued topography and on terraces, as well on marine, fluvial, and lacustrine terraces. These soils, earlier classified as Latosols or Ferralitic soils, are now called *Oxisols* (Soil Taxonomy, Staff, Soil Survey, 1975) or *Ferralsols* (FAO/UNESCO, 1971).

Ferralsols or Oxisols occur, in particular, over large areas in Africa and South America, as shown on the Soil Map of the World (FAO/UNESCO, 1971). They are strongly weathered deep soils, mostly low in fertility, with the exception of some of the Oxisols in basic parent rocks.

Related to the Oxisols are soils formerly known as Red–Yellow Podzolic Soils. These often have a low exchange capacity and a high percentage of exchangeable Al^{3+}. They are now classified as *Ultisols* (Soil Taxonomy) or Acrisols + Dystric Nitosols (FAO/UNESCO, 1971). They are also old and weathered with low base saturation. Ultisols mean ultimate soils, but this is not true for the humid tropics because they are not as old and often somewhat less weathered than the Oxisols. A subsurface horizon with clay illuviation is a diagnostic criterion of the Ultisols. Clay is transported from the topsoils to this horizon. These soils are found primarily in areas having a marked, but relatively short, dry period.

In the residual areas with more accentuated topography, younger soils can be found on acid or intermediate rocks. These belong most to the Dystropepts. They have low base saturation and high percentage of exchangeable Al^{3+} in common with the Ultisols but they lack an argillic horizon. Well-drained soils with a higher base saturation occur in the humid tropics on basic rocks and in areas with younger volcanic influences. The best known soils of this kind are the Reddish Brown Lateritic Soils, deep dark red, or dusky red soils with an argillic horizon (often not very well developed) and medium to high base saturation. These soils are now classified as *Rhodudalfs* (Soil Taxonomy) or Eutric Nitosols (FAO/UNESCO, 1971). An example of these soils is the well known "Terra Roxa estruturada" (structured Terra Roxa) of Brazil. The soils under discussion have predominantly 1:1 layer silicate clay minerals, mainly kaolinite. Red soils with 2:1 layer silicate clay minerals occur locally in the humid tropics. These soils do not have the same properties as those described in this chapter.

A. Oxisols

Oxisols are deep, permeable, well-drained soils (the soil is often 1–4 m deep) in which the subsurface horizon is composed of very fine granulars or of weakly developed blocks with a friable consistency. The coarser fractions in soils on acid parent materials (granites, sandstones, most sediments) are always dominated by quartz, and in soils on more basic parent materials (e.g., basalts) by iron oxides and titanium oxide together with quartz. Smaller percentages of kaolinite and aluminum oxide may also be found in the coarser fractions. Primary, easily weatherable minerals are absent or rare. The coarser fraction consists mainly of sand; the percent of silt is low. The mineral *colloidal fraction* consists of kaolinites, with varying amounts of iron and aluminum oxides. The molar ratio SiO_2/Al_2O_3 of the total clay fraction is often used as an indication of the composition and also of the weathering stage of the soil material of Oxisols. A ratio of about two is an indication that the clay fraction is dominated by kaolinites (kaolinite itself has a SiO_2/Al_2O_3 ratio of two). A ratio higher than two indicates that clay minerals which also have a higher cation exchange capacity (CEC) than kaolinite are present. However, ratios much higher than two are not found in Oxisols; they occur only in less weathered soils. A SiO_2/Al_2O_3 ratio well below two indicates the presence of free aluminum oxides (mainly gibbsite) in Oxisols. This often is the case in the residual Oxisols. Locally even very high amounts of free aluminum oxides may occur, e.g., in the cerrado Oxisols of central Brazil. The terrace Oxisols of the Amazon region have SiO_2/Al_2O_3 ratios that are never much below two. They are kaolinitic and free aluminum oxides do not play an important role.

Oxisols usually show a residual enrichment of iron oxides as a result of a loss of silica and bases, the main feature of so-called ferralitic weathering. The percentages of iron oxides in Oxisols are thus normally higher than in the original material. These percentages can, however, still be low in an absolute sense, if the soils originated from parent materials low in iron. Such soils may even have iron contents which are much lower than in young soils derived from parent materials high in iron. For example, most soils on the sediments of the Amazon region have relatively low iron content, usually between 2 and 8%. The residual Oxisols on the consolidated rocks often have medium to high iron contents, depending on the kind of parent rock.

The color of Oxisols is often related to their iron content. Soils low in iron content are often yellow, as is the case with many Oxisols on sediments. Soils high in iron usually are dark or dusky red (they may be yellow in cool and constantly wet conditions). The soils found on basic intrusions which are higher in iron have a dark red or dusky red color. Formerly the Oxisols were always described in terms of their color (Yellow Latosols, Red–Yellow Latosols, Dark Red Latosols).

The kaolinites in Oxisols have a low negative charge and CEC. The latter usually is not more than 6.5 mEq per 100 g of clay. This cation exchange capacity of the clay fraction and that of the organic matter often is only partly saturated with bases (low base saturation), whereas exchangeable Al^{3+} may be relatively high.

In its simplest form, an Oxisol consists of strongly weathered porous soil material well reworked by soil animals (chiefly by ants and termites) lying on strongly weathered, not reworked, parent materials, e.g., decayed rock or weathered sediments. Extensive areas of such Oxisols have been described in southern Brazil. However, in many other Oxisols, as well as in other tropical soils, plinthite—a mottled layer of red or brown iron in a gray matrix which hardens, if dried, to form iron gravel or an iron sheet—is present in the zone of fluctuating or seasonally perched water table, often with an underlying gray or whitish layer. In Oxisols it is most likely found deeper below the soil surface, e.g., at a 2–3 m depth, where there is no danger of hardening. However, hardening of phinthite is possible in soils where the plinthite is found higher in the profile, particularly in soils formerly known as Ground-Water Laterites. Fossil-hardened plinthite in the form of iron–gravel may form a part of the soil material in Oxisols. Iron–gravel and iron sheets can also occur in other soils.

In some regions of the Amazon the kaolinitic Yellow Latosols were used for agriculture many years ago. Large amounts of organic matter, often including bones, were brought to these places. These soils have deep, dark-colored topsoils and are known as "Terra Preta dos Indios"—Black soils of the Indians. The most typical of these soils have an "anthropogenic epipedon" (Soil Taxonomy). The "Terra Preta dos Indios" are considered to be fertile soils.

B. Other Red and Yellow Soils

Most of the younger red and yellow soils also consist primarily of clay minerals dominated by kaolinites and coarser fractions dominated by quartz, or locally, iron oxides.

The kaolinites of the younger soils often are more active than those in the Oxisols. Some accessory clay minerals with higher CEC may also be present, resulting in a somewhat higher CEC of the clay fraction and higher mobility of the clay. In somewhat older soils this higher mobility of the clay causes formation of a horizon with clay illuviation. Such a layer does not occur in the younger soils (Inceptisols).

The red and gray mottled material (plinthite) found in deeper layers of some Oxisols may also be present in the younger soils. In such soils it often occurs somewhat higher in the profile, suggesting less than optimal drainage conditions.

Most of the younger soils are also low in plant nutrients. However, some exceptions occur. As already mentioned, certain soils on basic rocks have much higher base saturation; for example, the "Terra Roxa estruturada" (an Alfisol). Significantly, these soils are not always as deep as most Oxisols, so that the nutrient cycle can be enriched by nutrients pumped up by the vegetation. It should, however, be stressed that these soils are also kaolinitic and that in addition to their higher base saturation, they have only very limited amounts of mineral reserves in the form of weatherable minerals. The extension of this kind of soil is very limited in the humid tropics.

Other exceptions are the soils in the humid and subhumid regions formed from volcanic ash. They cover greater areas, e.g., in Java, around the Rift Valley in East Africa, in the Andes, Central America, and Hawaii. The colloidal fraction of the younger members of these soils is often dominated by allophane, an amorphous aluminosilicate. If the soil is more mature, layer silicates become the dominant clay minerals, first halloysites, and later kaolinites. Most of these younger soils have a yellowish color, while the more mature soils are mostly dark red. The soils with higher allophane content have specific characteristics, such as a CEC strongly dependent on the pH (with low pH–low CEC) and a high phosphorus fixation.

The more mature soils are dominated by kaolinite in common with the old red and mostly yellow soils of the humid tropics. However, the amount of weatherable minerals often is still greater and the kaolinite often is more active and accompanied by halloysites or some allophane.

III. PHYSICAL CHARACTERISTICS

Under natural conditions most red and yellow soils in the humid tropics have a topsoil with good structure. If used in agriculture such soils are easy to work as long as the organic matter level does not decrease too much. Even the clayey soils are no exception. A farmer can work on his land shortly after rain, and these soils do not become very hard if dry. The preparation of a seedbed therefore is not difficult. However, if the organic matter level decreases, the soil structure deteriorates, a crust may form after heavy rains, and the soils may become hard if dry.

Although most tropical soils are deep, they do not necessarily favor root development for several reasons:

(1) A dense layer in the subsoil of the extremely developed Ultisols and Alfisols, in the form of a well-developed clay illuviation layer (well-developed argillic horizon). This relatively impermeable layer can be a mechanical barrier to root development. In extreme cases it may also impede water movement, leading to oxygen deficiency just above the dense layer.

(2) A compact subsurface layer that becomes hard when dried. Formation of this layer may be hastened by heavy machinery during land reclamation (Janssen and van der Weert, 1976). Such hardened layers are rather common in medium textured Oxisols with low iron content. Compact layers in finely textured Oxisols of the Amazon valley also occur (Sombroek, 1966).

(3) A layer of ironstone (hardened plinthite). Although not common in the humid tropics, it occurs more in savanna regions of subhumid to semiarid tropics. Plinthite (not hardened) occurs near the soil surface in some types of savanna soils (formerly called Ground-Water Laterites) and in some forest soils. It might be expected that following clearing of a forest, this layer will harden and impede root development. This is a subject of vigorous debate. The occurrence of plinthite near the surface in forest soils is, however, relatively rare.

(4) Low content of calcium and phosphorus and high content of aluminum in the subsoil. Under such conditions root growth is restricted to the topsoils, although the porosity of the subsoil may be favorable for root development. Soil management should attempt to influence chemical subsoil conditions in such a way that root growth is enhanced. It might also be useful to emphasize breeding of plant varieties better adapted to the special chemical conditions of these subsoils.

The unfavorable conditions for root growth in some subsoils may lead to very superficial root systems. Klinge (1973) showed, for example, that forest trees on a fine textured kaolinitic Yellow Oxisol in the Amazon valley had about 50% of their roots in the upper 20 cm. Sombroek (1966) indicated that if such soils had subsoils that were more compact than normal, root development was still more superficial and a rain forest with creeper developed. In Malaysia shallow-rooted rubber trees have been blown down (Chan and Pushpajajah, 1972).

In most soils of the humid and subhumid tropics, water deficits do not last long because there is no dry season, or it is short. The amount of available water, i.e., that between wilting point and field capacity, in most Oxisols and similar soils sometimes is, however, rather low. This is because these soils have many large and small pores but few medium-sized pores. The available water does not appear to depend on clay content. It is as high in more sandy soils as in very fine textured ones. The subsoils of the Yellow Oxisols of the Amazon valley have about 10% available water (Sombroek, 1966; see also Wessel, 1971 for related soils). The topsoil has more available water largely because of its higher organic matter content. However, not all the red and yellow tropical soils have such small amounts of available water. For example, Oxisols and similar soils with higher iron content, as well as Alfisols and some Ultisols, have more available water. The amount of water that plants can absorb within a certain time interval depends not only on available water in the profile but also on root development. As mentioned, conditions for root development are not always ideal as when the soil volume that supplies water to plants is limited. Some plants with low root–shoot ratios may undergo water deficits on moist soils because their capacity to absorb water is too low to keep up with a temporary high transpirational water loss.

The amount of water with a tension above 15 bars depends largely on clay content. It may be as high as 30% (weight percentage) in fine textured Oxisols. The large amount of water with tensions above the wilting point in fine textured soils is important for the conservation of organic matter within the granules. Soils with higher clay contents have a higher organic matter content, at least in part because of better conservation of the organic matter.

Oxisols and similar soils often are deep and highly permeable; thus runoff is substantially reduced. Erosion, especially sheet erosion, is not serious for such soils if the soil is covered with vegetation throughout the year so as to protect the topsoil from deterioration through rainsplash. Only under such conditions can deep soils be formed and maintained in nature. There is, however, some danger that the topsoil of Oxisols will deteriorate if used in agriculture without proper management. If the soils are exposed to direct influence of the rain, and especially if not enough organic matter is returned to the soil, the structure may deteriorate. This may result in sealing of the surface and consequently in runoff which causes both sheet and gully erosion on sloping terraces. It is also possible that more soil material illuviates to deeper horizons with agricultural use. In more coarsely textured soils the topsoil might become sandier due to these processes.

Not much is known about the differences in behavior of various Oxisols with respect to topsoil deterioration, but those low in iron, like the ones on sediments of the Amazon region, are most susceptible.

Typical Alfisols and Ultisols are generally more susceptible than Oxisols to erosion. They have topsoils which tend to seal, especially if they are low in iron. Moreover, they often have subsurface soils which are less permeable because of clay illuviation and related compaction. During the rainy season the topsoils may be saturated with water stagnating on the less permeable layer. This causes runoff and soil erosion even if the top layer is not properly sealed. This condition is further aggravated because many Ultisols and Alfisols occur in areas with more accentuated relief than where typical Oxisols are found. Nevertheless, many red and yellow soils of the humid tropics are relatively resistant to erosion if proper attention is given to this in their management. The soils should be protected as much as possible against the direct impact of rainfall. This is easier in the humid tropics than in regions with longer dry periods or with a cold period. Red and yellow soils of the humid and subhumid tropics can therefore be protected more easily against erosion than many soils in other parts of the world. An exception should be made in relation to mass movement, as slumps, which is not uncommon in some regions of the red and yellow soils. Erosion at the base of a hill by rivers may lead to very steep labile slopes.

IV. CHEMICAL CHARACTERISTICS

A. General Features

The availability of nutrients in the red and yellow soils is a many faceted subject. Two aspects will be emphasized: (1) the availability of bases, calcium, magnesium, and potassium; and (2) the phosphorus regime. Availability of other nutrients will be mentioned only briefly.

Deficiencies of Zn, B, Mo, and S often are reported for the strongly weathered red and yellow tropical soils (Drosdoff, 1972). Zinc deficiency appears to be related to fixation by crystalline sesquioxides. Boron deficiencies occur, in particular, on sands and also on neutral and alkaline soils. Molybdenum deficiency might occur in the process of nitrogen fixation in the strongly weathered soils of the humid tropics. This deficiency can often be alleviated by liming, but deficiencies of Zn and B are aggravated by addition of lime. Formerly, sulfur was added to fertilizers, mainly with phosphorus. However, with newer fertilizers, of greater purity this is not always done, and special attention should be given to providing plants with sulfur. In the natural ecosystems of the red and yellow soils sulfur appears to be present mainly in the organic cycle (Olson and Engelstad, 1972) it is partly lost by burning.

There is much speculation about the role of silica in plant nutrition. Some plants, especially rice, sugarcane, and corn, extract very large amounts from the soil. It is expected that available silica will be especially low in the old, strongly weathered soils because these have lost most of the silica which can be easily set free by weathering. However, as yet relatively little is known about silica deficiency (d'Hoore and Coulter, 1972).

The nitrogen regime of tropical soils is of special interest. Nitrogen does not seem to be a limiting factor in natural tropical ecosystems, but it is in agricultural production if no nitrogen is added in the form of fertilizers or by the use of leguminous crops. Some authors claimed that a substantial amount might become available by fixation other than that by leguminous crops (Dobereiner *et al.*, 1975), in particular with the production of plants belonging to the Gramineae.

Mineral toxicity may also occur. For example, Al has a negative effect on availability of Ca and P uptake. Manganese also has a negative influence on ion uptake in the more acid soils and in some hydromorphic ones.

B. Availability of Bases

Uptake of bases (Ca, Mg, K) is governed by (1) their supply on or near the surface of the plasmalemma in plant cell walls of the cortex in the apparent free space (AFC); (2) the mechanism by which nutrients are taken up by plant roots (e.g., by which they pass the plasmalemma).

The uptake of bases by plant roots varies among different groups of plants. An interesting point is that the ratio between cation and anion uptake is not the same for different groups of plants (de Wit *et al.*, 1963). However, this is not discussed here. Only the more general aspects of the uptake mechanism are considered.

At least two uptake mechanisms for bases exist. One is specific for a particular ion and the other is a common mechanism (Epstein, 1972; Schuffelen, 1974), which is especially effective at higher ion concentrations. The specific mechanism is particularly important for K. All cations, including K, can be absorbed by the common mechanism. Hence, K acts as a competitor for other ions, especially Mg. The other ions are weaker competitors for K because the specific uptake mechanism for K is well developed. For Ca and Mg the ratios of their concentrations, or activities with other ions (especially K and Al) with which they compete for passage through the plasmalemma, are important.

The nutrients, including bases, reach plant roots by mass flow and diffusion with an additional small contribution by root interception. The effect of the transport is greatly dependent on the concentration of nutrients in the soil solution.

Concentrations of bases in soil solutions vary with the moisture content. More stable and hence, more meaningful, soil characteristics are the ratios of the "reduced concentrations" (see Bolt, 1976) or of the "reduced activities" of the different bases in the soil solution. "Reduced concentrations" are the normal concentrations of the monovalent ions, the square roots of the concentrations of the divalent ions, and the cube root of the concentration of Al. The ratios of the "reduced concentrations" or activities are relatively stable; they are in equilibrium with the (normal) ratios of the bases on the CEC complex of the soil.

The cause of this equilibrium is that Ca and Mg are more strongly bound to the CEC complex than monovalent ions such as K. However, this situation applies only if the soil complex does not have a strong preference for the monovalent ions as is sometimes the case with clay minerals in relation to K. We may assume, however, that this is not the case for the top layers of the soils under discussion, in which the organic matter is the principal source of the CEC.

The lesser bondage of K to the CEC complex results in relatively easy transport to the roots. This, together with the easy passage through the plasmalemma, makes K a very mobile ion. The turnover of K as a part of the vegetation cycle is, as discussed later, also relatively fast—faster than the turnover of Ca and Mg. This mobility of K may explain why K is not strongly deficient in most natural ecosystems of the red and yellow tropical soils, although the reserve or total K can be extremely low.

The activity ratio $a_K/(a_{Ca} + a_{Mg})^{1/2}$ or, on acid soils, $a_K/(a_{Ca} + a_{Mg})^{1/2} + p(a_{Al})^{1/3}$, in which a indicates the activity in the soil solution, is sometimes used as a measure of the availability of K (for discussion see Boyer, 1972). Information on the ratios of reduced concentration or activities in soil solutions is, however, rather scarce. The data appearing in soil survey reports are mostly related to the bases of the cation exchange complex. They are often used as a measure of availability of the bases. Both absolute and relative amounts are important. For K, 0.10 mEq is seen as an absolute minimum for agricultural use, and it should be higher than 2% of the sum of exchangeable bases.

Ca and Mg uptake is, as previously mentioned, not only a function of the Ca or Mg present, but is also strongly influenced by competition with other ions. Earlier studies stressed the mutual influence of Ca, Mg, and K on uptake. For example, Mg might be deficient because of high levels of K. It now appears that in the acid red and yellow soils of the humid tropics competition with Al is also very important. This follows for Ca uptake from the work of Zandstra (1972), who studied Ca uptake in acid Oxisols from the Llanos in Colombia and in acid Oxisols and Ultisols from Puerto Rico.

Plant roots need small amounts of Ca for growth, particularly for plasmalemma functioning (Ca pectinates). Each root must absorb sufficient Ca for its needs because in most plants Ca is not translocated to the root tips (Harschner, 1974). Hence, reduced Ca uptake may inhibit root growth. Zandstra (1972) concluded that in the soils studied root growth was inhibited by Ca deficiency at the root tip.

Calcium uptake depended on the logarithm of the solution ratio $(Ca)^{\frac{1}{2}}/(Al)^{\frac{1}{3}}$ at least if higher concentrations of either K or Mg were absent. With higher K and Mg concentrations, (K) and $(Mg)^{\frac{1}{2}}$ must be added to $(Al)^{\frac{1}{3}}$ (parentheses indicate molar activities). He further related for practical purposes this ratio to ratios of bases and Al at the exchange complex.

The availability of bases in the red and yellow soils of the humid tropics varies from adequate to low. The reserves are, however, mostly low to very low. The bases at the CEC complex comprise a large portion of the total bases present in the soil. They act as a buffer and as a reserve. A balanced supply of bases to the plant is, however, easily upset, often due to the low amounts of bases at the CEC complex.

Different aspects of the exchange complex will be illustrated with samples of three different soils from the Amazon region. These soils have much in common but differ in relation to their exchange complex.

(1) A very fine textured kaolinitic Yellow Latosol from Belterra (Belterra clay), with low base saturation. This soil can be classified in Soil Taxonomy as a very fine kaolinitic isohyperthermic Haplic Acrorthox.

(2) A "Terra Preta dos Indios" developed in a soil similar to soil (1). Both base saturation and phosphorus content are high. The soil has an anthropogenic epipedon. The classification in Soil Taxonomy is, however, uncertain.

(3) A "Terra Roxa estruturada" developed on basic igneous rocks, with a relatively high base saturation. It is a fine, oxidic, isohyperthermic Oxic Tropudalf (most "Terra Roxa estruturada" are, however, Rhodudalfs).

A low CEC is an important characteristic of Amazon soils. The organic matter contributes importantly to CEC, because the CEC of the clay fraction is very low. This contribution of the organic matter decreases with soil depth. In Fig. 1, the influence of organic matter on the CEC in the three different soils is shown in the presence of 100 g clay.

The relation between mEq CEC and the percent of organic carbon (C) in the presence of 100 g clay can be expressed as (CEC determined at pH 7, see Vettori, 1969):

$$\text{CEC} = 2.0 + 5.0\ \text{C}\ (r^2 = 0.98)\ \text{(Belterra clay)}$$
$$\text{CEC} = 4.5 + 7.8\ \text{C}\ (r^2 = 0.998)\ \text{(Terra Preta dos Indios)}$$
$$\text{CEC} = 3.7 + 5.8\ \text{C}\ (r^2 = 0.999)\ \text{(Terra Roxa estruturada)}$$

Cation exchange capacity for both the clay and organic matter is lower in the soil with lower base saturation than in the one with higher base saturation, and still higher in the "Terra Preta dos Indios"; the values are, respectively, 2.0, 3.7, and 4.5 mEq for 100 g clay; 5.0, 5.8, and 7.8 mEq for 1 g C. The higher CEC might be related to less blockage of negative sites.

The CEC given in Fig. 1 is measured at a pH of 7. However, under field conditions the effective CEC is lower. The effective CEC can be expressed as the sum of bases + the exchange acidity (mostly exchangeable Al). In soils (2) and (3) the exchange acidity is zero. The effective CEC for these soils is thus the

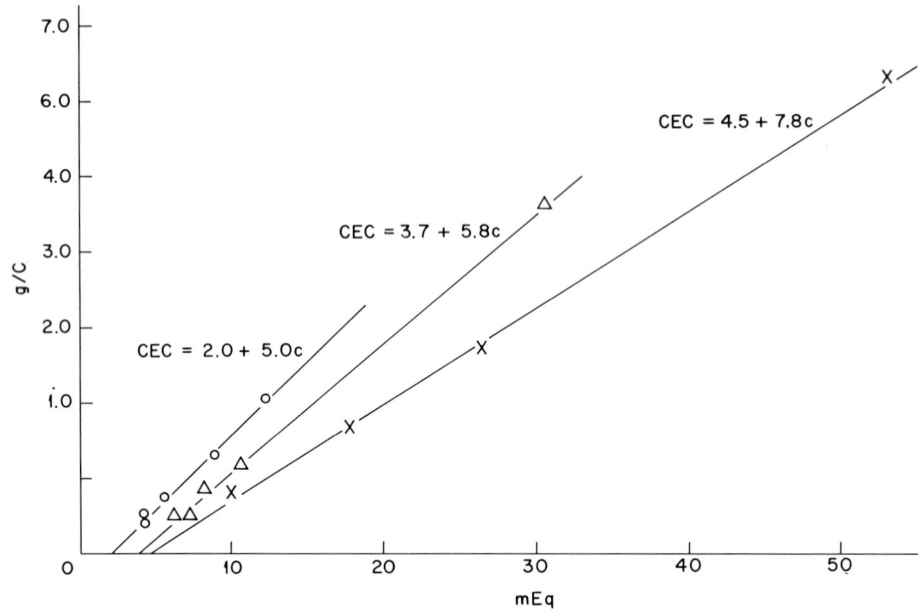

Fig. 1. The cation exchange capacity as a function of organic carbon (C) (in the presence of 100 g clay). ○—Very fine textured kaolinitic Yellow Latosol; △—Terra Roxa estruturada; ×—Terra Preta.

same as the sum of bases. In soil (1) appreciable amounts of exchange acidity (exchangeable Al) are present. Figures 2–4 show the exchange acidity (Al), if present, the sum of bases, and the components Ca and Mg. Particularly striking are the low amounts of exchangeable bases of the kaolinitic Yellow Latosol in relation to the other two profiles, and the high percentage of exchangeable Al. It is greater than the sum of the bases.

The amount of exchangeable K in the surface layers of the three soils varies from 0.10 to 0.28 mEq/100 g soil. The lowest amount is found in the kaolinitic Yellow Latosol; the highest in the "Terra Roxa estruturada." The amount of K in the "Terra Preta dos Indios" is not much higher than in the Yellow Latosol and, relative to Ca and Mg, is much lower. The amount of K, expressed as a percentage of the sum of the bases for the three profiles, was 10, 2, and 0.5%; the highest percentage was in the kaolinitic Yellow Latosol and the lowest in the "Terra Preta."

Data on exchangeable K were compared with the already mentioned absolute minimum requirements for agricultural use. The Yellow Latosol has 0.10 mEq, while the "Terra Roxa estruturada" has a saturation percentage of 2%. Hence, both of these ecosystems are functioning at a level of K which for different

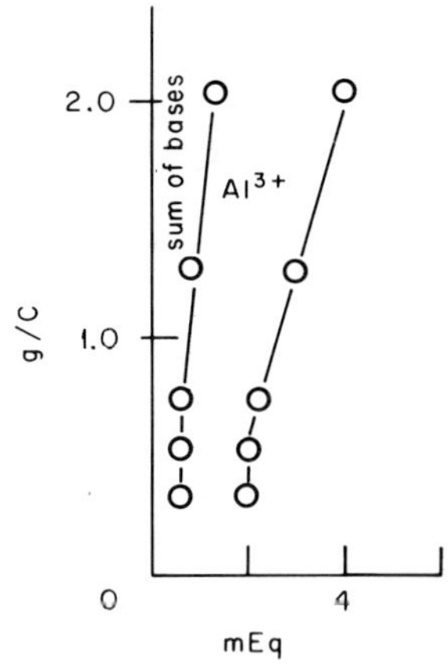

Fig. 2. Exchangeable bases and exchange acidity in a kaolinitic Yellow Latosol as a function of organic carbon (in the presence of 100 g clay).

reasons is regarded as minimal. The data on the activities of the equilibrium solutions (Fassbender *et al.*, 1970) indicate that these activities are low and the same for both soils. The major difference between the "Terra Roxa estruturada" and the "Yellow Latosol" is that the amount of available K, which is almost all the reserve present, is much smaller or negligible in the Yellow Latosol. The only reserve that can be used for agriculture is stored in the vegetation and can be easily lost by leaching after burning or by crop removal.

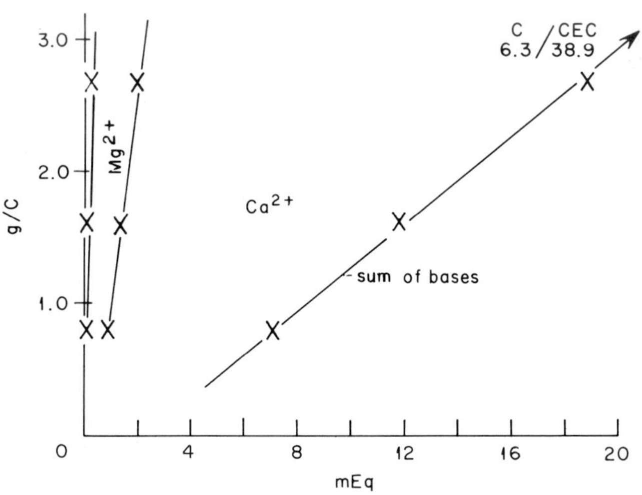

Fig. 3. Exchangeable bases in Terra Preta as a function of organic carbon (in the presence of 100 g clay).

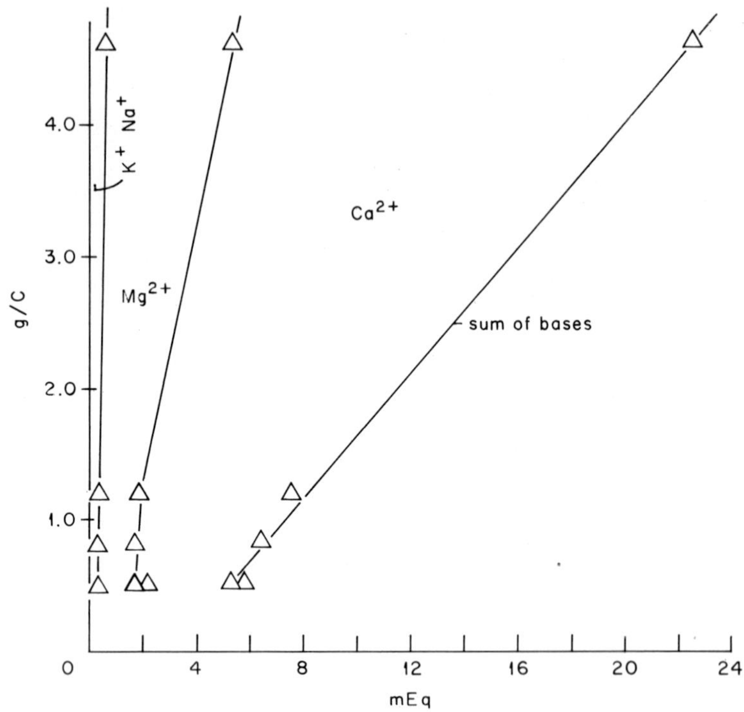

Fig. 4. Exchangeable bases in Terra Roxa estruturada as a function of organic carbon (in the presence of 100 g clay).

The amounts of exchangeable Ca and Mg are often low in dystrophic red and yellow soils as are the concentrations (or activities) of Ca and Mg in equilibrium solutions (see, e.g., data of Fassbender *et al.,* 1970). The concentrations (activities) are sometimes lower for these ions than for K, whereas the reverse is true in eutrophic soils. Exchangeable Ca and Mg are extremely low in the subsoil of the Yellow Latosol. It may be questioned whether Ca concentrations, in relation to Al, have reached such low levels that root growth of susceptible plants is greatly inhibited.

C. Fixation and Availability of Phosphorus

Much of the P in old weathered soils is present as iron phosphates (Chang and Jackson, 1958). In addition to iron phosphate, organically bound phosphorus and Al phosphates are present (see, e.g., Vierra, 1966). The iron phosphates may be so strongly bound that they are unavailable for plant growth. Phosphorus thus bound cannot be leached.

Bennema (1974b) showed the following relation of total P_2O_5 to Fe_2O_3 and organic carbon (C) content in Brazilian Oxisols:

$$P_2O_5(\%) = 0.002 + 0.010 \times \text{organic carbon } (\%) + 0.007 \times Fe_2O_3 \ (\%)$$

(r = 0.85 and the confidence limits at 95% level for the coefficient of Fe_2O_3 are 0.0065 and 0.0075). Figure 5 illustrates the relation of P_2O_5 to Fe_2O_3 for the means of different types of Oxisols. In addition to the relation of P_2O_5 to Fe_2O_3 and C, the relation of P_2O_5 to free Al_2O_3 was studied, but no significant relation was found. This does not necessarily mean that no Al phosphates were present, but that the amounts of Al phosphates were not correlated with the amount of Free Al_2O_3. Phosphorus concentrations in younger soils are in many cases considerably higher than in older soils, because of higher amounts of phosphate in weatherable minerals (mainly in apatite) and as secondary Ca and Al phosphates. When the soils become older and more strongly weathered, some P is obviously lost (leaching, erosion), as shown in the lower percentages of P_2O_5 in the older soils, and the P carried by the rivers to the ocean, estimated to be 3.5 million tons/year (Epstein, 1972).

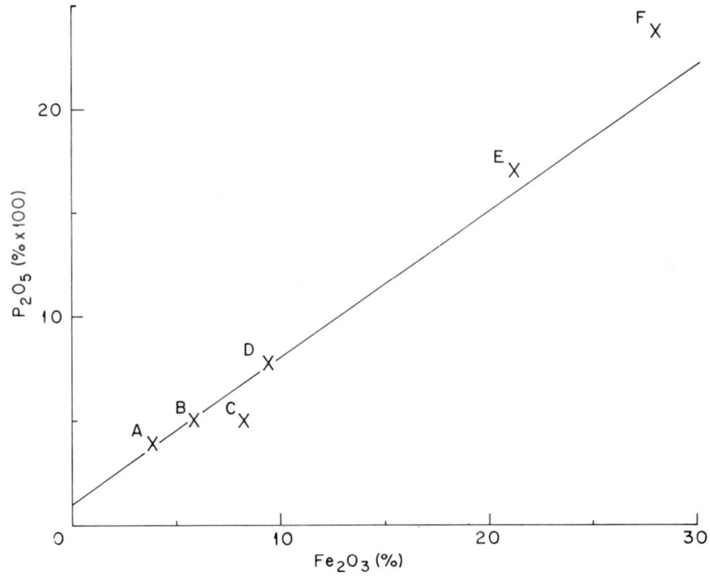

Fig. 5. The relationship of the average P_2O_5 contents with the average Fe_2O_3 contents of six different groups of Latosols. A–kaolinitic Yellow Latosol (on sediments); B–Red–Yellow Latosol; C–Humic Latosol; D–Dark Red Latosol; E–Red Latosol of the Rio Grande do Sul; F–Dusky Red Latosol (Terra Roxa legitima). The line represents the function with a constant (average) *C* content of 0.9%.

Much of the P that is not lost is transformed into iron phosphates and organically bound P. Transformation of P to iron phosphates in soils is to be expected because iron phosphates under acid conditions are less soluble than Ca or Al phosphates.

The amount of P_2O_5 in the upper 20 cm of the older tropical soils normally varies from 400 to 4000 kg/ha, depending largely on the iron content. In younger soils more than 4000 kg/ha may occur. Under natural conditions very old soils may have achieved a near equilibrium in which small amounts are gained annually through rainfall (±0.3 kg/ha) and small amounts are lost by leaching, erosion, and through the air. The small annual losses and gains add up to a large amount over the age of these soils. The soil and plants together act as a buffer with two competing subsystems, the mineral cycle and the organic cycle. The amount of P in the organic cycle depends on saturation of the total system and on the fixing power of the organic cycle in relation to that of the mineral soil. In the very old soils where saturation is low only small amounts of P can be expected in the natural vegetation.

If P is added to the soil, both the mineral cycle and the organic cycle will take their toll. It is not surprising that efficient use of fertilizers in the red and yellow soils is difficult because much of the added fertilizer is fixed by the soil.

To obtain higher yields, one might ask (1) whether the red and tropical soils should first be corrected with large amounts of P fertilizers, and later more P applied annually, or (2) whether management techniques should be developed in which the P fertilizer touches the mineral soil as little as possible so that P is absorbed by plants with minimal interference from the mineral soil. Different possibilities for such management techniques exist. In the tropics, large investments in fertilizers often are not economically feasible. Hence, correcting soil with P is practical only when relatively small amounts are needed.

The relation of fixation of added P and soil characteristics has been widely studied. The soil is generally mixed with solutions of different concentrations of P and the amounts of P that stay in solution or are bound are determined. Phosphorus fixation in red and yellow tropical soils is generally medium to high and lower than in volcanic ash soils with allophanes, and similar to or higher than in many soils of temperate regions. Variations in fixation of added phosphorus in different red and yellow tropical soils most often are a function of clay content, free Al_2O_3 and Fe_2O_3 content (see, e.g., Leal and Velloso, 1973a,b; Olson and Engelstad, 1972; Morelli, 1974).

Organic matter counteracts P fixation (Moshi *et al.*, 1974), while low pH favors it. The influence of pH is partly reversible, at least on a short-term basis (Leal and Velloso, 1973b). If the pH is increased some of the P fixed earlier can be set free. The influence of pH on fixation is related to the decrease in solubility of Fe and Al phosphates as pH is lowered. Aluminum phosphates may precipitate in the soil or in the root cortex (McCormick and Borden, 1974). The influence of the low pH and related high Al concentration in the soil solution is an important factor in soil fertility in the humid tropics because soils with low pH are so common.

In the laboratory, available P is often measured at a pH different than that in the field. Thus the effect of the field pH on the availability of P is obscured. For the interpretation of these data the field pH must be taken into account, as shown in Fig. 6, for cotton growing on lighter textured soils in Sâo Paulo State, Brazil. Three levels of response to P are evident: (1) an area of no response probably because additions of P were too small for these very low phosphate soils; (2) an area of response by most soils; and (3) an area of lack of response by most soils because the combination of P and pH is favorable for production of cotton. It follows from Fig. 6 that a pH between 6.2 and 6.6 is optimal for the supply of P. With lower pH higher "available P" is needed for the same level of supply.

From both short- and long-term experiments the following features of P

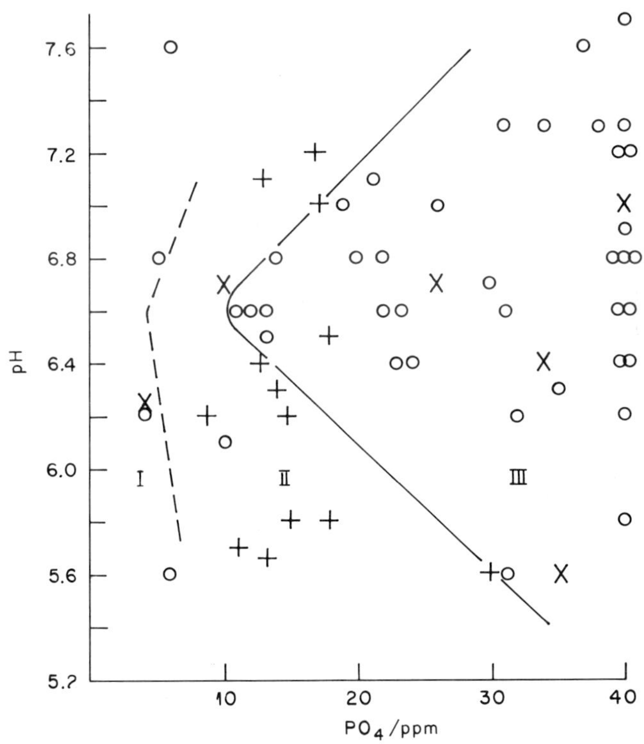

Fig. 6. Response of cotton to phosphorus as a function of soil phosphate extracted with oxalate and of pH. o–No response; X–significant; +–highly significant. Lighter textured soils, São Paulo, 1957–1958 (Campinas, 1960).

fixation have been deduced. Added P is first fixed on kaolinites and on the surface of the iron oxides and by free aluminum oxides. Some of the P begins to form iron phosphates, first in a nonoccluded form and later, in part, in an occluded form. The time change to iron phosphates can probably be measured in years. Wessel (1971) found in cacao soils of Nigeria that, after some years, much of the added P was in the form of iron phosphates. An additional part will be incorporated in the organic matter. It follows from these studies that in many soils the amounts of P necessary to "correct the soil" are very large, if an attempt is made to saturate the soil until P concentration in the soil solution is optimal for plant growth. In particular, large amounts of P must be added to soils with high contents of allophane, silicate–clay minerals, gibbsite, or iron oxides. The amounts needed may vary from tens to many hundreds of kilograms P per hectare. Therefore in most cases it is not feasible "to correct the soil" by heavy application of P. A reduction of P fixation by liming, if lime is available near the site, and by maintaining a sufficient supply of organic matter is often more practical. Some investigators hope that addition of silicate, e.g., in the form of silicate slag to the strongly weathered red and yellow soils, may be a practical way of lowering the phosphate fixation and thus rendering added phosphorus fertilizers more available to the crop (see Olson and Engelstad, 1972).

V. NUTRIENT RELATIONS

Most plant communities in the humid tropics are closed and efficient ecosystems which exist on a very small nutrient budget. In contrast to ecosystems of the temperate regions, the cycling nutrient capital is for a substantial part found in the living plants themselves. the shifting cultivation in the tropics is based on the use of these nutrients, which are freed by burning of the vegetation developed during the fallow period. Permanent or semipermanent agriculture also uses the nutrients derived from vegetation and decomposing organic matter in the early years. It is, therefore, important to discuss nutrient storage and cycling in different vegetation types.

The most common vegetation types of the humid and subhumid tropics are the rain forest, the semideciduous forest, the deciduous forest, and various savanna types. The savanna types are partly natural, such as the cerrado of Brazil. In other areas the savannas are anthropogenous. The amounts of available nutrients in the anthropogenous savanna system vary widely with the original eutrophy of the site and with the stage of degradation.

It sometimes is assumed that the lush tropical rain forests have high natural fertility. This is far from true, as will be shown by discussing nutrient cycling in a 40-year-old mature secondary forest in Ghana near Kade and in a virgin

Amazon forest. The latter is of special ecological interest because it represents vegetation of an almost flat surface that is about a million years old. With respect to nutrient cycling, this site is in or near equilibrium. Nutrients lost by leaching (erosion presumably is absent) are balanced by the amounts gained by the air from outside. The ecosystem thus represents a real climax vegetation. It is found in a tropical lowland with about 2000 mm rain a year and a dry period of about three months.

Such old and stable surfaces also occur in central Brazil under subhumid climate with a more marked dry period. The vegetation, in this case a savanna type (cerrado), must be regarded as an edaphic type. The soil is too infertile to maintain a forest, which in this climatic zone would be a deciduous forest. Obviously such a forest cannot maintain a sufficient fertility level and changes into a "cerrado." This is due to a loss of nutrients associated with occasional natural fires or heavy rains in the otherwise dry season, when much of the plant cover is still not active enough to fix the nutrients. It is understandable that primitive agriculture did not find any use for these infertile soils.

Nye and Greenland (1960) studied nutrient cycling in West-African ecosystems, including the Kade forest in Ghana, a 40-year-old mature secondary forest about 60 km from the sea. The soil was characterized as a rather eutrophic "Latosol." It probably would not be classified in the Soil Taxonomy as an Oxisol (these are rare in West Africa). The Kade forest is an interesting counterpart of the dystrophic Amazon forest. The amounts of nutrients received from rain in the Amazon forest and Kade forest are shown in Table I.

In the Amazon forest nutrients are in equilibrium because the amounts received approximately equal those lost by leaching.

The gain is partly from rainfall and partly from rainwash (through the canopy and along tree stems and branches). The additional nutrients returned in rain-

TABLE I

Nutrients Supplied Annually by Rain and Lost by Drainage in an Amazon Forest and Kade Forest in Ghana

	kg/ha					
	N	P	K	Ca	Mg	Ref.
Manaus (Amazon)						Stark (1972)
Received by rain	10.0	0.3	n.d.[a]	3.7	3.0	
Lost with drainage water	5.8	0.1	6.5	4.9	2.6	
Kade-Ghana received by rain	13.0	0.4	16.0	11.0	10.0	Nye and Greenland (1960)

[a]n.d.—not determined.

wash are derived from the vegetation, e.g., from leaves (important for K) and from dust. The amounts of nutrients returned by rain and rainwash in the Kade forest are higher than for German forests possible because of the desert dusts (Harmattan). However, K may be largely derived from leaves.

The importance of nutrients in rainwash in the Kade forest is shown when they are compared with those returned by litter. About 200 kg/ha K are returned annually by rainwash and only 60 kg/ha by litter. The amount of P returned by rainwash (3 kg) is about half of that returned by litter. It appears that the Kade forest receives a constant, but low nutrient supply from the desert. Although the contribution of nutrients from rain and dust is rather unimportant for agriculture, this is not necessarily true for a natural ecosystem. An annual contribution of 0.4 kg/ha of P from rain alone would mean 400 tons/ha (40 kg/m^2) in a million years. Although most soils are younger than 1 million years old, many are still old enough to have received large amounts of nutrients, but they have also lost large amounts. Hence, nutrient balance in old soils seems to depend largely on the "fixing capacity" of the system more than on the nutrient content of parent materials.

Table II shows stored nutrients and annual nutrient sources for the Walter Egler Reserve near Manaus in Brazil, the Kade forest in Ghana, and a mixed forest in the Congo.

Lower amounts of P, K, Ca, and Mg were returned in litter in the Walter Egler Reserve than in the Kade forest or Congo forest. The amount of litter in the Walter Egler Reserve also was low for a rain forest and lower than that of the other two forests. It is likely that net productivity of the Walter Egler Reserve was limited by nutrient availability.

In the Kade forest the amounts of K, Ca, and Mg stored in the vegetation were about the same as those present as exchangeable bases in the soil. There was about ten times as much P in the vegetation as was available in the soil. The P in vegetation will be made available for agricultural use by burning. Total nutrients in vegetation of the Amazon are not given, but low amounts may be expected because the nutrient content of the litter is very low, and a low P status is indicated by the low N/P ratio of the litter.

The data for Ghana indicate that K recycles faster than Ca and Mg, and much of the K in rainwash is derived from leaves and not from dust. About 35% of the K in the vegetation is returned annually to the soil. For Ca this amount is 12%; for Mg, 20%; for P, 12%; and for N, 14%.

Because of its high mobility in an ecosystem, K need not be limiting for plant growth, even if low amounts are present. However, low K reserves can be depleted by leaching or removal with crops (e.g., cassava). If the nutrient cycle is not functioning optimally, K is easily lost because it has only weak competitive capacity for exchange sites in most tropical soils.

TABLE II
Nutrient Cycling in Three Forest Ecosystems

		kg/ha					
		N	P	K	Ca	Mg	Ref.
Walter Egler Reserve, Manaus	yearly litterfall	105	2.2	12.7	18.4	12.6	Klinge and Rodrigues (1968)
	soil (± 25 cm depth)	total	avail.		exchangeable		Based on data by Equipe de Solos (1969)
		2000	10	150	200	100	
Mixed forest, Congo	yearly litterfall	200	6	43	94	46	Laudelout (1961)
	stored in vegetation	1100	110	850	1900	300	
Kade forest, Ghana	yearly litterfall	178	6.5	61	184	40	Nye and Greenland (1960)
	timberfall	32	2.6	5	73	7	
	rainwash	11	3.3	196	26	16	
	stored in vegetation (40-year-old forest)	1600	100	700	2300	300	
	soil (± 25 cm depth)	total	avail.		exchangeable		
		4000	10	600	2000	300	

The large amount of P in the vegetation of the Kade forest would ensure a high yield for many years after release by burning. However, some of the P is fixed and becomes unavailable. Hence, plants may become deficient in P. The low soil pH after leaching of Ca may further inhibit P uptake by plants.

In well-developed forests about 200 kg N/ha are returned annually to the soil in the litter.

In the secondary forest of Kade there was an average gain of 1600 kg N in 40 years in the vegetation store, or 40 kg a year. N was also accumulated in the soil organic matter. This amounted in total (vegetation + soil) to about 75 kg/year. Annual fixation then approximates 75 kg plus the annual losses. The N stored in the vegetation is, however, not available for agricultural use because it is lost by burning.

The number of years that crops can be grown after the forest has been cleared without added fertilizer depends on the site, the crop, and management practices. Nevertheless, fertilizers are necessary eventually because nutrients in the red and yellow soils, as well as the vegetation, are always limited. Shifting cultivation efficiently uses nutrients of the organic cycle on a low production level. An increase in yield, without additions of fertilizer, eventually leads to site degradation.

VI. ORGANIC MATTER

Organic matter is a very active component of the red and yellow tropical soils. The cation exchange complex and storage and availability of N and P are directly related to soil organic matter. This also is probably true for most minor elements. Hence, the changing amount of organic matter, as well as its influence on soil structure and erosion, are important in tropical agriculture. With a decrease in organic matter content, N, P, organically bound minor elements, and exchangeable bases can be released. As organic matter is increased some nutrients become fixed and rendered unavailable. Hence management practices that control the amount of organic matter can be the important factors in efficient use of nutrients. For these reasons, the factors that control the amounts and distribution of organic matter in tropical soils will be discussed briefly. Organic carbon content will be used as a measure of organic matter content.

Because the organic carbon content of soil is influenced by climatic and soil factors (Jenny, 1941), these should be evaluated with respect to their influence on the amount of carbon added to the soil and on the mean residence time of the organic carbon in the soil.

The organic carbon content in soils under equilibrium conditions is a linear function of the amount added (*Ad*) and its mean residence time (*R*). Organic carbon (%) = $Ad \times R$. Equilibrium, or near equilibrium conditions, are present in most red and yellow tropical soils under natural vegetation. In tropical lowlands the organic carbon contents of such soils are relatively low (high temperature, short residence time) and relatively high in tropical highlands (high temperature, long residence time). High clay contents tend to conserve organic carbon (long residence time). This is even more true for allophanes.

For equilibrium conditions the distribution of organic carbon with soil depth (organic carbon profile) can be considered a function of the amount of carbon added and its residence time at each depth. Under natural vegetation the organic carbon profiles of Oxisols and related deep soils can often be described as

$$C = ap^b \quad \text{or} \quad \log C = \log a + b \log p$$

in which C is the theoretical organic carbon content; p, soil depth; a and b are constants; and a is the content at 1 cm depth (Bennema, 1974a,b). Organic carbon profiles of a forest ecosystem in the Amazon valley and natural grasslands on dystrophic Oxisols are shown in Figs. 7 and 8.

Organic matter profiles of the very fine textured Amazonian Oxisols can be described approximately as $C = ap^{-0.5}$, which is valid to a soil depth of ±2 m. The value of a depends strongly on soil texture, and for the Amazonian very fine textured Oxisol is about 4.5. Thus, $C = 4.5/p^{\frac{1}{2}}$ at a depth of 1 cm the carbon content is thus about 4.5% and at a 16 cm depth about 1%. It thus decreases

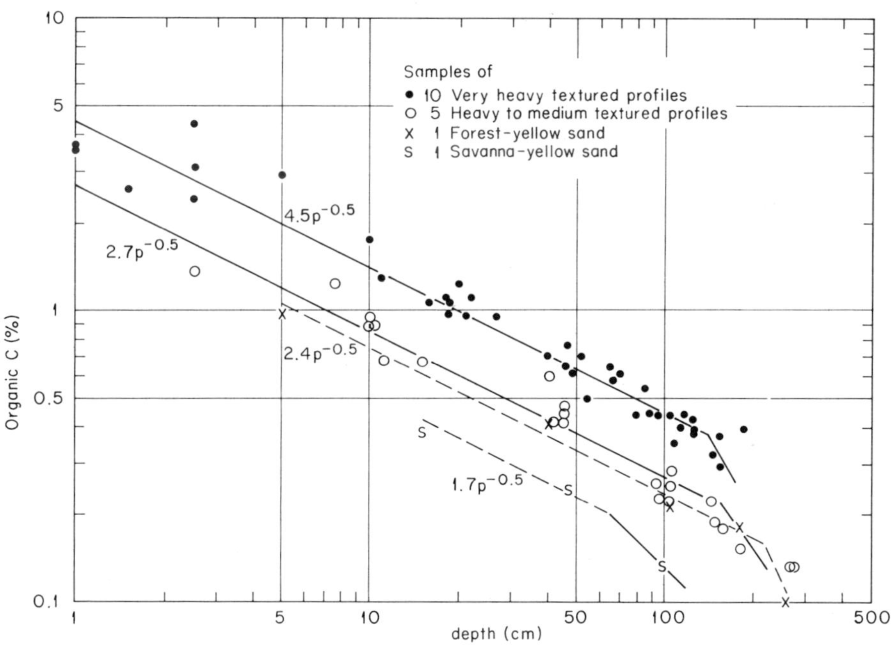

Fig. 7. The relation of organic carbon with depth in kaolinitic Yellow Latosols and sands in the Amazon region; p = depth (cm).

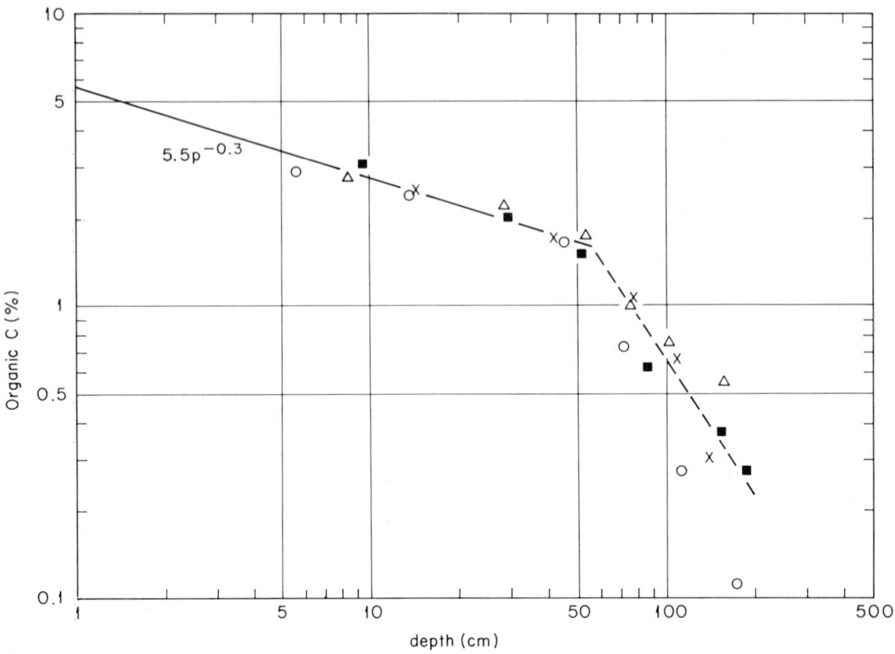

Fig. 8. The relation of organic carbon with depth in natural grasslands of Rio Grande do Sul. Vacaria 49 (○) and 14 (×), Durox 8 (■) and Bom Jesus 47 (△); p = depth (cm).

rapidly with depth. Figure 7 further shows that *a* is about 2.7 for the medium textured soils. In the sandy forest soil *a* is about 2.4, and still lower for sandy savanna soils.

Carbon profiles are different for soils supporting only grasses. In "grassland Oxisols" in southern Brazil (Rio Grande do Sul) $C = 5.5p^{-0.3}$. This relation is valid to a 90 cm depth (Fig. 8). A grassland profile from the Kenya–Kisii area shows almost the same function: $C = 6.1p^{-0.25}$ (van Wissen, 1974). The value of the exponent for p of −0.3 and −0.25 indicates that the decrease of organic carbon with depth in the surface layers is less than in most forest Oxisols.

A comparison of carbon profiles for grasslands and the dystrophic forest shows (1) a greater decrease in organic carbon with depth in the forest soil and (2) a relatively high total amount of organic carbon in the grassland profile in view of its relatively low production of organic matter. Net production of organic matter by grassland often is no more than one-half or one-third that of a forest ecosystem, but this is not reflected in total organic matter present in the profile. The residence time (e.g., conservation capacity) in dystrophic grassland soils is obviously higher than in dystrophic forest soils.

The residence time of carbon in tropical forest and grassland profiles increases with depth (Herrera and Tamers, 1971; Scharpenseel, 1971; Martel and Paul, 1971). Residence time at the soil surface can be counted in years and at a 30-cm depth in hundreds of years.

In tropical dystrophic forest ecosystems the organic matter is mainly deposited at the soil surface as litter and is not well mixed in the soil (Klinge and Fittkau, 1972; Fittkau and Klinge, 1973). Hence, most of the organic is decomposing in the zone where residence time is short so the total amount of organic matter in the soil is low. In grasslands a higher percentage of the organic matter is derived from grass roots. Grasses and herbs have many roots. These proliferate faster than tree roots and are deposited deeper in the soil where the conservation capacity is higher. Hence grassland profiles have higher organic matter content and their dark topsoils are deeper than in dystrophic forest profiles. It is further possible that the dead grass roots are more efficient than litter as in the building of humus. A similar effect on the amount of organic matter and development of dark topsoils occurs in other eutrophic forest profiles if earthworms are active.

It is important to recognize that the conservation capacity for organic carbon is strongly influenced by factors other than climate and soil. Maintaining a high organic matter content is important for sustained plant production. However, the amount of organic matter added to soils in most man-made ecosystems is lower than under natural conditions. The organic matter decomposes even faster in most agricultural soils than in natural ecosystems. This need not occur if proper techniques are used for conservation of organic matter. The conservation capacity of man-made ecosystems might be increased by adding less organic residues to the surface and more to the subsoil. In addition to other techniques it is important to pay attention to the kind of organic materials added and to influencing microbiological decomposition.

In most soils maintenance of an adequate amount of organic matter, together with an acceptable nutrient level, can be achieved only if a chemical fertilizer or organic matter is applied. However, this is not necessary during the initial agricultural occupation of previously forested area. The nutrients released by burning or by decomposition of organic matter become available to plants. If not used by crops they may be leached or strongly fixed. It is important, therefore, that during initial agricultural use, plants with dense and rapidly proliferating root systems be used. Grasslands following forests in the Amazon are a good example.

REFERENCES

Beek, K. J., and Bennema, J. (1972). Evaluacion de tierras para la planificacion del uso-rural–un metoda ecologico. *Bol. Lat.-Am. Fomento Tierras Aguas* FAO/PNUD No. 3.

Bennema, J. (1974a). Organic carbon profiles in Oxisols. *Pedologie* 119–146.
Bennema, J. (1974b). Oxissolos brasileiros. *An. Decimo. Q. Cong. Bras. Cienc. Solo.*, pp. 7–35. Soc. Bras. de Ciência de Solo.
Bolt, G. H. (1976). Adsorption of cations by soil. *In* "Soil Chemistry," Vol. 5A. Elsevier, Amsterdam.
Boyer, S. (1972). Soil potassium. *In* "Soils of the Humid Tropics," pp. 102–135. Natl. Acad. Sci., Washington, D. C.
Chan, H. Y., and Pushpajajah, E. (1972). "Productivity Potentials of Common Clones on Common Rubber Growing Soils." Rubber Res. Inst. Malaya, Kuala Lumpur.
Chang, S. C., and Jackson, M. L. (1958). Soil phosphorus fractions in some representative soils. *J. Soil Sci.* 9 (No. 1), 109–119.
de Wit, C. T., Dijkshoorn, W., and Noggle, J. C. (1963). Ionic balance and growth of plants. *Versl. Landbouwkd. Onderz.* **96.15**, 66.
d'Hoore, J., and Coulter, J. K. (1972). Soil silicon and plant nutrition. *In* "Soils of the Humid Tropics," pp. 163–174. Natl. Acad. Sci., Washington, D. C.
Dobereiner, J., Day, J. M., and von Bulew, J. F. W. (1975). Associations of nitrogen fixing bacteria with roots of forage grass and grain species. *Int. Winter Wheat Conf., 1975.*
Drosdoff, M. (1972). Soil micronutrients. *In* "Soils of the humid tropics," pp. 150–162. Natl. Acad. Sci., Washington, D.C.
Epstein, E. (1972). "Mineral Nutrition of Piants, Principles and Perspectives." Wiley, New York.
Equipe de Solos (1969). "Os Solos Da Area Manaus-Itacoatiara," Estudos e ensaios No. 1. IPEAN, Belem.
FAO-UNESCO. (1971). "Soil Map of the World," Vols. 1 and IV. UNESCO, Paris.
Fassbender, H. W., Vieira, L., and Stabili, M. E. (1970). Gleichgewichte der Kationen und Freisetzung von Kalium in einigen Boden von Amazonas in Brasilien. *Kali-Briefe,* Discipline 4, Soil Science, No. 46.
Fittkau, E. J., and Klinge, H. (1973). On biomass and thropic structure of the Central Amazonian rainforest ecosystem. *Biotropica* **5**(1), 2–14.
Harschner, H. (1974). Calcium nutrition of higher plants. *Neth. J. Agric. Sci.* **22,** 275–282.
Herrera, R., and Tamers, M. A. (1971). Radiocarbon dating of tropical soil associations in Venezuela. *Paleopedol.: Origin, Nat. Dating Paleosols, Pap. Symp., 1970* pp. 109–115.
Janssen, B. H., and van der Weert, R. (1976). The influence of fertilizers, soil organic matter and soil compaction on maize yields on the Surinam "Zanderij Soils." "Plant and Soil," Vol. 46. Netherlands.
Jenny, H. (1941). "Factors of Soil Formation." McGraw-Hill, New York.
Klinge, H. (1973). "Struktur und Artenreichtum des zentralamazonischen Regenwaldes," *Amazoniana* **4**(3), 283–293.
Klinge, H., and Fittkau, E. J. (1972). Filterfunktionen im Ekosystem des Zentralamazonischen Regenwaldes. *Mitt. Dtsch. Bodenkd. Ges.* **16,** 130–135.
Klinge, H., and Rodrigues, W. A. (1968). Litter production in an area of Amazonian Terra Firme forest. Part 1. Litter-fall, organic carbon and total nitrogen contents of litter. Part 2. Mineral nutrient content of the litter. *Amazoniana* **1**(4), 287–310.
Laudelout, H. (1961). "Dynamics of Tropical Soils in Relation to their Fallowing Techniques." FAO, Rome.
Leal, J. R., and Velloso, A. C. X. (1973a). Adsorcao de Fosfato em Latossolos sob Vegetacao de Cerrado. Part I. *Pesqui Agropecu. Bras., Ser. Agron.* **8,** 81–88.
Leal, J. R., and Velloso, A. C. X. (1973b). Dessorcao do Fosfato adsorvido em Latossolos sob Vegetacao de Cerrado. Part II. Reversibilidade da Isoterma de Adsorcao de

Fosfato em relacao ao pH da Solucao em Equilibro. *Pesqui. Agropecu. Bras., Ser. Agron.* **8**, 89–92.

McCormick, L. H., and Borden, F. Y. (1974). The occurrence of aluminum–phosphate precipitate in plant roots. *Soil Sci. Soc. Am., Proc.* **38**, 931–934.

Martel, Y. A., and Paul, E. A. (1974). The use of radiocarbon. Dating of organic matter in the study of soil genesis. *Soil Sci. Soc. Am., Proc.* **38**, 501–505.

Morelli, M. (1974). Fatores que influem na retencao e disponibilidade de Fosforo em Algumas. Unidades de Solo do Rio Grande do Sul. *An. Decimo Q. Congr. Cienc. Solo.*, pp. 191–205. Soc. Bras. de Ciência do Solo.

Moshi, A. O., Wild, A., and Greenland, D. S. (1974). Effect of organic matter on the charge and phosphate adsorption characteristics of Kikuyu Red Clay from Kenya. *Geoderma* **11**, 275–286.

Nye, P. H., and Greenland, D. J. (1960). "The Soil Under Shifting Cultivation," Tech. Commun. No. 51. Commonwealth Bureau of Soils, Harpenden, England.

Olson, R. A., and Engelstad, O. P. (1972). Soil phosphorus and sulfur. *In* "Soils of the Humid Tropics," pp. 82–101. Natl. Acad. Sci., Washington, D. C.

Sharpenseel, H. W. (1971). Radiocarbon dating in soils problems, troubles, hopes. *Paleopedol.: Origin, Nat. Dating Paleosols, Pap. Symp., 1970* pp. 77–89.

Schuffelen, A. C. (1974). Aspects of potassium and magnesium uptake by oats. *Neth. J. Agric. Sci.* **22**, 237–244.

Sombroek, W. G. (1966). "Amazon Soils." Pudoc, Wageningen, Netherlands.

Staff, Instituto Agronomico-Campinas. (1960). "Ensaios de Adubacao do Algodoeiro," Bol. No. 114. Instituto Agronomico-Campinas.

Staff, Soil Survey. (1975). "Soil Taxonomy." Agr. Handb. No. 436, Soil Conservation Service, U.S.D.A., Washington, D.C.

Stark, N. (1972). Nutrient cycling pathways and litter. *Bioscience* **22**, 355–360.

van Wambeke, A. (1974). "Management Properties of Ferralsols," Soils Bull. No. 23, FAO, Rome.

van Wissen, H. L. M. (1974). The influence of cultivation practices on the organic carbon content of some deep soil in Kisii District. M.S. Thesis, Wageningen, Netherlands.

Vettori, L. (1969). Metodos de analyse de solo. *Bol. Tec. no. 7, Equipe Pedol. Fertil. Solo (Braz.).*

Vieira, L. S. (1966). "Occorencia e formos de fosforo em solos da Amazonia, Brasil." Instituto Interam. de Ciencias Agricolas, Turrialba.

Wessel, M. (1971). Fertilizer requirements of cacao in South-Western Nigeria. Ph.D. Thesis, Wageningen, Netherlands.

Zandstra, H. G. (1972). Aluminum toxicity in some highly weathered soils of the tropics. Ph.D. Thesis, Cornell University, Ithaca, New York.

CHAPTER

3

Rice

SHOUICHI YOSHIDA

I. INTRODUCTION

Because the rice plant is highly adaptable to its environment and because man has succeeded in modifying the environment, rice can now be grown in widely different locations and under a variety of climatic environments.

Geographically, rice is grown in northeastern China at latitude 50°N, central Sumatra on the equator, New South Wales, Australia, and in Uruguay, South America at latitude 35°S. It is grown below sea level in Kerala, India, at sea level in many rice-growing areas, and at elevations above 2000 m in Kashmir and

Nepal. It can be grown under upland conditions, moderately submerged conditions, and in 150 to 500 cm of water.

Oryza sativa, the dominant rice species, is believed to have originated somewhere in Southeast Asia. Today, it is cultivated in Asia, Africa, Europe, North America, Central America, South America, and Oceania. Production statistics reveal that Asia is not only the "home area" of *Oryza sativa,* but also continues to be overwhelmingly the major rice-growing area of the world (Table I). On other continents, including Africa, where *Oryza glaberrima,* the other cultivated species of *Oryza,* originated, *Oryza sativa* is a newly introduced crop.

The average rice yield ranges from less than 1 to more than 6 tons/ha (Food and Agriculture Organization, 1972). There are a number of biological, environmental, and socioeconomic reasons for the large differences in rice yield. This chapter discusses some basic concepts that help establish a clearer understanding of the adaptability of rice cultivars to different environments, analyzes climatic factors that affect rice yield, and briefly discusses the growth pattern of rice as a crop and some cultural manipulations that increase yields.

II. ADAPTATION TO ENVIRONMENT

The duration of growth of a rice cultivar and the extent to which it is affected by daylength and temperature principally determine its specific adaptability to the regular crop season in a given location (Chang and Vergara, 1971).

Of the cultivated species, *Oryza sativa,* Indica variety group is being widely grown in tropical regions, such as Southeast Asia; the other variety group, Japonica, which is adapted to cooler areas, is largely grown in temperate

TABLE I
World Rice Production, 1971[a]

Region	Area (million hectares)	Production (million tons)	Average yield (tons/ha)
Asia[b]	124.5	286.6	2.30
South America	5.7	9.2	1.63
Africa	3.9	7.7	2.00
North and Central America	1.4	5.3	3.73
Europe[c]	0.8	3.2	4.30
Oceania	0.1	0.3	6.24
World	134.1	309.1	2.30

[a]Food and Agriculture Organization (1972).
[b]Including China.
[c]Including USSR.

countries, such as central and northern China, Korea, and Japan. Both Indica and Japonica varieties can also be grown in subtropical regions such as Taiwan.

Varieties adapted to temperate climate are insensitive or only weakly sensitive to daylength and can tolerate low temperatures, while varieties grown in the tropical climate have different degrees of sensitivity to daylength and are susceptible to low temperatures (Table II).

A. Response to Photoperiod

Photoperiodism of rice has held the interest of geneticists and physiologists for the past 50 years. Vergara *et al.* (1969) published a concise and useful review on this subject.

Growth of the rice plant can be divided into three stages—vegetative, reproductive, and ripening. The vegetative stage is further divided into a basic vegetative phase and a photoperiod-sensitive phase. The rice plant must attain a certain amount of growth—it usually must be 14 to 63 days old—before floral initiation can be triggered by short days. Because the very young plant is insensitive to photoperiod, the early growth stage has been termed the basic vegetative phase. Following this phase the plant enters a photoperiod-sensitive phase (psp) during which floral initiation can be triggered by short days. The psp of photoperiod-insensitive varieties varies between 0 and 30 days, while that of sensitive varieties lasts 31 days or much longer.

The length of the vegetative stage accounts mainly for the large differences (100 to 180+ days) in the time required to reach harvest among different rice varieties.

TABLE II
Responses by Rice Varieties Grown in Different Regions to Photoperiod and Temperature[a]

Region	Response to photoperiod	Response to temperature
Temperate	Insensitive to weakly sensitive	Tolerant to low temperatures at seedling stage and reproductive stage
Subtropical	Insensitive; weakly sensitive to strongly sensitive	Varying levels of high-temperature effect on basic vegetative phase. Varying degrees of tolerance to low temperatures at seedling stage. Basic vegetative phase generally unaffected by high temperatures
Tropical	Weakly to strongly sensitive in many varieties; insensitive in a few varieties	Development retarded by low temperatures at seedling stage; spikelet fertility reduced at reproductive stage. Basic vegetative phase unaffected by high temperatures

[a]Chang and Vergara (1971).

As is easily understood from the wide geographical distribution of rice, daylength during the growing season in rice-producing areas varies from 11 to 16 hours (Moomaw and Vergara, 1965). Response curves of seven varieties to photoperiod are shown in Fig. 1 (Chang and Vergara, 1971).

Information on duration of daylight at different latitudes and in different months is obtained, for instance, from the Smithsonian Meteorological Tables. In applying the results of laboratory studies to field performance, the usual "duration of daylight" may not be satisfactory since natural daylength is defined as the interval between sunrise and sunset, and since very low light intensities affect photoperiodic responses of rice in some experiments (Vergara *et al.*, 1969). For this reason, civil twilight may be added to natural daylength when laboratory studies need to be applied to field studies. A traditional Thai variety, Siam 29, strongly sensitive to photoperiod, does not flower when subjected to photoperiods longer than 13 hours. Improved varieties released by the International Rice Research Institute (IR8, IR5, IR20, IR22) and the University of the Philippines, College of Agriculture (C4-63) are weakly sensitive to photoperiod. Variety IR12-178 is completely insensitive to photoperiod and flowers in the same number of days after sowing under different daylengths. Most varieties in northern Japan are also completely insensitive to photoperiod (Sakamoto and Toriyama, 1967; Wada, 1952).

Plant breeders at the International Rice Research Institute have endeavored to select photoperiod-insensitive varieties (Beachell and Jennings, 1965), which

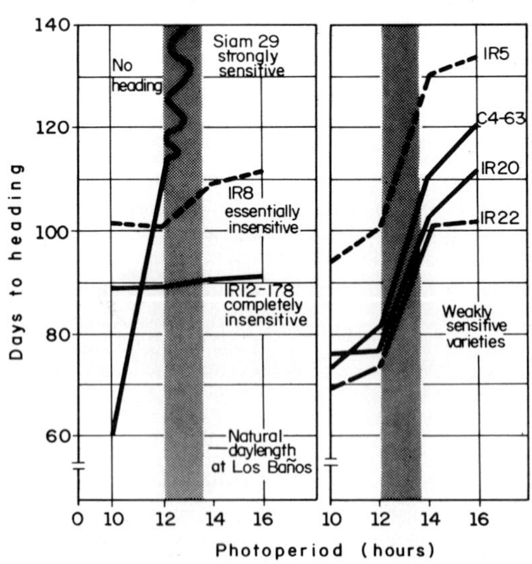

Fig. 1 Effect of four photoperiod treatments on the seeding-to-heading period of seven rice varieties (Chang and Vergara, 1971).

can flower and ripen throughout the year. The use of photoperiod-insensitive varieties makes the planning of rice cultivation more flexible and more suitable to the multiple cropping system characteristic of progressive agriculture. On the other hand, the photoperiod sensitivity of rice varieties continues to be a useful characteristic under certain environments.

Floating rice is planted early in the season, when the soil can still be worked, and usually is harvested when the flood waters have receded. It matures at the time the annual flood waters recede, about 180 to 200 days after sowing. Such a long growth period requires a photoperiod-sensitive variety. So far, there is no known tropical variety that has a long growth period and is not sensitive to photoperiod (Vergara *et al.*, 1969).

Photoperiodic sensitivity may function as a safety mechanism when the onset of the rainy season is considerably delayed (Oka, 1958). In most areas of tropical Asia, the onset of the rainy season is unpredictable and varies from year to year, whereas the end of the rainy season is fairly fixed. Thus, when planting is delayed because of delayed onset of the rainy season, a photoperiod-sensitive variety may still mature at its usual time, regardless of shortened growing periods. On the other hand, a photoperiod-insensitive variety requires a specified length of growing time for flowering and hence maturity. Thus, such a variety will suffer from drought at later growth stages after rainfall ceases.

Most traditional rice varieties in the tropics are sensitive to photoperiod, have long maturity periods, and are tall. When these varieties are grown with high levels of applied nitrogen they tend to lodge at later growth stages, resulting in very low grain yields. Takahashi *et al.* (1967) observed in Thailand that the popular photoperiod-sensitive varieties have a short growth period and usually are shorter when the farmers are forced by weather to transplant them later than at the usual time. In a delayed-planting experiment the September transplantings, two months later than usual, gave the highest yield for all varieties (Table III). By utilizing the photoperiodic sensitivity of the traditional Indica rice varieties "delayed planting" prevented excessive growth and lodging, resulting in high grain yields.

B. Response to Temperature

The response of the rice plant to temperature is much more complex and less understood than its response to daylength. The effects of temperature on growth and development of rice have been reviewed by Ishizuka *et al.* (1973), Nishiyama (1974), Owen (1971c), and Satake (1969).

Temperature affects growth of rice in two ways. First, a critically low or high temperature defines the environment under which the life cycle of the rice plant can be completed. Second, within the critically low and high temperature range, temperature influences the rate of development of leaves and panicles and the

TABLE III
Yield of Traditional Tropical Rice in Relation to Transplanting Time[a]

Variety	Yield (kg/ha) July	August	September	October	November
Bangkhen 293	229	3478	*4791*[b]	3386	2435
Luang Tawng 101	67	2455	*4121*	3098	2234
Nang Mon S-4	155	2222	*2971*	2010	1174
Puang Nahk 16	4136	4104	*4769*	3492	2388
Jao Luang 11	158	1791	*3571*	2120	2681
Khao Dok Mali 105	66	1219	*3031*	3020	1820
Bai Lod 104	388	1758	*3519*	2241	2019
Khao Pak Maw 17	265	1537	*3820*	2187	1528

[a]Takahashi *et al.* (1967).
[b]Italic figures indicate highest yields.

rate of ripening, thereby fixing the duration of growth of a variety under a given environment, and eventually determining the suitability of the variety to the environment.

The critically low or high temperatures vary from one growth stage to another (Table IV). These critical temperatures differ according to *variety, duration of critical temperature, diurnal change,* and *physiological activity of the plant.*

Subjecting the rice plant to temperatures below 20°C at about the meiotic stage of the pollen mother cells—about 11 days before heading—usually induces a high percentage of spikelet sterility (Satake, 1969). Differences among rice varieties in response to low temperatures at this stage of development have been clearly demonstrated. When a susceptible variety, Norin 20, was held at 15°C for 4 days, 51% of the spikelets were sterile, while a resistant variety, Hayayuki, under similar conditions produced only 5% sterile spikelets. Temperature as low as 12°C will not induce sterility if it lasts for only 2 days, but 6 days at 12°C will induce almost 100% sterility. Thus, not only the magnitude, but also the duration of ambient temperatures affects the response of the rice plant to low temperature. When the daily mean temperature was maintained at 17.5°C, the 10°C difference between day and night, i.e., 22.5°/12.5°C, decreased sterility considerably but the 5°C difference (20°/15°C) induced the same high sterility percentage as did the 0°C difference regime (17.5°/17.5°C). This suggests that a lower daytime temperature has a greater influence at the meiotic stage of pollen mother cells than does a lower nighttime temperature. On the other hand, nighttime temperature was more critical for floral initiation during the vegetative stage (Owen, 1969, 1972a,b). When variety IR8 was exposed to cool night

TABLE IV
Response of the Rice Plant to Varied Temperatures at Different Growth Stages

Growth stage	Critical temperature		Optimum temperature	References
	Low	High		
Germination	16°–19°C	45°C	18°–40°C	Chang and Vergara (1971); Nishiyama (1976)
Seedling emergence and establishment	12°–13°C	35°C	25°–30°C	Nishiyama (1976)
Rooting	16°C	35°C	25°–28°C	Nishiyama (1976)
Leaf elongation	7°–12°C	45°C	31°C	Nishiyama (1976)
Tillering	9°–16°C	33°C	25°–31°C	Nishiyama (1976)
Initiation of panicle primordia	15°C	–	–	Owen (1969, 1972a,b)
Panicle differentiation	15°–20°C	30°C	–	Nishiyama (1976); Satake (1969)
Anthesis	22°C	35°–36°C	30°–33°C	Poggendorft (1932); Kusanagi and Washio (1974); Sato *et al.* (1973); Tanaka and Wada (1955); Vergara *et al.* (1970)
Ripening	12°–18°C	>30°C	20°–29°C	Nishiyama (1976); Yoshida and Parao (1976)

temperature (15°C) during the vegetative stage and to a daytime temperature of 30°C, no panicle primordia were initiated. The same cool night temperature during the reproductive and ripening stages had an adverse effect, but did not have a similar dramatic effect during the vegetative stage (Owen, 1972a). More than 2 hours of cool night at 15°C delayed floral initiation of the variety Taichung Native 1 (Owen, 1972b).

Yatsuyanagi (1960) demonstrated that rice seedlings raised in upland beds could be transplanted when the daily mean temperature was about 13°–13.5°C, but seedlings raised in lowland nursery beds could be transplanted only when the daily mean temperature rose to 15.0°–15.5°C. Upland-grown rice seedlings have a higher starch and protein content, and thus have higher rooting capacity than lowland-grown seedlings.

The above examples illustrate the complexity of temperature effects on the different stages of rice growth and development. Hence, care must be taken to interpret critical temperatures, such as those shown in Table IV.

When optimum temperatures last for relatively short periods, it is important that the rice variety selected flower and mature before the temperature drops below the critical temperature, or that planting areas are suitable for rice cultivation. Temperature summation or heat sum is often used for this purpose (Chang, 1968; Went, 1957). Temperature summation is the sum of daily mean temperatures for an entire growing period or for the period from sowing to flowering. The temperature summation indexes of 14 Japanese rice varieties for the period from sowing to flowering range from 1000 to 3000 degree-days, depending on varieties, latitudes, and planting seasons (Toriyama *et al.*, 1969).

The predictive ability of temperature summation indexes for phasic development, and hence growth duration, of a rice variety under a given environment is based on the following:

(1) Temperature affects the rate at which a leaf develops (leafing rate). The higher the temperature, the faster the leafing rate (Komoda, 1958; Nagai, 1963; Yoshida, 1973). In addition, leafing rate is rapid in early stages of growth, and it slows down after initiation of panicle primordia. In terms of temperature summation indexes, the development of one leaf requires about 100 degree-days before panicle initiation and about 170 degree-days after panicle initiation (Komoda, 1958). Three to four leaves usually develop after panicle initiation. If a variety develops 16 leaves on the main culm before heading, 12 leaves develop before panicle initiation and 4 leaves develop later. Thus, 1200 degree-days are required for the first 12 leaves and 680 degree-days for the last 4 leaves. Additionally 250 degree-days are needed between flag leaf and heading. If a variety is grown at a mean temperature of 25°C, about 85 days are required to develop all 16 leaves.

(2) The number of leaves developed on the main culm before heading is fairly constant for a given variety. Most early- to medium-maturing varieties

develop about 10 to 18 leaves on the main culm. The number of leaves that develop on the main culm is constant for photoperiod-insensitive varieties. For example, Tainan-3 develops 15 leaves (Tanaka *et al.,* 1964) and IR747B2-6, 13 leaves (Yoshida and Parao, 1976) at Los Banos, Philippines, regardless of the planting season. On the other hand, photoperiod-sensitive varieties, Peta and BPI-76, develop 16 to 21 leaves and 13 to 23 leaves, respectively, depending on the planting season (Tanaka *et al.,* 1964).

(3) As a result of (1) and (2), the number of days from sowing to flowering becomes fairly constant for a photoperiod-insensitive variety under a given temperature regime.

(4) Temperature affects the rate of grain filling after flowering. Ripening persists for 30 days at a daily mean temperature of about 29°C to 53 days at a daily mean temperature of 18°C (Yamakawa, 1962). In terms of temperature summation indexes, the above figures correspond to 870 to 954 degree-days. In other studies, ripening requires about 700 to 800 degree-days (Ishizuka *et al.,* 1973; Komoda, 1958). For this reason the length of the ripening period varies from 64 to 66 days in cool regions such as Hokkaido, Japan, and Yanko, Australia to 30 days in warm regions such as Los Banos, Philippines, and Taichung, Taiwan. Temperature appears to be the predominant factor in the variation in length of the ripening period, varietal differences being small.

Generally an early-spring-planted crop in the temperate region requires larger temperature summation indexes and a variety adapted to warm climate, when planted in cool regions, requires larger temperature summation indexes. To explain these observations, Asakuma (1958) introduced "base temperature" into temperature summation indexes. The estimated base temperature of 30 varieties ranges from about 9° to 18°C. If the base temperature is subtracted from daily mean temperature, the temperature summation indexes for the sowing-to-heading period range from about 600 to 1000 degree-days, depending on variety. Varieties adapted to warm climates have higher base temperatures. Since temperature summation indexes are useful, primarily because they provide an estimate of the growth duration of a given variety under a given environment, a meticulous attempt to estimate an accurate base temperature for individual varieties may lower the value of temperature summation indexes; such an attempt will make the computation of temperature summation indexes more complex. As a compromise, a uniform base temperature, such as 10°C, may prove more useful in predicting the approximate growth duration under low temperature regimes. In this regard, Chang and Vergara (1971) found that a minimum temperature was better correlated than the temperature summation indexes with growth duration of IR8 rice planted at three locations. However, they obtained three different regressions for different locations, which indicated that the minimum temperature alone would not sufficiently account for the growth duration of IR8 at different locations.

Temperature summation indexes will probably be useful in predicting the growth duration of photoperiod-insensitive varieties or the feasibility of cultivating them under a given environment.

Ishizuka *et al.* (1973) discussed the use of temperature summation for rice cultivation in Hokkaido, Japan, where photoperiod-insensitive varieties are grown and where low temperature is the major factor limiting the growth of rice. In southern Japan, it is possible to grow two rice crops a year. Komoda (1958) estimated that two rice crops require 4400 to 4800 degree-days, and demonstrated that in regions below 36°N latitude in Japan, two rice crops can be successfully grown using early maturing, photoperiod-insensitive varieties.

C. Response to Water

One major characteristic of the rice plant is its ability to thrive in water, where the oxygen supply is limited. Because the rice plant has special tissues through which air can be transported from shoot to root (Arashi and Nitta, 1955; Arikado, 1953), it can be grown successfully in low-lying areas, such as swamps and other poorly drained areas. The transport of oxygen from shoot to root was demonstrated by the use of luminescence bacteria (Kumada, 1949), by oxidation of leucomethylene blue to methylene blue (Mitsui, 1960), and by the use of ^{15}O (Barber *et al.*, 1962).

Deep-standing water often inhibits seed germination and subsequent growth of the rice seedling. Some Indica varieties require more oxygen than some Japonica varieties for germination, rooting, and subsequent seedling growth (Takahashi, 1971). Consequently, seeds of these Indica varieties need exposure to air or to be under shallow standing water to ensure germination and rooting.

Depending on topography and rainfall pattern, low-lying areas may be subjected to different water depths and to different durations of high water. Vergara *et al.* (1976) proposed three categories for deep-water areas.

(1) Deep-water areas, where the water level is more than 150 cm up to 500 cm and where water usually remains in the field for three to four months. Bangladesh and Thailand have large acreages of deep-water rice. Special varieties known as "floating rice" are planted in these areas.

(2) Flooded areas, where the water level is less than 150 cm and the water usually remains in the field for several months. The margins of deep-water areas, low-lying areas, and tidal swamps are of this nature. India has a large flooded area in which the varieties used are usually the tall Indica type and are known as "flood rice."

(3) Submerged areas, where the water level is variable but the plants usually are completely submerged. The duration of submergence also varies. In some areas, submergence occurs almost every year while in many lowland rice fields

submergence occurs only during typhoons or unusually heavy continuous rains. No special varieties are used in these areas.

Floating rice has the ability to withstand an abrupt rise in water level and to grow rapidly above the water level. Floating rice can grow as tall as 600 cm. It normally elongates at 2–10 cm per day, but it can elongate by as much as 25 cm per day (Vergara *et al.*, 1976).

With respect to water regimes, rice varieties are often classified into lowland and upland varieties. It is not clear, however, if these two groups differ in some fundamental physiological functions. Upland rice varieties often are tall, early maturing, the resistant to blast: characterstics which are readily found in lowland varieties and, hence, are not specific to upland varieties.

III. CLIMATIC FACTORS AFFECTING GROWTH AND YIELD

A. Temperature

Even within the critical low and high temperatures, temperature has a more complex relationship with tillering, spikelet formation, ripening, and hence grain yield than does solar radiation since there is usually an optimum value for different physiological processes and since the optimum temperature varies with variety. Therefore, the results of an experiment depend on whether the range of temperatures studied is above or below the optimum temperature and on the variety used.

Temperature greatly influences growth rate just after germination (Fig. 2); the higher the temperature, the greater the growth rate. At later stages (3–5 weeks after sowing), however, temperature only slightly affects the tillering rate and the relative growth rate except at the lowest temperature studied. During the reproductive stage, within a temperature range of 22° to 31°C, the spikelet number per plant increases as the temperature drops. Thus, the optimal temperature appears to shift from high to low as growth advances from the vegetative to the reproductive stage (Yoshida, 1973). Low temperatures promote an increased number of spikelets in rice; this concurs with the findings from studies with other agronomic crops where relatively low temperatures have been found to increase the size of the inflorescence, the number of spikelets, the number of florets per spikelet, and the grain yield of perennial ryegrass (Ryle, 1965), wheat (Owen, 1971a,b; Thorne *et al.*, 1968), and barley (Tingle *et al.*, 1970).

The mean optimum air temperature for ripening of rice in Japan has been reported to be about 20° to 22°C (Aimi *et al.*, 1959; Matsushima *et al.*, 1957; Matsushima and Tsunoda, 1958; Tanaka, 1962). This optimum temperature agrees with the results of statistical analysis of effects of climatic factors on rice

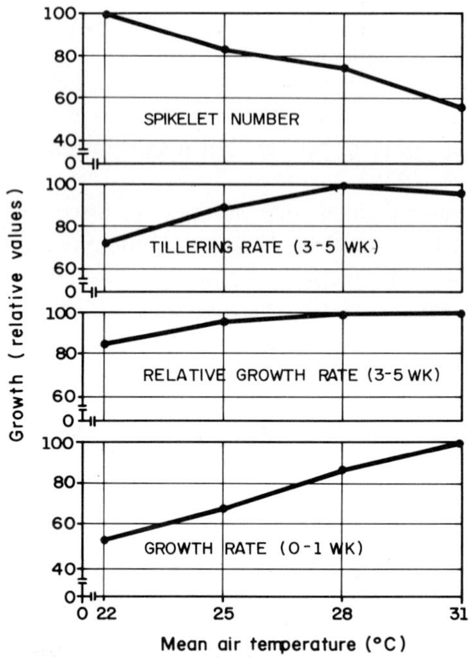

Fig. 2 Effects of temperature on growth of IR8 in a controlled environment (Yoshida, 1973).

yield (Hanyu *et al.*, 1966; Munakata *et al.*, 1967; Murata, 1964). There are optimum combinations of day and night temperatures for each stage of grain development of rice (Matsushima *et al.*, 1957, 1964; Matsushima and Tsunoda, 1958). A low nighttime temperature appears to favor ripening; this may be related to the effect of temperature on respiration (Yamamoto, 1954). The mean optimum air temperature decreases progressively from 21° to 14°C as the grain develops (Matsushima *et al.*, 1957, 1964).

In the tropics, however, a daily mean temperature as high as 28° or 29°C does not appear to be detrimental to ripening when solar radiation is high. High yields of from 8 to 10 tons/ha are obtained at Los Banos, Philippines, when rice crops ripen in April and May. These two months are characterized by high temperatures and high solar radiation. It is also possible that Indica varieties are better adapted to high temperatures while Japonica varieties require low temperatures for better ripening.

Temperature appears to influence the ripening of rice in two ways. First, low temperature favors an increase in weight per grain (Nagai, 1963; Murata, 1974). The 1000 grain weight in rice is generally a varietal character and, hence, is almost constant under different environments and different cultural practices.

However, the 1000 grain weight of the same variety has been shown to vary, in the 3 week period after heading, from about 24 g at a mean temperature of 22°C to 21 g at a mean temperature of 28°C (Murata, 1974).

Second, the length of the ripening period is inversely correlated with daily mean temperature (Yamakawa, 1962); the higher the temperature, the shorter the ripening period. Thus, persistent cloudy weather conditions will be more detrimental to grain filling under high temperatures because of a shorter ripening period. In fact, a combination of high temperature and low solar radiation can seriously impair ripening (Matsushima *et al.*, 1957).

In our recent studies on influence of temperature on ripening of one Indica (IR20) and one Japonica variety (Fujisaka 5), we found that upper spikelets matured in 13 days for IR20 and 18 days for Fujisaka 5 after anthesis at a mean temperature of 28°C (Fig. 3). Within the same panicle, date of anthesis among different spikelets usually varies by about 3 days. In the field, it takes about 14 days for all the panicles to emerge. Thus, in the tropics where the daily mean

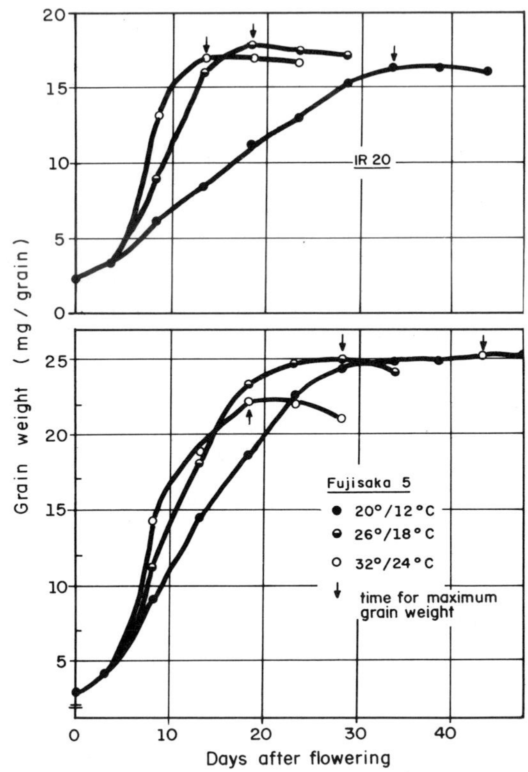

Fig. 3 Effect of temperature on grain growth of IR20 and Fujisaka 5.

temperature ranges from 25° to 30°C, duration of ripening becomes about 30 days after the beginning of panicle emergence. In rice agronomy, heading date is usually defined as the date when 50% of the panicles have emerged. If we take this as zero day of ripening, the duration of the ripening period becomes about 23 days after heading in the tropics.

The optimum temperature range for ripening of the whole panicle was 19° to 25°C for IR20 whereas it was 16° to 22°C for Fujisaka 5. Thus, IR20 appears to be better adapted than Fujisaka 5 to high temperatures.

B. Solar Radiation

Solar radiation is the only form of energy utilized in photosynthesis by green plants while temperature or heat is not used in the photosynthetic process. Thus, the amount of solar radiation affects rice growth through photosynthesis.

The solar radiation requirements of a rice crop differ from one growth stage to another (Stansel *et al.*, 1965; Yoshida and Parao, 1976). Shading during the vegetative stage only slightly affected yield and yield components (Table V). During the reproductive stage, however, shading had a very pronounced effect on spikelet number and, hence, on yield. The spikelet number per square meter

TABLE V
Effect of Shading at Different Growth Stages on Yield and Yield Components of IR747B2-6[a]

		Yield component			
Sunlight (%)	Grain yield (tons/ha)	Harvest index	Spikelet no./m^2	Filled grain (%)	1000 grain wt (g)
		Vegetative stage			
100	7.11	0.49	41.6	88.9	20.0
75	6.94	0.48	40.6	89.9	19.9
50	6.36	0.51	38.3	89.5	19.9
25	6.33	0.51	38.1	84.3	19.8
		Reproductive stage			
100	7.11	0.49	41.6	88.9	20.0
75	5.71	0.47	30.3	87.8	20.3
50	4.45	0.40	24.4	89.4	19.5
25	3.21	0.36	16.5	89.4	19.1
		Ripening stage			
100	7.11	0.49	41.6	88.9	20.0
75	6.53	0.49	41.1	81.1	20.0
50	5.16	0.44	40.6	64.5	19.5
25	3.93	0.38	41.7	54.9	19.1

[a]Yoshida and Parao (1976).

was positively and linearly correlated with dry matter production during the reproductive stage, implying that spikelet number may be linked with photosynthetic production during the reproductive stage. Shading during the ripening period also reduced grain yield considerably, because of a decrease in the percentage of filled grain. In comparing the relative importance of solar radiation at different stages of growth in terms of grain yield, the effect of solar radiation on grain yield can be seen to be the most marked at the reproductive stage, followed by the ripening stage (Fig. 4). The overall effect of solar radiation during the vegetative stage on grain yield is extremely small. A yield of 4 tons/ha can be obtained with 200 cal cm^{-2} day^{-1} during the reproductive stage (Fig. 4). To achieve the same yield, less solar radiation is required by the crop during the ripening period.

The amounts of solar radiation at eight locations (eight countries) in rice-growing areas are shown in Table VI. During the major rice cropping season, the monthly average of daily solar radiation ranges from about 300 to 700 cal cm^{-2} day^{-1} in most places. Thus, it is unlikely that incident radiation limits rice yield in most countries where a national average yield is below 4 tons/ha. A consistently high yield of 10 tons/ha is obtained in experimental fields in New South Wales, Australia (Boerma, 1964). The high incident solar radiation (Table VI) is considered one of the major reasons for this high yield.

When photosynthesis is restricted during ripening by cloudy weather, stored carbohydrates appear to be able to support growth of the rice grain despite the fluctuations in weather (Yoshida, 1972). A field study in which ^{14}C was supplied to a rice crop showed that 68% of the stored carbohydrates were translocated into the grain, 20% were respired during the ripening period, and

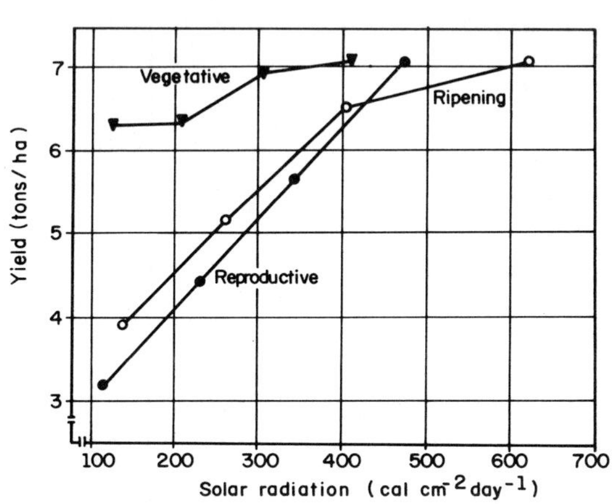

Fig. 4 Effect of solar radiation at different growth stages on grain yield of IR747B-26 (Yoshida and Parao, 1976).

TABLE VI

Solar Radiation Data ($cal \cdot cm^{-2} \cdot day^{-1}$) at Eight Locations of Rice-Growing Areas

Location	Jan.	Feb.	Mar.	Apr.	May	June	July	Aug.	Sep.	Oct.	Nov.	Dec.
Griffith, Australia[a]	700	670	520	380	260	250	240	340	460	560	720	710
Madras, India[b]	476	556	608	574	568	486	432	456	532	397	356	370
Djakarta, Indonesia[b]	361	382	400	396	374	361	388	426	451	422	397	370
Milano, Italy[b]	62	133	220	366	493	499	515	430	325	188	84	47
Saga, Japan[b]	206	295	330	393	381	322	375	444	349	349	298	228
Los Banos, Philippines	336	432	479	568	500	442	402	373	379	363	317	295
Bangkok, Thailand[b]	424	434	481	516	427	425	420	392	374	393	421	428
Davis, USA[b]	158	256	402	528	636	702	690	611	498	348	216	148

[a]C.S.I.R.O., Division of Irrigation Research, Griffith, N.S.W., Australia, meteorological data (1931–1967).

[b]Hayashi (1972).

12% remained in the vegetative tissues. The amount of the translocated carbohydrate was equal to about 21% of the grain carbohydrate, or equivalent to about 2 tons/ha of rough rice (Cock and Yoshida, 1972).

C. Rainfall and Water Requirements

Under rain-fed rice culture and where the temperature is within critical low and high temperatures, rainfall is perhaps the most limiting factor in rice cultivation. When irrigation is provided, growth and yield of rice are determined largely by temperature and solar radiation. Because of unpredictable variation in the frequency and amount of rainfall for a given location and from year to year, it is extremely difficult to find a simple relationship between rainfall and grain yield. At best, only the probable occurrence of drought can be predicted based on long-term records (Sastry, 1976). It is also difficult to generalize on the water requirements of the rice crop because of the large variations in topography, soil characteristics, and length of the crop-growing period among different locations. Currently, however, rice cultivation appears to be limited to areas where the annual rainfall exceeds 1000 mm (Kung, 1971).

The water requirements of rice crops in 43 locations in China, Japan, Korea, Philippines, Vietnam, Thailand, and Bangladesh are summarized in Table VII (Kung, 1971).

Transpiration is the loss of water through leaf surfaces and evaporation is the loss of water from a free water surface in an irrigated crop. Crop growth or leaf area growth is an increase in the "evaporation surface" in a physical sense. Transpiration loss increases with the increasing leaf area index (LAI) and reaches a plateau at an LAI of 3.5 to 4.0. At the plateau LAI, the transpiration loss is

TABLE VII
Water Requirement of an Irrigated Rice Crop[a]

By water loss	
Transpiration	1.5–9.8 mm/day
Evaporation	1.0–6.2 mm/day
Percolation	0.2–15.6 mm/day
	5.6–20.4 mm/day
By field operation	
Seed nursery	40 mm
Land preparation	200 mm
Field irrigation	1000 mm
	1240 mm/crop

[a]Kung (1971).

about 6 ± 2 mm/day (Kato *et al.*, 1965a), and it accounts for about 90% of the evapotranspiration loss (Kato *et al.*, 1965b).

Percolation loss is largely affected by topography, soil characteristics, and depth of the water table. Studies in Japan indicate that the water requirement of rice grown on sandy soils is about three times greater than that of rice grown on clay soils (Fukuda and Tsutsui, 1968). This difference is due largely to the difference in loss of water due to percolation. In fact, among the three ways water is lost from irrigated rice during the cropping season, the largest variation occurs in the percolation losses (Table VII). The total water loss ranged from 5.6 to 20.4 mm/day in irrigated rice, but most of the observed values for total water loss fell within the range of 6–10 mm/day. Thus, on the average, about 200–300 mm of rainfall per month are needed to produce a reasonably good crop of rice.

Culture of upland rice differs from that of irrigated rice in several aspects. First, the absence of levees allows water loss due to surface runoff during heavy rains; this is further increased by the sloping topography of most upland fields. Second, evaporation from the soil surface is much lower in upland fields than in flooded fields at early stages of growth. At later stages of growth, however, the difference may be negligible because water loss due to evaporation from the soil surface is only 10% of that lost by evapotranspiration. Third, transpiration per unit of the LAI is greater in upland crops than in lowland crops, presumably because of the drier environment in upland fields (Kato *et al.*, 1965a). If rainfall is evenly distributed, 200–300 m/month should be sufficient to produce a good rice crop. Under such conditions, the water losses due to percolation and surface runoff would be negligible. However, the rainfall pattern in most places is erratic and disproportionate. In lowland fields, the effective daily rainfall is defined as being not less than 5 mm and not more than 50 mm (Fukuda and Tsutsui, 1968). In upland fields, however, the upper limit of effective daily rainfall should be much less because the loss due to surface runoff would be greater in upland fields. Water stress, due to lack of rainfall or lack of irrigation, can retard the growth of rice. The most critical stage for grain yield is the period from about 10 days before flowering to flowering (Table VIII). Severe water stress at this critical stage causes high percentages of sterility, thereby reducing grain yield. Since the response is *irreversible,* an adequate supply of water at later stages is totally ineffective. On the other hand, water stress during the vegetative stage may reduce final plant height, reduce tiller number, and reduce leaf area, but the plant can recover from this retarded growth if water is supplied to the plant in sufficient time to permit its recovery before flowering.

D. Integrated Effects of Temperature and Solar Radiation

If the water supply is adequate, temperature and solar radiation appear to be the predominant factors affecting rice yield. The influence of climate, however, is apparent only when a good variety is grown with adequate spacing, on good

TABLE VIII
Effects of Drought on Yield and Yield Components of Rice[a]

Drought treatment (days from heading)	Yield (g/hill)	Panicle (no/hill)	Sterility (%)	Filled grains (%)	1000 grain wt(g)
−55	18.0	11	11	70	21.8
−51	16.8	11	9	66	22.0
−43	19.5	11	14	65	21.5
−35	20.0	12	11	60	20.5
−27	17.0	11	12	54	20.2
−19	15.7	11	34	52	20.8
−11	6.5	10	62	29	21.6
− 3	8.3	10	59	38	20.9
+ 5	16.5	11	10	59	21.9
+13	20.5	10	7	66	22.5
No stress	22.7	10	15	65	21.9

[a]Matsushima (1962).

soil, with an ample and timely nutrient supply, and with thorough control of pests and diseases. The yield difference between dry and wet seasons is pronounced only when the LAI is high or the grain yield is above 5 tons/ha (Fig. 5).

The grain yield of a rice crop can be expressed as

$$\text{yield (tons/ha)} = \text{spikelet number/m}^2 \times \text{grain weight (g/1000 grains)} \times \text{filled grain (\%)} \times 10^{-5} \quad (1)$$

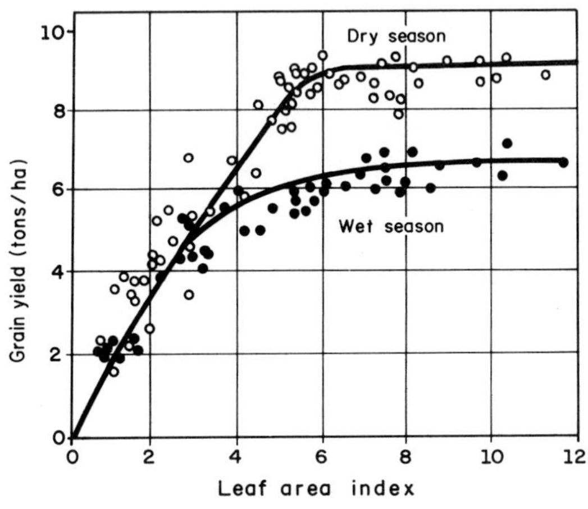

Fig. 5 Relationship between grain yield and leaf area index at heading in wet and dry season, 1966–1971, "Variety" IR8 (Yoshida and Parao, 1976).

Spikelet number per square meter is determined before flowering. Grain weight and filled grain percentages are determined during the period from just before flowering to maturity. Thus, weather conditions before and after flowering can affect the rice yield in different ways.

A number of studies in Japan indicated that yield per spikelet decreased with increased spikelet number per unit land area; this led many rice scientists working in temperate and tropical regions to study ripening as the most important stage affecting final crop yield.

In Japan, Murata (1964) found a close correlation between grain yield and solar radiation and daily mean temperature for August and September. Hanyu *et al.* (1966) also found a high correlation between grain yield and hours of sunshine and the daily mean temperature for the 40 days of the ripening period. Since these two studies were made on data collected from different localities where the soil and climate differed, the question of whether the effects of climatic factors on yield were adequately separated from those of soil factors remains unanswered. Murakami (1973), by growing rice in the same synthetic medium in different localities, also demonstrated close correlation between grain yield and daily mean temperature and hours of sunshine for 40 days of ripening period. These investigators developed formulas similar to the following:

$$Y = S[a - b\,(t - c)^2] \qquad (2)$$

where Y is grain yield, S is incident solar radiation or sunshine hours, t is average daily mean temperature, and a, b, c are constants. In spite of different sources of data, the optimum average daily mean temperatures for ripening were about the same; they ranged from 21.4° to 21.8°C. This optimum temperature for ripening agrees with findings obtained experimentally (Aimi *et al.*, 1959; Matsushima *et al.*, 1957; Matsushima and Tsunoda, 1958; Tanaka, 1962).

Munakata *et al.* (1967) derived a rather complex formula for a relationship between grain yield, solar radiation, temperature, and crop parameters, such as grain number and leaf blade weight. Most recently, Murata (1976) attempted an extensive analysis of the International Biological Program (IBP) experiments conducted in Japan, and derived the following formula:

$$Y = 382 + 0.627 W_0\, S \qquad (3)$$

where Y is yield in g/m^2, W_0 is dry weight in g/m^2 at heading time, and S is average daily solar radiation in cal cm^{-2} day^{-1} for the six weeks after heading. In this relation, W_0 was considered a better measure than spikelet number per square meter not only for representing the yield capacity but also for representing the average size of photosynthetic organs during the grain-filling period. Thus, the weather conditions before flowering affect the yield capacity of a crop and the size of the photosynthetic surface for ripening.

A high correlation was shown in the tropics between grain yield and solar radiation during the ripening period (Moomaw *et al.,* 1967) or during 45 days from 15 days before flowering to harvest (De Datta and Zarate, 1970). In addition, Tanaka *et al.* (1966) demonstrated a close association between grain yield and dry weight increase after flowering, suggesting that ripening is the most critical stage in grain production even in the tropics.

Since a major portion of grain carbohydrate comes from current photosynthesis during the ripening period (Yoshida, 1972), active photosynthesis during ripening is obviously important. However, whether ripening is more limiting to grain yield than any other growth stage under a given locality is yet another matter. In rice, the sink size of a crop is largely determined by spikelet number per square meter. Spikelets serve as a sink after fertilization and, therefore, sink size is determined not only before flowering, but also at flowering. Hence, the relative importance of climatic influence before and after flowering depends on whether the sink size is limiting grain yield.

Yoshida and Parao (1976) showed that spikelet number per square meter alone comprised 60% of yield variation while the combination of all the yield components accounted for 81% of yield variation. They derived the following relationship between grain yield and climatic influence for the experimental line IR747B2-6:

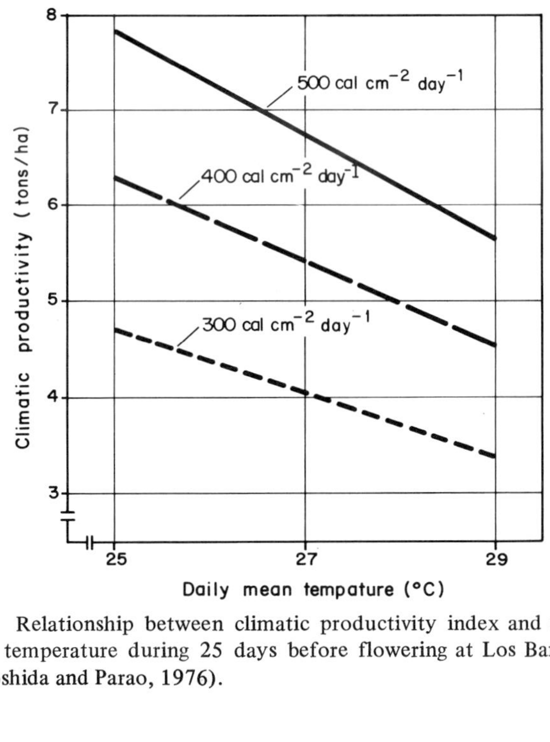

Fig. 6 Relationship between climatic productivity index and solar radiation and temperature during 25 days before flowering at Los Banos, Philippines (Yoshida and Parao, 1976).

$$Y = S(278 - 7.07t) \times 0.86 \times 18.1 \times 10^{-5} \tag{4}$$

where Y is yield in tons per hectare, S is solar radiation in cal cm^{-2} day^{-1}, t is temperature in °C during 25 days before flowering, 0.86 is the average filled grain percentage, 18.1 is the average 1000-grain weight, and 10^{-5} is a correction factor. The above formula implies that at Los Banos, Philippines, solar radiation and temperature during the reproductive stage (before flowering) have the greatest influence on rice yield because they determine the number of spikelets per square meter (sink size). Weather conditions during the ripening period are relatively stable and do not influence yield much. The relationship between daily mean temperatures and solar radiation during the reproductive stage and grain yield is shown in Fig. 6. A combination of low daily mean temperatures and high solar radiation gives a high yield. After examining the experimental results obtained at different locations in the tropics, however, Yoshida and Parao (1976) concluded that whether the weather conditions have a greater effect on rice yields before or after flowering is location specific or season specific. Thus, it is very important to examine the causes of yield variation, particularly the spikelet number per square meter and filled grain percentage under a given set of conditions.

IV. ANNUAL PRODUCTIVITY IN TEMPERATE REGIONS AND THE TROPICS

In temperate regions, one or two rice crops can be grown annually, while in the tropics, three to four crops a year are possible, provided irrigation is available. Thus, rice yield can be compared in two ways: on a per-crop basis and on an annual production basis.

The maximum recorded yield per crop in Japan is 10.5 tons/ha of brown rice, which is equivalent to 13.2 tons/ha of rough rice (Agricultural Policy Study Commission, 1971). Rice variety IR8 yielded about 10 tons/ha in the dry season in Los Banos, Philippines (De Datta *et al.*, 1968). On a yield-per-crop basis, the temperate regions appear to be more productive than tropical areas. In terms of annual production, however, the tropics have a greater potential than the temperate regions for rice production.

In Hokkaido, Japan, where only one crop a year is grown, about 6.5 tons/ha were obtained in a normal year (Table IX). In this region, low temperature is often a major cause of crop failure; hence, rice yield is unstable depending on weather conditions.

In southern Japan, where two crops can be grown in one year, a total of about 11 tons/ha was obtained from two crops. In Okinawa, the southernmost part of Japan, three crops a year are possible. One trial in the farmer's field

TABLE IX
Annual Productivity of Rice in Temperate Regions and in the Tropics

Location	Crop season	Yield (tons/ha)
Hokkaido, Japan[a] (43°N)	May–October	6.5
Kagawa, Japan[b] (34°N)	March–July	5.5
	July–October	5.4
		10.9
Okinawa, Japan[c] (27°N)	January–June	6.2
	June–August	4.9
	September–November	4.2
		15.3
Los Banos, Philippines[d] (14°N)	January–May	8.5
	May–July	4.9
	July–October	5.9
	October–December	4.4
		23.7

[a]Shiga *et al.* (1971), average of two years (1967 and 1968) for Sapporo.
[b]Komoda (1958), obtained for 1952.
[c]Shiraishi (1966), obtained for 1963.
[d]Yoshida and Parao (1976), average of 3 years (1971–1973).

recorded 15.3 tons/ha from three crops. In the Philippines, a total of about 24 tons/ha was recorded for four crops a year.

These data clearly indicate that the potential annual rice productivity is much higher in the tropics than in the temperate region. In practice, continuous rice cropping may not be advisable because it may cause severe disease and insect problems. However, the tropical environment for rice cultivation at any time of the year provides great flexibility in planning rice production.

V. GROWTH PATTERN AND GROWTH DURATION

As mentioned earlier, the life cycle of the rice plant can be divided into three growth stages: vegetative, reproductive, and ripening. The vegetative stage extends from germination to initiation of panicle primordia and is largely characterized by tillering. The reproductive stage is from initiation of the panicle primordia to flowering during which time the panicle primordia grow and develop into the panicle. Morphologically, the reproductive stage is characterized by internode elongation. During this stage, the spikelet number, and, hence, potential sink size of a rice crop, is determined. At this stage, the rice plant is very sensitive to such environmental stresses as low temperature, low solar radiation, and lack of water.

Ripening occurs from flowering to maturity. The major portion of grain carbohydrate is produced through photosynthesis during the ripening period, although some of the carbohydrate stored in the culms and sheaths before flowering is translocated into the grains. Thus, the life cycle of the rice plant includes sequential development of these three stages.

Tanaka (1976) proposed that the growth pattern of a rice crop be divided into three groups (types A, B, and C) with respect to the time of initiation of panicle primordia relative to tiller increase (Fig. 7a).

In type A, the initiation of panicle primordia begins soon after the maximum tiller number stage; the growth curve is generally type II or type IV (Fig. 7b); flowering is uniform and grain yield is high.

In type B of the growth pattern, tillering continues even after initiation of panicle primordia; the growth curve is often type III; flowering is irregular because of wide variation in the physiological age among tillers and grain yield is frequently low.

In type C, tillering ceases long before the initiation of panicle primordia; the

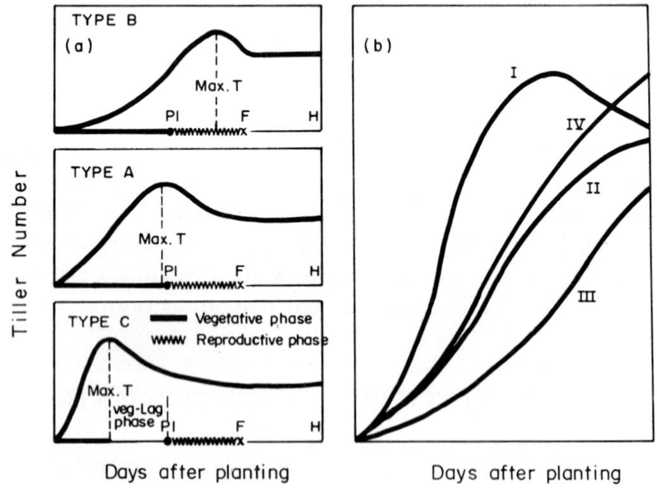

Fig. 7 (a) Diagram illustrating three types of phasic development. PI, Panicle-primordia initiation stage; F, flowering stage; H, harvest; Max. T, maximum tiller number stage. (b) Diagram illustrating various types of growth curves (adapted from Tanaka, 1976).

growth curve is generally type I; flowering is uniform but grain yield is often low.

Growth patterns of types A, B, and C result from the interaction among variety, environment, and management. Among the climatic influences on growth curves of a rice crop, temperature appears to be predominant (Ishizuka, 1971; Tanaka and Vergara, 1967). At high temperatures, the growth curve of a rice crop tends to be type I; at low temperatures, the growth curve is likely to be type III. Thus, the slowing of early growth in the tropics and its acceleration in cool regions may result in high yields with growth curve II or IV. The variety and cultural practices used under a given environment should be chosen with the objective of achieving type A growth pattern with a type II or type IV growth curve.

The above growth patterns are mainly concerned with duration of the vegetative stage which, in turn, largely determines growth duration (sowing to maturity). Hence, there appears to be a close relationship between duration of growth and rice yield (Tanaka and Vergara, 1967; Tanaka *et al.,* 1966; Kawano and Tanaka, 1968).

In the tropics, with conventional cultivation methods and ordinary commercial varieties, grain yields are highest when growth duration is about 130–140 days (Tanaka and Vergara, 1967). The optimum growth duration is affected by varietal character, cultural practices, and environment. Kawano and Tanaka (1968) found that the optimum growth duration for maximum yield is 120 days at high nitrogen levels and 160 days at low nitrogen levels. Spacing affects the grain yield of an early maturing variety because less time is available for vegetative growth of the crop (Table X). At the closest spacing, however, a

TABLE X
Effect of Plant Spacing on Grain Yield of Three Varieties During Wet Season[a]

Variety (line)	Spacing (cm)	Growth duration (days)	Grain yield (tons/ha)
IR747B2-6	5 × 5	95	5.62
	10 × 10	95	4.70
	20 × 20	95	4.15
IR22	5 × 5	114	4.92
	10 × 10	114	5.01
	20 × 20	114	4.57
IR8	5 × 5	124	5.32
	10 × 10	124	5.31
	20 × 20	124	5.14

[a]International Rice Research Institute Annual Report for 1970.

95-day rice variety can yield as much as a 124-day variety. Since it is difficult to achieve a spacing as close as 5 X 5 cm by the transplanting method, direct seeding would be preferable to transplanting for early maturing varieties.

VI. ENGINEERING FOR HIGHER YIELDS

A high grain yield of rice can be achieved only with the proper combination of variety, environment, and agronomic practices. Ishizuka (1969, 1971) attempted to synthesize currently available information on rice physiology into a fundamental concept of engineering for higher yields. He emphasized (1) designing a blueprint to obtain high yield, (2) tailoring an ideotype of a rice variety, and (3) manipulating desirable crop growth to achieve a target yield.

The concept of yield components (Matsushima, 1970) is useful in designing a blueprint for a desired rice yield and also in finding the "defects" of a given crop. For example, the following combination of yield components would result in a yield of 6 tons/ha.

$$6 \text{ tons/ha} = 32 \times 10^3 \text{ spikelets/m}^2 \text{ (400 panicles/m}^2 \times 80 \text{ spikelets/panicle)} \times 0.85 \text{ filled grain (\%)} \times 22 \text{ (1000-grain wt)} \times 10^{-5}$$

The required spikelet number could be obtained if 20 X 20 cm spacing is used, i.e., 25 plants/m^2, and each plant produces 16 panicles with 80 spikelets/panicle. Under reasonably good weather conditions, 85% of filled grains is normally expected, and 1000-grain weight is usually a varietal character; a similar example is described elsewhere (Chandler, 1969). Since the above computation is a kind of arithmetic, one cannot increase each of the yield components at his discretion. For instance, when the number of panicles per square meter is increased by using high plant density, the number of spikelets per panicle is decreased (Yoshida and Parao, 1972). A possible maximum yield is predetermined by the yield potential of a variety and environment. For designing a blueprint of target yield and examining defects of a given crop, a computation of yield using yield components would be meaningful if a comparison is made with a good crop already achieved under a comparable environment.

The tailoring of an ideotype variety and the choice of the right variety under a given environment are the keys to achievement of high yields.

Since Tsunoda's pioneering work in Japan and the concerted effort of breeders and physiologists at the International Rice Research Institute, considerable attention has been paid to the relationship between morphological characters and yielding capacity (Yoshida, 1972). The close association between certain morphological characters of rice varieties and yield response to nitrogen application led to the "plant type concept" as a guide for breeding high-yielding varieties. In summary, a variety of a good plant type has short stature, stiff culm, short and erect leaves, medium to high upright tillers, and a high harvest index (Chandler, 1968; Ishizuka, 1971; Jennings, 1964; Tanaka *et al.*, 1966; Yoshida, 1972).

Cultural manipulation of the growth of a rice crop usually includes changing the spacing both in plant density and in planting pattern, and regulating the rate of fertilizer used and its time of application. Since such manipulations are greatly affected by variety and environment, the best combination of cultural practices, usually location specific or season specific, must be found for a given set of conditions. Information on cultural practices is available from many publications (Matsubayashi *et al.*, 1963; UPCA, 1970).

In some areas, deficiencies and toxicities of nutrients and other chemical substances in soils cause poor plant growth. These can be identified by visual symptoms and chemical analysis of plant samples (Tanaka and Yoshida, 1970).

At extreme latitudes, where the summer season is short, as in Hokkaido, Japan, the greatest barrier to successful rice yield is the crop's failure to ripen because of a rapid decline in autumn temperatures. Under such climatic conditions, rice seedlings are raised in the early spring in a protected nursery bed covered with a plastic film, and transplanted into the main fields as soon as the temperature becomes favorable (Ishizuka *et al.*, 1973). This procedure extends the duration of the crop's growing season, and hence a crop can ripen in most years before the autumn temperature drops below a critical level.

The construction of irrigation and drainage facilities is man's most important effort to maximize and stabilize rice yields and to provide a fundamental requirement that affects the maximum realization of a rice variety's yield potential, cultural practices adopted, and climatic resources such as solar radiation and temperature. Although it is beyond cultural manipulation and requires a vast amount of capital investment, every effort should be directed toward providing irrigation and drainage facilities in rice-growing areas.

REFERENCES

Agricultural Policy Study Commission. (1971). "Technology of Japan No. 1 Winners in Rice Yield Contest." Fuji Publ. Co., Tokyo.

Aimi, R., Sawamira, H., and Konno, S. (1959). Physiological studies on mechanism of crop plants—the effect of the temperature upon the behavior of carbohydrates and related enzymes during the ripening of rice plant. *Proc. Crop Sci. Soc. Jpn.* **27**, 405–407.

Arashi, K., and Nitta, H. (1955). Studies on the lysigenous intercellular space as the ventilating system in the culm of rice and some other graminaceous plants. *Proc. Crop Sci. Soc. Jpn.* **24**, 78–81.

Arikado, H. (1953). Some considerations on the ventilating systems with special reference to anatomical observations in the rice plant. *Proc. Crop Sci. Soc. Jpn.* **22**, 49–50.

Asakuma, S. (1958). Ecological studies of heading of rice. I. Relations between the days from seeding to heading and several conditions. II. The basic vegetative growing habit, sensitivity to daylength, and sensitivity to temperature of Japanese rice. *Proc. Crop Sci. Soc. Jpn.* **27,** 61–66.

Barber, D. A., Ebert, M., and Evans, N. T. S. (1962). The movement of ^{15}O through barley and rice plants. *J. Exp. Bot.* **13,** 397–403.

Beachell, H. M., and Jennings, P. R. (1965). Need for modification of plant type. *In* "The Mineral Nutrition of the Rice Plant," pp. 29–35. Johns Hopkins Press, Baltimore, Maryland.

Boerma, E. B. (1964). "Sod-seeding of Rice in Southern Australia," Mimeogr. Pap. Int. Rice Comm. Meet., Manila, Philippines.

Chandler, R. F., Jr. (1968). Dwarf rice–a giant in tropical Asia. *In* Science for better living. *Yearb. Agric. U.S. Dep. Agric.,* pp. 252–255.

Chandler, R. F., Jr. (1969). Plant morphology and stand geometry in relation to nitrogen. *In* "Physiological Aspects of Crop Yield" (J. D. Eastin *et al.,* eds.), pp. 265–285. Am. Soc. Agron., Madison, Wisconsin.

Chang, J. H. (1968). "Climate and Agriculture. An Ecological Survey." Aldine, Chicago, Illinois.

Chang, T. T., and Vergara, B. S. (1971). Ecological and genetic aspects of photoperiod-sensitivity and thermo-sensitivity in relation to the regional adaptability of rice varieties. *Int. Rice. Comm. Newsl.* **XX,** 1–10.

Cock, J. H., and Yoshida, S. (1972). Accumulation of ^{14}C-labelled carbohydrate before flowering and its subsequent redistribution and respiration in the rice plant. *Proc. Crop Sci. Soc. Jpn.* **41,** 226–234.

De Datta, S. K., and Zarate, P. M. (1970). Environmental conditions affecting the growth characteristics, nitrogen response and grain yield of tropical rice. *Biometeorology* **4,** 71–89.

De Datta, S. K., Tauro, A. C., and Balaoing, S. N. (1968). Effect of plant type and nitrogen level on the growth characteristics and grain yield of indica rice in the tropics. *Agron. J.* **60,** 643–646.

Food and Agriculture Organization. (1972). "Production Yearbook," Vol. 26. FAO, Rome.

Fukuda, H., and Tsutsui, H. (1968). "Rice Irrigation in Japan." FAO, Rome.

Hanyu, J., Uchijima, T., and Suguwara, S. (1966). Studies on the agroclimatological method for expressing the paddy rice products. Part I. An agro-climatic index for expressing the quantity of ripening of the paddy rice. *Tohoku Nat. Agric. Exp. Stn., Bull.* **34,** 27–36.

Ishizuka, Y. (1969). Engineering for higher yields. *In* "Physiological Aspects of Crop Yield" (J. D. Eastin *et al.,* eds.), pp. 15–25. Am. Soc. Agron., Madison, Wisconsin.

Ishizuka, Y. (1971). Physiology of the rice plant. *Adv. Agron.* **23,** 241–315.

Ishizuka, Y., Shimazaki, Y., Tanaka, A., Satake, T., and Nakayama, T. (1973). "Rice Growing in a Cool Environment." Food and Fertilizer Technology Center, ASPAC, Taipei, Taiwan, Republic of China.

Jennings, P. R. (1974). Plant type as a rice breeding objective. *Crop Sci.* **4,** 13–15.

Kato, I., Naito, Y., Taniguchi, T., and Kamota, F. (1965a). Studies on the evapo-transpiration of crops. II. On the transpiration amount of paddy rice plants in paddy field and upland field conditions. *Bull. Tokai-Kinki Natl. Agric. Exp. Stn.* **12,** 39–63.

Kato, I., Naito, Y., Taniguchi, T., and Kamota, F. (1965b). Relation between transpiration amount and leaf area index on various crops. *Res. Prog. Rep., Tokai-Kinki Natl. Agric. Exp. Stn.* **2,** 14–18.

Kawano, K., and Tanaka, A. (1968). Growth duration in relation to yield and nitrogen response in the rice plant. *Jpn. J. Breed.* **18,** 46–52.

Komoda, K. (1958). "On the Flexibility of Rice Culture Period." Tokyo (unpublished), 58 pp.

Kumada, K. (1949). Investigation on the rhizosphere of rice seedling. Part 1. On the microscopic structure of the rhizosphere and oxidative power of root. *J. Sci. Soil (Tokyo)* **19,** 119–124.

Kung, P. (1971). "Irrigation Agronomy in Monsoon Asia." FAO, Rome.

Kusanagi, T., and Washio, O. (1974). The phytotron at Chugoku National Agricultural Experiment Station. *Bull. Chugoku Agric. Exp. Stn., Ser. A* No. 23, pp. 53–89.

Matsubayashi, M., Ito, R., Nomoto, T., Takase, T., and Yamada, N. (1963). "Theory and Practice of Growing Rice." Fuji Publ. Co., Ltd., Tokyo.

Matsushima, S. (1962). "Some Experiments on Soil Water Plant Relationship in Rice," Div. Agric. Bull. No. 112. Ministry of Agriculture and Cooperatives Federation of Malaya.

Matsushima, S. (1970). "Crop Science in Rice." Fuji Publ. Co., Ltd., Tokyo.

Matsushima, S., and Tsunoda, K. (1958). Analysis of developmental factors determining yield and its application to yield prediction and culture improvement of lowland rice. XLV. Effects of temperature and its daily range in different growth stages upon the growth, grain yield and its constitutional factors. *Proc. Crop Sci. Soc. Jpn.* **26,** 243–244.

Matsushima, S., Manaka, K., and Tsunoda, K. (1957). Analysis of developmental factors determining yield and yield prediction and culture improvement of lowland rice. XXXIV. On the mechanism of ripening (5) XL. On the mechanism of ripening (6). *Proc. Crop Sci. Soc. Jpn.* **25,** 203–206.

Matsushima, S., Tanaka, T., and Hoshino, T. (1964). Analysis of yield determining process and its application to yield-prediction and culture improvement of lowland rice. LXX. Combined effects of air-temperatures and water-temperatures at different stages of growth on the grain yield and its components of lowland rice. *Proc. Crop Sci. Soc. Jpn.* **33,** 53–58.

Mitsui, S. (1960). "Inorganic Nutrition, Fertilization and Soil Amelioration for Lowland Rice," 4th ed. Yokendo Ltd., Tokyo.

Moomaw, J. C., and Vergara, B. S. (1965). The environment of tropical rice production. *In* "The Mineral Nutrition of the Rice Plant," pp. 3–13. Johns Hopkins Press, Baltimore, Maryland.

Moomaw, J. C., Baldazo, P. G., and Lucas, L. (1967). Effects of ripening period environment on yields of tropical rice. *Int. Rice Comm. Newsl. Spec. Issue,* pp. 1–8.

Munakata, K., Kawasaki, I., and Kariya, K. (1967). Quantitative studies on the effects of the climatic factors on the productivity of rice. *Bull. Chugoku Agric. Exp. Stn., Ser. A* **14,** 59–95.

Murakami, T. (1973). Paddy rice ripening and temperature. *JARQ* **7,** 1–5.

Murata, Y. (1964). On the influence of solar radiation and air temperature upon the local differences in the productivity of paddy rice in Japan. *Proc. Crop Sci. Soc. Jpn.* **35,** 59–63.

Murata, Y. (1976). Productivity of rice in different climatic regions of Japan. *In* "Climate and Rice." Int. Rice Res. Inst., Los Banos, Laguna, Philippines.

Nagai, I. (1963). Growth and performance of rice varieties under controlled temperature and photoperiod conditions. *Int. Rice Comm. Newsl., Spec. Issue,* pp. 71–85.

Nishiyama, I. (1976). Effect of temperature on the vegetative growth of rice. *In* "Climate and Rice." Int. Rice Res. Inst., Los Banos, Laguna, Philippines.

Oka, H. I. (1958). Photoperiodic adaptation to latitude in rice varieties. *Phyton* **11,** 153–160.

Owen, P. C. (1969). The growth of four rice varieties as affected by temperature and photoperiod with uniform daily periods of daylight. *Exp. Agric.* **5,** 85–90.

Owen, P. C. (1971a). Responses of a semi-dwarf wheat to temperatures representing a tropical dry season. I. Non-extreme temperatures. *Exp. Agric.* **7,** 33–41.

Owen,. P. C. (1971b). Response of a semi-dwarf wheat to temperatures representing a tropical dry season. II. Extreme temperature. *Exp. Agric.* **7,** 43–47.

Owen, P. C. (1971c). The effects of temperature on the growth and development of rice. *Field Crop Abstr.* **24,** 1–8.

Owen, P. C. (1972a). Effects of night temperature on growth and development of IR8 rice. *Exp. Agric.* **8,** 213–218.

Owen, P. C. (1972b). Effects of cool periods (15°C) at night on Taichung Native 1 rice. *Exp. Agric.* **8,** 289–294.

Poggendorft, W. (1932). Flowering, pollination, and natural crossing in rice. *Agric. Gaz. N. S. W.* **43,** 898–904.

Ryle, G. J. A. (1965). Effects of daylength and temperature on ear size in S24 perennial ryegrass (*Lolium perenne* L.). *Ann. Appl. Biol.* **55,** 107–114.

Sakamoto, S., and Toriyama, K. (1967). Studies on the breeding of nonseasonal short duration rice varieties with special reference to the heading characteristics of Japanese varieties. *Bull. Chugoku Agric. Exp. Stn., Ser. A* **14,** 147–165.

Sastry, P. S. N. (1976). Climate and crop planning with particular reference to rainfall patterns. *In* "Climate and Rice." Int. Rice Res. Inst., Los Banos, Laguna, Philippines.

Satake, T. (1969). Research on cool injury of paddy rice plant in Japan. *JARQ* **4,** 5–10.

Sato, K., Inaba, K., and Tozawa, M. (1973). High temperature injury of ripening in rice plant. 1. The effects of high temperature treatment in different stages of panicle development on the ripening. *Proc. Crop Sci. Soc. Jpn.* **42,** 207–213 (with English summary).

Shiga, H., Miyake, M., Kurisaki, H., Sekiya, S., Kakimoto, A., Doi, Y., and Kataoka, T. (1971). Analysis of factors responsible for high yields in Hokkaido. *In* "Studies on Maximizing Rice Yield," Res. Rep. No. 49, pp. 117–122. Ministry of Agriculture and Forestry, Japan.

Shiraishi, K. (1966). Guides for improvement of rice cultivation and efficient utilization of paddy fields in Okinawa. *Ryuku Demonst. Farm.* 193 pp.

Stansel, J. W., Bollich, C. N., Thysell, J. R., and Hall, V. L. (1965). The influence of light intensity and nitrogen fertility on rice yield and components of yield. *Rice J.* **68,** 34–35.

Takahashi, H. (1971). Effect of depth of standing water on germination of some indica rice varieties. *Proc. Crop Sci. Soc. Jpn.* **40,** Suppl. 1, 143–144.

Takahashi, J., Danchanomai, P., Kanarengsa, C., and Krasesindhu, P. (1967). Increasing the yields of photo-sensitive varieties by modifying their cultural practices. *Int. Rice Comm. Newsl.* **26,** 39–44.

Tanaka, M. (1962). Studies on the growth of lowland rice caused by cool water irrigation and delayed heading. *Aomori Agric. Exp. Stn., Bull.* **7,** 1–107.

Tanaka, A. (1974). "Comparisons of Rice Growth in Different Environments," Mimeogr. Pap. Int. Rice Res. Inst., Los Banos, Laguna, Philippines.

Tanaka, A., and Vergara, B. S. (1967). Growth habit and ripening of rice plants in relation to the environmental conditions in the Far East. *Int. Rice Comm. Newsl., Spec. Issue* pp. 26–42.

Tanaka, A., and Yoshida, S. (1970). Nutritional disorders of the rice plant in Asia. *Int. Rice Res. Inst., Tech. Bull.* No. 10.

Tanaka, A., Navasero, S. A., Garcia, C. V., Parao, F. T., and Ramirez, E. (1964). Growth habit of the rice plant in the tropics and its effect on nitrogen response. *Int. Rice Res. Inst., Tech. Bull.* No. 3.

Tanaka, A., Kawano, K., and Yamaguchi, J. (1966). Photosynthesis, respiration, and plant type of the tropical rice plant. *Int. Rice Res. Inst., Tech. Bull.* No. 7.

Tanaka, M., and Wada, J. (1955). Effects of low temperatures on blooming of rice plants seeded and transplanted lately. *Proc. Crop Sci. Soc. Jpn.* **23,** 254–257.

Thorne, G. N., Ford, M. A., and Watson, D. J. (1968). Growth, development and yield of spring wheat in artificial climates. *Ann. Bot. (London)* [N. S.] **32,** 425–446.

Tingley, J. N., Faris, D. G., and Ormrod, D. P. (1970). Effects of temperature, light, and variety in controlled environments on floret number and fertility in barley. *Crop Sci.* **10,** 26–28.

Toriyama, K., Sakamoto, S., Iwashita, T., and Hsu, C. S. (1969). Heading characteristics of japonica varieties planted at the places of different latitudes. *Bull. Chugoku Agric. Exp. Stn, Ser. A* **17,** 1–16.

UPCA (University of the Philippines College of Agriculture). (1970). "Rice Production Manual." UPCA, College, Laguna, Philippines.

Vergara, B. S., Chang, T. T., and Lilis, R. (1969). The flowering response of the rice plant to photoperiod. A review of the literature. *Int. Rice Res. Inst., Tech. Bull.* **8.**

Vergara, B. S., Chu, T. M., and Visperas, R. M. (1970). Effect of temperature on the anthesis of IR8. *Int. Rice Comm. Newsl.* **19,** 12–17.

Vergara, B. S., Jackson, B., and De Datta, S. K. (1976). Response of the rice plant to deep water stress. *In* "Climate and Rice." Int. Rice Res. Inst., Los Banos, Laguna, Philippines.

Wada, E. (1952). Studies on the response of heading to daylength and temperature in rice plants. I. Response of varieties and the relation to their geographical distribution in Japan. *Jpn. J. Breed.* **2,** 55–62.

Went, F. W. (1957). "The Experimental Control of Plant Growth." Ronald Press, New York.

Yamakawa, H. (1962). Studies on the ecological variations of the growth of rice plants caused by the shifting of cultivation season in warm region in Japan. *Agric. Bull. Saga Univ.* **14,** 23–159.

Yamamoto, K. (1954). Studies on ripening of rice. Part 1. Part II. Part III. *Agric. Hortic.* **29,** 1161–1163, 1303–1304, and 1425–1427.

Yatsuyanagi, S. (1960). Scheduled cultivation of rice in Tohoku region. *Agric. Hortic.* **35,** 931–934, 1095–1098, and 1425–1427.

Yoshida, S. (1972). Physiological aspects of grain yield. *Annu. Rev. Plant Physiol.* **23,** 437–464.

Yoshida, S. (1973). Effects of temperature on growth of the rice plant (*Oryza sativa* L.) in a controlled environment. *Soil Sci. Plant Nutr. (Tokyo)* **19,** 299–310.

Yoshida, S., and Parao, F. T. (1972). Growth performance of improved varieties in the tropics with special reference to tillering capacity. *Exp. Agric.* **8,** 203–212.

Yoshida, S., and Parao, F. T. (1976). Climatic influence on yield and yield components of lowland rice in the tropics. *In* "Climate and Rice." Int. Rice Res. Inst., Los Banos, Laguna, Philippines.

CHAPTER

4

Sugarcane

LOUIS G. NICKELL

I. INTRODUCTION

Sugarcane (*Saccharum officinarum* L.) has been known from ancient times, long before the Christian era. The original home of sugarcane, for many years in dispute, is now believed to be New Guinea (Brandes, 1956). The dispersal of cultivated forms of sugarcane from New Guinea is closely related to the ancient migrations which covered a large part of the world.

Sugarcane was one of the first tropical crops to be adapted to large-scale farming. The length of the growing season varies from less than 9–10 months in Louisiana (frost dates setting the time limits), to two years in Peru and South Africa, and two years or more in Hawaii. Most of the rest of the sugarcane is grown in 14- to 18-month plant crops followed by 12-month ratoon crops.

Cane sugar is produced commercially in over 70 countries, territories, and island groups, generally within a band around the world bounded by 35°N and 35°S latitudes.

II. ENVIRONMENTAL EFFECTS ON GROWTH

In order to appreciate the importance of the Hawaiian breeding program for the sugar industry, one should keep in mind (1) that in Hawaii cane is grown from below sea level (behind sand dikes) to elevations over 3000 ft, (2) that rainfall in cane areas ranges from only a few inches to almost 300 inches per year, and (3) that because of the trade winds and mountains, cloud cover ranges from heavy to light, from all day to part of the day, giving wide variation in solar radiation. These and other climatic variables result in the need for almost a dozen varieties to take care of the "important" ecological zones. Thus, when a "variety is changed," it really means several varieties.

The varieties best for these several ecological zones differ in many characteristics. The studies referred to here were carried out (1) to determine how much variation in physiological characteristics existed among these varieties, and (2) to characterize the qualities of a variety best suited for a given environmental area to aid in selection of varieties for that area.

A. Solar Radiation

Sugarcane is well known as a "sun-loving" plant (Burr *et al.*, 1957). Being a C_4 plant (Kortschak *et al.*, 1965), sugarcane is not saturated by light as far as total photosynthetic production is concerned. However, in the case of photosynthetic efficiency, i.e., mg CO_2 fixed per unit of leaf area per unit of time, there are considerable differences among varieties. Figure 1 shows our concept (based on observed rates of photosynthetic efficiency) of limits of photosynthetic efficiency and how they might be used to select a variety for an ecological zone (at least as far as solar radiation is concerned).

We have found varieties which respond in their photosynthetic efficiency in the manner represented by the lower curve as well as those represented by the

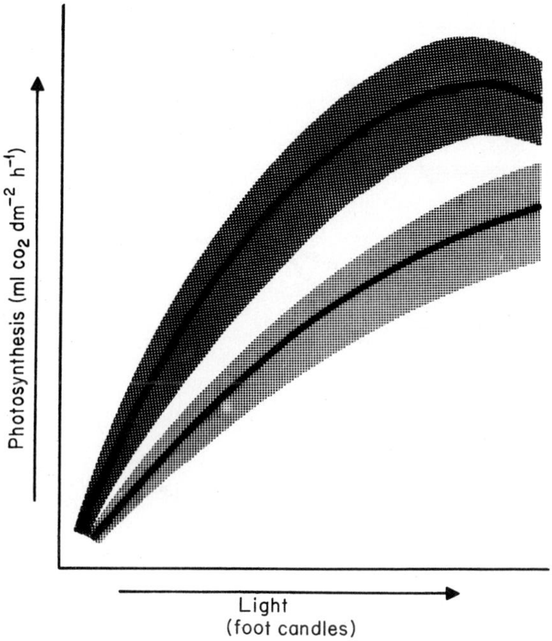

Fig. 1 Response (measured as photosynthetic efficiency) of sugarcane varieties to increasing light.

upper curve. Varieties represented by the lower curve have an increase in their efficiency that is directly related to the increase in solar radiation. Consequently, one would expect that if solar radiation in any given year decreases, both the efficiency and total output of a variety would decrease proportionately. Varieties represented by the upper curve increase in efficiency up to a certain amount of radiation and then decrease. For such a variety, a decrease in solar radiation would increase efficiency. Under field conditions we have found this to be the case. A severe decrease in radiation, e.g., 15–20%, significantly reduces the yield of a field-grown variety that has a photosynthetic efficiency curve such as that shown by the lower curve. A variety represented by the upper curve shows little or no decrease in yield under the same conditions.

Figure 2 shows yield data from experimental plots of four varieties. These results show the interaction between total radiation and amounts of nitrogen. There are increasing amounts of nitrogen on the horizontal axis. The vertical axis shows actual total yield. The four solid curves at the top of the figure represent yields in full sunlight; the dotted lines at the bottom show the yields under 50% of normal solar radiation. When sunlight is reduced, varieties H49-5 and H50-7209 still produce fairly well on a relative basis. On the other hand, H37-1933

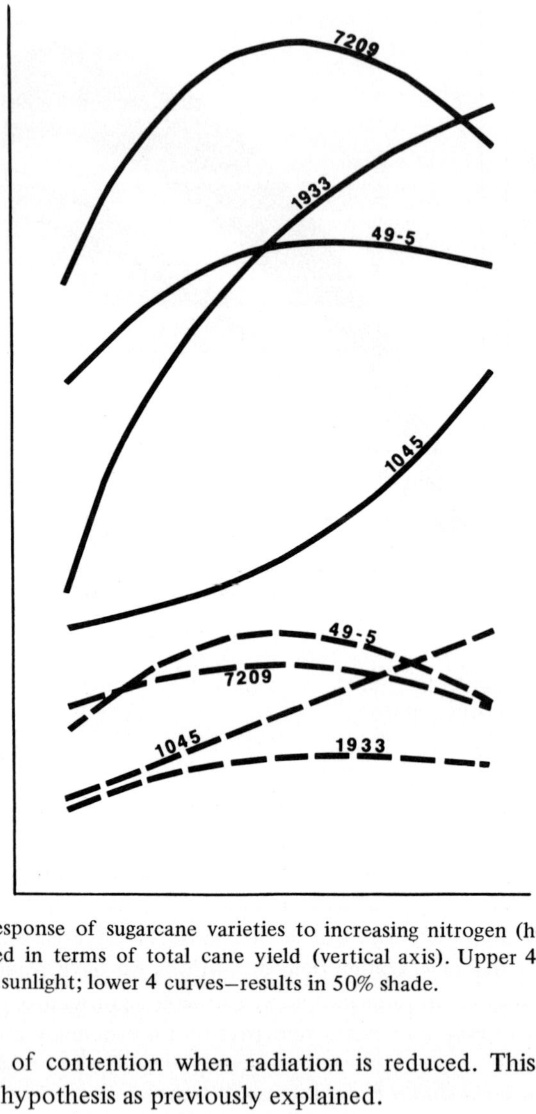

Fig. 2 Response of sugarcane varieties to increasing nitrogen (horizontal axis) measured in terms of total cane yield (vertical axis). Upper 4 curves—results in full sunlight; lower 4 curves—results in 50% shade.

almost drops out of contention when radiation is reduced. This is in keeping with our working hypothesis as previously explained.

B. Temperature

Our first experiment showing varietal differences in response to temperature is shown in Fig. 3. In this experiment, seed pieces of several varieties were planted individually in 1-gal cans and allowed to germinate and grow until the

Fig. 3 Response of sugarcane varieties to cold temperature (see text for explanation of experimental imposition of stress).

young plants were a few inches high. Growth during this initial period is represented by the part of each plant below the bottom ribbon. The plants were then placed in a cold chamber for a month. Growth during this cold period is represented by the part of each plant between the bottom ribbon and the one above it. The plants were then returned to ambient conditions for about 6 weeks–then replaced in the cold chamber for a month, etc. The alternate cold and normal periods are shown by the respective ribbons. The varietal differences to this treatment are readily apparent.

Figure 4 shows the root systems of cane plants taken from controlled temperature chambers (Brodie *et al.*, 1969). On the right, the plant at 90°F root temperature shows excellent growth; the decrease to 77°F does not have much

Fig. 4 Growth of root systems of sugarcane (variety H44-3098 in controlled temperature chambers). Effect of varying root temperatures. Left to right: 51°F, 64°F, 77°F, 90°F.

of an effect on root growth. A decrease to 64°F significantly reduces the yield of the root system, although the quality of it has not been impaired. A glance at the 51°F root temperature on the left shows that the roots are almost dead.

It is interesting that the response to temperature is expressed at the cellular level in cell suspensions of the several varieties grown at different temperatures.

Tissue cultures were started from three different varieties. From these tissue cultures, suspension cultures were developed. The suspension cultures were grown on rotary shakers at reduced temperatures, resulting in very poor growth. The cell suspensions were then grown under temperature conditions which were approximately 15°F higher. Figure 5 shows the effect of increasing these temperatures. Each test tube represents the settled cell volume from one of the Erlenmeyer flasks which had been shaken. In each group of tubes, the three on the left represent growth at the cold temperature, and the three on the right represent growth at the higher temperature. On the left is variety H49-5 which grows in the field under cold, rainy conditions. In this variety there is little increase in growth by increasing the temperature. The variety in the middle, NCo-310, shows an intermediate response, whereas the variety represented by tubes on the right, H50-7209, which is known to be adversely affected by cold temperatures under normal conditions, shows how much difference there is in the growth of these cell suspensions between the cold and the warm temperatures.

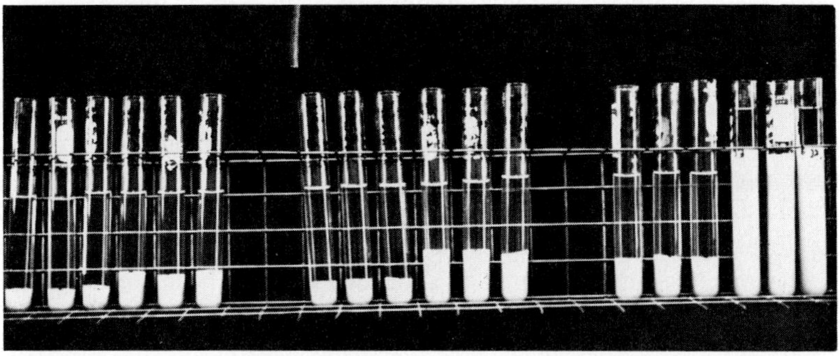

Fig. 5 Effect of temperature on growth of cell suspension cultures from 3 sugarcane varieties. (Demonstrated by settled cells from triplicate cultures). Left to right: H49-5 at cold temperature, H49-5 at increase of 15°F; NCo-310 (cold), NCo-310 (+15°F); H50-7209 (cold), H50-7209 (+15°F).

C. Water

Although it is well known that sugarcane requires an abundance of water (Meguro, 1963; Nickell, 1969), there are substantial varietal differences in the requirement or, more important, in the ability to produce under adverse water regimes (Tanimoto, 1969). Some of our early work on response of Hawaiian sugarcane varieties to salinity suggested by the way the varieties reacted to osmotic pressure of this type that there was a relationship between reaction to salinity and reaction in the field to drought. This was confirmed when we reviewed data from extensive field records showing crop performance as well as crop reaction throughout the growth cycle of cane under drought. After extensive experimentation, a test was developed (Tanimoto and Nickell, 1967a), using sodium chloride for osmotic shock, but adding it all at one time, rather than gradually increasing the level as had been done in salinity studies. This test permits a quick, simple, and cheap laboratory determination of drought resistance of sugarcane varieties. We are particularly interested in the level of salt at which the young cane plant is killed. In a typical test, sensitive varieties such as H53-1447 and H37-1933 were killed by adding 9–12 g of salt to the 1-gal cans in which the test plants had been grown. A tolerant variety, such as H57-5174, survived with 21 g of added salt. Over a period of years, there has been good correlation among varieties between reaction to this type of test and drought resistance in the field. Plant breeders now use this test routinely to screen varieties for drought resistance.

D. Varietal Responses to Environmental Factors

If one merely looks at the variation in varietal performance of sugarcane around the world, or, for that matter, even in nearby ecological zones of a small area, it becomes obvious that the environment plays a most important role in affecting the physiological expression of a given variety. As a practical matter, this variation is one of the reasons for the generous exchange of both information and cane varieties among sugar-producing areas of the world. Basic information, of course, is universally useful; field information from a given locality has limited direct usefulness elsewhere.

To emphasize the above points, data obtained in Hawaii will be discussed to show the differing responses to Hawaiian-bred varieties to a number of ecological factors. The practical use to which the information has been put will also be discussed.

E. Fertilizers

The response of almost any plant to increasing amounts of nitrogen is an increase in tillering, an increase in rate of leaf development, and quick coverage of the area in addition to greening of the foliage. This can be shown dramatically in sugarcane. Despite this generalization, there are tremendous varietal differences in response to nitrogen.

Table I shows, on a dry weight basis, the percentage of leaf punch nitrogen taken from the leaves of several varieties. The nitrogen levels in the leaves, as exhibited by the leaf punch nitrogen data, vary strikingly among these varieties. For example, after adding 500 lb of nitrogen per acre, variety H39-3633 does not have as much nitrogen in the leaf as does H56-432 at 100 lb of nitrogen when it is not growing well. Such information casts doubt on the widely accepted method of using nitrogen analysis as a means of fertilization prediction unless it is specifically designed for a given variety and its characteristics.

TABLE I
Varietal Differences in Leaf Punch Nitrogen at Increasing Amounts of Applied Nitrogen[a]

	Stimulated applied N levels		
Variety	100	300	500
56-432	0.995	1.792	2.134
49-3533	0.814	1.455	1.955
57-5174	0.946	1.438	1.899
50-7209	0.756	0.975	1.453
49-5	0.616	0.850	1.192
39-3633	0.659	0.715	0.856

[a]Values given are percent dry weight.

Sugarcane varieties show striking differences in response to various trace elements. For example, Fig. 6 shows the fourth leaf blades of two plants of variety H59-3775 grown in a nutrient culture solution for several months. Zinc was added to the culture solution of the plant on the right, but not to that of the plant on the left. It is very difficult to detect any difference between these two plants. When grown in soils which show low zinc levels, there apparently is no difference in field yields of this variety. On the contrary, when plants of variety H58-1545 are grown under the same conditions (Fig. 7), the minus zinc leaf on the left shows extremely severe symptoms of zinc deficiency, as opposed to the one on the right which was grown in an adequate supply of zinc. When this variety is grown in the field, it shows symptoms of this nature and will not grow in zinc-deficient soils unless zinc is added.

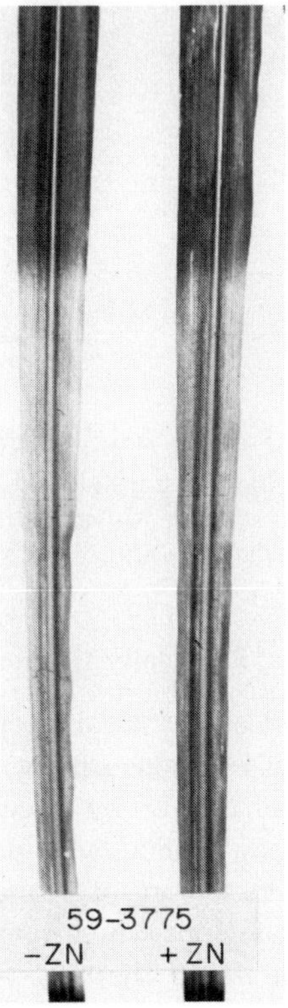

Fig. 6 Fourth leaf blade of variety H59-3775 (left, from plant in culture solution – Zn; right, from plant in culture solution + Zn).

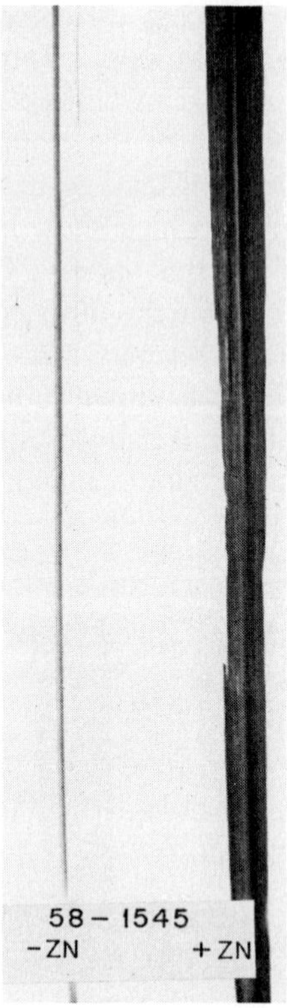

Fig. 7 Fourth leaf blade of variety H58-1545 (left, from plant in culture solution – Zn: only midrib has chlorophyll and shows in photograph; remainder of leaf is chlorotic, almost white; right, from plant in culture solution + Zn).

F. Hormones, Especially Gibberellic Acid

Commercial sugarcane varieties in Hawaii respond rather dramatically to additions of gibberellic acid. This hormone can be applied to the leaves, through the root system and nutrient culture, or by injecting it into the stalk of the plant. Because sugarcane increases in length by development from intercalary meristems in each internode, the effect of gibberellic acid is shown in the increased growth of those nodal regions which have not fully elongated—which were the number 4, 3, and 2 nodes at the time of application (Tanimoto and Nickell, 1968). All of our subsequent work with commercial varieties showed that this effect lasted for about one month (Tanimoto and Nickell, 1967b), because between 3 and $3\frac{1}{2}$ joints are affected. Commonly one joint is formed about every 10 days under Hawaiian conditions. Interestingly, elongation of noble canes was not confined to a one-month period but continued until the gibberellic acid was essentially "used up." This serves as the basis of investigations to determine the reasons for the difference between the noble canes and commercial types.

G. Salinity

As mentioned, early experiments showed the extreme variation among sugarcane varieties to increases in osmotic pressure. In these studies, the osmotic pressure increase was due to the addition of sodium chloride to the normal nutrient solution. After the seed pieces had been germinated in the nutrient solution, at weekly intervals one atmosphere of pressure in the form of sodium chloride was added. This was continued until the level had reached 6 atmospheres. One can readily observe the sensitivity of varieties such as H37-1933 and H53-1447 and the relative tolerance of varieties such as H50-7209 and H41-1045. Subsequent tests with salinity showed that two important varieties, H57-5174 and H50-7209, are fairly tolerant of saline conditions. However, there also is a difference between these two, with H57-5174 being more tolerant (Fig. 8). This photograph was taken at Pioneer Mill Company on the island of Maui in

Fig. 8 Varietal difference in response to soil salinity, Pioneer Mill Company, Island of Maui, Hawaii. Left, H57-5174; right, H50-7209. Both varieties planted at same time.

the Hawaiian chain. It shows a fairly saline field in which the two varieties are grown on either side of an irrigation supply pipe. H57-5174 on the left is growing much better than H50-7209 on the right. From these results a theoretical scheme shown in Fig. 9 was devised in which increased salinity is indicated on the horizontal axis and increased growth on the vertical axis. At extremely low levels of salinity, there is little difference among varieties. At extremely high levels of salinity, there is no difference—all of them will be killed. Nevertheless, these curves are fairly representative of tolerant varieties represented by the upper curve, sensitive varieties represented by the lower curve, and intermediate varieties which are somewhere in between.

H. Wind

Wind often affects sugarcane adversely. In Hawaii, the last three plantations, which have been abandoned, were on the tips of the islands in the head of the trade winds. These winds induce certain physical injury to the leaves, and also bring in salt spray. Through use of radioactive carbon dioxide, we were able to show the effects of different types of wind injury (Fig. 10). These effects were measured in terms of their inhibition of translocation. When the midrib alone was bent by wind, but not the lamina of the blade, there was about a 35% reduction in movement of photosynthate out of the leaf. However, when both the lamina and midrib were bent, there was almost complete inhibition of translocation. Surprisingly, when wind frayed the tips of the leaves, which is a very common form of damage, there was not as much injury as expected and translocation was inhibited by less than 50% (Hartt, 1964). The effects of wind are not all harmful. As in most other plants the rate of photosynthesis of

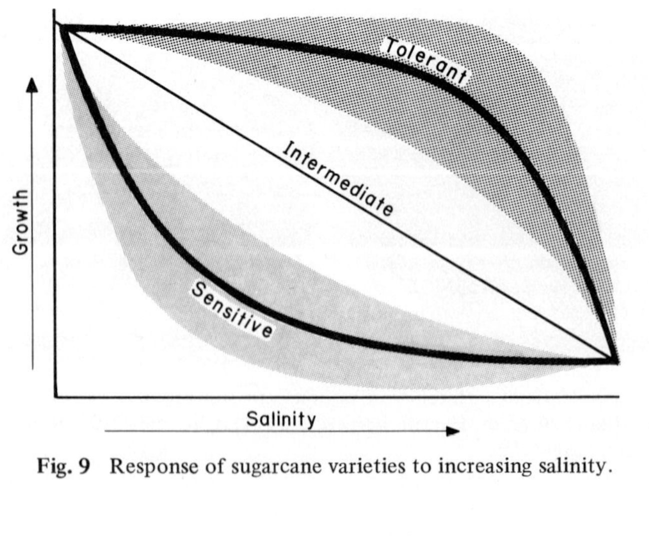

Fig. 9 Response of sugarcane varieties to increasing salinity.

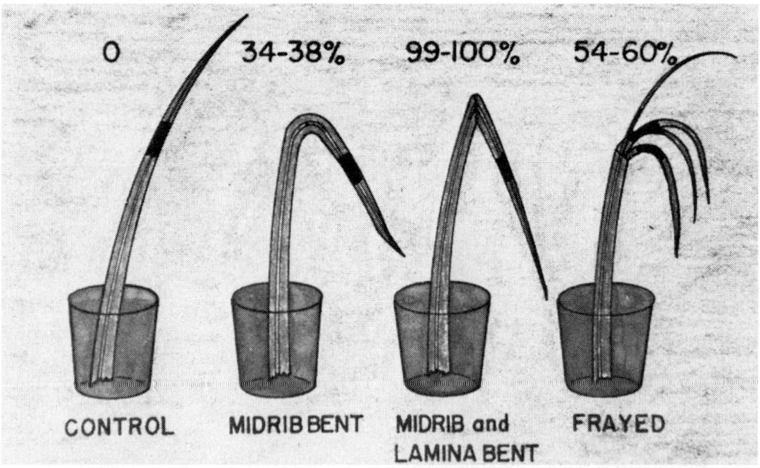

Fig. 10 The inhibition of translocation (of carbon-14) caused by various types of simulated wind injury.

sugarcane increases as the CO_2 level is increased. The trade winds, which normally are about 10 mph and occur throughout most of the year, probably ensure that there is no diminished supply of CO_2 in the middle of cane fields. Sensitive CO_2 monitors in cane fields have never detected reduction in CO_2 concentration. By comparison, studies in corn fields in the midwest show that on hot, sunny days growth is limited by low CO_2 concentrations.

I. Flooding

The problems caused by flooding result from a decreased oxygen supply. The cane plant is highly aerobic. When grown in nutrient culture, air must be continuously bubbled through the solution. Effects of aeration on roots of plants grown in long cylinders are shown in Fig. 11.

III. GERMINATION

If one-eye seed pieces are cut and planted in nutrient solutions, after several weeks there is a difference among some varieties in the manner in which germination occurs. This test was carried out by cutting single-eye seed pieces of two varieties, putting them into a trough of nutrient solution, covering them over with plastic so they would be in the dark, and allowing them to imbibe water and start germination. The cover was removed after about 10 days to 2

Fig. 11 Effect of decreased aeration on growth of sugarcane root systems. Numbers 1 and 2 on left, high aeration; 3 and 4, progressively decreased aeration.

weeks. Variety H37-1933 developed an extensive root system and hardly developed any shoot at all. H50-7209, on the other hand, developed an extensive shoot system with very little root system. These differences are apparent in the field. For example, H50-7209 tends to emerge very rapidly and after growing for a short while, to turn chlorotic. If left alone, the chlorosis disappears, because the root system will then have developed sufficiently for uptake of soil nitrogen.

A. Depth and Angle of Planting

The effect of planting depth on the percentage of emergence is shown in Table II. This test was carried out using single-eye seed pieces of variety H57-5174. Table III shows the effect of planting angle on percentage of emergence. This was done by using seed pieces of variety of H50-7209. Planting angle means the angle from the vertical of the "eye" of the seed piece—upright being 0°. These two sets of data suggest the value of proper planting depth and even the effect of the angle of the germinating shoot.

B. Effect of Chemicals, Especially Arginine

Experiments were conducted to find chemicals that would control the rate of germination and, ultimately, affect the rate of shoot emergence. Germination

TABLE II
Effect of Planting Depth on Percent Emergence[a]

	Planting depth (inches)				
Days after planting	1	2	3	4	5
8	20	10	–	–	–
9	50	20	–	–	–
10	50	20	10	–	–
11	50	40	30	–	–
12	60	60	50	10	30
13	100	70	50	10	30
15	100	80	60	60	70
17	100	80	70	70	90
19	100	80	70	80	100

[a]Variety H57-5174.

tests showed a strong stimulatory effect of adding 20 ppm of amino acid arginine to the water in which the seed pieces were immersed (Nickell and Kortschak, 1964). When arginine was added to the nutrient solution (replacing a small fraction of the nitrogen in the solution culture), it had a very strong positive effect on growth of sugarcane. Figure 12 shows the results of harvesting the millable cane from variety H44-3098 which had been grown in nutrient solution with and without added arginine for a period of 4 months. The cane on the right represents that grown in basic nutrient solution—the cane on the left represents that grown in basic nutrient solution to which 20 ppm of arginine had been added. Solutions in both these cases were replaced at weekly intervals.

TABLE III
Effect of Planting Angle on Percent Emergence[a]

	Planting angle (degree)				
Days after planting	0	45	90	135	180
13	40	30	–	–	–
14	50	40	–	–	–
15	60	40	10	–	–
17	70	80	10	–	10
19	80	80	50	–	20
20	100	80	100	30	30
22	100	100	100	70	60
25	100	100	100	100	100

[a]Variety H50-7209.

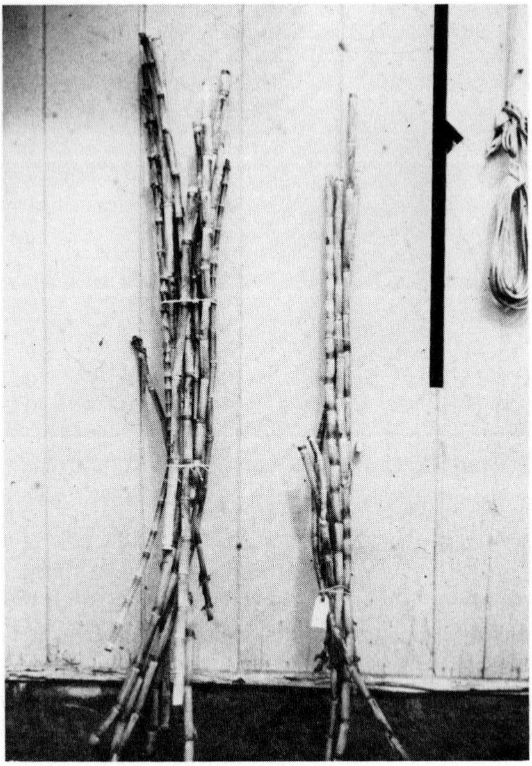

Fig. 12 Effect of arginine in culture solution on cane production by variety H44-3098. Left, arginine; right, basic nutrient solution.

Additional experiments were conducted to determine if arginine stimulation would occur at the cellular level (Maretzki and Nickell, 1968). Elimination of arginine from the nutrient medium for growing sugarcane cells caused a drastic reduction in growth. This prompted us to investigate rather thoroughly the arginine "story" in sugarcane, both at the cellular level and the possibility of its use in the field.

Figure 13 shows the growth of cell suspensions with and without arginine added. The striking effect of arginine at the cellular level, and on sugarcane growth, is well established (Nickell and Maretzki, 1972).

IV. TILLERING

Once germination of the seed piece has occurred and the young shoot has emerged from the ground, the grower of the cane is next interested in tillering of

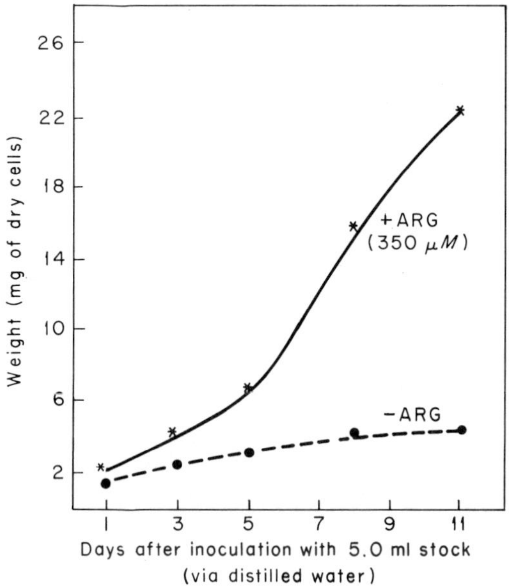

Fig. 13 Sugarcane cells grown either in the presence or absence of arginine in a synthetic medium.

his crop (Nickell, 1968). Probably as closely related to yield as any single factor is the number of stalks per acre at harvest. To obtain as large a number of stalks as can be supported by an acre of cane, maximum tillering early in the crop should be induced.

A. Effect of Sunlight

Sunlight or solar radiation is a major factor in production of shoots. The effect can be seen from controlled studies in which sunlight is excluded by 50% by using a Saran netting; other factors remaining equal. Such shading results in dramatic reduction in production of branches.

B. Effect of Pretreatment of "Seed"

For many years we have conducted tests under the code name VRN [Variety by Radiation (Solar) by Nitrogen]. Many sets of data discussed so far have come from tests of this nature. Samples were periodically harvested during growth of the crop and at final harvest at 24 months (normal crop age in Hawaii). At about 8 months of age, when we usually cut the seed for planting, we took samples of several stalks from a number of varieties which had been given three different levels of fertilizer, and two levels of solar radiation. We then divided these stalks into the first three upper joints which normally would be cut for seed cane and planting, and took three-node sections on down the stalk. The number of nodes, of course, is a function of varietal characteristics as affected by fertilization rates and amount of solar radiation. The effects of pretreatment of a cane stool on subsequent production of tillers by a given seed piece from such a stalk are shown in Figs. 14–18. Variety H49-3533 is well known for its poor tillering characteristic. No matter from which position on the stalk the three-joint seed pieces were cut, very little tiller production was demonstrated. Figures 15–18 are all of variety H50-2036. They were selected from numerous sets of data to illustrate the interaction between sunlight and nitrogen on effects of tillering. Figure 15 shows the effect of sun with no added nitrogen. A maximum of 70 tillers formed, and these were produced by the third seed piece from the top. If 200 lb of nitrogen were added at the same amount of solar radiation (Fig. 16), there was a tremendous increase in the number of tillers. With no nitrogen added and 50% sunlight, there was about the same amount of tillering as in full sun with no nitrogen (Fig. 17). However, each stalk gave only three good, three-joint seed pieces. With addition of 200 lb of nitrogen in 50% sunlight (Fig. 18),

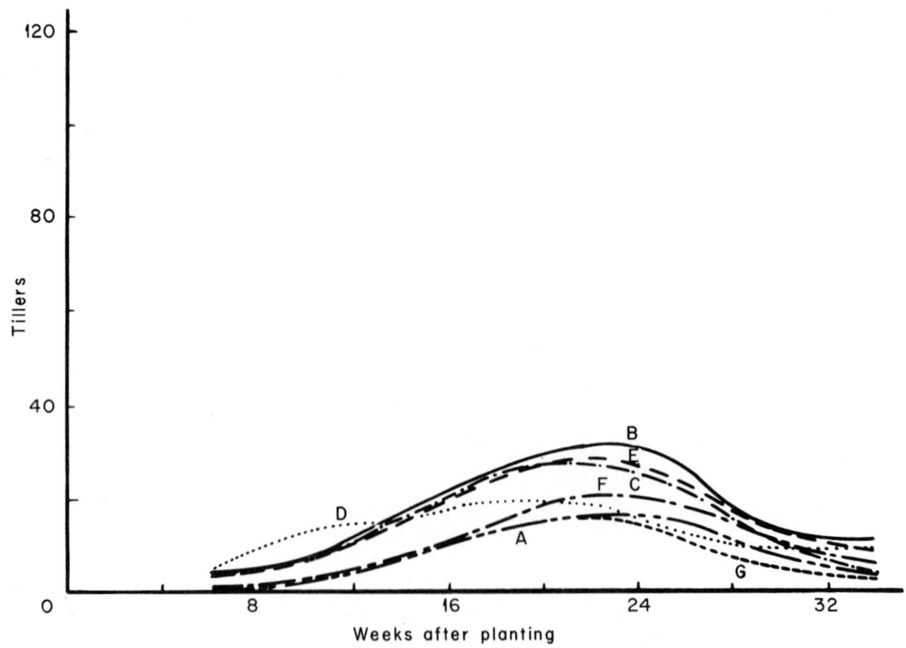

Fig. 14 Tillering of "poor tillering" variety H49-3533 (full sunlight, 200 lb N/acre). A–G, position of plantable three-eye seed pieces cut from stalk; i.e., A, topmost piece; B, next piece down, etc. (same for Figs. 15–18).

contrary to the effect in full sunlight where the addition of nitrogen increased the number of tillers dramatically, the number of tillers was actually reduced. This strongly supports our hypothesis that for a variety to be grown at its best, each of its individual agronomic characteristics must be determined separately, and suggests that this also is true for maximum production of a succeeding crop based on pretreatment of the cane plant stools which are to be used as seed.

V. RIPENING

A. Agronomic Practices

Traditionally, ripening of sugarcane has been achieved by taking advantage of the environment. In many countries low winter temperatures dictate the end of the growing season and the harvest period is adjusted accordingly. In tropical and semitropical areas, where either the temperature does not become low enough for excellent ripening or, as in Hawaii, where the economics of the industry dictate harvesting yearround, ripening has traditionally meant "drying off" or withdrawing water on the irrigated plantations.

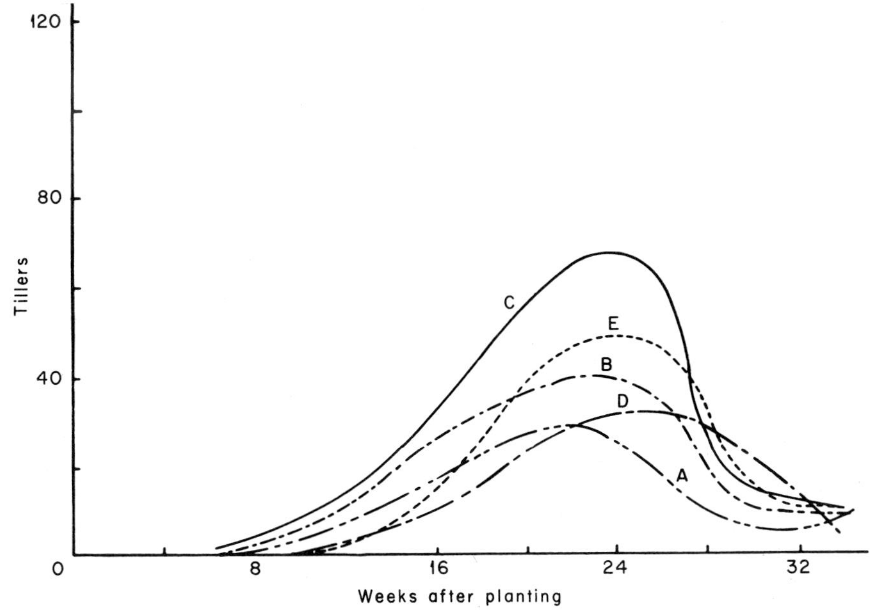

Fig. 15 Tillering of variety H50-2036 (full sunlight, 0 N/acre).

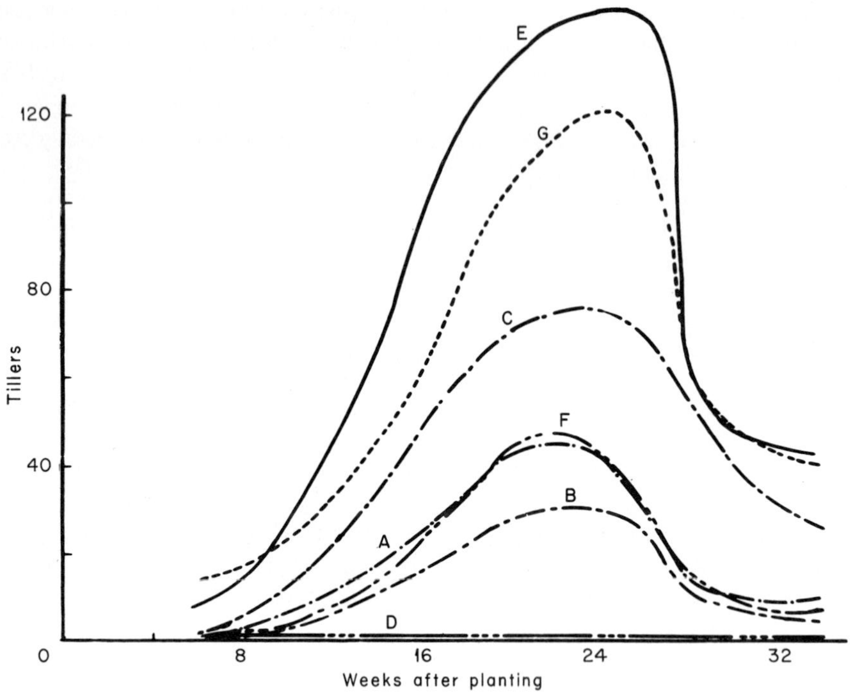

Fig. 16 Tillering of variety H50-2036 (full sunlight, 200 lb N/acre).

B. Chemicals

Basic work on translocation suggested that by affecting the youngest growing leaf or spingle or the growing point in defoliation studies, sugarcane might be artifically ripened (Hartt, 1964; Hartt *et al.*, 1964). Chemicals were sought that would kill either the young leaf or the growing point. Since only about half of these induced ripening, the technique was abandoned as a screening technique. Fortunately, however, very soon after we started such studies we found materials that were active (Nickell and Tanimoto, 1966). We based our study on the actual increase in the quality of the cane as expressed both in purity and sugar as percent of field weight. One of the most active compounds and certainly the most advanced in our testing is the Monsanto product "Polaris" (Nickell and Takahashi, 1972). This compound has been thoroughly tested under diverse ecological conditions, at different times of year, and on a number of our outstanding commercial varieties, as well as some of those that are candidates for commercial status. Our results to date show excellent effects in general in unirrigated regions with high rainfall and very little opportunity on the part of the growers to ripen their cane—there is much less of an effect under irrigated

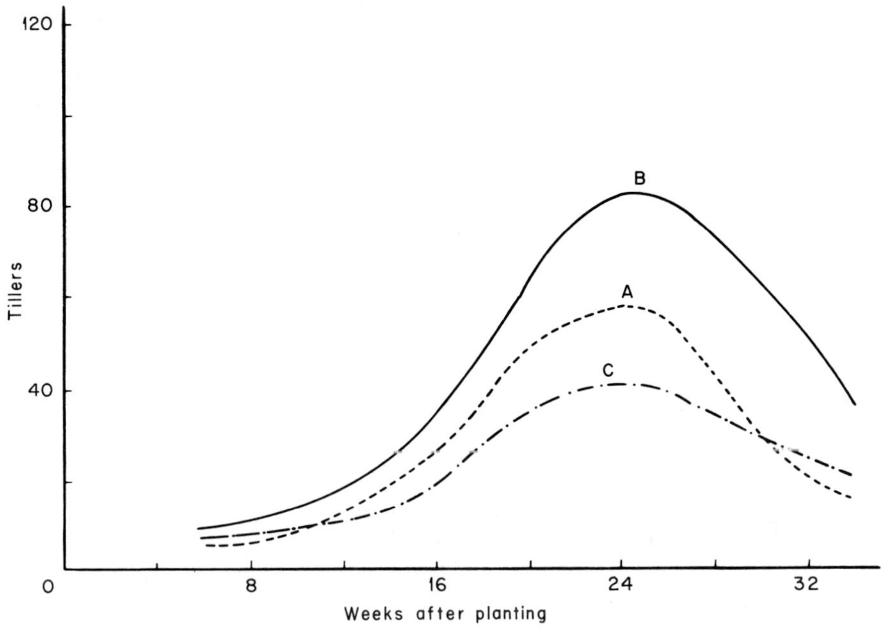

Fig. 17 Tillering of variety H50-2036 (50% shade, 0 N/acre).

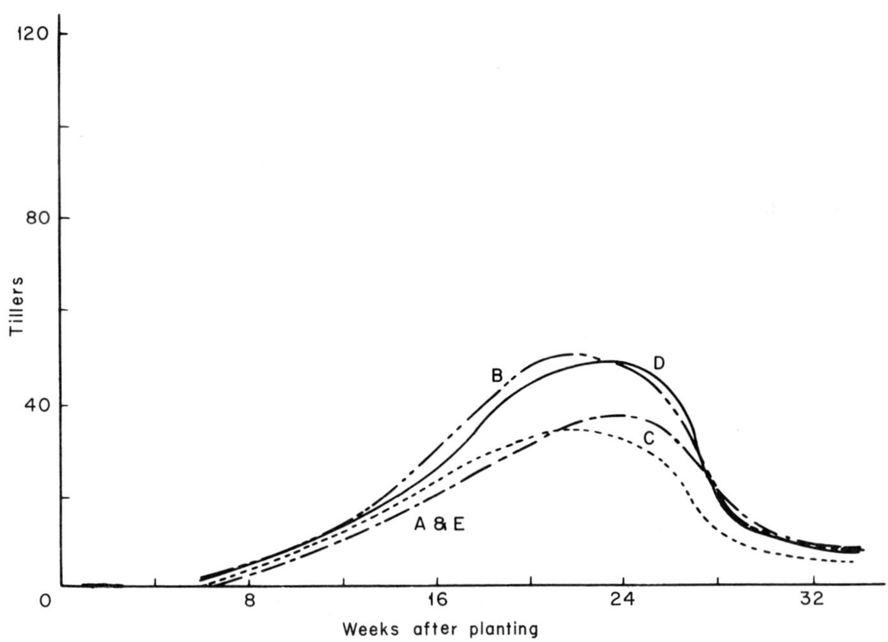

Fig. 18 Tillering of variety H50-2036 (50% shade, 200 lb N/acre).

conditions. The leading variety, H59-3775, responds very actively to this ripener (Nickell and Takahashi, 1974) and shows a 10–12% increase in recoverable sugar. The number two variety H50-7209 (which until last year was the number one variety) also shows a positive effect, particularly when unirrigated, but the effect is less than half of that on variety H59-3775. One of the next most important canes, variety H57-5174, does not respond to this ripener, or at least it did not until we found that apparently part of the problem on this variety is penetration. We have now found that adding extra surfactant to the formulation will induce and/or increase ripening in the latter two varieties (Nickell and Takahashi, 1975). Much effort will be given to searching for chemical ripeners which will be active under environmental conditions on irrigated plantations, i.e., low altitude, low rainfall, high sunlight areas.

REFERENCES

Brandes, E. W. (1956). Origin, dispersal and use in breeding of the Melanesian garden sugarcanes and their derivatives, *Saccharum officinarum* L. *Proc. Congr. Int. Sugar-Cane Technol., 9th* Vol. I, pp. 709–750.

Brodie, H. W., Yoshida, R., and Nickell, L. G. (1969). Effect of air and root temperatures on growth of four sugarcane clones. *Hawaii. Plant. Rec.* **58,** 21–52.

Burr, G. O., Hartt, G. E., Brodie, H. W., Tanimoto, T., Hortschak, H. P., Takahashi, D., Ashton, F. M., and Coleman, R. E. (1957). The sugarcane plant. *Annu. Rev. Plant Physiol.* **8,** 275–308.

Hartt, C. E. (1964). Translocation of sugar in the cane plant. *Hawaii. Sugar Technol., Rep.* pp. 151–167.

Hartt, C. E., Kortschak, H. P., and Burr, G. O. (1964). Effects of defoliation, deradication and darkening the blade upon translocation of C^{14} in sugarcane. *Plant Physiol.* **39,** 15–22.

Kortschak, H. P., Hartt, C. E., and Burr, G. O. (1965). Carbon dioxide fixation in sugarcane leaves. *Plant Physiol.* **40,** 209–213.

Maretzki, A., and Nickell, L. G. (1968). Dependence on arginine of sugarcane cells grown in submerged culture. *Proc. Int. Congr. Biochem., 7th, 1967* Abstracts, Vol. J-207.

Meguro, M. (1963). Economia d'agua de cana de azucar. *Univ. Sao Paulo, Fac. Filos., Cienc. Let., Bol.* **19,** 5–100.

Nickell, L. G. (1968). Agricultural aspects of transplanting and spacing. *Hawaii. Sugar Technol., Rep.* pp. 147–155.

Nickell, L. G. (1969). Water utilization. I. Basic plant-water studies. *Proc. Congr. Int. Soc. Sugar-Cane Technol., 13th,* pp. 38–48.

Nickell, L. G., and Kortschak, H. P. (1964). Arginine: Its role in sugarcane growth. *Hawaii. Plant. Rec.* **57,** 230–235.

Nickell, L. G., and Maretzki, A. (1969). Growth of suspension cultures of sugarcane cells in chemically defined media. *Physiol. Plant.* **22,** 117–125.

Nickell, L. G., and Maretzki, A. (1972). Developmental and biochemical studies with cultured sugarcane cell suspensions. *Proc. Int. Ferment. Symp., 4th,* pp. 681–688.

Nickell, L. G., and Takahashi, D. T. (1972). Ripening studies in Hawaii with CP-41845. *Hawaii Sugar Technol., Rep.* pp. 73–82.

Nickell, L. G., and Takahashi, D. T. (1974). Sugarcane ripeners in Hawaii. *Hawaii. Sugar Technol., Rep.* pp. 76–84.

Nickell, L. G., and Takahashi, D. T. (1975). Field studies with sugarcane ripeners in Hawaii. *Hawaii. Sugar Technol., Rep.* pp. 85–90.

Nickell, L. G., and Tanimoto, T. T. (1966). Effects of chemicals on ripening of sugarcane. *Hawaii. Sugar Technol., Rep.* pp. 152–166.

Tanimoto, T. T. (1969). Differential physiological response of sugarcane varieties to osmotic pressures of saline media. *Crop Sci.* 9, 683–688.

Tanimoto, T., and Nickell, L. G. (1967a). Estimation of drought resistance of sugarcane varieties. *Proc. Congr. Int. Soc. Sugar-Cane Technol., 12th,* 1965, pp. 893–896.

Tanimoto, T., and Nickell, L. G. (1967b). Re-evaluation of gibberellin for field use in Hawaii. *Hawaii. Sugar Technol., Rep.* pp. 184–190.

Tanimoto, T., and Nickell, L. G. (1968). Effect of gibberellin on sugarcane growth and sucrose production. *Hawaii. Sugar Technol., Rep.* pp. 137–146.

CHAPTER

5

Pineapple

D. P. BARTHOLOMEW and S. B. KADZIMIN

I. THE PLANT ENVIRONMENT

The relationship between crop varieties and climate is of considerable importance and many annual crops are bred for rather specific local conditions of temperature and daylength. For example, in temperate latitudes of the mainland United States, literally hundreds of corn and soybean varieties are selected on the basis of the number of heat units required to attain maturity within a specified photoperiod. Even in the subtropics and tropics one finds a relatively large number of varieties of some crops adapted to specific climatic zones. In Hawaii, for example, sugarcane varieties are screened for adaptability to higher elevation, high rainfall areas where irrigation is not required and low elevation areas where the crop is grown with supplemental irrigation.

Where plant breeding problems are more intractable, as they are with pineapple [*Ananas comosus* (L.) Merr.], it is not unusual to find a single crop variety grown over wide areas of the globe. The pineapple cultivar Smooth Cayenne is an excellent example of such a crop. Although three to five major groups of pineapple are recognized (Collins, 1960; Py and Tisseau, 1965; Samuels, 1969) and each has several subcultivars, Smooth Cayenne is the most widely planted cultivar, both in terms of acreage and range of latitude. The most northerly location for the field culture of pineapple appears to be the state of Assam, India at a latitude of 30°45′N (Hayes, 1960), and Strauss (1960) listed the most southerly location for the South African pineapple industry as being Port Elizabeth at 33°58′S. The range of latitudes and a climatic summary for the various pineapple growing areas are shown in Table I. Temperature and rainfall data for various areas are readily available. Despite the fact that the importance of solar radiation for the growth and development of pineapple has been recognized for many years (Nightingale, 1942), there appear to be few available records on solar radiation. Data on actual evapotranspiration are similarly lacking, although estimates have been made (Anonymous, 1962–1964).

Most pineapple is grown in or near coastal or island areas where climates tend to be more oceanic than continental, and temperature and humidity extremes are less severe than they would be for a continental climate at comparable latitudes. However, there are large differences in average annual temperature for various areas where pineapple is grown on a commercial scale. The lowest annual average temperature was 17.2°C at Port Elizabeth, South Africa and the highest was 26.9°C at Saigon, South Vietnam (Tkatchenko, 1947). Temperature extremes are equally as important as the annual average for tropical perennials such as pineapple. Pineapple will not tolerate temperatures near 0°C for prolonged periods. Additionally, high temperatures coupled with high insolation result in fruit and leaf temperatures in the range of 50°C (Nightingale, 1942; Aubert and Bartholomew, 1973) which pose sunburn hazards for both. The countries where temperature extremes detrimental to pineapple are most common are South Africa, Southern Australia, and Taiwan, where both fruit sunburn and frost damage can occur (Collins, 1960; Strauss, 1960; Su, 1969). The optimum daytime and nighttime temperatures for pineapple were reported to be about 30° and 20°C, respectively (Neild and Boshell, 1976).

The annual rainfall in various areas ranges from a low of 565 mm on the Island of Molokai, Hawaii (Noffsinger, 1961b) to 3550 mm at Neufchateau,

TABLE I
Climatic Characteristics of Some Pineapple-Growing Areas of the World

Location	Lat.	Elev. (m)	Temperature (°C) Ave.	Max.	Min.	Rain (mm)	Evaporation (mm)[a]	Reference
Nyombe, Cameroun	4.5°N	70				3000		Gaillard, 1970
Johore, Malaysia	1°39′N	5	26.9	35	18.9	2583	1742(p) 1742(a)	
Singapore, Malaysia	1°22′N	10	26.6			2881	1723(p) 1723(a)	Tkatchenko, 1947
Buenos Aires, Costa Rica	9°10′N		23	32	19.9	3078		Romero *et al.*, 1973
Rio Piedras, Puerto Rico	18°23′N	23	24.8			1866	1466(p) 1466(a)	Pico, 1974
Arecibo, Puerto Rico	18°23′N	15	25.3			2190	1473(p) 1431(a)	Pico, 1974
Rock Hampton, Australia	23°26′S	11.3	22.7	27.2	16.7	1002	1236(p) 1002(a)	
Brisbane, Australia	27°28′S	42	20.5	25.5	9.5	1143	1054(p) 1044(a)	Tkatchenko, 1947
Wahiawa, Oahu Hawaii	21°20′N	200	22.6	30.1	14.3	1062	1850(a)	
Touliu Expt. Sta., Taiwan	23°44′N	48				1984		Su, 1969
Karenko, Taiwan	23°58′N	19	22.2	27.1	17.3	1907	1242(p) 1196(a)	Tkatchenko, 1947
Thika, Kenya	1°01′S	1463	20.5	> 35	5.5	803	887(p) 787(a)	Lebedev, 1970
E. London, South Africa	33°02′S	125	18.6	22.8	14.4	824	858(p) 824(a)	Strauss, 1960
Port Elizabeth, South Africa	33°58′S	55	17.2	21.2	13.3	577	839(p) 579(a)	Tkatchenko, 1947
Malkerns, Swaziland	26°30′S		16.8	28	4	955		Dodson, 1968

[a]All potential (p) or actual (a) evaporation data and some of the rainfall data are from Thornthwaite, C. W. Associates Publications in Climatology, Centerton, New Jersey.

Guadeloupe (Py *et al.*, 1968). Optimum rainfall ranges from 1000 to 1500 mm (Collins, 1960; Teiwes and Grüneberg, 1963; Py and Tisseau, 1965). Successful production in low rainfall areas, such as the island of Molokai, Hawaii, requires good annual distribution of rainfall and water-conserving management practices such as the use of plastic and plant-litter mulches (Ekern, 1967; Py, 1968; Su, 1969). In high rainfall areas, good drainage is important because the pineapple plant root system is said to be very sensitive to waterlogging (Collins, 1960; Teiwes and Grüneberg, 1963). However, in Hawaii and elsewhere, growth decline and death associated with waterlogged soils are usually due to the presence of root fungal parasites (Py and Tisseau, 1965).

Solar radiation has long been recognized as an important factor for growth of pineapple (Nightingale, 1942) but its measurement on a wide scale has generally been precluded by lack of a suitable and inexpensive instrument. In Hawaii, 6-year monthly averages of solar radiation at Wahiawa ranged from 338 to 583 langleys (ly) day^{-1} with an annual average of 460 ly day^{-1} (calculated from Ekern, 1965b). Average total monthly hours of sunshine (instrument not specified) at Wahiawa for an 11-year period ranged from 127 to 200 with an annual total of 1931 hours (calculated from data of Collins, 1960). Recently, Aubert *et al.* (1973) reported hours of sunshine obtained with a Campbell–Stokes instrument ranged from 852 to 1577.5 hours per year for three locations in the Democratic Republic of Cameroun. In general, however, the limited data are inadequate to allow an assessment of solar radiation effects on growth and yield independent of temperature effects. The available data are discussed in a later section.

The specific soil requirements for pineapple appear to be minimal. Pineapple is grown in peat soils in Malaysia, quartz sands in Australia, Africa, and Taiwan, and highly weathered clays composed of kaolin and iron and aluminum oxides in Hawaii, the tropical Americas, Southeast Asia, and parts of Africa and India (Collins, 1960; Py and Tisseau, 1965; Teiwes and Grüneberg, 1963). The pH of the soils in which pineapple is grown ranges from a low of 3.3 for the peats of Malaysia to near 6.0 in other areas with the optimum reportedly being 4.5–5.5 (Collins, 1960; Py and Tisseau, 1965; Teiwes and Grüneberg, 1963). It would appear that almost any well-drained soil with a pH in the range of 4.5–5.5 would be suitable for the culture of pineapple. The type of soil seems to be a less important factor than the quality of management involved in the growing of the crop.

II. ANATOMY–MORPHOLOGY

A. General

The pineapple is a herbaceous perennial monocotyledonous plant that produces a single syncarpous fruit on a terminal inflorescence. The stem is short and thick, up to 35 cm in length with internodes of 1–10 mm at the time of floral differentiation (about 14 months after planting in Hawaii), and has a diameter of 6–7 cm. The leaf arrangement is spiral with a 5/13 or a 3,5 phyllotaxy (Collins, 1960; Ekern, 1968; Kerns *et al.*, 1936). Every thirteenth leaf is located directly above the initial leaf and five turns around the stem are made as the 13 leaves are formed. The leaves do not absciss as the plant grows and leaves produced after planting retain their chlorophyll and turgidity for many months. At an age of 12–14 months the plant has 60–80 fully developed, closely spaced leaves having a maximum length of about 100 cm. The leaves generally are semirigid, adaxially concave, and have angles with respect to the plant axis which range from near 90° (parallel to the soil surface) at the base of the plant to 0° at the top (Aubert and Bartholomew, 1973). This pattern of leaf arrangement is said to result in a velvet pile effect which tends to lower the reflectance of the pineapple plant community (Ekern, 1965a).

From a crop management standpoint, the most important whorl of leaves on the pineapple plant are the D-leaves. This group of leaves is the youngest physiologically mature whorl, usually representing the tallest leaves on the plant from the ground level. They are also the longest leaves present up to at least 8–12 months after planting. Their selection has been described in great detail (Sideris *et al.*, 1938; Krauss, 1949; Py, 1959). D-Leaves were first used as the index leaf for assessing the nutritional status of the pineapple plant by Nightingale (1942). They have been used for this purpose by other workers (Sanford, 1964; Gaillard, 1970; Su, 1969) and to evaluate the effects of environment on plant moisture status and on plant development (Gaillard, 1970; Su, 1969; Py *et al.*, 1968; Aubert *et al.*, 1973). The D-leaf weight has been shown to be highly correlated with the weight of the whole plant.

B. Leaf Anatomy

1. Trichomes

The leaf anatomy of pineapple is unique among crop plants in several respects. Both adaxial (upper) and abaxial (lower) leaf surfaces are covered by large peltate scutiform trichomes (Krauss, 1949; Fig. 1). The trichome heads overlap, forming a dense covering over the stomata. When leaves are fully expanded, the cells of the umbrella-shaped trichome heads are nonliving but the basal stalk cells retain their cytoplasmic organelles (Krauss, 1949). Several possible functional roles have been ascribed to the pineapple trichomes including absorption of water and aqueous nutrient solutions. Without doubt, pineapple leaves do absorb nutrients. However, no studies on pineapple were found which specifically demonstrated nutrient absorption by the trichomes. Most aqueous sprays primarily contact the distal three-quarters of the adaxial surface of the

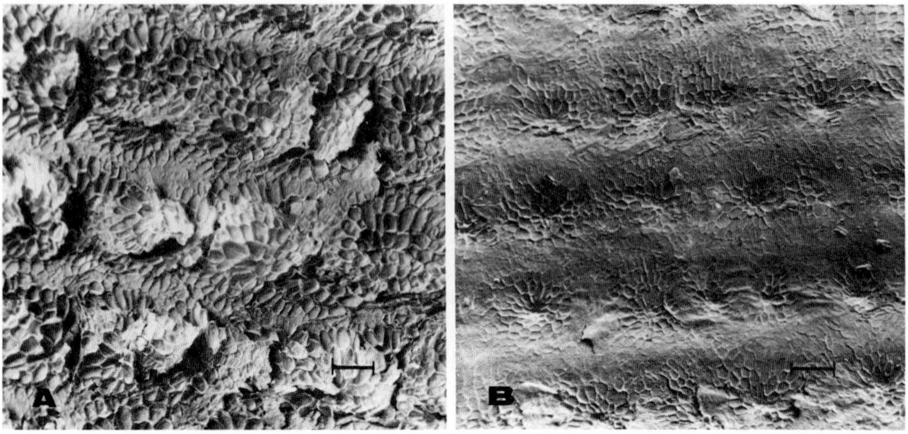

Fig. 1 Scanning electron micrograph of a pineapple leaf. Scale bar = 100 μm. (A) Abaxial leaf surface showing the large loosely attached trichomes located in valleys between the ridges. (B) Adaxial leaf surface with the trichomes in close contact with the leaf surface.

leaf and since the leaf is hypostomatous with the stomata on the abaxial surface, the only avenues of absorption are cuticular and trichomal. Trichomes on both surfaces of the proximal (stem) end of the leaf, where the leaf sheathes the stem, would be in almost constant contact with water and nutrient solutions which are funneled down the trough-shaped leaves to the leaf base. Trichomes located at the leaf base would have ample opportunity to play a role in nutrient and water absorption.

The complete covering of overlapping trichomes on the abaxial leaf surface may restrict loss of water vapor from the stomata. Benzing and Renfrow (1971) found desiccation resistance was greatest in *Tillandsia* species which had thick leaves and a dense trichome cover. We have been unable to measure the contribution of the trichomes to the resistance of the leaf to diffusion of water vapor (Bartholomew, 1975). It is possible that the trichomes are such a small part of the total leaf resistance that their effect is completely masked by the high stomatal resistance. One additional function attributed to trichomes is that they increase the reflectance of the leaf and thus reduce the heat load in an intense radiation environment (Krauss, 1949). In such an environment, leaf temperature can be 3° to 20°C above air temperature (Aubert and Bartholomew, 1973; Noffsinger, 1961a). While trichomes may enhance reflectance from a single leaf, it has been reported that reflectance from the whole plant was low because of the pattern of leaf arrangement (Ekern, 1965a).

2. *Epidermis and Hypodermis*

Leaf epidermal cells have undulated lateral walls which impart rigidity to the cells. This protects the cells from stresses resulting from collapse of the subepidermal water storage tissue during droughts (Nightingale, 1942; Sanford, 1962). Immediately below the epidermis is the hypodermal layer which, along with the thick inner-walled epidermis, provides protection against water loss, as in other xerophytes (Krauss, 1949). Johnson (1935) described a layer of cells below the epidermis which had reddish anthocyanin pigments and similar cells have been observed in many varieties of pineapple. It was hypothesized that this layer of cells acted as a screen against excessive light.

3. *Stomata*

The stomata occur only on the abaxial side of the leaf (Krauss, 1949), a general characteristic of the family Bromeliaceae. The abaxial side of the leaf is grooved longitudinally into parallel furrows and ridges. The stomata are located only in the furrows. The stomata are arranged irregularly (Fig. 2) in the furrows but the aperture is oriented parallel to the long axis of the leaf. The trichome mat described previously completely covers the furrows. Stomatal number per unit area varies from leaf to leaf, influenced perhaps by the growth status of the entire plant. Krauss (1949) indicated that counts made over a period of several years on the central portion of fully developed leaves averaged 70–85 stomata per mm^2. The stomatal apparatus is small with average dimensions of 26.5 μm perpendicular and 24.6 μm parallel to the long axis of the pore which itself is about 7 to 8 μm long (Bartholomew, 1975). Similar results were reported by Solereder and Meyer (1929). By contrast, maize is reported to have a stomatal aperture of about 20 μm (Zelitch, 1971). The depth of the pore and thus the diffusion path length is quite long, ranging from 20 to 40 μm (Aubert and Parcevaux, 1969).

4. *Water Storage Tissue*

The water storage tissue is colorless and translucent. It is adaxial to and contrasts sharply with the chlorophyllous mesophyll below (Krauss, 1949; Fig. 3). The relative amount of water storage tissue may vary with age, shading, planting density, temperature, evaporative demand of the atmosphere (Nightingale, 1936; Krauss, 1949), and location on the leaf. Tissue thickness ranging from one-quarter to two-thirds of the leaf cross section can thus be expected, and generally it is lacking toward the tip. Data obtained on a Smooth Cayenne cultivar grown in controlled environments showed that the relative proportion of water storage tissue midway between leaf tip and base appears to be unaffected by the temperature regime. However, at a light-period temperature of 30°C, the thickness and rigidity of the leaf were observed to increase as the night tempera-

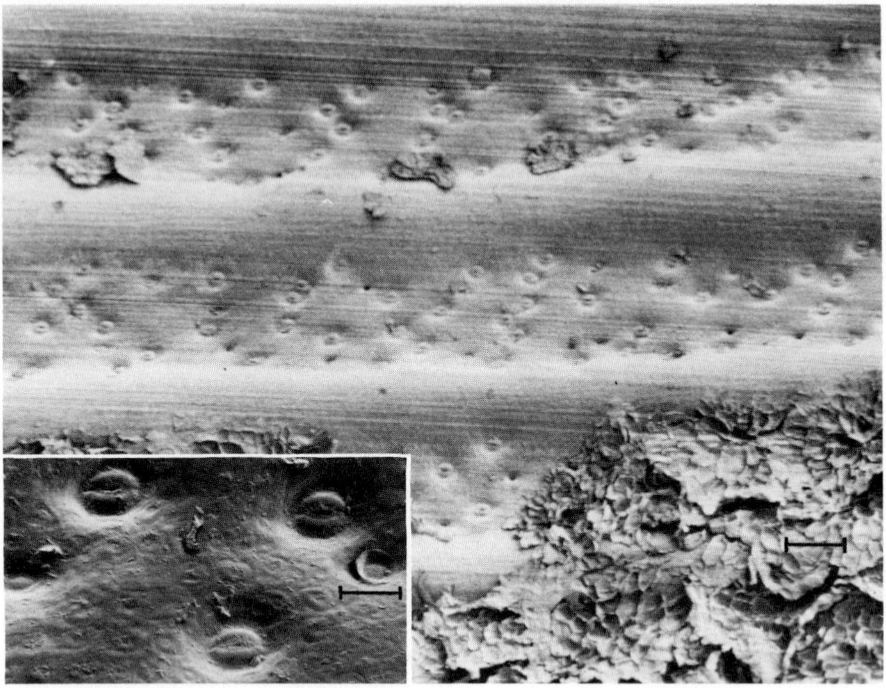

Fig. 2 Scanning electron micrograph of the abaxial surface of a pineapple leaf with the trichomes rubbed off. Scale bar = 100 μm. Inset: Micrograph showing the detail of the stomata. Note the trichome basal stalk remnant at the right margin of the picture. Scale bar = 30 μm.

ture was dropped from 30°C to 26° or 22°C (D. Bartholomew, unpublished). The leaves of plants grown at the warmer low elevations of Oahu, Hawaii similarly are less rigid than those on plants grown at higher elevations (W. G. Sanford and J. C. Fo, private communication, 1975). Black (1962) indicated that variation in water storage tissue thickness probably accounted for the wide variation in leaf water: dry matter ratios over the plant crop cycle for plants grown in Australia.

An important function of the water storage tissue is that it provides a replenishable reservoir of water for the plant. The tissue may also protect the underlying assimilatory tissue against excessive light, and stiffen the leaves when cells are turgid. With loss of water from the tissue, leaf blades lose some of their stiffness and the margins curl inward.

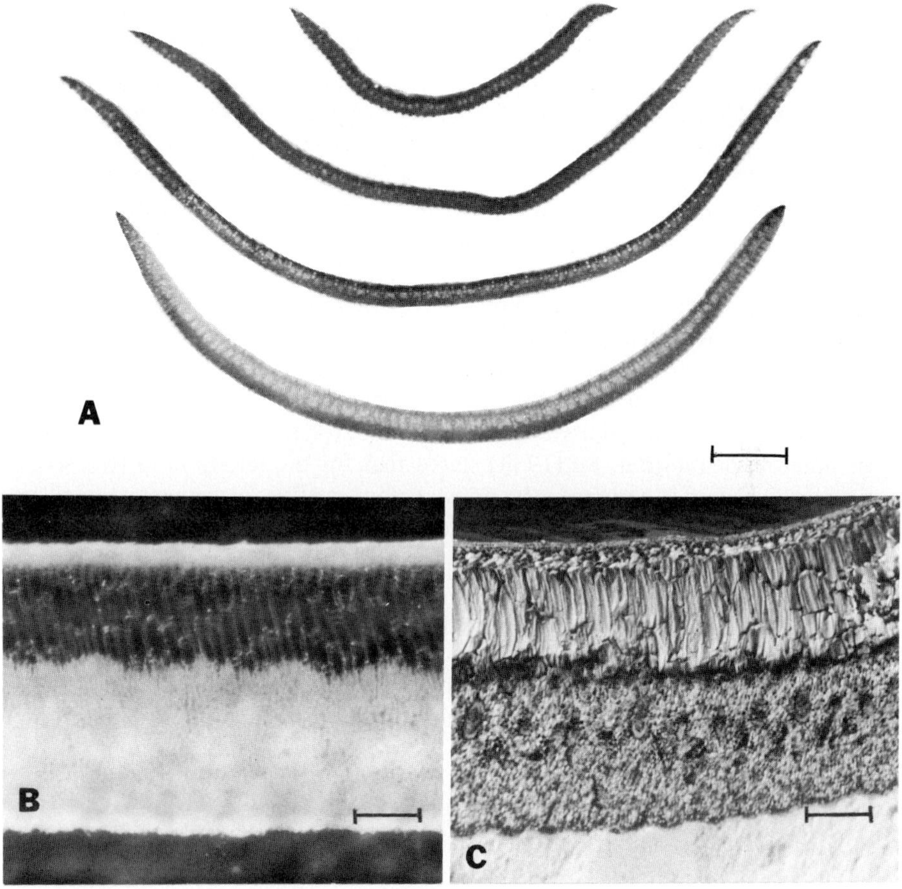

Fig. 3 (A) Cross sections through a pineapple leaf from near the tip (top of figure) to near the base. Note the increase in thickness of the water storage tissue on the upper surface of the leaf from the tip to the base of the leaf. Scale bar = 0.5 cm. (B), (C) Light and scanning electron micrographs, respectively, of a pineapple leaf cross section showing the detail of the very large water storage cells. Scale bar = 0.5 mm.

C. Root System

Two root systems are commonly associated with pineapple grown in Hawaii, the "soil" and "axillary" roots (Krauss, 1948). The soil root system, which arises as adventitious roots from the stem, has been reported to have a lateral spread of 1–2 m and to penetrate to depths of 85 cm (Krauss, 1948). Estimates of lateral spread based on root absorption of soil-applied ^{32}P have shown that the roots absorb nutrients up to 3 m away when soil tilth and nematode control are good (H. Y. Young, private communication, 1975). Above ground, adventitious roots develop in the leaf axils to form the axillary root system, probably in response to accumulation of water at the base of the leaves from dew, rain, or overhead irrigation. Axillary roots near the ground level grow into the soil as older leaves die off and decay. Axillary roots initiated at higher levels elongate within the leaf axils and extend around the stem for several centimeters. Su (1969) reported that significantly more ^{32}P was absorbed by pineapple plants when placed in the axils of mature leaves than when applied to the soil. Work in Hawaii has shown that both the leaves and the axillary root system are important avenues of absorption of foliarly applied nutrient solutions (H. Y. Young, private communication, 1975).

The activity of the root system after the first (plant crop) fruit is harvested is an unresolved question. Su (1969) stated that for the ratoon crop (the second crop produced on shoots from the original crop), axillary placement of fertilizers was more advantageous than soil placement because the "roots of the mother plant are senescent." In Hawaii, shoots generally develop too high above the soil level to establish independent root systems. In addition, the benefits of soil fumigation after plant-crop harvest support the contention that, in Hawaii at least, development of the ratoon crop depends on an active mother-plant root system. Experience has shown that while a poor ratoon crop yield can follow a good plant-crop yield, good ratoon yields are never obtained from plants having a low plant-crop yield.

III. CARBON ASSIMILATION

The pineapple is one of the few commercially cultivated crop plants which possesses Crassulacean acid metabolism (CAM). The subject of CAM generally has been well covered in various reviews (Ranson and Thomas, 1960; Ting *et al.*, 1972; Queiroz, 1974) but there are special implications specifically for pineapple that have not been adequately covered elsewhere.

Plants having CAM are characterized by large diurnal changes in organic acids, primarily malic, in the leaf. Accumulation of the acid occurs in the dark and is associated with nonautotrophic assimilation of CO_2. Pineapple was first demonstrated to have CAM by Sideris *et al.* (1948) (Fig. 4) and this result was later confirmed by Seshagiri and Suryanarayanamurthy (1957). In pineapple, organic acids accumulated during the night in leaves which had previously been illuminated. During the day, the acid level decreased with a concomitant increase in sugars. Sideris *et al.* (1948) suggested that the sugars were precursors of the

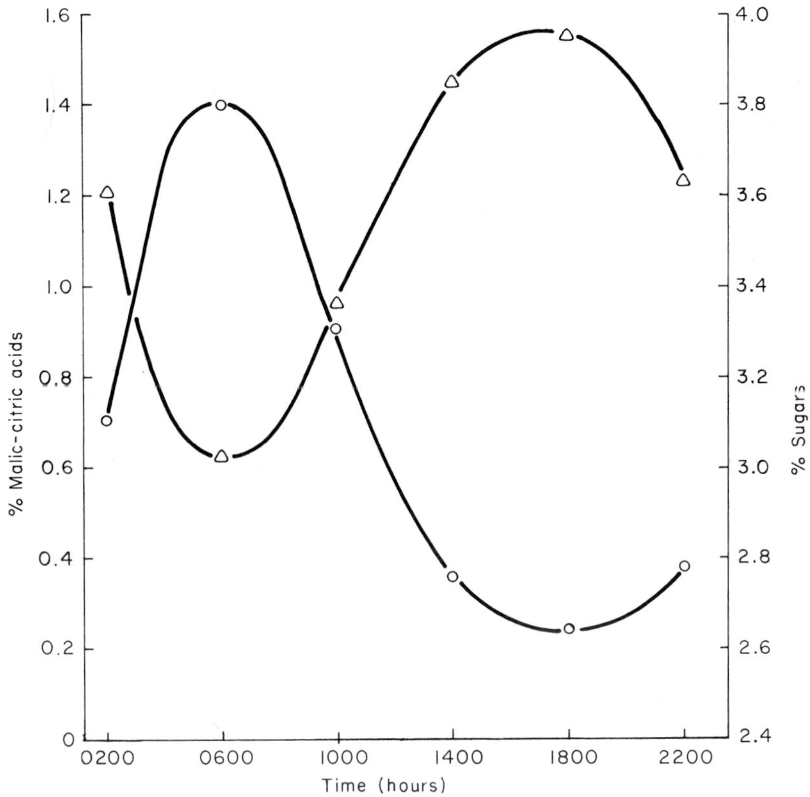

Fig. 4 Diurnal variation in malic plus citric acid (circles) and sugars (triangles) in pineapple D-leaves (Sideris *et al.*, 1948, with permission).

accumulated organic acids. They also noted that low sunlight resulted in reduced levels of acid.

An inverted pattern of gas exchange (suspension of transpiration during the day) was observed to be a characteristic of desert plants early in the twentieth century (Shreve, 1916) and Krauss (1930) reported similar results for pineapple. The first measurements of direct stomatal involvement in the inverted rhythms of gas exchange observed for plants having CAM appear to be those of Nishida (1963). Since that time, additional reports have demonstrated conclusively that most succulents and xerophytes show an inverted pattern of stomatal opening (Ehrler, 1969; Neales *et al.*, 1968; Szarek *et al.*, 1973; Coutinho, 1969). Measurements of leaf diffusive resistance to water vapor transfer for pineapple (Aubert, 1970, 1971; Neales *et al.*, 1968; Bartholomew, 1975; Bartholomew and

Kadzimin, 1976) also demonstrate an inverted stomatal rhythm for this plant. However, pineapple leaf resistances are low early in the day, increase progressively to a maximum, which corresponds approximately to the minimum level of titratable acidity, and then decrease (Fig. 5; Aubert, 1971; Neales *et al.*, 1968). The maximum daytime leaf resistance attained has been shown to be influenced by the quantity of malic acid accumulated the previous night and by the prevailing weather (Aubert, 1971; Bartholomew and Kadzimin, 1976). The inverted stomatal rhythm has significant adaptive advantages for pineapple since the crop is often grown where annual rainfall is low.

The prevailing thermoperiod markedly influences the observed pattern of stomatal opening in CAM plants. Ting *et al.* (1967) reported that when *Kalanchoe blossfeldiana* plants were conditioned in different thermoperiods, the stomatal opening pattern shifted from normal daytime opening to apparent nighttime opening as day temperature was increased from 15° to 30°C. Night temperature was kept constant at 15°C. Yoder (1969) and Yoder and Bartholomew (1969) found that transpiration rates of pineapple were markedly affected by thermoperiods. The highest transpiration rates occurred in the light when day temperature was 25° to 30°C and was less than or equal to night temperature. Nighttime transpiration rates remained constant and daytime rates decreased progressively as the night temperature was dropped 5°, 10°, or 15°C below the

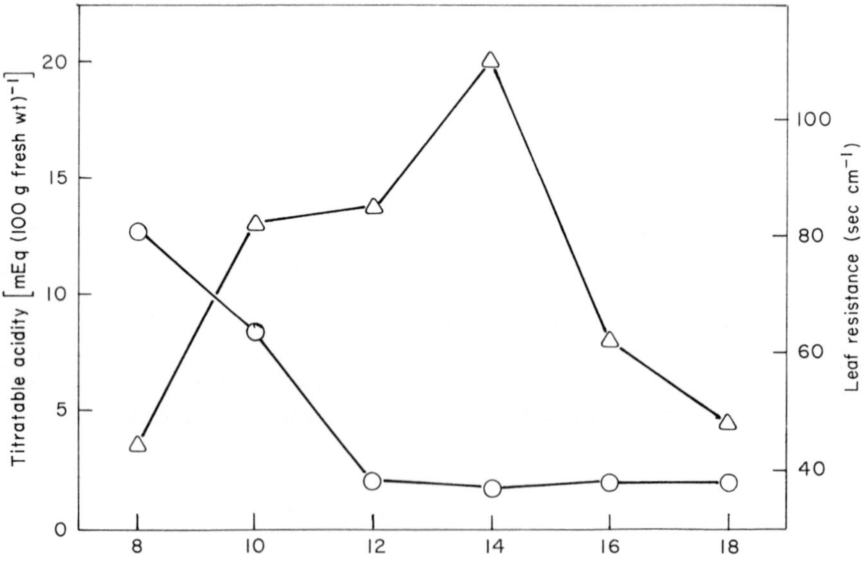

Fig. 5 Change in titratable acidity (circles) and leaf diffusive resistance (triangles) of a pineapple D-leaf with time of day.

day temperature. Evaporation rates decreased with decreasing night temperature, while transpiration remained approximately constant, suggesting that the nighttime stomatal resistance decreased as the day/night temperature differential was increased.

The earliest study of carbon dioxide assimilation by pineapple (Joshi *et al.*, 1965) showed a diurnal uptake rhythm which was similar to the diurnal change in organic acids observed by Sideris *et al.* (1948). Although titratable acids in the leaf were not measured, the relationship between nonautotrophic assimilation of CO_2 and accumulation of malate in the dark is well documented (Ransom and Thomas, 1960; Bruinsma, 1958; Ting and Dugger, 1968). Reported rates were very low, approximately 6–7 mg CO_2 dm^{-2} 24 hr^{-1}; more recent results (Neales *et al.*, 1968; Connelly and Bartholomew, 1971; Connelly, 1972) were somewhat higher but similar in magnitude.

The biochemistry of CO_2 assimilation by CAM plants is relevant to this discussion since the peculiarities of the circadian rhythm of CO_2 assimilation by pineapple have important ecological implications. The subject of photosynthetic carbon fixation in plants was recently reviewed by Black (1973). On the basis of the first products of photosynthesis or dark carboxylation, plants have been characterized as assimilating CO_2 by the C_3 (Calvin cycle), C_4 (dicarboxylic acid cycle), or CAM pathways. In C_3 plants, the primary carboxylating enzyme is ribulose diphosphate (RuDP) carboxylase and the first detectable product of CO_2 assimilation is 3-phosphoglyceric acid, a three carbon compound. In C_4 plants, on the other hand, the primary carboxylating enzyme is phosphenolpyruvate (PEP) carboxylase and the first detectable product present in appreciable amounts is malic or aspartic acid, depending on the species. In C_4 plants, it has been proposed that the malate (or aspartate) is transported from the site of fixation in the mesophyll cell to the bundle sheath cell where decarboxylation of the 4-carbon acid takes place. The released CO_2 is then reassimilated via the C_3 pathway and ultimately converted into starch. Plants with CAM utilize PEP carboxylase when fixing CO_2 in the dark and, as with C_4 plants, the first detectable product is malate. The newly formed malate is transported to a closed pool (Ting, 1971; Kluge, 1969) where, in pineapple, it can accumulate to levels as high as 1% of the fresh weight of the green leaf tissue (H. Y. Young, private communication). In the light, malate is moved from the storage pool, probably by a light activated process (Kluge, 1971a; Queiroz, 1974), converted to oxaloacetate and decarboxylated by PEP carboxykinase (Dittrich *et al.*, 1973). The released CO_2 is reassimilated via the C_3 cycle (Dittrich *et al.*, 1973). The fate of the 3-carbon residue from decarboxylation has not been elucidated but it has been suggested that it may be oxidized via the tricarboxylic acid cycle or converted directly to starch (Dittrich *et al.*, 1973).

Two additional metabolic features of C_3, C_4, and CAM plants which give evidence of environmental effects on carbon assimilation, and ultimately on plant development, are the CO_2 compensation point (the equilibrium CO_2 concentration between a brightly illuminated plant leaf and the surrounding atmosphere within a sealed chamber to which no additional CO_2 is supplied) and the isotopic composition of the carbon in plant tissue.

At optimum temperatures, the carbon dioxide compensation point (Γ) of C_4 plants is at or very near zero. For C_3 plants, on the other hand, Γ is near 50 ppm CO_2. The optimum temperature for CO_2 assimilation for C_3 plants is near or below 25 C and Γ increases with increasing temperature above the optimum.

Few data on Γ are available for CAM plants. Both Jones and Mansfield (1972) and Connelly (1972) found Γ of CAM plants to be extremely variable in the light but more nearly constant in the dark. When Γ is determined in the dark, it does not fit the above definition. However, for pineapple, we will use Γ to mean the equilibrium CO_2 concentration between the leaf and its environment in a sealed chamber, and additionally, note whether the leaf is in the dark or the light. Both Jones and Mansfield (1972) and Connelly (1972) found Γ values of *Bryophyllum fedtschenkoi* and pineapple, respectively, to be at or very near 0 ppm when the plant is actively assimilating CO_2 in the dark (for pineapple, this was at or below 20°C). An increase in the day/night temperature from 35°/20° to 35°/30°C resulted in a much higher maximum value of Γ in the dark for pineapple. High nighttime temperature apparently influences Γ by inhibiting PEP carboxylase activity, by increasing the respiration rate, or both. At a constant 30°C, no net CO_2 uptake was observed in the dark and Γ approached 50 ppm (Connelly, 1972).

In the light, Γ was influenced both by light intensity (Jones and Mansfield, 1972) and temperature (Connelly, 1972). The maximum value of Γ in the light for pineapple was obtained when appreciable amounts of CO_2 were assimilated in the dark. With a nighttime temperature of 20°C, the maximum daytime Γ decreased from about 500 ppm CO_2 to 60 ppm CO_2 when the light-period temperature was reduced from 35° to 25°C (Connelly, 1972). With a nighttime temperature of 30°C, the maximum daytime Γ was 200 ppm at 35°C. The high nighttime temperature resulted in negligible net CO_2 assimilation in the dark and the maximum dark-period Γ was 40 ppm CO_2. Γ was also found to be near zero in the light during the first hour or so after the lights were turned on (Jones and Mansfield, 1972) suggesting that CO_2 was being assimilated via the C_4 pathway. Γ then increased with time, reaching a maximum value about 4 hours after the lights were turned on. A probable explanation is inhibition of PEP carboxylase by light-activated transport of malate from the storage pool to the cytoplasm where decarboxylation is believed to occur (Kluge, 1971; Querioz, 1968, 1974).

Some time after the malate storage pool was depleted, the stomatal resistance declined (Fig. 5; Aubert, 1971). Autotrophic assimilation of CO_2 then commenced, apparently by the C_3 pathway. Evidence for autotrophic CO_2 assimilation by the C_3 pathway was a burst of CO_2 given off at the time the lights were turned off (Connelly, 1972; Crews *et al.,* 1975). Similar results have been reported for other CAM plants (Wynn *et al.,* 1973) and the burst disappeared when pineapple leaves were exposed to low oxygen tensions (Crews *et al.,* 1975). Such an O_2 sensitive CO_2 burst is indicative of photorespiration (Zelitch, 1971), a process which has been demonstrated by this technique to occur mainly in C_3 plants. High daytime values of Γ and the CO_2 burst at the end of the light period indicate that carbon assimilation in the light by pineapple occurs via the C_3 pathway. It has been reported that malate is heavily $^{14}CO_2$ labeled in the light in *Bryophyllum* once the malate pool has become depleted (Kluge, 1971a) suggesting that C_4 photosynthesis is initiated when accumulated malate has been decarboxylated. Other work (Osmond *et al.,* 1973), based on $^{14}CO_2$ pulse-chase experiments utilizing *Kalanchoe daigremontiana,* suggests that direct carboxylation of ribulose diphosphate occurs in the light and thus provides additional evidence for the operation of the C_3 pathway in the light. Thus, the results of several workers suggest that depending on the environment, some CAM plants including pineapple can assimilate CO_2 by either the C_3- or C_4-photosynthetic pathway.

The short-term effects of thermoperiod on CO_2 assimilation in the light and dark by pineapple are considerable (Connelly, 1972; Neales, 1973). When nighttime temperatures were 20°C or below, dark fixation of CO_2 was at a maximum and uptake in the light was greatly reduced or completely eliminated. At a constant 25° or 30°C, uptake of CO_2 in the light as well as total uptake were significantly enhanced. The available data show that the proportion of carbon assimilated in the light (presumably by the C_3 pathway) by pineapple increases as the night temperature increases above 20°C and as the difference between day and night temperature decreases. Similar results were recently reported for *Aloe vera* (Crews *et al.,* 1976), a plant which also decarboxylates accumulated organic acids via PEP carboxykinase. The data for *Aloe vera* show that when the night temperature was high and CAM activity low, the activities of PEP carboxylase, and especially PEP carboxykinase, were also low. Where high CAM activity prevailed, the activity of the carboxylase and decarboxylase showed concomitant increases.

The isotopic carbon discrimination ratio also has been used as a diagnostic criterion to distinguish between plants having C_3 and C_4 photosynthetic pathways. C_3 plants discriminate against the heavier ^{13}C to a greater degree than do C_4 plants so that the isotopic carbon composition of C_3 plants is quite unlike that in the atmosphere, whereas that of C_4 plants is more nearly similar (Smith and Epstein, 1971). Recent reports show that CAM plants can have a variable carbon isotope composition within an identical group of plants (Allaway *et al.,* 1974; Bender *et al.,* 1973; Lerman *et al.,* 1974) and that isotopic composition can be varied by altering the plant environment (Bender *et al.,* 1973; Crews *et al.,* 1976; Osmond *et al.,* 1972). Plant temperature, light intensity, day length, and moisture stress all cause the isotopic carbon composition to shift. The results obtained from carbon assimilation studies with pineapple (Connelly, 1972; Neales, 1973; Neales *et al.*, 1968; D. P. Bartholomew, unpublished) and studies cited above of the isotopic carbon composition of CAM plants other than pineapple support the assumption that cool night temperatures and water stress, conditions under which dark assimilation of CO_2 predominates, cause CAM plants to operate basically in the C_4 photosynthetic mode. Warm night conditions, on the other hand, promote autotrophic assimilation of CO_2 and would cause CAM plants to function more like C_3 plants. Additional reviews of CAM have appeared recently (Kluge, 1976; Osmond, 1976), and a number of models of CAM regulation have been proposed. Thus far, no one model seems able to account for all of the data.

IV. WATER RELATIONS

It has long been known that pineapple may be grown under semiarid conditions and it has been reported to be nearly as hardy as cactus under drought conditions (Sideris and Krauss, 1928). Pineapple is one of a very few crop plants which can be classified physiologically as a true xerophyte. The crop is grown without irrigation in many areas where rainfall is 1000 mm or less (Table I), and in the tropics and subtropics potential evaporation often greatly exceeds this value. Pan evaporation at Wahiawa, Hawaii averaged 1850 mm per year (Ekern, 1965b), while annual rainfall was in the range of 1000 mm per year (Noffsinger, 1961b). For perennial mesophytic crops, this would result in an annual water deficit. This water deficit aspect is further accentuated by the fact that rainfall distribution tends to be seasonal. In Hawaii, the 4-month summer dry season when rainfall averages 25 mm per month is the period when water supply is most likely to limit growth. The seasonality of rainfall can restrict the growth of pineapple even in humid climates (Py *et al.*, 1957; Py and Tisseau, 1965). Survival during dry months may be related in part to the fact that the spiral array of semierect to erect leaves provides an efficient collection mechanism for light rains and dew (Ekern, 1965c). In addition, the relatively large volume of water stored in the leaves allows dark CO_2 assimilation to continue relatively unabated after the onset of stress (Neales *et al.*, 1968). The water storage tissue collapses as water is depleted and expands when water again becomes available (Nightingale, 1942; Collins, 1960; Sanford, 1962).

The relationships between water deficits and plant growth and evaluation of methods to assess plant water deficits have received much attention (Kozlowski, 1968a,b, 1972, 1976; Wiebe *et al.*, 1971; Hsiao, 1973). Water deficits in pineapple have been indexed by the relative thickness of water storage tissue of the youngest physiologically mature leaf (termed D-leaf) and by the percentage of white root tips visible on the roots in the soil (Sanford, 1962). Both methods rely on visual estimation of the plant water status and thus are likely to suffer from the handicap that growth may have been reduced by the time stress is visually detectable. Recently, estimates of plant tissue water potential (Ψ_p) for pineapple have been made (Wambiji and El-Swaify, 1974; Kadzimin, 1975) using the dye technique (Barrs, 1968). Normal values of Ψ_p for pineapple were −4 to −8 bars and were comparable to those obtained for other crop plants (Hsiao, 1973). Growth of pineapple plants in soils salinized to various levels (Wambiji and El-Swaify, 1974) or dried to −15 bars or lower (Kadzimin, 1975) resulted in a decrease in Ψ_p to values in the range range of −20 to −30 bars. The relationship between Ψ_p and growth has not yet been adequately ascertained for pineapple.

There are few quantitative data available on the relationship between water supply and growth or yield of pineapple. Py and Tisseau (1965) indicated that application of 60 mm of irrigation water per month during the dry season in Guinea was sufficient to meet plant water requirements. However, there do not appear to be any data which allow the prediction of growth and yield reductions as a result of soil water deficits which occurred during the development of pineapple crops.

Successful pineapple culture in areas where potential evapotranspiration is mugh greater than annual rainfall is related to the fact that pineapple has a very low consumptive use rate of water relative to mesophytic crops (Fig. 6) (Ekern, 1965c). Transpiration rates of pineapple range up to 0.075 g dm^{-2} h^{-1} (Joshi *et al.*, 1965; Neales *et al.*, 1968; Yoder, 1969) and are about one-tenth to one-twenty-fifth of those for mesophytic plants (Neales *et al.*, 1968; Boyer, 1969). At full cover, the fraction of sunlight used for evaporation from a pineapple canopy was 0.08–0.28 (Ekern, 1965c); this contrasts with values of 0.6–0.8 for Bermuda grass sod exposed to the same conditions (Ekern, 1966).

Water loss from the leaves of any plant is directly related to cuticular and epidermal characteristics, including stomatal features. Morphological and anatomical features of the pineapple leaf responsible for the low transpiration rates of pineapple have been described (Section II). Control of water loss from plant leaves is exercised only at the stomata and the coincidence, or lack thereof, of stomatal opening with peak atmospheric evaporative demand primarily determines plant transpiration rates. The inverted stomatal rhythm of pineapple results in a lack of such coincidence and is in part responsible for the low water use rates observed for this plant (Ekern, 1965c; Joshi *et al.*, 1965; Neales *et al.*, 1968). However, in addition to their inverted stomatal rhythm, plants with CAM are also characterized by very low transpiration rates when the stomata are open during periods of relatively high evaporative demand (e.g., see Fig. 6). This result is due to an almost infinite cuticular diffusive resistance to water vapor and a high stomatal diffusive resistance. We have estimated normal cuticular resistances

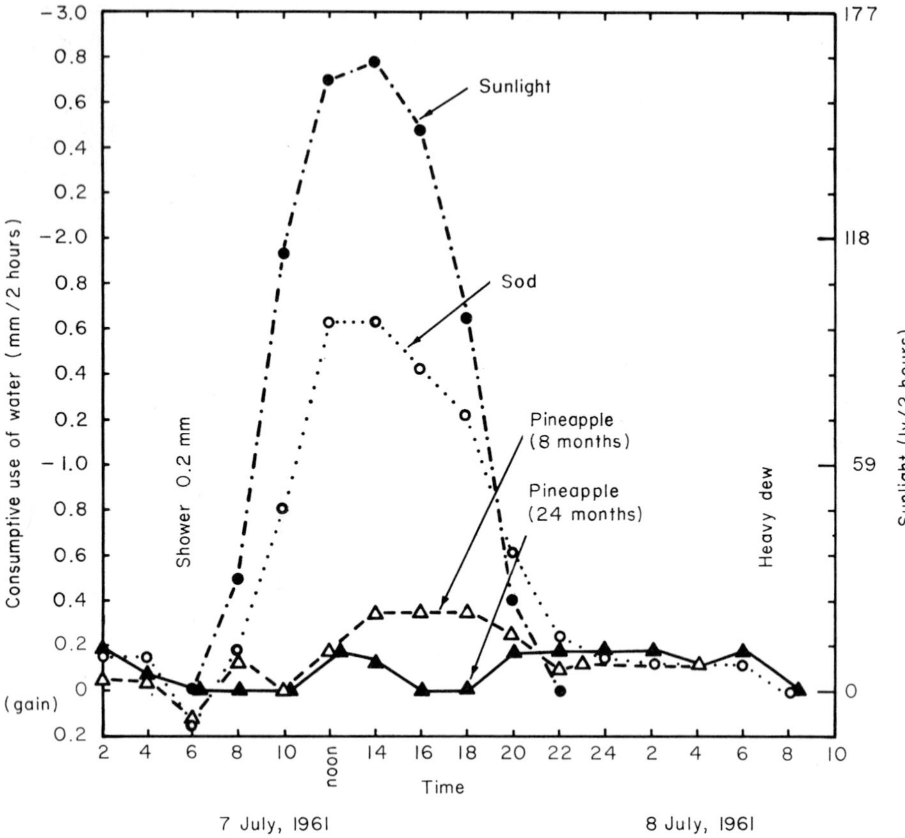

Fig. 6 Relationship between sunlight [closed circles, langleys (ly) per 2 hours] and consumptive use of water by Bermuda grass sod (open circles) and by pineapple 8 (open triangles) and 24 (closed triangles) months after planting (Ekern, 1965c, with permission).

(adaxial surface with no stomata) in excess of 800 sec cm^{-1} with a specially designed and calibrated diffusion porometer (Bartholomew and Kadzimin, 1976). Water loss from stressed pineapple plants can be almost completely suspended as evidenced by the fact that fresh weight losses from severed vegetative shoots of pineapple over periods of weeks to a few months generally amounted to no more than 60% (Sideris and Krauss, 1928, 1955). This contrasts with measurable cuticular resistances in the range of 10 to 80 sec cm^{-1} for conventional vegetation (Kramer, 1969). Stomatal diffusive resistances of pineapple also are high relative to mesophytic plants. Minimum resistances are in the range of 7–10 sec cm^{-1} (Aubert, 1970, 1971; Bartholomew, 1975) and maximum values can exceed 300 sec cm^{-1} (Neales *et al.*, 1968; Kadzimin, 1975). By contrast, stomatal resistances for mesophytes range from a low of one or less to highs in the 20–30 sec cm^{-1} range (Holmgren *et al.*, 1965; Whitman and Koller, 1967). Where pineapple plants were stressed for water but were not visibly desiccated, midday maximum diffusive resistances greater than 600 sec cm^{-1} were not uncommon (Kadzimin, 1975). High midday resistances presumably estimate abaxial cuticular resistance. It is interesting to note the somewhat lower value of this resistance on the abaxial leaf surface where stomata are present than on the adaxial surface.

The efficiency with which water is used by crop plants to produce dry matter is of particular importance where water supplies can limit productivity. While no data are available on the efficiency of water use by pineapple in terms of fruit yield, transpiration ratios for this plant are among the lowest on record. The transpiration ratio ranges upward from 200 for conventional vegetation (Kramer, 1969). Laboratory and greenhouse results for pineapple yield values of 50 to 60 (Sideris and Krauss, 1928; Joshi *et al.*, 1965; Neales *et al.*, 1968). While pineapple water use rates are low, Joshi *et al.* (1965) suggested that carbon assimilation rates were insufficient to allow the plants to be agriculturally competitive in well-watered areas. Results presented in Section V, A of this chapter show that crop growth rates (g m^{-2} day^{-1}) for pineapple can be comparable to those of alfalfa and sugarbeets over a period of several months. Relatively high growth rates for pineapple coupled with its efficient use of water makes the crop an extremely attractive one in areas where water supplies may be marginal for mesophytic crop plants.

V. CROP ECOLOGICAL ASPECTS

A. General Growth

Growth, development, and fruiting of pineapple are determined by a complex of plant and environmental factors. Growth has been indexed by whole plant weights, but perhaps more frequently by the weight and/or approximate area of the D-leaves (Aubert *et al.*, 1973; Gaillard, 1969, 1970; Py *et al.*, 1968; Py, 1959, 1968; Su, 1969; Black, 1962). In addition, theoretical estimates of total leaf mass have been obtained through the use of the product of D-leaf weight times the number of new leaves formed since the last data were collected (Gaillard, 1970; Py *et al.*, 1968; Py and Tisseau, 1965). One study of pineapple growth based on dry matter yields has been reported (Tay and Tan, 1971) and dry weights have been estimated from plant moisture contents of a subsample and the fresh weights of test plants (Joshi *et al.*, 1965). However, most pineapple growth studies have been conducted on the basis of fresh weights (Sideris and Krauss,

1937, 1938; Sideris and Young, 1950; Py, 1959) so there is little opportunity to compare dry matter production rates of pineapple grown in different environments.

Within a given environment, the increase in fresh weight for Smooth Cayenne is determined primarily by the weight of the plant at the time of planting. Commonly used sources of planting material are as follows:

(1) Suckers (shoots), which develop from buds located in leaf axils along the stem soon after floral differentiation occurs. Sucker weights generally range from 0.5 to 1.0 kg.

(2) Slips, which form at the base of the fruit and morphologically are the crowns of vestigial fruits (Jacques-Felix, 1950). Slip weights generally range from 0.3 to 0.5 kg.

(3) Crowns, which develop above the fruit after all flower parts have been formed. Crown weights generally range from 0.075 to 0.2 kg.

Growth, as indicated by an increase in fresh weight, is sigmoid for both the whole plant and its component parts (Sideris and Krauss, 1937, 1938; Teisson, 1973) and recent data (D. P. Bartholomew, unpublished; Pineapple Research Institute, unpublished) show that a similar relationship holds for dry weight. A plot of the logarithm of fresh or dry weight against time is approximately linear up to about three months after floral differentiation. The use of planting material of different weights results in different amounts of initial capital (leaf area, storage reserves, etc.). Since the logarithm of plant weight at some interval of time after planting is linearly related to initial weight at the time of planting, where no competition exists the size of planting material is a primary determinant of plant weight at some future time. Where competition for environmental factors exists, initial weight differences would be maintained until competition between plants in the community limits the growth of the individual.

There are no data to show what combination of size of planting material and plant population results in interplant competition to the point where individual vegetative growth is suppressed. Black (1962) suggested that the large variability in plant weights observed at the later stages of his experiment was the result of interplant competition. The implication was that large plants were in a position to obtain environmental resources at the expense of adjacent smaller plants. Although no statistics were presented, such an effect should result in increased coefficients of variation as competitive pressure increased with vegetative development. In Black's study, both planting density (34,600 plants ha^{-1}) and final average plant fresh weight (2496 g at 14 months) were low when compared with similar data collected in Hawaii (42,978 plants ha^{-1}, 6069 g at 14 months; Pineapple Research Institute, unpublished). In the Hawaii experiment, the logarithm of plant fresh and dry weights increased linearly with time for about 350 days. While standard deviations increased with increasing plant weight and time, no consistent trend in coefficients of variation was observed. Also, no consistent decrease in plant weight or D-leaf weight with increasing planting density to levels greater than 60,000 plants per hectare has been measured in Taiwan or Hawaii (Su, 1969; W. G. Sanford, private communication). Small decreases in leaf number have been reported when density was increased from 29,000 to 104,000 plants ha^{-1}, but D-leaf weight was not altered significantly (Wee, 1969).

Competition between pineapple plants apparently does exist even at relatively low densities, but its effect is primarily on the growth of reproductive parts and propagative structures such as slips and suckers. Linear regression equations calculated from the data of Dodson (1968) and Wee (1969) show that fruit weight decreases approximately linearly with increasing planting density and correlation coefficients were 0.95 or better for both sets of data. Py and Tisseau (1965) cite similar results for Martinique and Hawaii. In all instances cited, the decrease in fruit weight per increase of 1000 plants ha^{-1} was both a function of variety and of environment (presumably including quality of management). As planting density was increased, slips and suckers per plant, percent of plants fruiting, fruit weights, and fruit sugars all decreased while fruit acid increased (Wee, 1969; Mitchell and Nicholson, 1965; Dodson, 1968; Su, 1969).

The apparent absence of competition between plants during vegetative growth is surprising when one considers the tremendous amounts of leaf area displayed by the pineapple crop community. Py (1959) gives data showing a leaf area index (LAI, m^2 of leaf per m^2 of ground) of 9.3 with a planting density of 38,461 plants ha^{-1} at 14 months after planting. In an experiment conducted in Hawaii, green leaf weight of the mother plant increased to a maximum at about 14 months after planting, remained nearly constant for about 4 months, and then began to decline. No leaf area data were taken in the experiment but we have since shown that the relationship between green leaf weight and area is linear (Yoder, 1969; D. Bartholomew, unpublished). However, data obtained from plants grown in controlled environment chambers showed that the specific leaf weight (mg cm^{-2} fresh weight basis) increased significantly from 114 in a 30°C night temperature regime to 153 in a 18°C night temperature regime. Therefore, a single relationship between leaf weight and leaf area is not universally applicable. Calculations of LAI for the Hawaii data, based on a conservative relationship between leaf area and leaf weight, show that at a planting density of 43,000 plants ha^{-1}, the LAI could remain at or above 10 for at least 7 months. Sustaining such a high LAI for several months suggests that light distribution in the pineapple leaf canopy is uniform or the light compensation point is very low. Observations of leaf display support the former alternative. The spiral phyllotaxy of pineapple leaves results in rather wide vertical separation between leaves. If spatial separation between leaves is of the order of $d > 2w$, where d is the distance between leaves and w is leaf width, then the angle of occlusion of skylight γ, where $\gamma = 2 \tan^{-1} (w/d)$, is near a minimum (Loomis *et al.*, 1971). Values of $(w/2d)$ for Smooth Cayenne pineapple in Hawaii are of the order of 0.13 to 0.15 about midway between the leaf base and the tip. These values would result in relatively low occlusion angles (γ) of 15° to 17°. Foliage angle also affects the distribution of light over plant leaves and Loomis *et al.* (1971) state that simulations show little difference in productivity due to varying leaf angle when LAI > 2. However, a canopy where leaf angle increased progressively from 0° at the top of the canopy to 90° at the base, as is the case for pineapple (Fig. 7), was shown to be more efficient than the opposite arrangement. Thus the leaf display features of pineapple support the contention that maintenance of very high LAI over long periods of time results from uniform distribution of light in the canopy.

Net assimilation rates (NAR) for pineapple range from about 0.4 to greater than 2.0 g m^{-2} day^{-1} (Joshi *et al.*, 1965; Tay and Tan, 1971; D. Bartholomew, unpublished). The highest rates are one-tenth, or less, of those reported for mesophytic crop plants (Wilson, 1966, 1967). However, the low rates for pineapple gain significance when dry matter production is evaluated on a land area basis. At 8 months of age, field-grown plants in Hawaii have approximately 0.7 m^2 of leaf area and leaf area per plant continues to increase for another 4–6 months, remains about constant for an additional 4–5 months, and then begins to decline (Pineapple Research Institute of Hawaii, unpublished

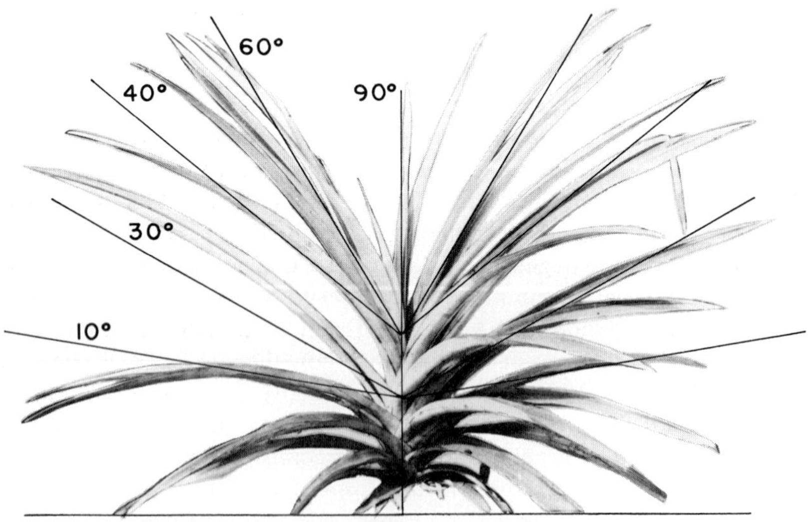

Fig. 7 A Smooth Cayenne pineapple plant showing the spirally arranged leaves and the wide range of leaf angles.

data). At an average planting density of 45,000 plants per hectare, LAI would increase to a maximum in the range of 12. An average NAR of 1.5 g m^{-2} day^{-1} was obtained for greenhouse-grown pineapple (Joshi *et al.,* 1965) and higher values have been obtained for field-grown plants in Hawaii. An NAR of 1.5 g m^{-2} day^{-1} and an LAI of 10 or greater could result in a sustained pineapple crop growth rate (CGR) of 15 g m^{-2} day^{-1}, a value which Hawaiian field data show would be sustained for a period of several months. The maximum CGR for alfalfa and sugarbeets was reported to be somewhat higher (Zelitch, 1971) but maximum CGR's generally are not sustained for long periods of time. Thus, low NAR values for pineapple appear to be compensated for by high LAI's over relatively long periods of time. In one Hawaiian experiment, about 390 tons/ha of fresh plant material and 62 tons/ha of dry matter, including the fruit, were produced in 656 days (Pineapple Research Institute of Hawaii, unpublished data).

B. Effects of Temperature on Growth

Few data on the specific effects of temperature on pineapple growth are available. The optimum temperatures for root and leaf elongation were about 29° and 32°C, respectively (Sanford, 1962). In Hawaii, reduced solar radiation and temperature levels from about November to April resulted in slower vegetative growth rates (the slope of a logarithmic plot of plant fresh weight vs. time decreased). With the onset of warmer weather in the spring, the steeper slope of the previous summer's growth curve was reestablished (Pineapple Research Institute, private reports). The observed reduction in growth was probably due in part to lower soil temperatures which prevailed during the winter months. Average soil temperatures at Wahiawa, Hawaii from December to April were near 21°C and about 0.5°C above air temperature (Ekern, 1967). It was reported that the average winter soil temperature was increased 1.5°C with the use of black or clear plastic mulch, and leaf weights of pineapple grown with mulch were about 1.4 times those of plants grown on unmulched soil. The projected increase in growth based on the temperature–growth curves for pineapple corresponded well with the observed result (Ekern, 1967). The few data available would seem to suggest that low temperatures, rather than reduced solar radiation, are primarily responsible for the reduced growth observed during the winter season.

The temperature of the soil or rooting medium can have a profound influence on the rate of growth of pineapple. Ravoof (1973) examined the effects of root temperature on the growth and nutrient uptake of pineapple slips grown in solution culture in a short-term (40-day) experiment. Root temperatures ranged from 15° to 30°C in 5° increments while tops were maintained at ambient greenhouse temperatures. At 15°C, fresh weight of the slips was less than at the outset of the experiment. Above 15°C, fresh weight increased with temperature from 20° to 30°C. However, the growth increment was less from 25° to 30°C than from 20° to 25°C as would be predicted from the published temperature–growth curves. Leaf dry matter decreased from 16.5% at 15°C to about 12 to 12.5% at 30°C. The interaction between temperature and nitrogen carrier was interesting. Root weights were greatest at 30°C when ammonium nitrate and sodium nitrate were the nitrogen sources while root weights were maximal at 25°C in ammonium sulfate. Leaf weight, as a percentage of total plant weight, increased significantly from 77 to 80% as temperature increased from 25° to 30°C.

The short-term effects of temperature on growth (Sanford, 1962) or on CO_2 uptake (Connelly, 1972; Neales, 1973) do not correspond with long-term growth data, however. The maximum rate of CO_2 uptake for a single attached pineapple leaf was observed at constant temperatures of 25° and 30°C and essentially all of the CO_2 was taken up in the light (Connelly, 1972). Neales (1973) reported similar data for whole plants maintained at a 35°C night–27.5°C day temperature, i.e., total CO_2 assimilation was greater when dark fixation was suppressed by high night temperatures. An experiment conducted in the South Eastern Plant Environment Laboratories phytotron at Raleigh, North Carolina showed that plant growth was maximum in a 30°C day–26°C night or 26°C day–22°C night temperature regime (Table II). On the basis of CO_2 assimilation data and the temperature–growth curves available for pineapple, we had anticipated that dry matter production would also be optimum at or near 30°C. The data in

TABLE II
Effects of Thermoperiod on the Relative Rates of Leaf Elongation and Total Plant Dry Matter Production of Smooth Cayenne Pineapple as a Percent of That Obtained on a 30°C Day/26°C Night Regime[a]

Day/night temperature regime (C)	Average temperature (°C)	Relative heart leaf elongation rate (cm/week)	Relative total dry wt. at 8 months
30/26	28	100.0[b]	100.0[c]
30/30	30	78.1	48.2[d]
26/22	24	63.9	89.5[c]
22/22	22	61.9	75.5
22/18	20	47.7	80.8

[a]The study was conducted in the South Eastern Plant Environment Laboratory phytotron, Raleigh, North Carolina, during 1974.

[b]Results are weekly averages of four replicates measured over a period of 6 weeks. The data were not analyzed statistically.

[c,d]Results followed by *c* were significantly greater (0.05 level of probability) than those followed by *d* as determined by student's *t* test.

Table II show that such was not the case, however. The relative weight data of Table II are based on unadjusted totals. Crown weights were measured prior to the installation of the experiment and the length of the tallest leaf was measured two months after planting at the time the treatments were imposed. Adjustment of dry weights on the basis of original crown weight or on length of the tallest leaf permitted discrimination between treatments which were not significantly different on an unadjusted basis. On the basis of adjusted totals, the relative yields at the different thermoperiods were $30°/26° > 26°/22° \cong 22°/18° \cong 22°/22° > 30°/30°C$.

The biweekly increase in length of small heart-leaves has been used to index plant response to nutrients in Hawaii and a similar technique has been used elsewhere (Mitchell and Nicholson, 1965). In the phytotron study mentioned above, weekly measurements of heart-leaf growth were made. The maximum leaf elongation rate was about 9 cm per week for plants in the 30°/26°C thermoperiod. The values were similar to midsummer values obtained on well-watered field-grown plants in Hawaii. Two interesting results were evident from the data: (1) there is poor correspondence between increase in leaf length and final dry weight; (2) dry matter production can be considerable at temperatures which are suboptimal on the basis of short-term leaf and root elongation measurements, e.g., in a 22°/18°C thermoperiod where the average temperature is only 20°C.

The discrepancy between leaf elongation and dry weight can be accounted for by the fact that at night temperatures of 30°C, specific leaf weights (g cm^{-2}) were significantly lower than they were at night temperatures of 26°C and below. Thus, a high rate of leaf elongation was not necessarily reflected in high dry matter production. On the basis of plant appearance, we are of the opinion that plants at 22°/22°C were clearly superior to those grown at 22°/18°C. The lack of a clear distinction between plants at several of the temperatures is no doubt related in part to the small sample size necessitated by the limited growing space available in the phytotron. The relatively small decrease of 20% in average dry weight with a drop in average temperature of 8°C (from 30°/26° to 22°/18°C) certainly attests to the adaptability of the Smooth Cayenne plant to a wide range of thermoperiodic regimes. The discrepancy between the long-term growth results obtained in the phytotron and short-term CO_2 uptake studies can only be accounted for by the lack of adequate adaptation at a given temperature prior to data collection. To the extent that the data have been analyzed, studies of CO_2 assimilation by a single attached D-leaf for the phytotron-grown plants generally support the dry matter accumulation data.

C. Effects of Light on Growth

Relationships between light intensity and the rate of CO_2 uptake have been established for large numbers of C_3 and C_4 plants. The leaves of C_3 plants generally are light saturated at 30 to 55 klux while C_4 plant leaves saturate at 100 klux or greater (Zelitch, 1971). Net photosynthesis of CAM plants was reported to saturate at about 1000 ft-c (10.8 klux) (Black *et al.,* 1971). However, more recent data of Crews *et al.* (1975) show that both light and dark CO_2 assimilation by pineapple increase with light intensity to 80.7 klux. The rate of deacidification of *Bryophyllum tubiflorum* leaves in light increased as light intensity was increased from 8 to 17 klux (Kluge, 1971). Aubert (1971) reported that both acidification and deacidification rates in pineapple leaves were reduced on days when energy receipt was reduced. The few data available would suggest that an estimate of light saturation of CAM leaves at about 11 klux is too low, but definitive data are unavailable.

It has been amply demonstrated that even though the leaf of a C_3 plant is light saturated at less than half of full sunlight, dry matter production by a crop community increases with increasing light intensity to levels of full sunlight (Zelitch, 1971; Loomis *et al.,* 1971). We could find no studies for pineapple in which the relationship between light intensity and CO_2 assimilation or dry matter production has been adequately examined. As was implied earlier, the collection of reliable CO_2 assimilation data is much more difficult for CAM plants than it is for mesophytes. In order to obtain a realistic estimate of temperature and light effects on CO_2 assimilation the CAM plant must be examined for an entire 24-hour period. This is because dark CO_2 fixation and the synthesis of large amounts of malate have been demonstrated to inhibit the activity of PEP carboxylase (Ting and Dugger, 1968; Kluge, 1969; Queiroz, 1968) and this is qualitatively indicated by the observed variation in Γ early in the light period (Jones and Mansfield, 1972; Connelly, 1972). A value of 0 ppm CO_2 at the onset of the day suggests that CO_2 is being assimilated via PEP carboxylase. A gradual increase in Γ suggests inhibition of PEP carboxylase resulting in an increase in the intercellular-space CO_2 concentration which triggers an increase in stomatal resistance.

It is logical to assume that daytime conversion of malate to starch and additional autotrophically assimilated CO_2 will determine, at least to some degree, the level of available substrate and therefore the quantity of malate assimilated during the following night. Such an assumption is supported by the observation that low levels of solar radiation greatly reduced the diurnal amplitude of titratable acids in pineapple leaves (Sideris *et al.,* 1948; Aubert, 1971). The maximum titratable acid level attained after a low-sunlight day was about 42% of that after a high-sunlight day (Aubert, 1971). Other results showed that titratable acid levels decreased with increasing shade for plants supplied with adequate amounts of nitrogen (Connelly, 1969).

In addition to the shorter-term effects of light on carbon assimilation, temperature also affects the rate and pattern of CO_2 assimilation with carryover effects lasting for periods of a week or more (Connelly, 1972; D. Bartholomew, unpublished). An additional carryover effect of unknown duration has recently been demonstrated. The similarities in CO_2 assimilation patterns for the data of Connelly (1972) and Neales (1973) are striking. The lack of correspondence between their short-term results and the longer-term growth response of pineapple plants grown under controlled conditions (see Section V, B) suggests that great care must be exercised when drawing conclusions about the longer-term responses of pineapple to environmental factors from short-term experiments.

The effects of solar radiation on pineapple growth and yield have been recognized for quite some time. Nightingale (1942) emphasized that the carbohydrate–nitrogen relationship of the pineapple plant, which is determined in part by the energy available for photosynthesis, was of greater importance than the amount of nitrogen applied to the plant. Several workers have reported that pineapple leaves were darker green at reduced light levels and a constant rate of nitrogen application, indicating that the rate of nitrogen utilization for growth was restricted by carbohydrate production at low-sunlight levels (Nightingale, 1942; Sanford, 1964; Connelly, 1969; Su, 1969). In a shading experiment in which plants were widely spaced, it was reported that 25% shade had no effect on growth, while growth was reduced by 50% shade (Connelly, 1969). Aubert *et al.* (1973) concluded that the primary factor limiting yield at 1000 m elevation in Cameroun was a low level of solar radiation. However, average temperature was 20°C which was 2.6°C lower than that at 500 m elevation. The changes in both hours of sunshine and temperature with elevation make isolation of specific effects more difficult.

The only reported effect of sunlight on yield was an observation that a 20% reduction in sunlight decreased yield by 10% (Sanford, 1962). The scarcity of data on the effects of light on pineapple growth and yield makes it difficult to interpret much of the available data on season and planting density. A decrease in fruit weight with increasing density (Dodson, 1968; Tan and Wee, 1973; Su, 1969) seems to implicate light as a possible limiting factor during development of reproductive parts. However, the lack of any well-demonstrated effect on vegetative development, even at very high densities (Wee, 1969) where the LAI must be very high, is perplexing to say the least. It seems an understatement to say that much work is left to be done in this important area of pineapple physiology.

VI. PHYSIOLOGY OF FLOWERING

A. Natural Differentiation

Vegetative growth and reproductive development of pineapple are difficult to characterize as a specific sequence of events associated with the progression of climatic change through the year. Numerous reports state that the length of time from planting to harvest of a fruit of a given size is determined by the weight or size of the material planted (Cerighelli, 1955; Collins, 1960; Hayes, 1960; Py and Tisseau, 1965; Py *et al.*, 1968; P. Mitchell, 1953; A. R. Mitchell, 1962). In many areas where pineapple is grown, the crop is planted in the fall, winter, or early spring. Prior to the discovery of chemical forcing agents which could be used to induce flowering in pineapple, natural differentiation of the apex occurred during the late fall or early winter of the year following planting.

In Hawaii during the fall–winter season, the rate of reproductive development and particularly the period from differentiation, either natural or induced with hormones, to appearance of the inflorescence requires about 60 days. The further development of the inflorescence and fruit to the time of harvest requires an additional 5 months for a total period from induction to harvest of 7 months. When plants are induced to flower in late spring or summer, or at other times in warmer climates such as Ivory Coast, the period between floral induction and appearance of the inflorescence is 40–45 days. Further fruit development requires slightly less than 5 months resulting in a total period from induction to harvest of about 6 months (W. G. Sanford, personal communication).

A number of experiments have been conducted to determine what environmental factors are primarily responsible for the natural induction of flowering (Py and Guyot, 1970; van Overbeek and Cruzado, 1948a,b; Py *et al.*, 1968; Teisson, 1972; Gowing, 1961). There is general agreement that plant size determines to a great extent the susceptibility of Smooth Cayenne to flowering and that flowering occurs predominantly in the fall. In Hawaii, crowns and slips planted in the fall will differentiate naturally the following year, usually in the last week of November or the first week of December at the latest (J. C. Fo, personal communication). Some flowering also occurs at other times of the year, however. Small plants planted in the spring usually will not flower in the fall of the same year while large suckers planted at the same time often will flower (Gowing, 1961). However, in some cases very large plants have a lower percentage of flowering than medium-sized plants at the normal time for natural differentiation (Py and Guyot, 1970).

Varietal differences in susceptibility to flower induction have also been noted. In Puerto Rico, nearly 100% of Red Spanish plants flowered in the fall in response to shorter days and cooler temperatures, irrespective of plant size. Under the same conditions, less than 50% of the Cabezona variety flowered naturally regardless of plant size and age (van Overbeek, 1946). Additional experimentation showed Red Spanish did not respond to reduced daylength but did flower if exposed to lower night temperatures (van Overbeek and Cruzado, 1948a). Daylength is considered to be a factor in the flowering of Smooth Cayenne and Cabezona (Py and Guyot, 1970; van Overbeek and Cruzado,

1948a; Py *et al.*, 1968); Gowing (1961) indicated the Smooth Cayenne grown in Hawaii is a quantitative, but not an obligate, short-day plant. In the Ivory Coast (about 4°N latitude) the change in daylength is very small with little or no change in night temperature. Under such conditions, the stimulus to flower is thought to occur as a result of a reduction in the hours of sunshine (Teisson, 1972).

Rapid vegetative growth rates tend to inhibit differentiation. Growth stimulated by nitrogen fertilizers and irrigation or rainfall (Py and Guyot, 1970; Su, 1969; Nightingale, 1942; Evans, 1959; Gaillard, 1969) inhibits flowering. Thus, once a plant has reached sufficient size to be susceptible to floral induction, environmental factors which promote flowering are those which tend to retard vegetative growth. Those factors would include a decrease in nutrition or water supply and reduced temperature, daylength, and solar radiation.

B. Flower Induction with Growth Regulators

The pineapple is unique among crop plants in being susceptible to flower induction by the application of growth regulators. It has been known for many years that pineapple could be induced to flower by exposure to smoke (Collins, 1960). Rodriquez (1932) found that the active ingredient in smoke responsible for flower induction in pineapple was ethylene. Somewhat later, it was discovered that auxin and auxinlike compounds could be used to induce pineapple to flower (Clark and Kerns, 1942). Gowing (1956) explained the effects of naphthaleneacetic acid (NAA) on flowering on the basis of antagonistic effects on native auxin levels in the pineapple stem apex. Although there apparently were no published observations, it has been recognized for some years that pineapple flowered a week to 10 days later if forced with NAA than if forced with ethylene, acetylene, or ethephon (2-chloroethylphosphonic acid). The delay in flowering when plants are forced with NAA appears to be explained by the results of Burg and Burg (1966) who showed that the application of NAA to pineapple resulted in ethylene production by the plant. Ethylene production peaked about 5–7 days after application of the NAA and it was proposed that the ethylene produced as a result of the NAA treatment was the factor responsible for stimulating differentiation. With the recognition of ethylene as an important plant growth regulator (Warner and Leopold, 1967), it is now generally accepted that artificial induction of flowering in pineapple is intimately associated with ethylene. However, no more complete explanation of the physiological role of ethylene in this morphogenic process has been forthcoming.

Despite the fact that the pineapple plant can be induced to flower by the application of growth regulators, chemicals differ in their effectiveness and susceptibility of the plants varies. In general, acetylene, ethylene, ethephon, and β-hydroxyethylhydrazine are more effective forcing agents at most times of the year than NAA, indoleacetic acid, 2,4-dichlorophenoxyacetic acid, and related compounds and the two groups of chemicals produce somewhat different results (Terry, 1975). According to Terry, a fruit produced as a result of forcing with NAA forms a more pointed cylinder, the top of the fruit is immature when the base is completely ripe, and the ripening peak in the field is fairly board. Fruit from plants forced with ethylene and related compounds form a more-or-less uniform cylinder which ripens more uniformly than NAA-fruit, and the ripening peak is narrowed.

Differences in plant susceptibility to hormones have been observed also. As with natural differentiation, within limits, the larger the plant the more susceptible it is to floral induction (Cooper, 1942; Das *et al.*; 1965; Py, 1958). Plants are more susceptible to chemical forcing near the season when natural differentiation occurs (Das, 1965; Gowing, 1961). Rapid growth makes forcing more difficult (Py and Guyot, 1970; Wee and Ng, 1968; Evans, 1959; Gaillard, 1969; Py, 1958) as does a period of water stress sufficiently severe to cause cessation of growth (Py, 1958). Differences in susceptibility among cultivars have also been noted (Cooper, 1942; Das, 1965; Groszmann, 1950; Randhawa *et al.*, 1970; Wee and Ng, 1968; van Overbeek, 1946). Unsusceptible plants often can be induced to flower by applying higher concentrations of ethephon (E. J. Anderson, personal communication) or by multiple applications of acetylene (Aldrich and Nakasone, 1975) or NAA at intervals of a few days, a common practice in Hawaii.

A recent study of the effects of temperature on pineapple growth and development under controlled conditions (D. Bartholomew, unpublished) was described in Section V, B. In that study, Smooth Cayenne plants grown in physiological long days with high nitrogen were forced with ethephon at both 7 and 9 months after planting. Although only two or four plants were forced in each treatment, ethephon was 100% effective in all thermoperiods where the night temperature was below 26°C with day temperatures ranging from 34° to 22°C. Since all conditions except temperature were held constant, the data suggest that night temperature can play an important role in determining plant susceptibility to forcing. We suspect carbohydrate/nitrogen relationships were associated with the degree of susceptibility to forcing, but tissue analyses have not yet been completed.

One additional factor associated with forcing is the time of day. In a study conducted in Mexico, it was found that the percentage of flowering induced with acetylene was higher in the early morning and late afternoon than it was at midday (Aldrich and Nakasone, 1975). It has also been known for some time that ethylene is more effective if applied at nighttime than in the daytime. These results suggest that ethylene and acetylene are entering the plant leaf through the stomata. The inverted stomatal rhythm of pineapple would account for the greater effectiveness of gaseous forcing agents at night and in the early morning.

VII. EFFECTS OF ENVIRONMENT ON GROWTH CYCLE, FRUIT YIELD, AND FRUIT QUALITY

As was mentioned previously, the size of material planted determines the weight of the plant at some later point in time. It has been shown conclusively for several pineapple cultivars growing in a number of areas that fruit weight is related to the size of planting material. Fruit weight is also highly correlated with plant weight, estimated leaf mass, and D-leaf weight at the time of floral differentiation (either natural or forced) (Gaillard, 1969; Py and Lossois, 1962; Py and Tisseau, 1965; Tan and Wee, 1973; van Overbeek, 1946; Mitchell, 1962). For commercial fruit, two fruit sizes generally are desired. The fresh fruit market demands a fruit weighing approximately 1.5 kg while a somewhat larger fruit, usually 2.0 kg or larger, is needed for processing. Since fruit size for processing must of necessity be approximately a constant, the determinants of fruit size at some point in time are size of planting material and length of the vegetative growth period. Fruit yield then becomes primarily a function of planting density and percentage of plants bearing fruit.

The only criteria by which environments (physical and chemical characteristics of the soil, quantity and distribution of rainfall, temperature, solar energy, and management factors such as fertilizer application, and weed, pest, and disease control) can be compared is on the basis of tons of fruit per hectare per month. The data of Table III give some indication of the tremendous variation in yield obtained in the tropics. In general, these data represent the highest yields reported in the literature. The yield data for various areas incorporate many variables, some of which are amenable to management and some of which are not. On the basis of published data, it would be difficult to develop a predictive relationship that could be used to evaluate the yield potential of a new area or the low yield realization of some areas currently producing pineapple. For instance, there was no indication why yields were so low in South Africa, especially when contrasted with the yield from Swaziland, an area having a similar climate. Management is one possibility, but more likely it is the lack of an adequate data base resulting from the fact that private research organizations or corporations do not publish their results.

Some generalizations about relationships between location (and thus to some extent climate) and length of time from planting to harvest can, however, be made. Although temperature data were not available for all locations, the data of Table III are arranged approximately in order of decreasing average temperature. The months from planting to plant-crop harvest generally increased with decreasing average temperature from 11 months at Nyombe, Cameroun to 32 months at Malkerns, Swaziland. Temperatures in Taiwan tend to be warmer than those in Hawaii in the summer and cooler in the winter. Although average temperatures and months to plant-crop harvest were similar at Touliu Experiment Station and

TABLE III
Length of Crop Cycle and Yield Data for Pineapple at Various Locations in the Tropics and Subtropics

Location	Cultivar	Planting material and wt. (g)	Date planted	Months to plant crop		Months to ratoon crop		Density[b] (plants ha^{-1})	Yield				Reference
									Plant-crop		Plus ratoon (tons ha^{-1})	Plant-crop fruit wt (kg)	
				forcing[a]	harvest	forcing	harvest		tons ha^{-1}	tons ha^{-1} $month^{-1}$			
Nyombe, Cameroun	Cayenne	suckers 350	3/1966	6 (A)	11			61,500	98.4	8.94		1.60	Gaillard, 1970
Pekan Nenas, Malaysia	Singapore Spanish	slips	7/1964	10 (A)	15–17	Var.	Var.	71,729	59.4	3.71	89.9	0.95	Wee, 1969
Buenos Aires, Costa Rica	Montelira	suckers	7/1969	13 (NAA)	20			53,280	103.4	5.17		1.94	Romero *et al.*, 1973
Rio Piedras, Puerto Rico	Red Spanish	slips	11/1956	n.d.	n.d.	n.d.	n.d.	23,100	45.4		74.1	n.d.	Samuels and Gandia-Diaz, 1960

Rio Piedras, Puerto Rico	Cayenne	slips	11/1956	n.d.	n.d.	n.d.	n.d.	23,100	73.9		117.7	n.d.	
Yeppoon, Australia	Cayenne	slips	9/1967	8 (BOH)	16–17	21	28–29	36,000	59.4	3.60	108.2	1.65	Jorgenson, 1973
Wahiawa, Hawaii	Cayenne	slips	8/1958	14 (BOH)	23	28	36	43,000	101.5	4.41	188.2	2.36	Pineapple Research Institute, unpublished data
Touliu Expt. Sta., Taiwan	Cayenne	slips	8/1958	16	23	n.d.	40	44,444	64.8	2.82	134.5	1.62	Su, 1969
Thika, Kenya	Cayenne	crowns	4/1959	n.d.	22–30	n.d.	51	24,829	30.8	1.18	61.8	1.97	Evans, 1957
E. London, S. Africa	n.d.	n.d.	n.d.	n.d.	24–36			33,839	24.4	0.81		n.d.	Strauss, 1960
Malkerns, Swaziland	Cayenne	crown 85–199	3/1959	22 (NAA)	32	32	n.d.	56,800	84.2	2.63	152.7	1.68	Dodson, 1968

[a]Forcing agent: NAA, napthaleneacetic acid; A, acetylene; BOH, β-hydroxyethylhydrazine; n.d., no data given.
[b]Where more than one treatment was reported, only the highest densities and yields are given.

Wahiawa, the difference in average fruit weight suggests a much smaller plant was forced in Taiwan. The cooler winters and warmer summers could account for the observed difference in fruit weight. Su (1969) has indicated that plant-crop fruit were harvested about 25 months after planting and ratoon fruit at 38 months after planting at Tashu and Kukeng, Taiwan (average temperatures for the two areas were 21.3° and 23.7°C, respectively). At other locations in Taiwan, it was stated that plant-crop fruit were harvested in the third year and ratoon-crop fruit about $1\frac{1}{2}$ years later. Uniform plantings at four locations in Australia had quite large variations in a leaf growth index (Cannon, 1957). The lowest rate of growth and the lowest yield was at Flaxton, an area described as having cooler conditions than the other sites. The results, however, were confounded with water supply which was noted to have retarded plant establishment at Flaxton and one other site. Since growth-rate reductions at latitudes where temperatures are cooler are often compensated for by an increased period of vegetative growth, the effects of location can be summarized by the following observations from Brazil (Anonymous, 1962). The cool climate of the subtropics or high altitudes retards growth and reduces yield (on a tons ha^{-1} $month^{-1}$ basis, not necessarily on a tons per plant- or ratoon-crop basis, e.g., see Table III). Two years are required to mature fruits in southern Brazil, while near the equator in northern Brazil, fruits are harvested approximately 1 year after planting.

An additional effect of climate on the growth cycle and yield is of importance. In warm climates as in the Ivory Coast, no second or ratoon crop is obtained from the established planting because suckers do not develop (Lacoeuilhe, 1975; W. G. Sanford, private communication). Shoots do develop high on the stem (called "hapas" in Hawaii) and, on becoming large, break off. It has been observed in Martinique that the higher the elevation, the better is the development of shoots (C. Py, personal communication, via W. G. Sanford). Thus, ratoon cropping may not be practical at low elevations near the equator, especially with Smooth Cayenne. If ratoon crops are harvested, ratooning may be labor intensive as in Malaysia, for example, where suckers are forced individually as they reach optimum size (Wee, 1969).

Fruit quality is affected by climate also. Fruits which mature during the winter or at high altitudes where cooler temperatures and low levels of solar radiation prevail often are of inferior quality because of poor shape for canning, low brix, and/or high acid (Huang, 1956; Mitchell, 1953; Groszmann, 1950; Cannon, 1957; Collins, 1960; Aubert *et al.*, 1973). Collins (1960) also indicated that fruit produced in warm climates such as Cuba, the Philippines, Haiti, and parts of Mexico have lower acid than comparable fruit in somewhat cooler climates. Fruit brix tends to decline and acid increases with increasing planting density, probably due to competition for sunlight. The above observation implicates solar radiation equally with temperature as a determining factor for fruit yield and quality. As was mentioned previously, the available data are inadequate to allow the determination of the relative contribution of each of these factors to pineapple plant growth and fruit development.

VIII. PESTS AND DISEASES

Crop pests and diseases usually accumulate when a crop is monocultured. Pineapple is no exception in this regard and the incidence of problem pests and diseases is no doubt associated with the fact that pineapple production on a relatively large scale often is a monoculture based on a single asexually propagated cultivar. The buildup of certain pests may also result from ecological changes associated with the repetitive use of a particular set of cultural practices. Two prominent examples are available in Hawaii which would have eliminated the pineapple industry if control measures had not been developed. Mealy bug wilt caused by *Pseudococcus brevipes* (Ckll.) was a serious problem in the 1930's. Dr. Walter Carter discovered the cause of and control measures for this threat to the industry (Carter, 1973). Dr. Carter (1943) and Dr. C. T. Schmidt (Carter, 1973), respectively, discovered the soil fumigants D-D (dichloropropane:dichloropropene) and DBCP or Nemagon (dibromochloropropene). These two fumigants made it possible to inexpensively control the root knot [*Meloidogyne javanica* (Treub) Chitwood] and reniform (*Rotylenchulus reniformis* Linford and Oliviera) nematodes, populations of which had built up to near disastrous levels.

The incidence of pests and diseases varies widely in importance. Pests unknown or of little importance in one area may pose serious hazards to pineapple culture in other areas. In general, the ecological aspects of pineapple pests and diseases have been adequately discussed in the two most recent books on pineapple culture (Collins, 1960; Py and Tisseau, 1965). The discussion here is, therefore, restricted to some recent findings associated with plant pest problems in Hawaii that may in some cases have ramifications beyond the Islands.

Root rot and heart rot of pineapple caused by *Phytophthora cinnamomi* Rand and *P. parisitica* Dastur can cause large losses of plants in localized areas where soil drainage is poor and/or during periods of heavy rainfall (Collins, 1960; Py and Tisseau, 1965). It has been observed that the incidence of rot caused by *P. parisitica* was frequently associated with areas where soils are alkaline. Damage associated with *P. cinnamomi* is also somewhat dependent on soil pH. In areas where pineapple is grown in highly weathered soils, when climate favors disease development, a relatively small increase in soil pH due to the application of lime can result in a relatively large increase in the percentage of plant damage (W. G. Sanford, private communication).

Soil pH apparently also influences nematode populations. A recent pot study showed that an increase in soil pH as a result of the application of $CaCO_3$ resulted in a decrease in the number of extractable reniform nematodes (*Ro-*

tylenchulus reniformis Linford and Oliviera) (Englerth, 1969). A recently completed experiment corroborates the above results but indicates that the relationship between nematode population and soil pH is not a straightforward pH effect (W. J. Apt and J. A. Silva, private communication). Further studies on this interesting finding are currently underway.

Two fruit diseases, pink disease caused by *Acetomonas* sp. and interfruitlet corking (IFC) caused by *Penecillium funiculosum* Thom (Hepton and Anderson, 1968), have received recent attention. Pink disease is a problem in Hawaii, Australia, Taiwan, and the Philippines. The most evident symptom of the disease is a browning of the fruit, which occurs during canning. Infection occurs through the open flower but the sporadic incidence of the disease had baffled researchers. The incidence of infection is likely to be highest when temperatures are cool and relative humidity is high (Rohrbach and Pfeiffer, 1971). On the basis of observed relationships between disease incidence and climate it is believed that pink disease occurs when temperatures are within a few degrees of 17°C at the time of flower opening and the temperature does not exceed 29°C during the period from flowering to harvest (K. G. Rohrbach, private communication). Temperatures above 29°C apparently are lethal to the organism. Disease incidence thus likely would be restricted to areas of the subtropics and at higher elevations in the tropics.

Interfruitlet corking has been reported only in Hawaii where the disease was primarily responsible for eliminating a promising new hybrid pineapple. The disease reportedly invades the noncutinized flower parts about 3 months after floral differentiation (Hepton and Anderson, 1968). Recent work indicates infection may take place much closer to or at the time of flower differentiation (K. G. Rohrbach, private communication). If this finding is supported by additional data, corrective measures would need to be taken much earlier than indicated by the results of Hepton and Anderson (1968). Symptoms similar to those of interfruitlet corking have also been observed in the pineapple growing areas of Africa (W. G. Sanford, private communication).

ACKNOWLEDGMENTS

We gratefully acknowledge the technical assistance of Sally S. Bartholomew, Dr. R. J. Downs and his staff at the North Carolina State University phytotron, and Glenn K. Masukawa and Judith Koja in the collection of the unpublished data reported herein. Unpublished data of the Pineapple Research Institute of Hawaii were collected by a large number of people over a period of several years and their efforts are also acknowledged with thanks. We also wish to thank Dr. B. Cooil, Dr. W. G. Sanford, and Dr. A. S. Whitney for critically reviewing the manuscript.

This paper was published with the approval of the Director of the Hawaii Agricultural Experiment Station as Journal Series No. 1979. Some of the unpublished results reported here were obtained while the senior author was partially supported by NSF Grant 28951 to the Southeastern Plant Environment Laboratories, North Carolina State University, Raleigh, North Carolina.

REFERENCES

Aldrich, W. W., and Nakasone, H. Y. (1975). Day versus night application of calcium carbide for flower induction in pineapple. *J. Am. Soc. Hort. Sci.* **100,** 410–413.

Allaway, W. G., Osmond, C. B., and Troughton, J. H. (1974). Environmental regulation of growth, photosynthetic pathway and carbon isotope discrimination ratio in plants capable of Crassulacean acid metabolism. *In* "Mechanisms of Regulation of Plant Growth" (R. L. Bieleski, A. R. Ferguson, and M. M. Cresswell, eds.), Bull. No. 12, pp. 195–202. R. Soc. N. Z., Wellington.

Anonymous. (1962). "Queen of Tropical Fruits," Vol. 2, pp. 36–37. Crop Protection Courier of Farbenfabriken Bayer AG, Lenerkusen, West Germany.

Anonymous. (1962–1964). "Average Climatic Water Balance Data of the Continents," Publ. Climatol., Vols. XV–XVIII. Drexel Inst. Technol., Canterton, New Jersey.

Aubert, B. (1970). Etude de la résistance à la diffusion gazeuse au niveau de l'épiderme foliaire du bananier (*Musa acuminata* Colla cv. *sinensis*) et de l'ananas (*Ananas comosus* (L.) Merr) en conditions naturelles. *Fruits* **25,** 495–507.

Aubert, B. (1971). Effet de la radiation globale sur la synthèse d'acides organiques et la régulation stomatique des plantes succulentes. Exemple d'*Ananas comosus* (L.) Merr. *Oecol. Plant.* **6,** 25–34.

Aubert, B., and Bartholomew, D. P. (1973). Mesures en champ de la température de limbes d'ananas par radiométrie infrarouge. *Fruits* **28,** 623–629.

Aubert, B., and de Parcevaux, S. (1969). Résistance à la diffusion gazeuse au niveau de l'épiderme foliaire de quelques plantes fruitères tropicales et subtropicales. *Fruits* **24,** 177–190.

Aubert, B., Gaillard, J. P., Py, C., Lossois, P., and Marchal, J. (1973). Influence de l'altitude sur le comportement de l'ananas "cayenne lisse": Essais réalisés au pied du Mont Cameroun. *Fruits* **28,** 203–214.

Barrs, H. D. (1968). Determination of water deficits in plant tissues. *In* "Water Deficits and Plant Growth" (T. T. Kozlowski, ed.), Vol. 1, p. 267. Academic Press, New York.

Bartholomew, D.P. (1975). Use of diffusion porometer on xerophytic plants. *Wash., Agric. Exp. Stn., Bull.* **809,** 26–28.

Bartholomew, D. P., and Kadzimin, S. B. (1976). Porometer cup to measure leaf resistance of pineapple. *Crop Sci.* **16,** 565–568.

Bender, M. M., Rouhani, I., Vines, H. M., and Black, C. C. (1973). $^{13}C/^{12}C$ ratio changes in crassulacean acid metabolism plants. *Plant Physiol.* **52,** 427–430.

Benzing, D. H., and Renfrow, A. (1971). The significance of photosynthetic efficiency to habitat preference and phylogeny among Tillandsioid bromeliads. *Bot. Gaz. (Chicago)* **132,** 19–30.

Black, C. C. (1973). Photosynthetic carbon fixation in relation to net CO_2 uptake. *Annu. Rev. Plant Physiol.* **24,** 253–286.

Black, C. C., Edwards, G. E., Kanai, R., and Mollenhauer, H. H. (1971). Photosynthetic assimilation of carbon in certain higher plants. *Photosynth., Two Centuries After Its Discovery by Joseph Priestley, Proc. Int. Congr. Photosynth. Res., 2nd, 1971* pp. 1745–1775.

Black, R. F. (1962). Pineapple growth and nutrition over a plant crop cycle in south-eastern Queensland. I. Root development and general growth features. *Queensl. J. Agric. Sci.* **19**, 435–451.

Boyer, J. S. (1969). "Physiological Limitations on Crop Production under Temperature and Moisture Stress." Nat. Acad. Sci., Washington, D.C.

Bruinsma, J. (1958). Studies on the crassulacean acid metabolism. *Acta Bot. Neerl.* **7**, 531–590.

Burg, S. P., and Burg, E. A. (1966). Auxin-induced ethylene formation: Its relation to flowering in the pineapple. *Science* **152**, 1269.

Cannon, R. C. (1957). Pineapple fertilizer investigations in south-eastern Queensland. *Queensl. J. Agric. Sci.* **14**, 93–110.

Carter, W. T. (1943). A promising new soil amendment and disinfectant. *Science* **97**, 383–384.

Carter, W. T. (1973). "Insects in Relation to Plant Diseases," 2nd ed., pp. 285–303, 715. Wiley (Interscience), New York.

Cerighelli, R. (1955). La production commerciale de l'ananas on Malaisie britannique. *Fruits* **10**, 499–513.

Clark, H. E., and Kerns, K. R. (1942). Control of flowering with phytohormones. *Science* **95**, 536–537.

Collins, J. L. (1960). "The Pineapple." Leonard Hill, London.

Connelly, P. R. (1969). The relative response of the pineapple plant, *Ananas comosus* (L.) Merr., under varying nitrogen rates and carriers and different levels of light intensity. M.S. Thesis, Library, University of Hawaii, Honolulu.

Connelly, P. R. (1972). The effects of thermoperiod on the carbon dioxide uptake and compensation point of the pineapple plant, *Ananas comosus* (L.) Merr. Ph.D. Dissertation, Library, University of Hawaii, Honolulu.

Connelly, P. R., and Bartholomew, D. P. (1971). Effects of temperature on carbon fixation by pineapple. *Plant Physiol.* **47**, Annu. Suppl., Abstr. No. 178.

Cooper, W. C. (1942). Effect of growth substances on flowering of the pineapple under Florida conditions. *Proc. Am. Soc. Hortic. Sci.* **41**, 93–98.

Coutinho, L. M. (1969). Novas observações sôbre a ocorrencia do "efeito de de saussure" e suas relações com a suculência, a temperature folhear e os movimentos estomaticos. Bol. Faculdade Filosofia, Ciências et Letras, U. São Paulo, *Botanica* **24**, 77–102.

Crews, C. E., Vines, H. M., and Black, C. C. (1975). Postillumination burst of carbon dioxide in crassulacean acid metabolism plants. *Plant Physiol.* **55**, 652–657.

Crews, C. E., Williams, S. L., Vines, H. M., and Black, C. C. (1976). Changes in the metabolism and physiology of crassulacean acid metabolism plants grown in controlled environments. *In* "CO_2 Metabolism and Plant Productivity" (R. H. Burris and C. C. Black, eds.), pp. 235–250. Univ. Park Press, Baltimore.

Das, N. (1965). Control of flower and fruit formation of pineapples by application of auxin and anti-auxin alone and in mixtures. *J. Indian Bot. Soc.* **43**, 498–507.

Das, N., Baruah, S. N., and Baruah, A. (1965). Induction of flowering and fruit formation of pineapples with the aid of acetylene and calcium carbide. *Indian Agric.* **9**, 15–23.

Dittrich, P., Campbell, W. H., and Black, C. C. (1973). Phosphoenolpyruvate carboxykinase in plants exhibiting crassulacean acid metabolism. *Plant Physiol.* **52**, 357–361.

Dodson, P. G. C. (1968). Effects of spacing, nitrogen and hormone treatment on pineapple in Swaziland. *Exp. Agric.* **4**, 103–115.

Ehrler, W. L. (1969). Daytime stomatal closure in *Agave americana* as related to enhanced water-use efficiency. *In* "Physiological Systems in Semi-Arid Environments" (C. C.

Hoff and M. L. Riedesel, eds.), pp. 239–247. Univ. of New Mexico Press, Albuquerque.

Ekern, P. C. (1965a). The fraction of sunlight retained as net radiation in Hawaii. *J. Geophys. Res.* **70,** 785–793.

Ekern, P. C. (1965b). Disposition of net radiation by a free water surface in Hawaii. *J. Geophys. Res.* **70,** 795–800.

Ekern, P. C. (1965c). Evapotranspiration of pineapple in Hawaii. *Plant Physiol.* **40,** 736–739.

Ekern, P. C. (1966). Evapotranspiration by bermudagrass sod, *Cynodon dactylon* L. Pers., in Hawaii. *Agron. J.* **58,** 387–390.

Ekern, P. C. (1967). Soil moisture and soil temperature changes with the use of black vapor-barrier mulch and their influence on pineapple (*Ananas comosus* (L.) Merr.) growth in Hawaii. *Soil Sci. Soc. Am., Proc.* **31,** 270–275.

Ekern, P. C. (1968). Phyllotaxy of pineapple plant and fruit. *Bot. Gaz. (Chicago)* **129,** 92–94.

Englerth, E. J. (1969). Nitrogen nutrition of the pineapple plant, *Ananas comosus* (L.) Merr., soil nitrogen status and dynamics of the reniform nematode population *Rotylenchulus reniformis* Linford and Oliveira in relation to form of nitrogen fertilizer, soil acidity, and fumigation. Ph.D. Dissertation, Library, University of Hawaii, Honolulu.

Evans, H. R. (1957). The influence of fertilizers on pineapple replant areas. *East Afr. Agric. J.* **22,** 122–126.

Evans, H. R. (1959). The influence of growth-promoting substances on pineapples. *Trop. Agric. (London)* **36,** 108–117.

Gaillard, J. P. (1969). Influence de la date de plantation et du poids des rejets sur la croissance des ananas au Cameroun. *Fruits* **24,** 75–87.

Gaillard, J. P. (1970). Recherche d'un équilibre K/N dans la production de l'ananas frais au Cameroun. I. Résultats agronomiques. *Fruits* **25,** 11–24.

Gowing, D. P. (1956). An hypothesis of the role of naphthaleneacetic acid in flower induction in the pineapple. *Am. J. Bot.* **43,** 411–418.

Gowing, D. P. (1961). Experiments on the photoperiodic response in pineapple. *Am. J. Bot.* **48,** 16–21.

Groszmann, H. M. (1950). Hormones and flowering in pineapples. *Queensl. Agric. J.* **70,** 88–89.

Hayes, W. B. (1960). "Fruit Growing in India." India University Press, Allahabad.

Hepton, A., and Anderson, E. J. (1968). Interfruitlet corking of pineapple fruit, a new disease in Hawaii. *Phytopathology* **58,** 74–78.

Holmgren, P., Jarvis, P. G., and Jarvis, M. S. (1965). Resistances to carbon dioxide and water vapour transfer in leaves of different plant species. *Physiol. Plant.* **18,** 557–573.

Hsiao, T. C. (1973). Plant responses to water stress. *Annu. Rev. Plant Physiol.* **24,** 519–570.

Huang, Y. J. (1956). A comparative experiment between the planting of summer seedlings and winter seedings of pineapple. *J. Agric. Assoc. China* [N. S.] **14,** 75, English summary. in *Hortic. Abstr.* (Taipei), 1954–1958, No. 0079. Chin. Hortic. Soc., Taipei, Taiwan.

Jacques-Felix, H. (1950). Discussion sur la fascisation de l'ananas. *Fruits* **5,** 39–51.

Johnson, M. O. (1935). "The Pineapple." Paradise of the Pacific Press, Honolulu, Hawaii.

Jones, M. B., and Mansfield, T. A. (1972). A circadian rhythm in the level of carbon dioxide compensation in *Bryophyllum fedstchenkoi* with zero values during the transient. *Planta* **103,** 134–146.

Jorgenson, K. R. (1973). Fertilizing pineapples in a soil high in potassium. *Queensl. J. Agric. Anim. Sci.* **30**, 213–223.
Joshi, M. C., Boyer, J. S., and Kramer, P. J. (1965). Growth, carbon dioxide exchange, transpiration, and transpiration ratio of pineapple. *Bot Gaz. (Chicago)* **126**, 174–179.
Kadzimin, S. B. (1975). Effects of water stress on pineapple (*Ananas comosus* (L.) Merr.). M.S. Thesis, Library, University of Hawaii, Honolulu.
Kerns, K. R., Collins, J. L., and Kim, H. (1936). Developmental studies of the pineapple *Ananas comosus* (L.) Merr. I. Origin and growth of leaves and inflorescence. *New Phytol.* **35**, 305–317.
Kluge, M. (1969). Zur Analyse des CO_2-Austausches von *Bryophyllum.* II. Hemmung des nächtlichen Stärkeabbaus in CO_2 verarmter Atmosphäre. *Planta* **86**, 142–150.
Kluge, M. (1971). Studies on CO_2 fixation by succulent plants in the light. *In* "Photosynthesis and Photorespiration" (M. D. Hatch, C. B. Osmond, and R. O. Slatyer, eds.), pp. 283–287. Wiley (Interscience), New York.
Kluge, M. (1976). Models of CAM regulation. *In* "CO_2 Metabolism and Plant Productivity" (R. H. Burris and C. C. Black, eds.), pp. 205–216. Univ. Park Press, Baltimore.
Kozlowski, T. T. (1968a). "Water Deficits and Plant Growth," Vol. 1. Academic Press, New York.
Kozlowski, T. T. (1968b). "Water Deficits and Plant Growth," Vol. 2. Academic Press, New York.
Kozlowski, T. T. (1972). "Water Deficits and Plant Growth," Vol. 3. Academic Press, New York.
Kozlowski, T. T. (1976). "Water Deficits and Plant Growth," Vol. 4. Academic Press, New York.
Kramer, P. J. (1969). "Plant and Soil Water Relationships: A Modern Synthesis." McGraw-Hill, New York.
Krauss, B. H. (1930). The transpiration of pineapple. M.S. Thesis, Library, University of Hawaii, Honolulu.
Krauss, B. H. (1948). Anatomy of the vegetative organs of the pineapple, *Ananas comosus* (L.) Merr. I. Introduction, organography, the stem, and the lateral branch or axillary buds. *Bot. Gaz. (Chicago)* **110**, 159–217.
Krauss, B. H. (1949). Anatomy of the vegetative organs of the pineapple, *Ananas comosus* (L.) Merr. II. The leaf. *Bot. Gaz. (Chicago)* **110**, 333–404.
Lacoeuilhe, J. J. (1975). Le problème de la deuxième récolte en culture d'ananas en Côte d'Ivoire. *Fruits* **30**, 83–89.
Lebedev, A. N. (1970). "The Climate of Africa." Israel Program for Scientific Translations, Jerusalem.
Lerman, J. C., Deleens, E., Nato, A., and Moyse, A. (1974). Variation in the carbon isotope composition of a plant with crassulacean acid metabolism. *Plant Physiol.* **53**, 581–584.
Loomis, R. S., Williams, W. A., and Hall, A. E. (1971). Agricultural productivity. *Annu. Rev. Plant Physiol.* **22**, 431–468.
Mitchell, A. R. (1962). Plant development and yield in the pineapple as affected by size and type of planting material and times of planting and forcing. *Queensl. J. Agric. Anim. Sci.* **19**, 453–466.
Mitchell, A. R., and Nicholson, M. E. (1965). Pineapple growth and yield as influenced by urea spray schedules and potassium levels at three plant spacings. *Queensl. J. Agric. Sci.* **22**, 409–417.
Mitchell, P. (1953). The pineapple. *Queensl. Agric. J.* **77**, 62–84 and 125–138.

Neales, T. F. (1973). Effect of night temperature on the assimilation of carbon dioxide by mature pineapple plants, *Ananas comosus* (L.) Merr. *Aust. J. Biol. Sci.* **26**, 539–546.

Neales, T. F., Patterson, A. A., and Hartney, V. J. (1968). Physiological adaptation to drought in the carbon assimilation and water loss of xerophytes. *Nature (London)* **219**, 469–472.

Neild, R. E., and Boshell, F. (1976). An agroclimatic procedure and survey of the pineapple production potential of Colombia. *Agric. Met.* **17**, 81–92.

Nightingale, G. T. (1936). The growth status of the pineapple plant. *Pineapple News* **10**, 68–78.

Nightingale, G. T. (1942). Nitrate and carbohydrate reserves in relation to nitrogen nutrition of pineapple. *Bot. Gaz. (Chicago)* **103**, 409–456.

Nishida, K. (1963). Studies on stomata movement of crassulacean plants in relation to the acid metabolism. *Physiol. Plant.* **16**, 281–298.

Noffsinger, T. L. (1961a). Leaf and air temperature under Hawaii conditions. *Pac. Sci.* **15**, 304–306.

Noffsinger, T. L. (1961b). Monthly probabilities of different rainfall levels. *Univ. Hawaii, Land Study Bur., Tech. Pap.* No. 4, pp. 1–44.

Osmond, C. B. (1976). CO_2 assimilation and dissimilation in the light and dark in CAM plants. *In* "CO_2 Metabolism and Plant Productivity" (R. H. Burris and C. C. Black, eds.), pp. 217–233. Univ. Park Press, Baltimore.

Osmond, C. B., Allaway, W. G., Sutton, B. G., Troughton, J. H., Queiroz, O., Luttge, U., and Winter, K. (1973). Carbon isotope discrimination in photosynthesis of CAM plants. *Nature (London)* **246**, 41–42.

Pico, R. (1974). "The Geography of Puerto Rico." Aldine, Chicago, Illinois.

Py, C. (1958). La production d'ananas de novembre–décembre. *Fruits* **13**, 105–110.

Py, C. (1959). Etude sur la croissance de l'ananas en Guinée. *Fruits* **14**, 3–24.

Py, C. (1968). Intéret, dans la culture de l'ananas en zone humide, d'une couverture du sol en polyéthylène. *Fruits* **23**, 139–148.

Py, C., and Guyot, A. (1970). La Floraison controlée de l'ananas par l'ethrel, nouveau régulator de croissance (lère partie). *Fruits* **25**, 253–262.

Py, C., and Lossois, P. (1962). Prévision de récolte en culture d'ananas: études de correlations. Part 2. *Fruits* **17**, 75–87.

Py, C., and Tisseau, M. (1965). "L'ananas." G. P. Maisonneuve & Larose, Paris.

Py, C., Tisseau, M. A., Oury, B., and Ahamada, F. (1957). "La culture de l'ananas en Guinee," Manual du planteur. de L'Institut français de Recherches Fruitières Outre-Mer, Paris.

Py, C., Lossois, P., and Karamkam, M. (1968). Contribution à l'étude du cycle de l'ananas. *Fruits* **23**, 403–413.

Queiroz, O. (1968). Sur le métabolisme acide des Crassulacees. III. Variations d'activité enzymatique sous l'action du photopériodisme et du thermopériodisme. *Physiol. Veg.* **6**, 117–136.

Queiroz, O. (1974). Circadian rhythms and metabolic patterns. *Annu. Rev. Plant Physiol.* **25**, 115–134.

Randhawa, G. S., Dass, H. C., and Chacko, E. K. (1970). Effect of etherel, NAA (naphthaleneacetic acid) and NAD (naphthaleneacetamide) on the induction of flowering in pineapple (*Ananas comosus*). *Curr. Sci.* **39**, 530–531.

Ranson, S. L., and Thomas, M. (1960). Crassulacean acid metabolism. *Annu. Rev. Plant Physiol.* **11**, 81–110.

Ravoof, A. A. (1973). Effects of root temperatures and nitrogen carriers on nutrient uptake, growth, and composition of pineapple plants, *Ananas comosus* (L.) Merr. Ph.D. Dissertation, Library, University of Hawaii, Honolulu.

Rodriguez, A. G. (1932). Influence of smoke and ethylene on the fruiting of the pineapple (*Ananas sativus* Shult). *J. Dep. Agric. P. R.* **16**, 5–18.

Rohrbach, K. G., and Pfeiffer, J. (1971). Temperature and moisture requirements for infection of detached pineapple inflorescences by pink disease bacteria. *Phytopathology* **61**, 908 (abstr. annu. meet.).

Romero, A. C., López, C., and Gurdián, G. R., (1973). Nutricion mineral de la piña cultivada en el campo vs. producción y calidad de la fruta. *Proc. Trop. Region Am. Soc. Hortic. Sci.* **17**, 203–219.

Samuels, G. (1969). Pineapple cultivars 1970. *Proc. Trop. Region Am. Soc. Hortic. Sci.* **14**, 13–24.

Samuels, G., and Gandia-Diaz, H. (1960). A comparison of the yield and nutritional requirements of the Red Spanish and Smooth Cayenne pineapples. *Proc. Caribb. Region Am. Soc. Hortic. Sci.* **4**, 41–47.

Sanford, W. G. (1962). Pineapple crop log—concept and development. *Better Crops Plant Food* **46**, 32–43.

Sanford, W. G. (1964). Factors influencing the interpretation of the pineapple crop log. *In* "Plant Analysis and Fertilizer Problems" (C. Bould, P. Prevot, and J. R. Magness, eds.), Vol. 4, pp. 255–267. W. F. Humphrey Press, Geneva, New York.

Seshagiri, P. V. V., and Suryanarayanamurthy, A. (1957). Studies in the organic acid metabolism of *Ananas sativus*. *J. Indian Bot. Soc.* **36**, 254–261.

Shreve, E. B. (1916). An analysis of the causes of variations in the transpiring power of cacti. *Physiol. Res.* **2**, 73–127.

Sideris, C. P., and Krauss, B. H. (1928). Water relations of pineapple plants. *Soil Sci.* **26**, 305–315.

Sideris, C. P., and Krauss, B. H. (1937). The growth of pineapple plants in complete water cultures with either ammonia or nitrate salts. *Growth* **1**, 204–210.

Sideris, C. P., and Krauss, B. H. (1938). Growth phenomena of pineapple fruits. *Growth* **2**, 181–196.

Sideris, C. P., and Krauss, B. H. (1955). Transpiration and translocation phenomena in pineapples. *Am. J. Bot.* **42**, 707–709.

Sideris, C. P., and Young, H. Y. (1950). Growth of *Ananas comosus* (L.) Merr. at different levels of mineral nutrition under greenhouse and field conditions. I. Plant and fruit weights and absorption of nitrate and potassium at different growth intervals. *Plant Physiol.* **25**, 594–616.

Sideris, C. P., Krauss, B. H., and Young, H. Y. (1938). Assimilation of ammonium and nitrate by pineapple plants grown in nutrient solutions and its effects on nitrogenous and carbohydrate constituents. *Plant Physiol.* **3**, 489–527.

Sideris, C. P., Young, H. Y., and Chun, H. H. Q. (1948). Diurnal changes and growth rates as associated with ascorbic acid, titratable acidity, carbohydrate and nitrogenous fractions in the leaves of *Ananas comosus* (L.) Merr. *Plant Physiol.* **23**, 38–69.

Smith, B. N., and Epstein, S. (1971). Two categories of $^{13}C/^{12}C$ ratios for higher plants. *Plant Physiol.* **47**, 380–384.

Solereder, H., and Meyer, F. J. (1929). Bromeliaceae. *In* "Systematische Anatomie der Monokotyledonen" (F. J. Meyer, ed.), Vol. IV, pp. 80–129. Verlag von Gebrüder Borntraeger, Berlin.

Strauss, E. B. (1960). "Pineapples in the Eastern Cape." Inst. Soc. Econ. Res., Rhodes University, Grahmstown, South Africa.

Su, N. R. (1969). "Research on Fertilization of Pineapples in Taiwan and Some Associated Cultural Problems," Spec. Publ. No. 1. Society of Soil Scientists and Fertilizer Technologists of Taiwan.

Szarek, S. R., Johnson, H. B., and Ting, I. P. (1973). Drought adaptation in *Opuntia basilaris:* Significance of recycling carbon through crassulacean acid metabolism. *Plant Physiol.* **52,** 539–541.

Tan, K. M., and Wee, Y. C. (1973). Influence of size of pineapple slips on plant growth, fruit weight and quality in graded and mixed plantings. *Trop. Agric. (London)* **50,** 130–142.

Tay, T. H., and Tan, K. M. (1971). Leaf area and growth studies of pineapple. *Malays. Pineapple* **1,** 11–21.

Teisson, C. (1972). Etude sur la floraison naturelle de l'ananas en Côte D'Ivoire. *Fruits* **27,** 699–704.

Teisson, C. (1973). Développement et croissance de l'inflorescence d'*Ananas comosus* (cv Cayenne lisse). *Fruits* **28,** 433–439.

Teiwes, G., and Grüneberg, F. H. (1963). "Pineapple," Green Bull. No. 3. Verlagsges. Ackerbau mbH, Hannover, Germany.

Terry, R. M. (1975). Use of plant growth regulators in Hawaii on pineapple. *Hawaii, Agric. Exp. Stn., Misc. Publ.* **124,** 37–39.

Ting, I. P. (1971). Nonautotrophic CO_2 fixation and crassulacean acid metabolism. *In* "Photosynthesis and Photorespiration" (M. D. Hatch, C. B. Osmond, and R. O. Slatyer, eds.), pp. 169–185. Wiley (Interscience), New York.

Ting, I. P., and Dugger, W. M. (1968). Non-autotrophic carbon dioxide metabolism in cacti. *Bot. Gaz. (Chicago)* **129,** 9–15.

Ting, I. P., Thompson, M. D., and Dugger, W. M. (1967). Leaf resistance to water vapor transfer in succulent plants: Effect of thermoperiod. *Am. J. Bot.* **54,** 235–251.

Ting, I. P., Johnson, H. B., and Szarek, S. R. (1972). Net CO_2 fixation in crassulacean acid metabolism plants. *In* "Net Carbon Dioxide Assimilation in Higher Plants" (C. C. Black ed.), pp. 26–53. Proceedings of the Symposium of the Southern Section of the American Society of Plant Physiology, 1972. Department of Biochemistry, University of Georgia, Athens, Georgia.

Tkatchenko, B. (1947). Ecologie de l'ananas dans le Sud-Indochinois. *Fruits* **2,** 206–212.

van Overbeek, J. (1946). Control of flower formation and fruit size in the pineapple. *Bot. Gaz. (Chicago)* **108,** 64–73.

van Overbeek, J., and Cruzado, H. J. (1948a). Note on flower formation in the pineapple induced by low night temperatures. *Plant Physiol.* **23,** 282–285.

van Overbeek, J., and Cruzado, H. J. (1948b). Flower formation in the pineapple plant by geotropic stimulation. *Am. J. Bot.* **35,** 410–412.

Wambiji, H., and El-Swaify, S. A. (1974). Effects of soil salinity status on pineapple. I. Growth parameters. *Hawaii, Agric. Exp. Stn. Dep. Pap.* **22,** 1–14.

Warner, H. L., and Leopold, A. C. (1967). Plant growth regulation by stimulation of ethylene production. *BioScience* **17,** 722.

Wee, Y. C. (1969). Planting density trials with *Ananas comosus* (L.) Merr. var. Singapore Spanish. *Malays. Agric. J.* **47,** 164–174.

Wee, Y. C., and Ng, J. C. (1968). Some observations on the effect of month of planting on the Singapore Spanish variety of pineapple. *Malays. Agric. J.* **46,** 469–475.

Whitman, P. C., and Koller, D. (1967). Species characteristics in whole plant resistances to water vapor and CO_2 diffusion. *J. Appl. Ecol.* **4,** 363–377.

Wiebe, H. H., Campbell, G. S., Gardner, W. H., Rawlins, S. L., Cary, J. W., and Brown, R. W.

(1971). Measurement of plant and soil water status. *Utah, Agric. Exp. Stn., Bull.* **484,** 1–71.

Wilson, J. W. (1966). High net assimilation rates of sunflower plants in an arid climate. *Ann. Bot. (London)* [N.S.] **30,** 745–751.

Wilson, J. W. (1967). Effects of seasonal variation in radiation and temperature on net assimilation and growth rates in an arid climate. *Ann Bot. (London)* [N.S.] **31,** 41–57.

Wynn, T., Brown, H., Campbell, W. H., and Black, C. C. (1973). Dark release of $^{14}CO_2$ from higher plant leaves. *Plant Physiol.* **52,** 288–291.

Yoder, R. C. (1969). Effects of thermoperiod on the stomatal opening and transpiration of pineapple (*Ananas comosus* (L.) Merr.). M.S. Thesis, Library, University of Hawaii, Honolulu.

Yoder, R. C., and Bartholomew, D. P. (1969). Effects of temperature on pineapple transpiration. *Plant Physiol.* **44,** Annu. Suppl., Abstr. No. 113.

Zelitch, I. (1971). "Photosynthesis, Photorespiration, and Plant Productivity." Academic Press, New York.

CHAPTER

6

Grasslands

G. O. MOTT and H. L. POPENOE

I. INTRODUCTION

Tropical and subtropical grasslands encompass a large floristic diversity, a wide spectrum of environmental circumstances, and include, for the purposes of this chapter, man's domestic livestock. Most other chapters in this volume deal with only one species, some of which occupy a very limited niche in the environment. However, this chapter will take into account the heterogeneity of plant species, principally those in the grass and legume families and the wide range of environments in which they grow between 30°N and 30°S latitudes. A brief summary of the natural grassland ecosystems of the tropical zone is given to acquaint the reader with their floristic composition, their similarities and differences, and some factors which influenced the development and production of certain ecosystems.

Any particular natural tropical grassland is composed of a multitude of species, and any change in a single component of the environment may bring about drastic changes in the ecosystems. Components of the environment such as light, temperature, and water are dealt with only in a general way, since quantiative data for natural grasslands do not exist and only recently have data become available on a few cultivated forage species. The main emphasis is on the soil factors which limit production of the primary producers in soil–plant–animal systems. The interface between the soil and plant is of primary concern, although we are well aware of the effect of grazing pressure on the ecophysiology of tropical grasslands. The subjects of forage utilization by grazing animals, the effects of quantity and quality of forage on offer, and grazing and animal management are omitted from this discussion.

II. FLORA OF NATURAL GRASSLANDS

A. Distribution of the Grass Family

There is a phylogenetic similarity among the grasses associated with natural grasslands throughout the tropical regions of the world. If the classification of Stebbins and Crampton (1959) is accepted, most grasses occur in the subfamilies of Panicoideae and Eragrostoideae. We must be aware, however, that a wide spectrum of environments occurring in the tropics suggests that a variety of genera and species have evolved, some of which fit into very narrow niches of ecosystems. If we define the tropical and subtropical region to include, in many parts of the world, up to 30°N and 30°S latitudes, we greatly expand the environments to which tropical grasses have become adapted. Within this zone there are large variations in light intensity and duration, oscillations in temperature, large differences in altitude, amount and distribution of rainfall, and tremendous variations in all aspects of soil environment.

The sensitivity of certain genera and species of grasses to specific components of the environment is becoming better understood. Kikuyu grass (*Pennisetum clandestinum*) grows well at an altitude of 2500 to 3000 m near the equator but descends to near sea level at 30°N and 30°S latitudes. Certain grasses are adapted to aquatic conditions and a complete spectrum of moisture conditions is found in the Amazon basin from almost continuous flooding to very dry desert conditions. Even with this very broad spectrum of conditions, the primary subfamily of grasses represented is Panicoideae. It is in this subfamily that most tropical grasses occur, with a few grasses from Eragrostoideae. On the other hand, as we move out of the tropical regions of the world, Festucoideae become the dominant subfamily.

Speciation in the grass family is extremely difficult because changes in environmental circumstances result in morphologically distinguishable differences within species. We are also dealing in many cases with extreme differences in chromosome numbers without any morphological manifestations. To add to the difficulty, apomictic swarms occur in many genera of the Panicoideae subfamily which are difficult to classify. In recent years, plant geneticists have discovered sexual types in *Cenchrus ciliaris* and *Panicum maximum,* and this has provided a means of releasing the gene pool. Many new types of grasses are being developed which differ greatly from parent plants both morphologically and agronomically. This promises to be one of the most important developments for production of new cultivars which, no doubt, will increase the range of adaptation of many tropical grasses. Many other species of tropical grasses also display apomictic behavior and the development of new cultivars awaits discovery of sexual plants within these species.

A classic example of improvement which can be realized by development of new cultivars is the selection of a hybrid by Dr. Glenn Burton at Tifton, Georgia. This is a hybrid between a South African introduction and a naturalized *Cynodon dactylon* growing in Georgia. This hybrid is almost completely sterile, so it depends on vegetative methods for propagation. It is much more productive than naturalized forms of *Cynodon* introduced in the southeastern area of the United States and at least five million hectares of "Coastal" Bermuda grass are now being grown in the southern part of the United States. Another cultivar, "Midland" Bermuda grass, has extended the range of the genus *Cynodon* several degrees further north in the temperate region of the United States. This grass provides a very important source of feed during periods of the year when temperate species are of low production or low quality. Coastcross-1 is another hybrid *Cynodon* which was selected specifically for its high rate of digestibility. These examples serve to illustrate some potential improvements which may be made by the genetic manipulation of grasses available in the tropics.

B. The African Grasslands

African grasslands have been the source of most tropical grasses used in improved pastures throughout the world. A description of these grasslands is too complex to include in this chapter, but they are well described by Rattray (1968). African grasslands cover a wide spectrum of environments from dry deserts of the Sahara to the tropical rain forest of Zaire. Ecologically, most grasslands of this continent must be considered disturbed grasslands because of the long influence of man. Another unique feature of African grasslands has been the presence of ungulate animals which have been an important factor in development of grasslands of high productivity in terms of animal products.

Our interest in African grasses relates to their migration to other tropical regions of the world and the superiority of many of these grasses to indigenous species of the grasslands of South America, Australia, and Asia. In Central and South America, as well as Australia, most artificial pastures are being planted to grasses which have their origin on the African continent. Forested areas in the Amazon basin are being cleared and planted to perennial African grasses. Similar procedures are being practiced in Central America, in Australia, the cerrados of Brazil, the savannas of Venezuela, and the llanos of Colombia. The superiority of African species is well recognized, and research is now progressing at a rapid pace to explore the gene pools of a large number of African genera and species for the purposes of developing superior grasses for animal production which fit a particular niche in the environment. The migration of African grasses was discussed by Parsons (1972). Some of the most important migrants have been species of *Panicum, Pennisetum, Melinis, Hyparrhenia, Cynodon, Brachiaria, Cenchrus,* and *Chloris.*

C. South and Central American Grasslands

For purposes of this chapter, major areas of tropical natural grasslands as represented by the cerrados and caatingas of Brazil, the savannas of Amapa and Roraima in Brazil, as well as the savannas of Venezuela, Guyana, Surinam, and the llanos of Colombia will be considered. The cerrados of Central Brazil are mixed formations of trees, shrubs, and an understory of grasses. They occupy extensive areas in the central plateau region of Brazil, primarily in the states of São Paulo, Minas Gerais, Goiás, and Mato Grosso. Most natural grasses are very low in quality and belong to the genera *Panicum, Paspalum, Andropogon,* and some less important genera. Ecologists do not entirely agree as to the reasons for this vast area of grasslands in the central plateau. Three primary factors have been associated with the existence of this grassland region. It has been suggested that cerrados represent a disclimax because of the occurrence of annual burning. It has also been suggested that the forest has been prevented from developing because of moisture deficits due to the relatively long dry season, while others accept the theory that edaphic factors associated with extremely low levels of nutrients in the soil are responsible for the grassland type of vegetation. A most important feature of the flora of the cerrados is the number of species and diversity of the legumes which occur in this region of extremely low soil fertility. This region is a very important source of the genera *Stylosanthes, Arachis, Desmodium, Galactia,* and many other tropical legumes peculiarly adapted to this particular climatic and soil environment. Since there are regions of the world which have similar environmental circumstances, this area may serve as a source of genes for plant breeding programs around the world. Without doubt, one of the most important research efforts which should be undertaken in Brazil is to make extensive collections of this treasure of legume germ plasm and to make provisions for its preservation for future generations of plant scientists. Most legumes indigenous to this region are very tolerant of acid soils, extremely low fertility, high aluminum and manganese levels, and moisture stress.

The savannas of Northern Brazil are dominated by the grass genera *Trachypogon, Andropogon, Paspalum, Panicum,* and *Axonopus.*

Since the soils of the Guyanas, Venezuela, and Colombia are heavily leached and the region is characterized by a wet–dry season pattern of rainfall, a thread of similarity exists among the genera of grasses represented throughout this region. There are local differences which are easily recognized by taxonomists and ecologists, but the procedures for development and improvement of these grasslands for animal production are very similar. Many genera of legumes also occur in the savannas and representatives of the genera *Stylosanthes, Desmodium, Aeschynomene, Calopogonium, Centrosema, Clitoria, Galactia, Indigofera, Phaseolus, Teramnus, Vigna,* and *Zornia* are to be found. Very little if any exploitation of these herbaceous legumes has been made and researchers need to avail themselves of the opportunity of developing these legumes for livestock production.

D. The Tropical Australian Grasslands

The characteristic species of humid and subhumid tropical grasslands of Australia are *Heteropogon contortus* and *Themeda australis.* There are many other species in the genera *Sorghum, Dichanthium,* and *Bothriochloa,* but the first two genera dominate the vegetation. As rainfall diminishes toward the interior of the continent and conditions become more arid, species of *Aristida, Triodia,* and *Astrebla* dominate the vegetation. As in the American tropics, there has been a migration of African species into Australia including *Panicum maximum, Digitaria decumbens, Melinis minutiflora, Setaria anceps, Cenchrus ciliaris, Pennisetum purpureum,* and *Chloris gayana* for improved pasture. Australian scientists have made great strides in developing tropical legumes which are adapted to a wide range of conditions in the Australian grassland environment. For the humid and subhumid tropical region, *Centrosema pubescens, Desmodium intortum, Glycine wightii, Pueraria phaseoloides,* and *Macroptilium atropurpureum* are being planted as rapidly as seed stocks become available. In drier regions, three different species of *Stylosanthes* are finding a place in their pasture programs. Most of these legumes have been imported from tropical America, although some have been introduced from other tropical regions of the world.

Only a very small area of tropical Australia exceeds 1600 mm of annual rainfall. The relatively low summer rainfall characteristic of tropical Australia has provided an incentive to seek tropical grasses and legumes which will tolerate a long dry season. Many of the same problems, associated with low fertility levels, high acidity, and moisture stress found in the savannas of tropical America also occur in Australia.

III. THE TROPICAL ENVIRONMENT AND GRASSLANDS

A. Light

Radiant energy is one of the greatest assets of tropical environments since the variation in the amount of solar radiation received at the top of the atmosphere during the year is quite small. Many tropical species of grasses have a unique CO_2 fixation pathway which, at higher light intensities, results in a much higher rate of photosynthesis when compared with temperate species. Currently, two major pathways of photosynthesis are recognized and plants are classified as either C_3 or C_4 plants. Some physiological features of C_3 and C_4 grasses are summarized in Table I.

Rates of photosynthesis in tropical grasses (C_4) are of the order of two or three times those of temperate grasses (C_3) and tropical legumes (C_3). Such high rates of CO_2 fixation are partially accounted for by a more complicated biochemical photosynthetic pathway and are also partially due to lack of photorespiration. These differences in rate of CO_2 fixation between C_3 and C_4 plants become much more apparent at higher light intensities. Our present knowledge suggests that C_4 grasses may not reach light saturation even at highest levels of solar radiation (Fig. 1), whereas assuming that temperature, CO_2 levels, and other factors of the environment are not limiting, light saturation is achieved for C_3 legumes at solar radiation levels of less than one-fourth of maximum. C_4 grasses have the capacity of converting up to 5 to 6% of incoming solar radiation at the earth's surface at relatively high temperatures giving them a comparative advantage over C_3 tropical legume species (Fig. 1). Since variation in solar

TABLE I
Physiological Features of Temperate (C_3) and Tropical (C_4) Grasses

Characteristics	Temperate	Tropical
Photosynthetic capacity	20–30 mg CO_2 dm^{-2} h^{-1}	40–60 mg CO_2 dm^{-2} h^{-1}
Growth rate at optimum	X^a	2–3X
Photorespiration	Present	Absent
Efficiency of water use	X	2–3X
Optimum growth temperature	15–22°C	22–35°C
Low temperature limit	5°C	10°C
High temperature limit	30°C	45°C
Light saturation	20–30 klux	60 klux or more

[a] X, level of attribute.

energy is much smaller during the year under tropical than temperate climates, if other environmental conditions are near optimum for forage growth, high rates of production can be realized. Yields of over 50,000 kg dry matter per hectare per year have been reported for such grasses as *Pennisetum purpureum* and *Panicum maximum.* On the other hand, a high yield of pure tropical legume swards would be of the order of 12,000 to 16,000 kg/ha.

1. Light Response and Legume–Grass Association

Establishment and maintenance of legume–grass swards in tropical environments have been difficult to achieve. Many factors affect the balance between these two plant families, and it is apparent that the difference in their photosynthetic mechanisms may be one of these factors. Since C_4 grasses have growth

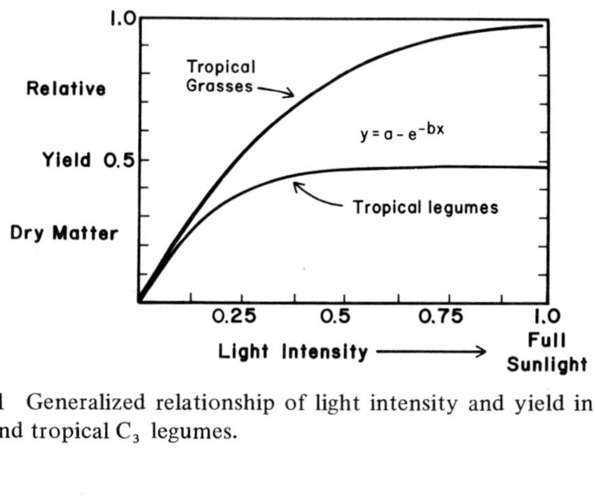

Fig. 1 Generalized relationship of light intensity and yield in tropical C_4 grasses and tropical C_3 legumes.

rates at least twice those of C_3 tropical legumes, optimum nutrient levels for legumes must be provided, and defoliation systems devised to utilize the faster growth of tropical grasses, if the association is to be maintained. Research is being concentrated on this subject to develop production and management systems which will maintain both components in the sward at levels satisfactory for high production and forage quality. Legume–grass associations are very complicated ecosystems in which the environmental factors, light, temperature, nutrient supply, and moisture, play important roles. Growth habits of species, their geometry, and recovery mechanisms may also be critical factors in their persistance.

Many tropical legumes are viny plants which maintain a high level of exposure to solar energy by climbing on tall tropical grasses; these are *Pueraria phaseoloides, Centrosema pubescens, Macroptilium purpureum,* and many others. If environmental conditions are optimum for the growth of legumes, it is not uncommon for legumes to become dominant and eliminate associated grasses, even though grasses may have a comparative advantage of a faster growth rate.

2. *Plant Geometry and Light Interception*

Species of tropical grasses differ greatly with respect to their ability to trap solar radiation, and thus, their net assimilation rates. The optimum leaf area index (LAI) varies from 2 to 3 to over 15 among species, and there are also extreme differences among strains within species. Some strains of *Pennisetum purpureum* and *Panicum maximum* have nearly vertical leaves during much of their growth period, which permits development of large leaf areas and greater efficiency of intercepted light. Other species of *Cynodon, Digitaria,* and *Brachiaria* have their leaves oriented more nearly horizontal and develop much lower leaf areas. Most species of viny legumes have mostly horizontal leaves and relatively low LAI's when compared with grasses. This marked difference between leaf structure of tropical legumes and some tropical grasses adds another complicating factor to management of tropical legume–grass associations. The foliage area required to intercept most incoming radiation is very important in maximizing production of pasture swards. Canopies which display a more horizontal leaf surface probably have an advantage under more frequent defoliations by grazing animals than those which are more nearly vertical. The latter will reach their maximum production level with less frequent defoliation so that the canopy can better attain its optimum LAI. Herein lies one of the difficulties of maintaining legume–grass associations with C_3 legumes having a high extinction coefficient, which attain their maximum growth rates at an LAI of about 4, and associated C_4 grasses having low extinction coefficients which attain their maximum growth rate at an LAI of about 12. The concept is illustrated in Fig. 2. Although the curves do not indicate any specific legume–grass combination, the relationship is obvious. Grasses have a comparative advantage, assuming that other

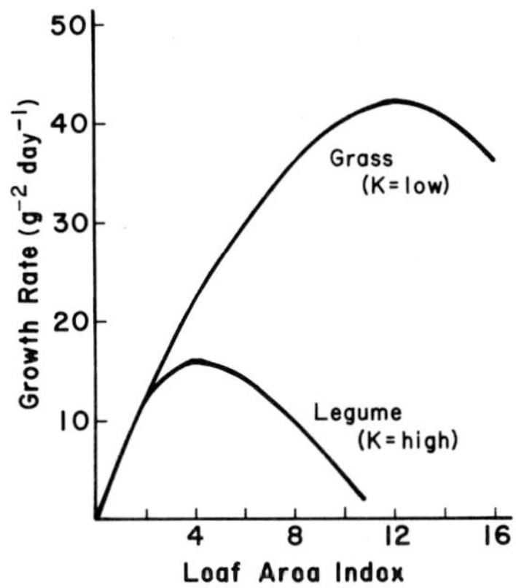

Fig. 2 Leaf area index and growth rates in grasses and legumes as related to leaf architecture. *K* = extinction coefficient.

conditions of the environment are optimum for each component of the mixture. However, legumes may occupy about the same competitive position as grasses under low light intensities—rates of CO_2 fixation are about the same for C_3 legumes and C_4 grasses at light intensities of about one-fourth full sunlight—if they have a viny growth and are able to maintain a high level of leaf exposure to solar energy. The fact that mixtures of *Macroptilium, Centrosema,* and *Pueraria* in combination with *Panicum maximum, Pennisetum purpureum,* and *Brachiarias* sp. are found under high rainfall conditions, where a heavy cloud cover is common, suggests that grasses and legumes under this light environment are about equally competitive and form compatible mixtures. Areas where such combinations seem to thrive are the wet coastal areas of Queensland, the high rainfall areas of western Ecuador and Colombia, the piedmont region on the eastern slopes of the Andes, and the southeastern littoral areas of the Brazilian coast. No doubt, many areas of the Amazon basin will also support such legume—grass associations.

3. Photoperiodic Response

Since relatively short daylengths prevail in the tropics, it would be expected that most tropical grasses and legumes would be short-day plants. It is well known that some species and cultivars of tropical forages are indeterminate over a wide range of daylengths and show very little photoperiodic response, while others are very sensitive to daylength. Some cultivars of *Paspalum notatum* produce abundant seed at higher latitudes, but never flower at low latitudes. A collection of ecotypes of *Panicum maximum* in Florida (ca. 30°N) exhibits a wide spectrum of photoperiodic responses ranging from types which do not flower during the growing season to those that flower profusely even during the longest days of the year. A similar collection of *Pennisetum purpureum* in Venezuela (ca. 10°N) has both abundant-flowering and nonflowering types. As a pasture grass, *Melinis minutiflora* performs quite differently at about 20°S latitude in Brazil than it does at lower latitudes in Colombia. Initiation of reproductive tillers in many tropical grasses results in interruption of growth which is frequently followed by a period of dormancy or relatively slow growth. Maintaining grasses in a vegetative state is usually favorable to sustaining a high nutritive value for grazing animals, as well as preserving high growth rates over a longer period of time. The above statements are from observations made by the authors; few critical observations are available on flowering responses in tropical grasses. The conflict of interest between plant breeders, who strive for high yielding types which produce an abundance of seed, and pastoral agronomists, who prefer nonflowering types with high nutritive value, suggests that new cultivars should include the combined efforts of the two disciplines. Future development of new cultivars for the tropics may require close interregional cooperation with seed being produced of a superior strain in a seed producing environment, which then can be grown for pasture and forage in an environment where no flowering occurs, so that a high nutritive value may be attained. If the development of new cultivars of tropical species is to progress at a rapid pace, there must be simultaneous screening and selection of superior types under a broad spectrum of environments. This will provide exposure to all kinds of environmental stresses, including differences in solar radiation, temperature, water, and nutrient supply, as well as biological factors such as diseases, insects, weeds, and the stress imposed by the grazing animal. Today, the grassland husbandman in the tropical and subtropical regions of the world is dependent upon his seed supply from an entirely different set of environmental circumstances from those in which he operates. Many failures have occurred and will continue until the breeder tests his material under a much broader environment. The photoperiodic response is only one aspect of the total development problem, but a very important one when considering seed propagated cultivars.

Most of the principles with respect to photoperiodic response suggested for grasses also apply to tropical legumes. Photoperiodic response in the genus *Stylosanthes* (Table II) was studied by t'Mannetje (1965). The strain of *S. guyanensis* which he studied flowered only at the 8- and 10-hour photoperiod, while *S. mucronata* flowered at all photoperiods observed including 14 hours. The yield of dry matter continued to increase in both species as daylength increased. Since natural distribution of *S. guyanensis* extends beyond 25°N and 25°S latitudes, a broad spectrum of photoperiod, i.e., responses, would be expected within the species. Much more research is needed to determine the strains best adapted to a particular environment.

TABLE II
Photoperiodic Response of *Stylosanthes guyanensis* and *S. Mucronata*[a]

Photoperiod (hr)	8	10	12	14
No. of days to flowering				
S. guyanensis	52	52	–	–
S. mucronata	51	53	54	62
Yield of dry matter (g)				
S. guyanensis	9.60	8.77	12.30	12.08
S. mucronata	2.78	2.69	2.18	4.10

[a]t'Mannetje, 1965.

B. Temperature

Temperature is an important factor in determining the general distribution of both grasses and legumes in the tropics, and no doubt had a major role in the evolutionary development of both families. Two important aspects of temperature are seasonal changes, which occur during the year and are usually expressed as average daily means, and diurnal variations which are indicated by the oscillation of day and night temperatures. The amplitude of these variations increases with increases in latitude and altitude. Low temperatures restrict the growth of both tropical legumes and grasses and, when temperature during any period of the day drops to 10°C, growth rates approach zero. General relationships between temperature and growth rates of temperate grasses and legumes, tropical legumes, and tropical grasses are shown in Fig. 3. Optimum tempera-

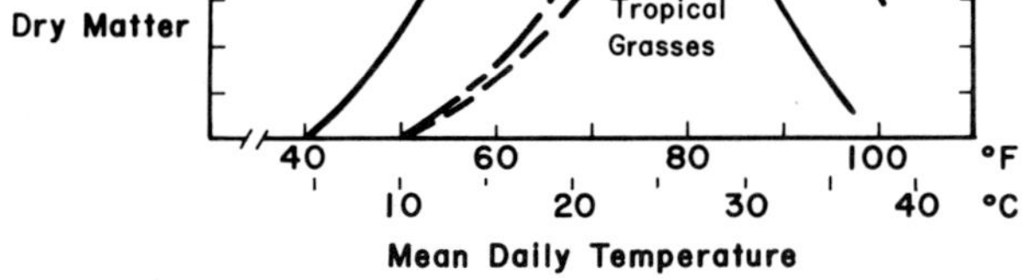

Fig. 3 Temperature and fractional dry matter yield in tropical and temperate grasses and legumes.

tures for maximum production of temperate species are in the range of 20°C, for tropical legumes 30°C, and for tropical grasses may be as high as 40°C. From physiological studies on other crop plants, increases in temperatures are known to increase rates of respiration, and this general relationship probably also applies to tropical forages. With legume–grass associations, high temperatures above the optimum for tropical legumes seem to favor the tropical grasses, since tropical legumes reach their maximum growth rate at lower temperatures.

C. Moisture Regimes

Moisture relationships and their effect on distribution of tropical forage species and their seasonal production are beyond the scope of this chapter. However, it is well known that water balance in a particular environment may be the dominant factor for pasture production. The water available to forages takes into account the rainfall, the rate of evaporation, and the amount of water which the soil is capable of storing within the root zone. Since the root zone differs for different plants, this means that, in addition to physical factors associated with water balance, water utilization is also species dependent. Australian investigators (Fitzpatrick and Nix, 1970) have developed a model for expressing a moisture index which takes into account long-term weekly rainfall and observed or estimated evaporation. The decline of the ratio of actual to potential evapotranspiration as the available water in the root zone is reduced was estimated by a continuous function. A moisture index was derived from the ratio of actual evapotranspiration to potential evapotranspiration. The model has given satisfactory relationships with production of forages under Australian conditions.

The pattern of rainfall, especially the length of the dry season, is an overriding factor in production of forages in the tropics. The ability of many species to withstand moisture stress determines to a large extent their adaptation to a particular region. Areas in which moisture supply is adequate throughout the year are limited, but length and severity of dry periods are extremely variable. Wide spectrums of moisture regimes in the regions of natural distribution of many species have resulted in development of large numbers of ecotypes of both grasses and legumes.

It has already been mentioned that C_4 grasses use water much more efficiently than C_3 plants (see Table I). Various cultivars and species of grasses give different responses to the amount of water available and nitrogen levels as indicated by Burton *et al.* (1957). Three cultivars of *Cynodon dactylon* and one each of *Paspalum notatum* and *Digitaria decumbens* were studied at three nitrogen levels under field conditions. In the first year, the rainfall during the growing season was 1007 mm and during the second year 347 mm. Results are given in Table III. All these factors had large effects upon water efficiency.

TABLE III
Effect of Different Grasses, Rainfall, and Nitrogen Levels on Water Use Efficiency[a]

	kg water/kg dry matter					
	N levels (kg/ha) 1953 (1007 mm rain)			N levels (kg/ha) 1954 (347 mm rain)		
	56	112	224	56	112	224
Cynodon dactylon, Common	6812	2896	1546	9738	5028	4336
Cynodon dactylon, Coastal	2478	1431	803	1547	982	641
Cynodon dactylon, Suwannee	1923	1104	692	1107	725	452
Paspalum notatum, Pensacola	2200	1321	870	3103	1945	1239
Digitaria decumbens, Pangola	2249	1585	2240	2843	2513	3016

[a]Burton *et al.* (1957).

"Suwannee" had the greatest water use efficiency and "Common" the lowest efficiency of the grasses. With the exception of "Pangola," increased nitrogen levels increased water use efficiency. "Coastal" and "Suwannee" showed greater efficiency during the dry year than during the wet year, while the reverse was true for the other three grasses.

IV. THE SOIL ENVIRONMENT

A. Classification of Soils

This discussion of the soil environment is focused on soil as a medium for production of forage plants and how it can be modified to provide for more efficient food production for man. Based on the most recent taxonomic nomenclature, the following orders of world soils account for the major land surface dedicated to forage and pasture production in the tropics: Oxisols, Ultisols, Alfisols, Inceptisols, and Entisols. These are arranged in the descending order of age or degree of weathering. Inceptisols, being of more recent origin, are extremely variable, depending upon the origin of the parent materials, and may or may not be useful for crop production. Oxisols represent the most extreme weathering of parent materials and Ultisols are for the most part highly weathered. McLean (1971) suggests that there is no clear line of demarcation with respect to chemical and physical properties of soil orders, but that they represent a continuum from most recent deposits, with little or no weathering to extreme weathering of the Oxisols (Table IV). Since soil acidity and the nutrient supplying power of soils are of primary concern in forage production, the

TABLE IV
Acid Soil Continuum: Soil Attributes Pertinent to Liming Effects[a]

Attribute	Mollisols	Alfisols	Ultisols	Oxisols
1. Permanent charge acidity	H+ (some $OH\text{-}Al^{2+}$)	$OH\text{-}Al^{2+}$	Al^{3+}	$Al^{3+}(Fe^{3+})$
2. Type of clays	Illite, Mont.	Chloritized Mont. and Illite	Kaolinite hydrous oxides	Hydrous oxides kaolinite
3. Chrystallinity of clays	Chrystalline	Chrystalline	Chrystalline and amorphous	Amorphous and Crystalline
4. Ionic exchange cap.	High cation	High cation (OH–Al blocked)	Low cation Low anion	High anion Low cation
5. Base (Ca and Mg) saturated	Relatively high	Medium to low	Low	Very low
6. Phosphate-fixing tendency	Low	Medium	High	Very high
7. Potassium-release tendency	High	Medium	Low	Low
8. Structure	Excellent (Ca aggregates)	Good to fair (Ca + Al aggreg.)	Fair to poor (Fe + Al aggreg.)	Good to excellent (Fe + Al aggreg.)

[a]McLean (1971).

characteristics discussed by McLean are pertinent. As the degree of weathering increases, the predominant source of soil acidity shifts from the H-ion concentration to Al^{3+}. Cation exchange capacity shifts from a high level in a temperate zone Mollisol to a very low cation exchange capacity with a high anion exchange capacity in an Oxisol. Base saturation (Ca and Mg) shifts from high to very low, K release from high to low, and P fixation from low to high, as degree of weathering increases.

B. Water Relations

Physical relationships of interest are the structural stability of soil aggregates (Table IV), rainfall acceptance of the soil as measured by infiltration rate, permeability and rate of percolation, and soil retention of moisture. Mollisols and Oxisols seem to have an advantage because of their more stable aggregates, formed as a result of binding by organic matter and hydrous oxides, respectively, although there is considerable variation within the orders. Most heavily weathered Oxisols from both the savannas and forested areas have high infiltration and percolation rates. A grass cover has a modifying effect upon these physical characteristics by intercepting raindrops, formation of a mulch on the soil surface, and increasing organic matter content of the soil.

C. Response to Plant Nutrients

The adaptation and production of tropical forages in many cases will depend on the nutrient supplying power of the soil and the input of artificial fertilizers into soil–plant–animal systems. If a mixture of tropical legumes and grasses is the objective for a pasture ecosystem, nutrient requirements of legumes must first be met for them to survive. It does not necessarily follow that all tropical legumes have a higher nutrient requirement than grasses growing in the same environment. Some species of *Desmodium, Stylosanthes,* and *Arachis* are unusually effective in extracting nutrients which are at very low levels of availability in the soil. Oxisols and Ultisols are the natural habitats of many species of these genera. One should not conclude that these legumes will not respond to applications of nutrients which are very deficient. Levels of critical nutrients may be the key to the persistence of legumes in an association. A useful mathematical function for expressing the response to a nutrient or combination of nutrients is the following:

$$Y = A - Be^{-CX}$$

where Y = yield; A = sensitivity of the plant to a nutrient or level in nutrient pool; C = curvature of the response; and X = nutrient added.

Appropriate coefficients are species dependent and this is probably also true for strains within species, so that the values of *A*, *B*, and *C* will have to be determined experimentally for each species and combination of species.

D. Recycling of Nutrients

The efficiency with which nutrients are utilized in any agricultural system depends on the amounts and frequency of recycling of those nutrients. Mott (1974) has discussed some key ideas in recycling of nutrients in soil–plant–animal systems. A simplified concept of the recycling in such a system is illustrated in Fig. 4. The three main compartments are the soil, plant, and animal. The soil is in an equilibrium state with residues which are mainly organic in nature and serve as the principal reservoir of nutrients for plant growth. The sizes of unavailable nutrient pools vary considerably for different nutrients and soils with different chemical and physical properties. In Oxisols and Ultisols, with their low content of total cations—Ca, Mg, and K—and their extremely low level of P, only very small amounts of these elements enter the recycling stream and become available for plant growth. These are nutrients which are harvested in shifting agriculture in one or two crops out of every 15 to 25 years. Many years of solar energy are required to generate these nutrients during a forest fallow, which is quickly destroyed by a good burn. In a soil–plant–animal system, the animal is a very important component, continually converting solar energy captured by photosynthesis in forage plants into useful products and

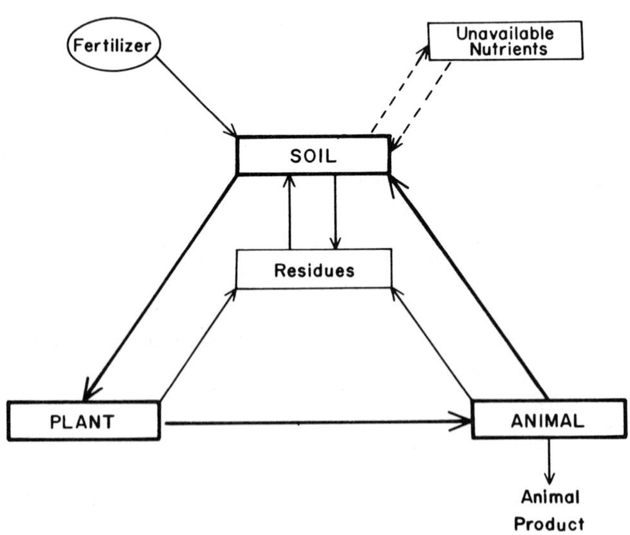

Fig. 4 Nutrient recycling in grazed grasslands.

recycling inorganic nutrients back into the system for production of more forage. Up to 90% of mineral nutrients and N are returned in excreta of grazing animals, making the grazing animal an effective vehicle for retention and frequent reutilization of nutrients in the system.

Many factors influence the efficiency of the recycling mechanism, but only a few will be considered. The more intensively the system is managed the more rapid will be the turnover of nutrients. As biomass yield is increased there will also be an increase in the amount of organic matter in the soil which will increase amounts of nutrients in the residue pool and in the recycling system. Since nutrient levels in many tropical soils are inherently very low, some input of artificial fertilizers may be needed to set the mechanism in motion and maintain a high level of nutrient flow. If a suitable environment for biological N fixation can be created, this can become the source of N in the systems. Intensive systems with high stocking rates, now common on the humid grasslands of Africa, can be achieved in other areas with the technology now available.

Little attention has been given to the role of insect fauna in accelerating rates of transport or retention of nutrients in the recycling stream. Lindquist (1933) studied the activity of several different species of coprophagous beetles in native and cultivated pastures of Kansas (Table V). He found that beetles were not only effective in disposing of cow dung at a very rapid rate, but they also excavated large amounts of soil. Recently, Australian scientists have observed that dung tends to accumulate in their pastures where beetles associated with cattle do not occur. This is in contrast to South African pastures where dung beetles are abundant in excreta of domesticated cattle. Several thousand species of dung beetles are known, and steps are being taken to introduce species into Australia which are adapted to that environment. Where dung beetles are present, dung dropped in the pasture by grazing animals may disappear within 24 hours, thus accelerating rates of recycling and retaining most of the nutrients in the nutrient pool.

TABLE V
Quantities of Dung Buried and Soil Excavated by Coprophagous Beetles[a]

	Quantity/burrow (g)	
Species	Dung buried	Soil excavated
Pinotus carolinus	48.5	287.1
Copris tulius	7.3	37.8
Phanaeus sp.	9.6	93.4

[a]Lindquist (1933).

E. Nitrogen

For grasses in a pasture sward, N is frequently the most important nutrient limiting factor. Only rarely does a lack of response occur to an application of N fertilizer in a cultivated pasture. A review by Vicente-Chandler (1974) indicates some of the potential production of humid tropical grasslands. Studies have been made on *Panicum maximum, Digitaria decumbens, Brachiaria mutica, Cynodon nlemfuensis, Pennisetum purpureum,* and many other species in Puerto Rico, Brazil, Venezuela, Colombia, and Nigeria. Responses to N fertilizer have been almost linear up to 400–600 kg of N/ha/year. At higher levels, the rate of response diminishes until no further increase occurs. Within this range of application, increases in dry matter production are from 30 to 50 kg for each kilogram of N applied. Under grazing conditions, an increase of 1.2–2.0 kg of liveweight of beef was obtained for each kilogram of N applied. Gross annual yields of dry matter have exceeded 50,000 kg/ha under favorable humid tropical conditions at high levels of fertilizer inputs. These responses occurred under conditions of minimum moisture stress and where other nutrients in the system were near optimum. These rates of increase were obtained on an Ultisol with from 30 to 50% slopes, with an initial pH of 4.8–5.2 and 4–8 mEq of exchangeable bases/100 g of soil. Annual rainfall ranged from 1500 to 2000 mm, and mean monthly temperatures from 21° to 27°C. These data are cited to indicate levels of production which are possible on highly weathered soils. It is not suggested that these ecosystems are economically feasible or desirable, but the results under experimental conditions reveal some of the biological potentials and limitations associated with the soil.

Possible sources of N to the soil–plant system include biological N fixation by both legumes and grasses. Fixation of atmospheric N by legume–*Rhizobium* symbiosis is well known and does not need elaboration. Reports of annual N fixation rates in excess of 200 kg of elemental N/ha have been recorded for tropical legumes. This is of the same order of magnitude as for temperate legumes and should be an incentive to exploit tropical legumes for forage, for cover crops, and as green manure crops in rotational and multiple cropping systems. In intensive grazing management systems, at least a portion of the N generated by the legumes is recycled, and the general level of N in the nutrient pool is raised.

It has been known for some time that certain organisms associated with members of the grass family fix N. In Brazil Dobereiner and Day (1974) described such associations in *Paspalum notatum, Pennisetum purpureum,* maize, and sugarcane.

Since 1972, new evidence shows that N fixation in a much broader spectrum of grasses may be occurring in the tropics, and this discovery could have far-reaching implications in the future development of forage–livestock feeding systems (Dobereiner and Day, 1974). Nitrogenase activity has been demonstrated in a number of grasses including *Digitaria decumbens.* The organism has been identified as *Spirillum lipoferum* and is concentrated in cortex cells. There is a suggestion that it is associated with C_4 grasses which have malate as one of its primary products of photosynthesis. *S. lipoferum* is known to utilize malate as an energy source, and it seems reasonable to assume that malate may be available as a substrate. In any symbiotic relationship there is a partnership between the host plant and the microorganism. In this case, the *Spirillum* appears to be genotype specific, and the matching of the proper strain of *Spirillum* with the grass genotype is of utmost importance.

F. Phosphorus

In highly weathered soils of the tropics, total and available P levels are very low. Fixation of P is extremely high due to formation of Fe and Al phosphates. Small amounts of P released in ash from burning the forest in shifting cultivation or in the savanna are either quickly utilized by crops planted by farmers or by regrowth of savanna vegetation. Very little P in available form is in the soil for an extended period of time. The adaptation of savanna grasses to these very low levels of P is related to their ability to extract P and to their physiological tolerance. Many African introductions will not tolerate these low levels of P and require phosphate fertilizers to be productive.

Applications of lime to raise pH to a level of between 5 and 6 may neutralize exchangeable Al sufficiently to reduce fixation of P and increase rates of decomposition of organic matter, but great care must be taken to avoid overliming which may have an adverse effect on availability of other nutrients. The possibility also exists that, as more P is released, the decrease in P levels will be more rapid. In this respect, liming of highly weathered soils may have a long-term deleterious effect on the savanna and on shifting agriculture. Application of 100–250 kg/ha of molybdenized superphosphate in seedings of several tropical legumes in Australia is standard practice followed by a maintenance application every one to three years.

This amount is not sufficient to attain maximum yields initially, but as production increases and with constant recycling of an increment of the P previously applied in a grazed pasture, levels of P can be increased over a period of time. Similar phenomena also occur with other nutrients which are not easily leached.

G. Sulfur

Sulfur deficiency is of very frequent occurrence in highly weathered soils of the humid tropics. This deficiency is frequently not evident since other nutrients, especially N and P, are at such a low level that symptoms of S deficiency do not appear. Raising N levels in cultivated pastures by N fertilizer application may result in increasing organic matter sufficiently to create an imbalance in the N:S ratio so that S becomes unavailable and symptoms become strikingly evident. With the disappearance of fertilizers containing sulfur, such as ammonium sulfate and simple superphosphate, the S problem with both tropical crops and cultivated pastures promises to become more acute. Sulfur is a critical element for both tropical grasses and legumes.

H. Potassium

There are at least five different forms of K in the soil–plant system: K in the organic residues, exchangeable K, fixed K, K in the crystalline matrix, and K in the soil solution. In soil–plant–animal systems, the major pool of K available to plants and in the recycling stream may be found in the organic residues. Residue K is in equilibrium with exchangeable and soil solution K, and the key to a sufficiency of K in the system depends on applying K if it is not already present in the soil. Potassium responses on well-established productive pastures are seldom observed because of effective recycling of K in the system.

I. Critical Plant Percentage of Phosphorus, Potassium, and Sulfur

Andrew and his co-workers (1969a,b, 1974) in Australia have established critical percentages of P, K, and S for several of the more common tropical legumes. The critical percentage is defined as the "percentage of the element in the plant tissue corresponding with optimum growth, above which no further yield response may be expected and below which additional growth could be expected if the element concerned was supplied." The critical percentages were obtained under standardized growth conditions, sampling procedures, and with all other nutrients except the one being tested supplied at adequate levels. The results are presented in Table VI. *Stylosanthes humilis* and *Centrosema pubescens* reached optimum levels of production at lower levels of P, K, and S than most other legumes. The requirements of alfalfa (*Medicago sativa*) are obviously higher than for most tropical legumes except for P. Five of the seven legumes appear to have as high a requirement for P as *Medicago sativa*. This seems to indicate that P may be a key element in the satisfactory production of tropical legumes especially on highly weathered soils which are very low in available P. Levels as high as those provided for *Medicago sativa* may be needed for some tropical legumes.

TABLE VI
Critical Percentage Levels of Phosphorus, Potassium, and Sulfur in the Tops of Eight Legume Species[a]

	Critical percentage of dry matter		
	P[b]	K[c]	S[d]
Macroptilium lathyroides	0.20	0.75	0.15
Macroptilium atropurpureum	0.24	0.75	0.17
Desmodium intortum	0.22	0.72	0.17
Desmodium uncinatum	0.23	0.80	0.17
Stylosanthes humilis	0.17	0.60	0.13
Glycine wightii	0.23	0.80	0.18
Centrosema pubescens	0.16	0.75	0.15
Medicago sativa	0.24	1.20	0.21

[a]Preflowering stage (Andrew and Robins, 1969a,b; Andrew, 1973–1974).
[b]Andrew and Robins (1969a).
[c]Andrew and Robins (1969b).
[d]Andrew (1973–1974).

J. Calcium

In temperate zones of the world, lime is applied as a soil additive to adjust pH to 6.5 or above for crop production. Research during the past 25 years by soil chemists suggests that lime should, on some soils, be considered as a source of nutrient Ca rather than as a soil additive. Table IV indicates that the source of soil acidity in a Mollisol is the H^+ ion concentration, while in the Ultisols and Oxisols, Al^{3+} is the primary source. It is now known that the poor growth of many crop plants on the highly weathered soils of the humid tropics is due to toxic levels of exchangeable Al and in some cases Mn. A comparison of tropical and temperate legumes and their response to increasing ratios of Al to Ca + Mg is given in Fig. 5. Very little quantitative data are available for forage legumes and grasses as to their Al tolerance and it is very difficult to separate the effects of pH, nutrient Ca, toxic levels of Mn, and availability of micronutrients from Al toxicity in a soil–plant system. There are some indications that the yield of some tropical legumes is reduced to about 70% of maximum at high levels of exchangeable Al, whereas temperate legumes may not survive under similar conditions. It is not clear, however, whether the effect is due solely to toxic Al or whether a combination of factors is operating. Some examples of cation–Al relationships in Brazilian soils are given in Table VII. The pH of all soils is low and Al saturation of the exchange capacity ranges from 2 to 90%. Ratios of

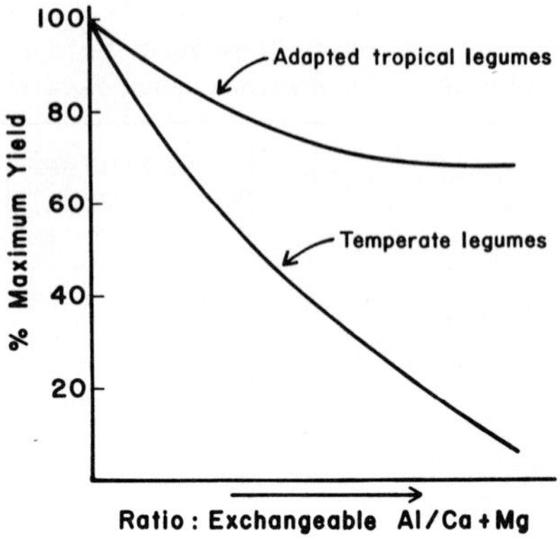

Fig. 5 Exchangeable Al/Ca + Mg ratios and fractional yield of tropical and temperate legumes.

Al/Ca + Mg found in some of these soils will depress the yield of some of the most Al-tolerant legumes. Under these conditions, a small amount of calcium carbonate applied to the soil may have a marked effect on yields.

Spain (1973) studied the effect of lime application rates on yield and composition of tropical legumes on an Oxisol in the llanos of Colombia. The soil had a pH of 4.5, 3.5 mEq/100 g of Al^{3+}, 0.5 of Ca^{2+}, and 0.3 of Mg^{2+}. His results indicate that only 150 kg/ha of lime were required to give nearly maximum yields and that up to 4 metric tons gave no further increase. Percent Ca in plant tops increased markedly for the first 150 kg/ha and continued with a small incremental increase up to 4 tons of added lime. Percent Mn decreased drastically for the first 150 kg/ha and continued to decrease as the rate of lime increased up to 4 tons. The results for *Stylosanthes guyanensis* (Fig. 6) suggest that tropical legumes respond to nutrient Ca but probably are quite tolerant of Al and Mn. Further research is needed to more clearly define the effect of Ca on the other nutrients in the system and their interaction with various tropical legume and grass species.

In solution culture, Andrew and Hegarty (1969) studied the effect of Mn concentration on yield of several legumes and established their tolerance levels to this potentially toxic element (Table VIII). The seven legumes are listed in order of their tolerance of Mn in the solution. At the highest concentration of Mn studied—40 ppm—the percent of maximum yield ranged from 70% for *C. pubescens* to 9% for *M. atropurpureum,* indicating a far greater tolerance of *C. pubescens* than *M. atropurpureum.* Toxicity threshold was defined as the concentration of Mn in plant tops which corresponded to a 5% reduction in yield. *C. pubescens* had the highest concentration, followed by *S. humilis, D. uncinatum,* and *M. sativa,* with the lowest concentration at the toxicity threshold.

TABLE VII

pH, Exchangeable Cations, Percent Aluminum Saturation, and Exchangeable Al/Exchangeable Cation Ratios of Some Highly Weathered Tropical Soils

Soil	pH	Exchangeable Ca + Mg (mEq/100 g)	Exchangeable Al (mEq/100 g)	Aluminum saturation (%)	Al/Ca + Mg ratios	Source
Red–yellow latosol	4.3	0.3	1.7	83	5.7	Pratt and Alvahydo, 1966
Red–yellow latosol	5.5	4.2	0.1	2	0.02	Pratt and Alvahydo, 1966
Red–yellow latosol	4.5	1.3	0.8	36	0.6	Pratt and Alvahydo, 1966
Red–yellow podzolic	3.7	0.3	3.8	90	13	Pratt and Alvahydo, 1966
Hydromorphic laterite	4.2	0.1	2.24	49	22	Organização dos Estados Americanos, 1974
Hydromorphic laterite	5.0	0.14	4.14	77	30	Organização dos Estados Americanos, 1974
Yellow latosol	4.7	0.24	1.83	24	8	Organização dos Estados Americanos, 1974

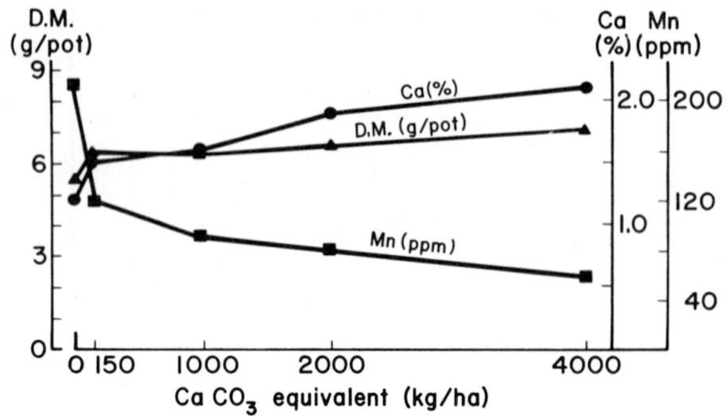

Fig. 6 Effect of lime on dry matter (D.M.) production and Ca and Mn concentration of *Stylosanthes guyanensis.*

K. Micronutrients

With the low level of macronutrients in highly weathered soils of the tropics, micronutrient deficiencies have been obscured or do not exist. As the fertilizer input has been increased for more intensive crop production, deficiencies of one or more of the micronutrients have appeared. Another factor which has increased the likelihood of micronutrient deficiencies is the use of higher analysis fertilizers with low micronutrient impurity. Under natural savanna conditions, micronutrient deficiencies are seldom observed, but, as more legumes and grasses

TABLE VIII
Manganese Toxicity Associated with Yield of Legumes[a]

Species	Rank	Yield b[b] (g)	% Yield at 40 ppm[c] (5 ppm = 100)	Toxicity threshold[d] (ppm)
Centrosema pubescens	1	−0.0023	70	1600
Stylosanthes humilis	2	−0.0038	50	1140
Macroptilium lathyroides	3	−0.0066	29	840
Desmodium uncinatum	4	−0.0080	24	1160
Medicago sativa	5	−0.0102	18	380
Glycine wightii	6	−0.0128	11	560
Macroptilium atropurpureum	7	−0.0159	9	810

[a]Adapted from Andrew and Hegarty (1969).
[b]Unit decrease in yield (log) per unit increase of Mn in solution culture.
[c]Percentage yield when 40 ppm Mn supplied in solution culture.
[d]Toxicity threshold level = concentration of Mn in dry matter of the tops, which corresponds to a 5% reduction in yield.

which have a higher nutrient requirement are introduced, micronutrient deficiencies begin to appear.

There are many difficulties in assessing the micronutrient situation in tropical pastures because of extreme variations in local soil situations and the variation among species and cultivars in their ability to extract nutrients from the soil. The situation becomes clear in the case of Mn (Table VIII), where the concentration of Mn in the dry matter of seven legumes, which results in a 5% reduction in yield, varies from 380 to 1600 ppm (Andrew and Hegarty, 1969). Documentation of micronutrient deficiencies in tropical forages is rare, but deficiencies of zinc, copper, boron, molybdenum, iron, and manganese are certain to receive greater attention as pasture–livestock feeding systems become more intensive. It is also well known that legumes have greater sensitivity to micronutrient deficiencies than grasses. As scientists and farmers attempt to include legumes in pastoral systems, micronutrients seem certain to enter as limiting factors for production.

V. FIRE AS AN ENVIRONMENTAL FACTOR

Burning of grasslands and use of fire as a management tool have both been the subject of much discussion for many years. Differences of opinion stem from personal experiences of scientists working with various grassland types in different parts of the world under a wide variety of circumstances. There seems to be little justification for burning a grassland which remains acceptable to the grazing animal during most of the year, as is the case of many improved pastures in the humid tropics. Solar energy stored in the pasture sward, which can be utilized to produce a useful animal product, justifies the utilization of that forage if productivity of the grassland is maintained. Some ecologists have difficulty in accepting man and his domesticated animals as components of grassland ecosystems, and any change which occurs as a result of man's intrusion is considered to be a degradation of the ecosystem. Fire is also one of those components of the environment which is considered in a negative sense, since many ecologists regard a natural grassland savanna as a disclimax. It does not matter that all evidence points to the existence of fire as an environmental factor without the presence of man. These scientists prefer to include fire as a component of grasslands ecosystems over which man may or may not exercise a degree of control.

Fire in wet–dry tropics has resulted in the evolution of grasses and other plants which have a high degree of resistance to burning. Very little if any burning occurs in the wet–humid tropical environment subjected to a dry season of less than two months. In areas where a longer dry season is characteristic of the environment, burning the natural grassland savannas every one to three years is a common practice. The necessity and advantages of burning have been reviewed by Tothill (1971). He quotes West (1965) who gave six main reasons for using fire regularly as a management practice in a pastoral system.

1. To remove top hamper in order to increase the availability of new season's growth and to remove patchiness in unevenly grazed pastures.
2. To stimulate growth at a time when it might otherwise not occur.
3. To control regrowth of trees and shrubs and to clean up fallen timber.
4. To attract animals to areas that might otherwise be left ungrazed.
5. To remove the hazard of wild fires and to establish fire breaks.
6. To control diseases and pests such as ticks.

If the objective is to maintain the species already present in the native grassland, then burning near the end of the dry season will probably accomplish this objective. The indigenous plants have evolved over the ages from a similar burning regime. Burning at periods other than at the end of the dry season may adversely effect some of the native grasses. Considerable shifts in botanical composition have been reported in Africa and Australia from off season burning. Since less than 50% of the biomass is utilized on most native tropical grasslands, there is usually plenty of fuel for fire.

Burning removes unpalatable accumulation from the previous wet season's growth and contributes to more uniform utilization of forages by grazing animals. It also keeps in check invasion of trees and shrubs which frequently reduce production of desirable forage species. There are numerous examples of the invasion of trees and shrubs into grasslands which were originally grasslands. Much of western Texas was covered with short grasses prior to settlement by European immigrants. When the incidence of fire was reduced from that set by the American Indians, *Prosopis juliflora* moved in from stream banks and completely invaded the grasslands. Similar invasions have been observed on the margins of the gallery forests and tall grass prairies of the central United States and the forest–savanna margins in Africa.

Burning of the savanna is charged with reducing the organic matter content of the soil but this has been extremely difficult to document. However, two very deficient nutrient elements are almost completely volatilized in the burning process–nitrogen and sulfur. Since sulfur is present at such extremely low levels in the atmosphere over the tropics, loss of this element in the burning process may be a serious loss. A large proportion of mineral nutrients are in plant tops and litter; nonvolatile ones are released in the ash and become more readily available to growing plants early in the next rainy season. It is evident that fire provides a means of accelerating the recycling of nutrients in soil–plant–animal systems. Extreme care must be exercised, however, to prevent burning of the savanna after new growth has started in order to prevent injury to the desirable species. A period of rest during the previous wet season to allow for accumulation of fuel and buildup of food reserves, and another period of rest after the burn to prevent the depletion of reserves by overgrazing, are sound pre- and postfire management practices. Further effects of fire on ecosystems are discussed by Kozlowski and Ahlgren (1974).

VI. CONCLUSIONS

Grasses and legumes occupying the highly weathered Oxisols and Ultisols of the South American grasslands are adapted to very low fertility levels, high levels of exchangeable aluminum and, in some cases, manganese, periods of moisture stress, and frequently to abuse from overgrazing. Not enough attention has been given to these indigenous plants to determine the potential of selected species as feed for domestic animals. Scarcity and high cost of fertilizers and other physical inputs used in pasture–livestock systems dictate that improvement in production must come from the systems which make maximum use of the natural grasslands and the species which are best adapted to the environment. A better understanding of the physiology of these forage species, their responses to various components of the environment, and to changes which can be imposed by man will provide a basis for development of production systems which are economical and which will preserve the ecosystem. However, they have been almost completely neglected. Most of the exotic tropical forages introduced from other parts of the world require high levels of inputs for their maintenance. Under certain circumstances, their strategic use in forage livestock systems may be justified to provide higher quality feed than that provided by the natural grasslands, but benefits obtained must be balanced with the additional costs. A combination of a relatively large area of natural pastures for the breeding herds and a small area of high producing, high-quality pasture for the young stock and for flushing the females during the breeding season appears to be a viable system of management.

The shift from forest to grassland, as is occurring at a very rapid pace in the Amazon region, presents another problem. Native forages are almost entirely absent, and the only species available for planting are the exotics. Since almost the entire nutrient supply of the tropical forest ecosystem is retained in the above-ground biomass, the relatively small quantity of nutrients released by burning is quickly leached and may be lost beyond the root zone of the forage species. This is especially true if a few years of intertilled crops are grown before pasture establishment. Unless legumes which are adapted to the climatic environment and particularly to the soil environment are included in pasture mixtures, the chances of a productive pasture for any long period are remote. Even the indigenous legumes, which are well adapted to the highly weathered soils, will require additional nutrients to persist in association with the exotic grasses.

The nutrient demand varies greatly between species, strains, and cultivars of tropical grasses and legumes. Research efforts should be focused on development of forages which are highly efficient in their extraction of soil nutrients and accelerate their turnover rate in the system. Most evaluation of forages by the introduction specialist and plant breeder is conducted under higher fertility levels than those to which the released cultivar will be exposed. Under such a testing regime, superior germ plasm may be discarded because it fails to respond to high levels of fertilizer.

Central and South America provide the natural habitat of almost all the known genera and species of promising tropical forage legumes. The conservation of this treasure of germ plasm is urgent, since large areas of natural grassland and forest fringe areas, where these legumes occur, are being destroyed by settlement schemes. If an extensive program of plant collection is not initiated soon, a large proportion of the gene pool will be lost. This will narrow the genetic base available in the future for improvement and development of cultivars adapted to a wide spectrum of environments. Before it is too late, germ plasm banks for these promising species should be established in several countries of Latin America.

Additional research is needed to determine the net primary production of tropical grasslands and their effectiveness in controlling erosion. The role of the grazing animal as a component in the system for recycling nutrients needs further investigation. It is of interest in this connection that the most productive natural grasslands occur in regions where there is a long history of large populations of ungulate animals, which suggests that the animal may have been an important agent in reducing loss of nutrients and may have contributed to the retention of small increments of N added to the system by biological and nonbiological N fixation. The possibility of accelerating this rate of soil improvement through the use of legumes and selected grasses, biological N fixation, and fertilizer N deserves intensive study.

The use of animal products to alleviate the food deficiency situation in certain countries has been criticized on the basis of the inefficiency of the grazing animal in converting above-ground plant biomass to an acceptable food product. The critics do not seem to understand that the biomass which is converted to animal product has no alternative use for food except through fermentation by the ruminant animal. Meat is often considered to be a food product for only the affluent, developed countries, but it is well known that millions of people subsist on foods generated by their herds and flocks. Perhaps the grassland scientist can make a contribution to mankind by developing techniques for increasing the productivity of grasslands and their conversion to animal-based foods.

REFERENCES

Andrew, C. S. (1973–1974). Critical sulfur concentration levels in pasture legumes. *CSIRO, Div. Trop. Agron., Annu. Rep.*, p. 82.

Andrew, C. S., and Hegarty, M. P. (1969). Comparative responses to manganese excess of eight tropical and four temperate pasture legume species. *Aust. J. Agric. Res.* **20**, 687–696.

Andrew, C. S., and Robins, M. F. (1969a). The effect of phosphorus on the growth and chemical composition of some tropical pasture legumes. I. Growth and critical percentages of phosphorus. *Aust. J. Agric. Res.* **20**, 665–674.

Andrew, C. S., and Robins, M. F. (1969b). The effect of potassium on the growth and chemical composition of some tropical and temperate pasture legumes. I. Growth and critical percentages of potassium. *Aust. J. Agric. Res.* **20**, 999–1007.

Burton, G. W., Prine, G. M., and Jackson, J. E. (1957). Studies of drouth tolerance and water use of several southern grasses. *Agron. J.* **49**, 498–503.

Dobereiner, J., and Day, J. M. (1974). Associative symbioses in tropical grasses: Characterization of microorganisms and dinitrogen fixing sites. *Int. Symp. N_2 Fixation, 1970.* Interdiscip. Discuss. (in press).

Fitzpatrick, E. A., and Nix, H. A. (1970). The climatic factor in Australian grassland ecology. *In* "Australian Grasslands" (R. M. Moore, ed.). Aust. Natl. Univ. Press, Canberra.

Goodland, R. (1966). On the savanna vegetation of calabozo, Venezuela and Rupununi, British Guiana. *Soc. Venez. Scienc. Nat., Bol.* No. 110, Tomo XXVI.

Kozlowski, T. T., and Ahlgren, C. E., eds. (1974). "Fire and Ecosystems." Academic Press, New York.

Lindquist, A. W. (1933). Amount of dung buried and soil excavated by certain coprini. *J. Kans. Entomol. Soc.* **6**, 109–125.

McLean, E. D. (1971). Potentially beneficial effects from liming: Chemical and physical. *Soil Crop Sci., Soc. Fla., Proc.* **31**, 189–196.

Mott, G. O. (1974). Nutrient recycling in pastures. *In* "Forage Fertilization" (D. May, ed.), pp. 323–339. Am. Soc. Agron., Madison, Wisconsin.

Organização dos Estados Americanos. (1974). "Marajo–um estudo para o seu desenvolvimento." Secretaria-Geral, OEA, Washington, D.C.

Parsons, J. J. (1972). Spread of African pasture grasses to the American tropics. *J. Range Manage.* **25**, 12–17.

Pratt, P. F., and Alvahydo, R. (1966). Cation exchange characteristic of soils of São Paulo, Brazil. *IRI Res. Inst.* [*Bull*] **31.**

Rattray, J. M. (1968). The grass cover of Africa. *FAO Agric. Stud.* **49.**

Spain, J. M. (1973). Effect of lime upon dry matter production and forage composition on an oxisol. *CIAT Annu. Rep.*, pp. 22–24.

Stebbins, G. L., and Crampton, B. (1959). A suggested revision of the grass genera of temperate North America. *Recent Adv. Bot.* **1**, 133–145.

t'Mannetje, L. (1965). The effect of photoperiod on flowering, growth habit, and dry matter production in four species of the genus *Stylosanthes* SW. *Aust. J. Agric. Res.* **16**, 767–771.

Tothill, J. C. (1971). A review in the management of native pasture with particular reference to North-Eastern Australia. *Trop. Grasslands,* **5**, 1–10.

Vicente-Chandler, J. (1974). Fertilization of humid tropical forages. *In* "Forage Fertilization" (D. Mays, ed.), pp. 277–298. Am. Soc. Agron., Madison, Wisconsin.

West, O. (1965). Fire in vegetation and its use in pasture management with special reference to tropical and subtropical Africa. *Commonw. Bur. pastures Field Crops (G. B.), Mimeo* **1**.

CHAPTER

7

Root Crops

L. A. WILSON

I. INTRODUCTION

This chapter discusses comparative aspects of the ecophysiology of tropical root crops and of the edible aroids in relation to the environments in which they are normally cultivated. *Solanum* potato will be used as a reference for comparison with tropical root crops because of its long history of genetic, agronomic, and physiological investigations, and also because considerable effort has recently been given to the reintroduction of *Solanum* cultivation to tropical latitudes.

The term ecophysiology is interpreted in this chapter to mean the performance of crop plants in relation to environmental conditions. Plant performance is defined as the growth, development, productivity, and yield characteristics of crop species as well as the functional attributes, e.g., physiological and metabolic processes, which determine performance characteristics. Three aspects of the comparative ecophysiology of root crops are considered here: (a) macroecophysiology, i.e., crop performance in relation to ecological zones with characteristic macroclimates; (b) microecophysiology, i.e., crop performance in relation to the immediate environment, including the microclimate created by the crop community; (c) species dependent ecophysiology, i.e., the design and performance of ideal crop ideotypes (Donald, 1968) to maximize crop performance in selected macroclimates. Other factors known to affect crop performance, e.g., cultural practices, i.e., soil type, fertilizer use, water management, etc., and control of pests, diseases, and weeds will be referred to only in so far as they relate to the subject matter of ecophysiology, as defined here.

II. COMPARATIVE MACROECOPHYSIOLOGY

This section summarizes the climatological characteristics of ecological zones in which tropical root crops occur as well as the origin and distribution of individual crops within the zones described. Some ecobotanical characteristics of root crops in relation to their current and potential distribution are also discussed. Yield, production, and producitivty of the crops are compared and related to selected aspects of the macroclimate in which cultivation occurs.

A. Macroclimatology of Root Crop Ecozones

The climates of the world may be broadly classified into tropical, subtropical, intermediate or temperate, and polar types; generally these climate types coincide with well-defined ecozones. In addition, mountain areas may also be considered as separate ecozones since the climatological and ecological characteristics of such areas are often distinctive. Separation of mountain ecozones is especially significant in considerations of root crop ecophysiology because the most widely cultivated root crop, *Solanum* potato, originated in an Andean mountain climate. Also, after several decades of selection and breeding for cultivation in intermediate climates, *Solanum* potato is now being reintroduced to its mountain climate of origin in several tropical countries, e.g., Kenya, Venezuela, Jamaica, and Ceylon.

Like most other crop species, e.g., cereals and legumes, the major proportion of root crop cultivation occurs in tropical, subtropical, and intermediate climates, but there is considerable potato production in mountain regions, and the crop is also cultivated in the short summer season of polar climates in Norway and Finland. Here, the characteristics of tropical and subtropical climates are discussed and only brief references are made to intermediate climates for purposes of comparison.

The major climatological factors which might limit agricultural production are temperature, precipitation, and daylength, while windspeed merits consideration as an important limiting factor in certain circumstances. Of these, temperature is the only absolute limiting factor and the classification of climates on the basis of temperature is therefore most important. Accordingly, tropical climates are defined as those with an average annual temperature greater than 25°C (77°F) and an average for each month exceeding 18°C (65°F). In tropical climates, therefore, temperature conditions suitable for crop growth occur throughout the year. Subtropical climates are those with short, mild winters in which there is a period of 1–2 months when freezing temperatures may occur, but the average temperature of the coldest month is above 6°C (43°F). Summer temperatures are, however, frequently as high as those in tropical climates. Intermediate climates, i.e., those between polar and subtropical, have cold winters and warm to hot summers.

On the basis of prevailing precipitation, tropical and subtropical climates may be further subdivided into humid and dry types as shown in the classification of these climates (Table I) adopted by Van Royen and Bengston (1964). Thus, of the humid tropical climates, tropical rain forests have no pronounced or prolonged dry season and there is little seasonal variation either in temperature or in precipitation (200–400 cm/year). In contrast, the tropical monsoon climate exhibits marked daily and seasonal temperature changes and there is abundant precipitation (100–200 cm/year) during a wet season, which alternates with a dry season of 4–6 months or longer. In the tropical savannas, precipitation (100 cm/year) is well spread over 4–6 months, with a prolonged dry season often lasting 6–7 months. The dry tropical climates are subdivided into semiarid or arid types according to whether annual precipitation is greater or less than 20 cm, respectively. Of the subtropical climates, humid types have precipitation (75 cm/year) which is evenly spread throughout the year with no pronounced dry season and temperatures are below 18°C for 4–6 months. Alternatively, in the dry subtropical or Mediterranean climates, precipitation (50–75 cm/year) occurs mainly during the 6–8 months of the cool season in which temperatures are below 18°C. However, in some regions there is moderate summer precipitation. Frost may occur in both subtropical climates during the coldest period.

The mountain climates have temperatures which are inversely related to altitude, and generally high, but seasonal precipitation. Thus, in tropical and

TABLE I
Average Annual Temperatures and Precipitation Values of Tropical and Subtropical Climatic Types[a]

Climates	Annual averages	
	Temperature (°C)	Precipitation (cm)
Tropical climates		
Humid		
Rain forest	25–27	200–400
Monsoon	18–27	100–200
Savanna	18–27	80–120
Dry		
Semiarid	18–27	20–50
Arid	18–27	<20
Subtropical climates		
Humid	6–27	75–100
Dry	*6–27*	50 – 75

[a]After Van Royen and Bengston (1964).

subtropical latitudes, mountain climates at high altitudes are often intermediate in character except for prevailing short days and high light intensities.

Using the climatic diagrams proposed by Walter and Lieth (1960) some typical tropical, subtropical, and mountain climates of South America are described (Fig. 1). Four tropical and subtropical climate types are characterized by these authors:

Type I Equatorial humid
Type II Tropical summer rains
Type III Subtropical hot and arid
Type IV Mediterranean winter rains

The intermediate and polar types include:

Type V Warm temperate humid
Type VI Humid with cold season
Type VII Arid with cold season
Type VIII Boreal
Type IX Arctic

The tenth climatic type includes mountain areas in the other regions. The climatic diagrams clearly distinguish arid, humid, and very wet periods of the year by means of temperature and precipitation curves in which 10°C corresponds to 20 mm precipitation (Gaussen, 1955). When this fixed proportion

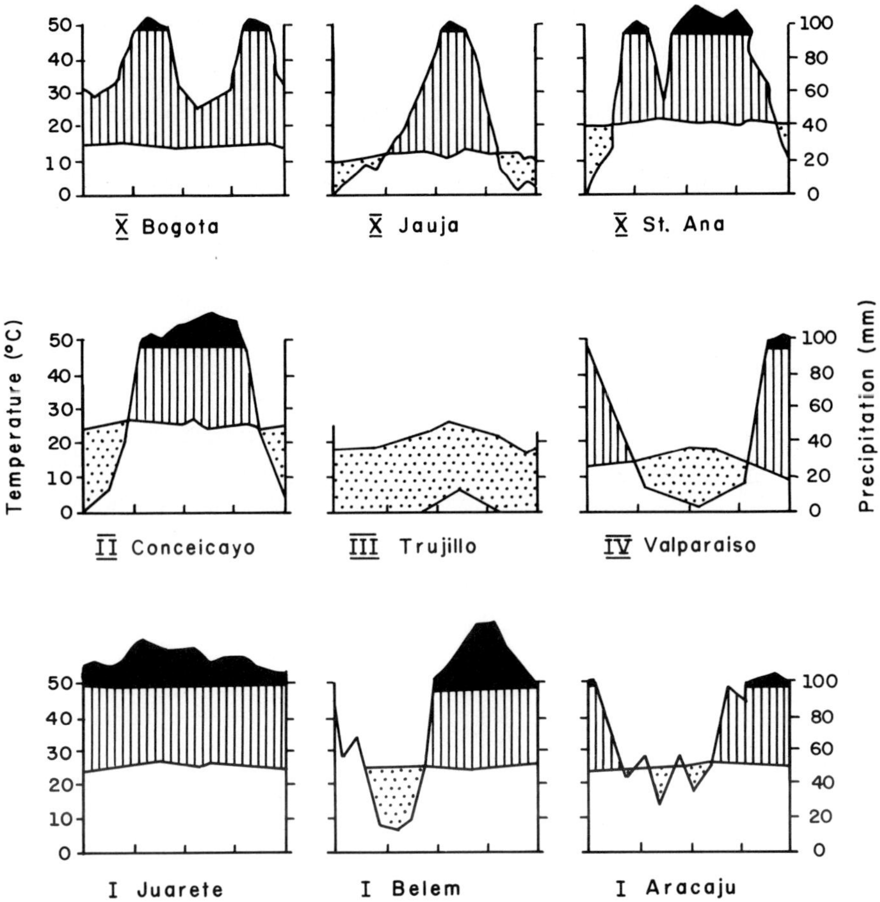

Fig. 1 Climatic diagrams of selected regions of South America. Drought periods (dotted); humid periods (vertical hatching); precipitation exceeding 100 mm (black). The year is from July to June. Roman numbers before locations refer to climatic types.

between the curves is used, an arid period prevails whenever the precipitation curve falls below the temperature curve. Conversely, humid and very wet periods occur when the precipitation curve is above the temperature curve. Climatic types similar to those in Fig. 1 can be recognized in all of the major land masses of the world. It is proposed, therefore, to consider the macroecophysiology of root crops in relation to climatic types found in current and potential root crop producing areas of the world.

Daylength is an important climatic factor in the ecophysiology of root crops because it may influence tuber initiation and thereby limit tuber yield. Critical daylengths are of importance only in *Solanum* potato cultivation, which extends from equatorial daylengths of 12 hours to polar daylengths of 24 hours. However, although tuber initiation is earlier and early tuber growth more rapid under short-day conditions, the effect is quantitative rather than qualitative and *Solanum tuberosum* cultivars exist which would give economic yields under all daylength conditions. In the tropical root crops, there is some evidence that similar quantitative interrelationships between tuber initiation and daylength may exist, but these are not economically important because of the restriction of tropical root crop cultivation to short daylength tropical ecozones. Sweet potato cultivation in Japan and the southern area of the United States are exceptions to this generalization, but possible daylength effects on tuber initiation are apparently overcome by the use of suitable cultivars.

There are two major effects of windspeed on crop growth: (1) effects of windspeeds of very high velocity, e.g., monsoons and hurricanes on the complete destruction of the crop; (2) effects of slower windspeeds in increasing evapotranspiration to the point where growth and yield of the crop are adversely affected.

The first effect is not dealt with in this chapter and little information exists on effects of windspeed on yield in tropical root crops, although the use of windbreaks in root crop cultivation on exposed sites, e.g., on hilly slopes in Jamaica, is well established.

B. Origin and Distribution

In this section, the origin of tropical root crops is first described and their distribution outlined by consideration of areas under root crop cultivation in major producing countries and continents.

The origin of tropical root crops shown in Table II indicates that five of the six genera under consideration originated in a relatively small region of tropical America, in northern South America and Central America; in the sixth genus, *Dioscorea,* the species *trifida* is also American in origin. Thus, sweet potato, cassava, the aroid, *Xanthosoma* or tannia, *Solanum* potato, and *Dioscorea trifida* originated either in tropical or North Andean America, and taro, *Colocasia esculenta,* in Southeast Asia. The yams differ from the other cultivated species in their apparent genetic diversity and origin in widely separated centers, e.g., in tropical America (*D trifida*), West Africa (*D. rotundata, D. cayenensis, D. bulbifera, D. dumetorum*), and South Asia (*D. alata, D. esculenta, D. opposita*).

The distribution of tropical root crops is illustrated (Table III) by data showing areas cultivated in the various continents and in the major producing countries of each continent. The total area under tropical root crops and potato amounts to 49.3 million hectares (M ha) of which the tropical root crops account for 26.8 M ha or 54%. Cassava cultivation on 9.8 M ha represents 45% of the

TABLE II
Origin of Tropical Root Crops[a]

Root crop species	Suggested origin
American species	
Ipomoea batatas	Tropical America, Mexico, North Brazil
Manihot esculenta	Tropical America, South Mexico/Guatemala, Northeast Brazil,
Xanthosoma sagittifolium	Tropical America, North Brazil
Solanum tuberosum	Andean South America, Colombia/Peru
Dioscorea trifida	Tropical America, Guyana/Surinam, North Brazil
African species	
Dioscorea rotundata	Tropical West Africa
Dioscorea cayenensis	Tropical West Africa
Dioscorea dumetorum	Tropical West Africa
Dioscorea bulbifera	Tropical West Africa
Asian species	
Dioscorea alata	South Asia
Dioscorea esculenta	South Asia
Dioscorea opposita (syn. batatas)	South Asia
Colocasia esculenta	Southeast Asia

[a]From Purseglove (1968, 1972).

area under tropical root crops and sweet potato and yams 55%. It is estimated that approximately equal acreages (8.5 M ha) are under sweet potato and yam cultivation. The edible aroids (*Colocasia* and *Xanthosoma*) excluded from Table III are also estimated to occupy 1.5 M ha or 6% (approximately) of the area cultivated to cassava, sweet potato, and yams.

The percentage distribution of root crops according to continent (Table IV) shows that Old World areas, i.e., Africa (16.9%), USSR (15.9%), Europe (14.2%), and Asia (10.3%), together account for 57.3% of the total root crop cultivation area, whereas the area under root crop cultivation in the New World continents, i.e., Latin America (8.3%) and North America (1.5%), amount to only 9.8% of the total root crop area. The area under root crop cultivation is particularly low in North America where cereal cultivation is preferred. The data indicate, therefore, that distribution of root crops is also determined by factors other than ecological suitability, since regions with macroclimates suitable for root crop cultivation exist on all the continental land masses in Table IV. Analysis of root crop cultivation according to country further illustrates country preferences for certain tropical root crops, over and above the ecological restriction of root crop cultivation in countries with intermediate climates. Thus, Brazil, Tanzania, Zaire, Indonesia, and Thailand may be considered cassava producing countries; in South Korea and Taiwan, sweet potato is preferred, while *Colocasia* is cultivated in many Pacific Islands. In Nigeria and Japan, root

TABLE III
Distribution of Root Crop Species (Area Cultivated)[a]

	Area (1000 ha)			
Continent/country	Cassava	Yam and sweet potato	Potato	Total
Latin America	2,563	454	1,126	4,143
Brazil	2,025	181	220	2,426
Peru			315	315
Argentina			179	179
Africa	4,942	3,271	299	8,512
Nigeria	1,100	1,800		2,900
Tanzania	285			285
Zaire	800			800
Kenya			55	55
Asia	2,249	1,587	1,304	5,140
India	353	230	514	1,097
Indonesia	1,434	356		1,790
Japan		129	148	277
South Korea		127		127
Taiwan		288		288
Thailand	130			130
Oceania		18		18
Europe			7,099	7,099
West Germany			554	554
Holland			154	154
United Kingdom			256	256
Yugoslavia			330	330
North America		55	691	746
U.S.		55	563	618
Canada			128	128
USSR			7,960	7,960
World	9,783	17,001	22,466	49,250

[a]From FAO Production Yearbook, 1971.

TABLE IV
Distribution of Root Crops according to Continent

Crop	Percentages of total area cultivated to root crops[a]					
	Latin America	North America	Africa	Asia	Europe	USSR
Cassava	5.1		9.8	4.5		
Sweet potato and yams	0.9	0.1	6.5	3.2		
Potato	2.2	1.4	0.6	2.6	14.2	15.9
	8.2	1.5	16.9	10.3	14.2	15.9

[a]Calculated on a total acreage of 50,750 M ha, i.e., including root crops (Table III), 49,250 M ha and edible aroids, 1500 M ha.

crop cultivation is distributed between cassava, sweet potato, and yams, and sweet potato and *Solanum* potato, respectively (Table III). This later distribution of root crops best illustrates the ecological suitability of current tropical root crop species. Accordingly, cassava, sweet potato, and yams can be cultivated in humid tropical climates, but sweet potato cultivation extends into the subtropics. Also, although *Solanum* potato can be cultivated in most ecological zones, considerations of productivity (limited by disease) restrict cultivation to intermediate, subtropical, and mountain climates. The distribution of root crops in different climatic zones shown in Table V illustrates the relative climatic suitabilities of the crops to different macroclimates, including indications of the use for supplementary water supply in the form of irrigation.

Table VI shows that the latitudinal and altitudinal distribution of potato is greater than that in the tropical root crops, and that sweet potato is the most

TABLE V
Distribution of Root Crops on a Climatic Basis

Climate	Cassava	Sweet potato	Yam	Taro	Tannia	Potato
Tropical rain forest	*	*	(*)	*	*	
Tropical monsoon	*	*	*	(*)I	*	(*)
Tropical savanna	*	*	*	(*)I	*I	(*)
Dry tropical, i.e., steppe or desert		(*)I				
Humid subtropical		*		(*)	(*)	*
Subtropical (Mediterranean)		*				*
Humid intermediate						*
Dry intermediate						

[a]After Kay (1973). ( ), limited cultivation, I, irrigation required.

TABLE VI
Latitudinal Distribution and Macroclimate Requirements of Root Crops[a]

Crop	Distribution (latitude/altitude)	Macroclimate requirements					
		Length of wet season	Annual precipitation (cm)		Temp (°C)		Daylength (hours)
			Optimal	Minimal	Optimal	Minimal	
Cassava	30°N–30°S (1000 m)	Long or short	100–150	50–75	25–29	710	10–12
Sweet potato	40°N–32°S (2100 m)	Medium to short	75–100	50	24	710	11.5–13.5
Yams (except *D. opposita*)	20°N–20°S (1000 m)	Medium to long	100–150	60	30	720	10–11
Taro	30°N–30°S (2100 m)	Long	200–250	175	21–27	710	
Tannia	30°N–30°S (1500 m)	Long	140–200	100	24	710	
Potato	70°N–40°S	Long	75 (2.5 cm/week)	50	15–13	510	12–17

[a]Adapted from Kay (1973).

widely distributed of the tropical crops. However, the Asian yam *Dioscorea opposita* has the potential for a distribution similar to *Solanum tuberosum*. This root crop distribution is due mainly to the following species characteristics: (a) shorter growth periods of potato and sweet potato compared with other root crops; (b) capacity for tuberization under daylengths longer (i.e., up to 13.5 hours in sweet potato and 24 hours in potato) than the remaining root crops; (c) capacity of potato for tolerating lower temperatures than the tropical root crops. Within these limits of temperature and daylength, each of the crops has different water requirements for growth and tuberization. These range from paddy field conditions in the *Colocasia* aroids, through the small, but evenly distributed water requirements of *Solanum* potato, to the minimal requirements of cassava, after germination and establishment of the crop. It is also interesting to note that the ecodistribution of root crops described is achieved in different ways in the various crop species. Thus, in potato, different varieties suitable for cultivation under different daylength and temperature conditions exist. In sweet potato, there is a range of varieties with growth periods ranging from 3 to 8 months, suitable for cultivation under different water and temperature regimes. However, in cassava, similar varieties have been used for cultivation both under dry and wet conditions, since types suitable for specific ecosystems have not been rigorously selected. In yams, distribution is achieved through the existence of different species adaptable to different ecosystems and hence the considerable genetic potential for development of varieties within these species is yet to be exploited. The same situation exists in the edible aroids in which tolerance of low water supply might be considerably improved by breeding.

C. Ecobotanical Characteristics

The adaptability of a root crop species to cultivation in different macroclimatological zones depends either on (a) the availability of different varieties suitable for cultivation under various environmental conditions or (b) the existence of individual varieties capable of successful cultivation in different ecosystems. Therefore, species adaptability depends either on the existence of a range of botanical or physiological characteristics in different varieties within the species or on the existence of plant characteristics in single varieties of the species capable of adaptive expression in different ecosystems. Thus, different *Solanum* potato varieties are adapted for growth under a wide range of daylength conditions, while single cassava varieties can grow successfully under both wet and dry conditions. Some of these characteristics are compared in Table VII. These are discussed with particular reference to the ability of the plant to survive unfavorable climatic conditions, e.g., temperature, excessive or limited supply of water, and wind which may prevail for varying periods in the course of the

TABLE VII
Some Botanical Characteristics of Tropical Root Crops[a]

Common name	Cassava	Sweet potato	Yams	Taro	Tannia	Potato
Habit	Perennial woody shrubs	Perennial prostrate to semierect herb	Annual vining herb	Perennial cormous herb	Perennial cormous herb	Annual, erect to semiprostrate herb
Stem Height and thickness	0.9–4 m thick	0.9–4.2 m, slender–thick	2–3 m (Asian), 9–15 m (African), slender	Cormous	Cormous	0.3–0.6
Stem branching	Little to much branched	Little to much branched	Moderately branched			Little branching
Leaf size and shape	Large, simple, palmate, 5-lobed	Medium, simple, entire or lobed	Small–medium, simple, ovate, entire or lobed	Very large, simple, heart-shaped	Very large, simple, sagittate to hastate	Small–medium compound, odd-minnate
Tubers	Root	Root	Stem	Stem (corm)	Stem (corm)	Stem
Number	1–15	1–10	2–20 (Asian), 1–5 (African)	1 corm (+ cormels)	1 corm + cormels	5–10
Shape	Fusiform	Globular–fusiform	Lobed (Asian), cylindrical (African)	Globular	Globular–fusiform	Globular–oval
Daylength response	Short-day	Indifferent	Short day	Syort day	Unreported	Short day–long day
Growth period (months)	9–24 (7)	3–8 (3)	6–12 (5)	6–18 (3)	9–12 (6)	3–5 (3)

[a]Numbers in parenthesis indicate growth periods of early maturing types.

growth cycle of the crop. Such survival may be of major importance in species adaptability.

1. Growth Period

Growth periods varying in length from 3 months in *Solanum* potato to up to 24 months in cassava have been recorded in the root crops under consideration (Table VII). Thus, ability of *Solanum* potato and, to a lesser extent, sweet potato to be cultivated in climates with short growing seasons and long unfavorable seasons, due either to low temperatures or low precipitation, depends to a large extent on short growth cycles (3–5 months) in these crops. Cultivation of most of the other root crop species outside the tropics is limited by their longer growing period (9–12 months) as well as by the inability of the crops to withstand unfavorably low temperatures or dry conditions. A major exception in this respect is *Dioscorea opposita,* which can tolerate cold conditions including frost and is widely cultivated in Korea, China, and Japan. The species was also grown experimentally in Europe during the potato famine in the nineteenth century. *Dioscorea rotundata* with growth periods as short as 6 months is also cultivated in northern savanna regions of West Africa. However, the existence of early types in all root crop species suggests that selection and breeding programs may lead to considerable reduction of growth periods and consequent increase in the climatic adaptation in tropical root crops.

2. Habit

Root crop habits range from the relatively tall (up to 4 m) woody shrubs of cassava to the short (30–50 cm) semierect herbs of *Solanum* potato. The former cassava habit is associated with perennial growth in which the secondarily thickened stem from which the majority of leaves are shed is capable of withstanding unfavorable dry periods, while the latter habit is characteristic of quick growing annuals, which complete development in a short growing season. Cassava also grows well under wet conditions. The sweet potato is a perennial herb in which the foliage never senesces completely in the humid tropics, but senescence does occur in climates with dry or cold periods. Tubers are, however, capable of perennation and subsequent germination under favorable conditions. In the humid tropics, stem cuttings are used as planting material. The prostrate habit of sweet potato is also particularly resistant to damage by high winds, e.g., hurricanes and typhoons, and has been thought to be one of the reasons for the success of the crop in Japan (Fujise, 1970). Yams are annual herbs with relatively long growth periods and are, therefore, exposed to dry conditions of varying intensity, depending on the tropical climates in which they are cultivated. They apparently originated in forest regions and their long twining stems are apparently adaptations to growth under conditions of high relative humidity and low light intensity. However, all of the known commercial species seem to grow successfully under the higher light intensity and lower relative humidity conditions of tropical cultivation, provided that staking is practiced, but cultivars of both *Dioscorea alata* and *Dioscorea rotundata* can also be grown without staking in drier climates, e.g., in Barbados.

The *Colocasia* aroids grow most successfully under wet conditions and are normally cultivated in paddy fields on imperfectly drained soils, but cultivation also occurs under drier upland conditions (de la Pena and Plucknett, 1967). By contrast, *Xanthosoma aroids,* particularly the early maturing cultivars, are more tolerant of dry conditions. Attention is drawn to the four types of edible aroids in cultivation in the Caribbean:

Colocasia aroids		
Dasheen	*C. esculenta* var esculenta–	one large main corm; few small cormels
Eddoe	*C. esculenta* var *globulifera*–	small main corm; many well-developed cormels
Xanthosoma aroids		
Tannia	*X. sagittifolium*–	elongated cormels
Barbados nut eddoe	*X. sagittifolium*–	globular cormels similar to eddoe

The Barbados nut eddoe is probably best adapted for growth under dry conditions, while of the *Colocasia* aroids, the eddoe is said to be more drought tolerant than the dasheen (Kay, 1973). *Colocasia* aroids have also been satisfactorily used in Japan and Egypt as the first crop in reclamation of saline sandy soils.

Although in cassava, sweet potato, and yam species, varieties with leaf shapes ranging from entire through lobed to finely divided exist, there is no evidence that such leaf shapes are related to the climatic adaptation of these species. The occurrence of physiological adaptations to dry conditions, e.g., few stomata sensitive to high evapotranspiration and with short periods of opening during the day, low cuticular transpiration, and deep root systems, has not been investigated in root crops. However, it should be noted that while the yam and aroids have rather shallow root systems unsuitable for withstanding dry conditions, there is some evidence that in cassava and sweet potato roots concerned with water and ion uptake are deep seated, but tuberous roots are shallow (Fujise and Tsuno, 1967). Such root distribution would seem to be a desirable drought-tolerant characteristic in root crops. None of the root crops under consideration including *Solanum* potato is resistant to prolonged exposure to frost. However, the optimal temperature for growth is much lower in potato (15°–18°C) than for the tropical root crops (25°–29°C). These temperature responses also reflect the adaptations to altitude found in the root crops under consideration. Thus, the maximum altitudes recorded for successful root crop cultivation are as follows:

cassava and yams, up to 1000 m; sweet potato and the aroids, 1500–2400 m; *Solanum* potato, higher than 2100 m.

3. Tuberization

A major botanical characteristic restricting the adaptation of root crops to different macroclimatic zones is the daylength requirement for tuberization. Thus, in cassava, tuberization is promoted under short-day conditions and reduced at daylengths greater than 10–12 hours. The crop is, therefore, most successfully cultivated between latitudes 15°N and 15°S (Kay, 1973). A similar short-daylength requirement for tuberization has been reported for yams (Njoku, 1963), although *Dioscorea opposita* which is cultivated as far north as Korea (30°N) produces tubers under longer daylengths. A positive influence of short days on tuberization in the edible aroids (*Colocasia antiquorum*) has also been reported by Tsukamoto and Inaba (1961). *Solanum* potato originated under short-daylength Andean conditions and, in general, indigenous South American cultivars will only produce reasonable yields with a daylength of 12–13 hours (Kay, 1973). However, there are numerous temperate cultivars which will tuberize under longer daylengths and indeed, tuberization also occurs under the 24-hour daylengths of polar summers. The response of tuberization to soil water conditions is also an important characteristic of the process. Thus, in the aroids, tuberization is indifferent or encouraged by high soil water contents, whereas in sweet potato, cassava, and yams, waterlogged conditions restrict tuber initiation and development. Also, temperature conditions necessary for tuberization in *Solanum* potato (i.e., night temperatures of 10°–14°C) inhibit the process in the tropical root crops, but reduced nighttime temperature (20°C) promotes tuberization in sweet potato (Kim, 1961). Thus, the high nighttime temperatures prevalent in the humid tropics are not conducive to the highest productivity in tropical root crops. However, short daylengths and temperatures suitable for year-round cultivation in such climates allow for the possibility of obtaining more than one crop in a calendar year, e.g., sweet potatoes.

4. Reproductive Biology

The reproductive biology of tropical root crops is a significant feature of their ecophysiology because of the vast potential for improving both productivity and distribution by breeding, as has been done with the *Solanum* potato. There have, however, been suggestions that the capacity for flowering is inversely related to the capacity for tuberization, e.g., in yams and cassava. Although these suggestions have not been experimentally substantiated, inhibition of tuberization is known to lead to improved flowering in *Solanum* potato.

The root crops are largely short-day flowering plants, but there are serious problems in flower production for hybridization in all the species except *Solanum* potato, in which temperate varieties are long-day flowering plants.

Thus, in sweet potato, plants must be staked to improve flowering; problems of self- and cross-incompatibility (Williams and Cope, 1967), as well as seed viability, are encountered in hybridization programs. Yam species normally produce dioecious flowers, but hermaphrodite flowers do occur. There are also staminate and pistillate plants in many species, e.g., *Dioscorea trifida* and *Dioscorea rotundata.* However, flowering is never profuse and comparatively few hybridizations have been achieved. Thus, *Dioscorea trifida* and *Dioscorea bulbifera* are the only species that flower regularly in the Caribbean and even in these species flowering is restricted to certain cultivars. Recently, flowering, open pollination, and production of viable seed have been observed in *D. rotundata* in Nigeria (Sadik and Okerere, 1975). Similarly, although there are cassava varieties that flower adequately, flowering is restricted in many types which are desirable parents for breeding programs. Flowering is rare in the edible aroids (McDavid and Alanu, 1976). Therefore, there is a great need for improving flower production in the root crops by better understanding of the reproductive biology of these species.

D. Productivity

Data on the comparative production of tropical root crops shown in Table VIII indicate a total world production of 443 million tons (M tons), to which may be added edible aroid production estimated at 10 M tons by Coursey and Haynes (1970). Coursey (1967) also estimated that yam production accounted for 41% or 18.2 M tons and sweet potato for 59% or 26.2 M tons of the joint total (44.4 M tons) for sweet potato and yam production. Nigerian yam production (12 M tons) accounts for more than half of the world's production. The genetic diversity of cultivated yam species is used to advantage in ensuring a fairly evenly distributed monthly production of this crop in Jamaica (Fig. 2), where cultivation occurs on steep hillsides (Rankine and Ferguson, 1974). *Solanum* potato production amounts to approximately 67% of the world's production of root crops; more than 80% of this production occurs between latitudes 40°N and 60°N in Europe, the USSR, and North America; European potato production is almost half of the world's potato production, and one-third of the world's root crop production. Production of tropical root crops occurs mainly between latitudes 30°N and 30°S and including edible aroid production, amounts to 150 M tons (approximately) or one-third of the world's root crop production. Cassava accounts for about two-thirds of the tropical root crop production and the South and Central American and African continents each produce 35% (approximately) of this amount. Brazilian cassava production accounts for some 86% of the South and Central American production, while Nigeria and Zaire together account for some 57% of African cassava production. Sweet potato and yam production is approximately half that of cassava produc-

TABLE VIII
Production of Tropical Root Crops and Potato in Selected Countries[a]

	Production (million metric tons)			
Country/continent	Cassava	Sweet Potato and yams	Potato	Totals
Latin America	34.82	4.12	9.13	48.07
Brazil	29.46	2.13	1.59	33.18
Peru	0.50	0.18	1.93	2.61
Argentina	0.29	0.44	1.96	2.69
Africa	37.02	23.41	2.74	63.17
Nigeria	7.30	13.50		20.80
Tanzania	1.50	0.31		1.81
Zaire	10.00	0.35		10.35
Kenya		0.46	0.21	0.67
Asia	20.12	15.99	13.63	49.74
India	5.22	1.60	4.64	11.46
Indonesia	10.45	3.03		13.48
Japan		2.56	3.16	5.72
South Korea		2.14		2.14
Taiwan		3.44		3.44
Thailand	1.90	0.25		2.15
Europe			134.70	134.70
North America		0.63	16.94	17.57
USSR			92.30	92.30
World	92.22	44.43	306.45	443.10

[a]From FAO Production Yearbook (1971).

tion, 15% of *Solanum* potato production, and only 10% of the world's root crop production. Highest production (30% approximately) occurs in Africa, of which Nigerian production (13.5 M tons) accounts for 50%. Production (mainly sweet potatoes) is evenly distributed in South and Central American and Asian countries. De Witt (1967) compared the potential photosynthetic productivity of different latitudes and showed that the tropical zone between 30°N and 30°S had a productivity of 120×10^3 kg carbohydrate/ha/year or 1680×10^2 kg potato tubers/ha/year, which was twice that at latitude 50°N and 1.3 times that at latitude 40°N (Table IX). Despite this, root crop productivities in tropical

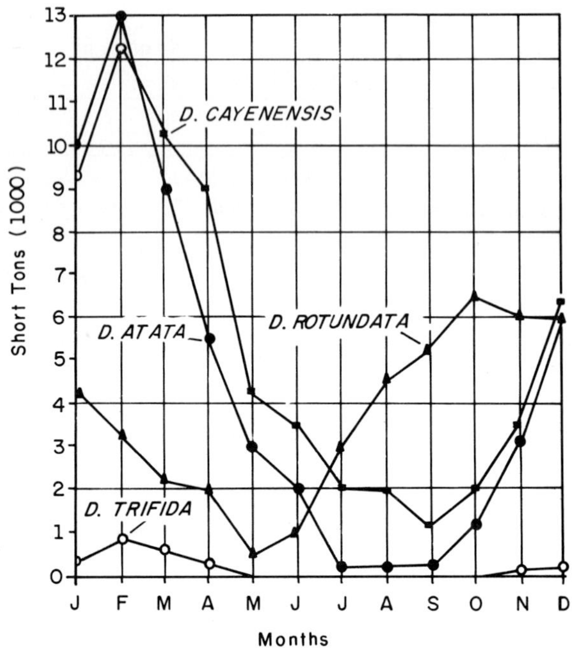

Fig. 2 Total monthly yam production by species for 1972–Jamaica. From Rankline and Ferguson (1974). Short ton = 2000 lbs.

TABLE IX
Comparative Potential Productivity of the Land Surface in Different Latitudinal Zones on the Earth's Surface[a]

°N Latitude	Land surface: Total (ha × 10^6)	Land surface: % Agricultural land	Land surface: Agricultural land (ha × 10^6)	Months above 10°C	Yield (ha/yr): Carbohydrate (kg × 10^3)	Yield (ha/yr): Tubers (kg × 10^2)
70	800	52	420	1	12	168
60	1400	38	530	2	21	294
50	1600	18	290	6	59	826
40	1500	13	200	9	91	1274
30 to –30	7700	10	770	12	120	1680
–40	100	14	14	8	89	1246
–50	100	53	53	1	12	168

[a]Adapted from de Witt (1967).

latitudes (Table X) for cassava (94×10^2 kg/ha) and sweet potato/yams (87×10^2 kg/ha) are only 69% and 64%, respectively, of *Solanum* potato productivity at 136×10^2 kg/ha. The highest cassava productivities (above 120×10^2 kg/ha) occur in Brazil, India, and Thailand; the highest sweet potato productivity (above 150×10^2 kg/ha) is found in Japan, South Korea, and Taiwan. *Solanum* potato productivities in the Netherlands, Great Britain, and West Germany were, however, some 2- to 3-fold higher at 380, 280, and 274×10^2 kg/ha, respectively. Nevertheless, the highest potato productivity in the Netherlands is still only 46% of the potential photosynthetic productivity (Table XI) of latitude

TABLE X
Comparative Productivity of Tropical Root Crops and Potato in Selected Countries[a]

	Productivity (100 kg/ha)			
Country/continent	Cassava	Sweet potato and yams	Potato	Averages
Latin America	135	91	81	102
Brazil	146	118	72	112
Peru	129	130	61	97
Argentina	113	99	110	107
Africa	75	72	82	76
Nigeria	66	75		70
Tanzania	53	78		65
Zaire	125	66		95
Kenya		84	38	61
Asia	90	101	105	98
India	148	70	90	102
Indonesia	73	85		79
Japan		199	213	206
South Korea		168		168
Taiwan		151		151
Thailand	152	53		107
Europe			190	190
North America		115	226	170
USSR			116	116
World	94	87	136	105

[a]From FAO Production Yearbook (1971).

TABLE XI
Root Crop Productivities of Selected Countries as Percentages of Potential Latitudinal Productivities[a]

Root crop/country	Productivity (kg/ha × 10^2)	% Potential latitudinal productivity
Cassava		
Thailand	152	9
Brazil	146	9
Zaire	125	7
Nigeria	66	4
Sweet potato/yams		
Japan	199	16
Taiwan	151	9
India	70	4
Brazil	118	7
Peru	130	8
Solanum potato		
The Netherlands	380	46
North America	226	27
USSR	116	14
Brazil	72	4
Peru	61	4
India	90	5
Kenya	38	2

[a]All % potential latitudinal productivities are calculated from potential photosynthetic productivities for 30°N–30°S, except for Japan (40°N), the Netherlands, North America, and USSR (50°N).

50°N and the highest cassava and sweet potato/yam productivities are 9% and 16% of the potential photosynthetic productivity of latitudes 30°N–30°S and 40°N, respectively.

Although there are many reasons for the low productivity of root crops cultivated in tropical latitudes (Coursey and Haynes, 1970; Wilson, 1974), the following sections of this chapter deal with some microecophysiological considerations of tropical root crop productivity.

III. COMPARATIVE MICROECOPHYSIOLOGY

In Section II, on the macroecophysiology of tropical root crops, it was shown that although the potential photosynthetic productivity of tropical ecosystems was up to twice that of intermediate ecosystems, actual *Solanum* potato productivity in the latter ecosystem was greater than cassava, sweet potato, and yam productivities in the former. This section discusses physiological and metabolic processes of root crops in response to different aspects of the immediate crop environment in order to explain the low productivity of root crops in tropical ecosystems.

The ultimate expression of interactions between the environment and the crop involves: (a) the size and form of individual plants in the crop community at any given stage in the crop growth cycle and (b) the pattern of growth and development of the crop community. The expressions are effected both through (1) quantitative processes, e.g., cell division and expansion, photosynthesis, protein synthesis, leading to increase of the total mass (dry weight) of the crop, and (2) qualitative processes, i.e., morphogenetic changes, e.g., tuber initiation, leaf senescence leading to changes in the quality of growth. In turn, these processes are affected by changes in the total environment of the crop which consists of the following components: (i) the immediate external aerial environment; (ii) the soil environment; and (iii) the internal crop environment, e.g., hormonal, nutritional, and assimilate balances, which are affected both by environmental factors, as well as plant factors, such as spacing, mutual shading of leaves, and sink capacity of tubers. Therefore, in order to compare the effects of different aspects of the environment on root crop growth and development, the following approach is adapted: (a) characteristic growth patterns of individual root crops are compared in relation to selected aspects of the environment; (b) effects of different components of the total environment on growth and development of root crops are compared; (c) effects of the environment on quantitative and qualitative processes leading to tuber yield are examined.

A. Growth Pattern–Environment Interactions

Leaf area index (LAI) and tuber weight are chosen as parameters for expressing differences in the growth patterns of root crops because LAI is a measure of the effective leaf surface exposed to sunlight for photosynthesis while tuber weight measures the ultimate distribution of assimilate to the yield organ. Typical growth patterns for sweet potato, *Xanthosoma* aroids, sugar beet, *D. alata, D. trifida,* and *D. esculenta* yams, as well as *Solanum* potato, grown both under temperate and tropical conditions are compared in Figs. 3 and 4. Major differences between the growth cycles of different root crop species emerge on the basis of the following characteristics: (1) overall length of the crop growth period and (2) relative timing, duration, and magnitude of shoot and tuber growth. Differences in productivity between root crop species in response to environmental conditions are the result of these characteristics of crop growth patterns.

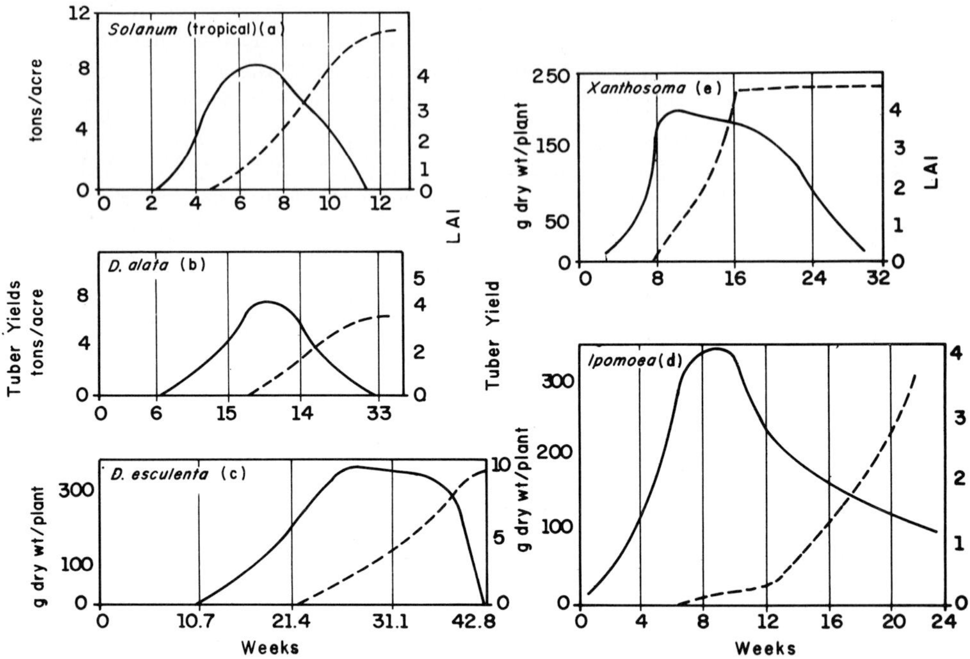

Fig. 3 Growth patterns of root crops: (a) adapted from Chapman (1965a); (b) Chapman (1965b); (c) Ferguson and Haynes (1969); (d) Walter (1966); and (e) Enyi (1973).

1. Length of Crop Growth Period

Lengths of root crop growth cycles may be classified as short (3–4 months), e.g., *Solanum* potato under tropical conditions, intermediate (4–6 months), e.g., sweet potato, *Solanum* potato in intermediate climates, or long (9–12 months), e.g., yams, edible aroids, and cassava, but there is considerable intervarietal variation in crop growth cycles within individual species. Consideration of lengths of crop growth cycles in relation to the length of the growing season also leads to the recognition of three types of crop cycles as follows: (i) cycles which occupy most of the growing season, e.g., late potatoes, aroids, and yams, and in which growth occurs in more than one climatic season, e.g., spring, summer, autumn, or wet and dry seasons; (ii) cycles which can occupy one climatic season, e.g., wet season or dry season, sweet potato, early potatoes; (iii) cycles which are curtailed by unfavorable weather, e.g., unirrigated yams in the savanna regions of northern West Africa. Crop cycles which occupy the entire growing season are twice as long in the humid tropics (9–12 months for yams, aroids, and cassava) as in intermediate climates (5 months for *Solanum* potato). The major

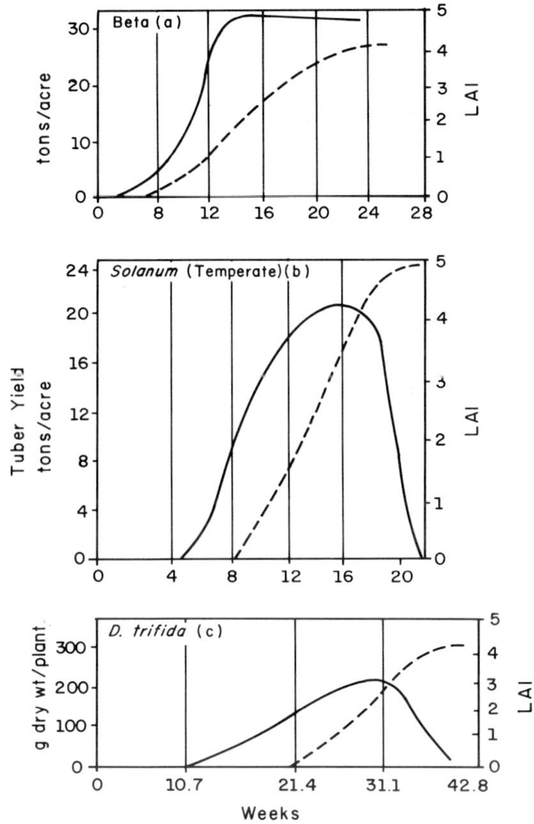

Fig. 4 Growth patterns of root crops: (a) adapted from Milthorpe (1967); (b) Radley (1963); and (c) Ferguson and Haynes (1968).

environmental factors which affect such crop cycles are early onset of the unfavorable season toward the end of the cycle, e.g., cold weather and dry season in intermediate and tropical climates, respectively; factors which postpone onset of tuber bulking, e.g., waterlogged soils resulting from heavy precipitation in tropical root crops (Lowe and Wilson, 1974), high temperatures in *Solanum,* or heavy nitrogen fertilization (Haynes *et al.,* 1967). Reduction of the growth period in *Solanum* potato from 22 weeks in intermediate climate cultivation to 14 weeks in the tropics (Figs. 3 and 4) is an example of the effect of short daylengths in reducing growth cycles through its stimulatory effect on tuber initiation in this species. High light intensity, high temperature, and low water supply also shorten growth cycles through similar effects (Bodlaender, 1963). Late planting is an obvious, but very common factor responsible for low yields due to reduction in the length of crop growth cycles. Thus, Boyd and Dyke (1950) reported a 0.7 ton/acre/week reduction in yield for every week's delay in *Solanum* potato sowing in a nationwide survey of this crop in England. Reductions in tuber yield (up to 74%) in yam farms in Barbados were also to a great extent accounted for by late planting (Rankine and Ferguson, 1974).

The early growth rate is a major determinant of the lengths of crop growth periods. Since all root crops are propagated vegetatively, a major environmental factor restricting early shoot growth is the availability of water at crop germination, e.g., in dry compared with wet season sweet potato crops (Walter, 1966; Lowe, 1971). However, the weight of planting material (tuber sett), particularly in potato and yams, is a most important cultural practice affecting early growth. Thus, both in yams (Ferguson, 1973) and *Solanum* potato (Bremner and El Saeed, 1963), increased tuber sett weight is known to increase early growth and yield of the crop. There is an optimal tuber sett weight for maximum yield but this weight is influenced by the spacing used. The early growth of yam whereby several cataphylls are produced before mature leaf types is characteristic of the species and is no doubt determined by either hormonal or nutritional balances in the tuber sett. Severing of the young plant from the mother tuber hastens the appearance of mature leaves. Similar effects of size of planting material (stem cuttings) on early growth have been established for cassava (Enyi, 1973) and sweet potato (Godfrey-Sam-Aggrey, 1974). Lowe and Wilson (1975b) also showed that the number of subterranean nodes in sweet potato stem cuttings influenced the number of tubers subsequently initiated. When tuber setts are used for sweet potato propagation, exposure of the mother tuber to light (Akita *et al.,* 1962) leads to improved tuberization. Species differences in early growth are, however, quite considerable in the root crops. Thus, the attainment time of a crop LAI of one varies from 4 weeks in *Solanum* grown under tropical conditions to 17 weeks in *Dioscorea trifida* (Figs. 3 and 4). Such variations no doubt lead to differences in the efficiency of utilization of incident radiation for photosynthesis and low efficiency, e.g., in yams, which in turn leads to slower growth and longer crop cycles.

Tuber initiation and early development also occur [even in root crop species with late bulking (Fig. 3)] during the period of early growth (to LAI = 1). However, factors which stimulate early shoot growth, e.g., high nitrogen, water supply, and long daylength, usually inhibit tuber initiation and early tuber growth. Therefore, environmental conditions which optimize these opposing effects are desirable during the early development of root crops. In species in which tuber growth does not occur during early crop growth despite the initiation of tubers, the alternative shoot sink (Wilson, 1967) competes successfully with the tuber sink to postpone the advent of tuber bulking. Dry conditions approximately 40 days after planting in sweet potato (Hernandez and Hernandez, 1967), moderate supply of nitrogen fertilizers for all root crops

(Haynes *et al.,* 1967), and moderate soil temperature (15°–20°C) in *Solanum* potato (Kay, 1973), as well as early planting during short-day spring conditions are all factors encouraging tuber initiation and early tuber growth during the period of early development of tuberous crops. However, consideration of environmental effects on tuberization, e.g., daylength in *Solanum* potato and soil oxygen tension in sweet potato, is discussed in Section III, B.

2. *Relative Timing, Duration, and Magnitude of Shoot and Tuber Growth*

Although the relative timing, duration, and magnitude of shoot and tuber growth as shown in the growth patterns described in Figs. 3 and 4 are characteristic of individual root crop species, growth patterns can be influenced by factors which have opposing effects on shoot and tuber growth. Thus, in sweet potato, high levels of nitrogen supply increase the rate of shoot growth and advance the time to attainment of maximum LAI, but postpone the initiation of rapid tuber bulking (Haynes *et al.,* 1967) (Fig. 5). Similar observations have been made with other root crops, e.g., *Solanum* potato (Watson, 1963). However, postponement of tuber bulking in sweet potato by high nitrogen supply also resulted in more rapid late bulking either because of increased capacity for tuber growth (Wilson, 1969) or increased availability of photosynthate due to higher LAI (Haynes *et al.,* 1967). There is also considerable intervarietal variation in the timing, duration, and magnitude of shoot and tuber growth in sweet potato (Fig. 6).

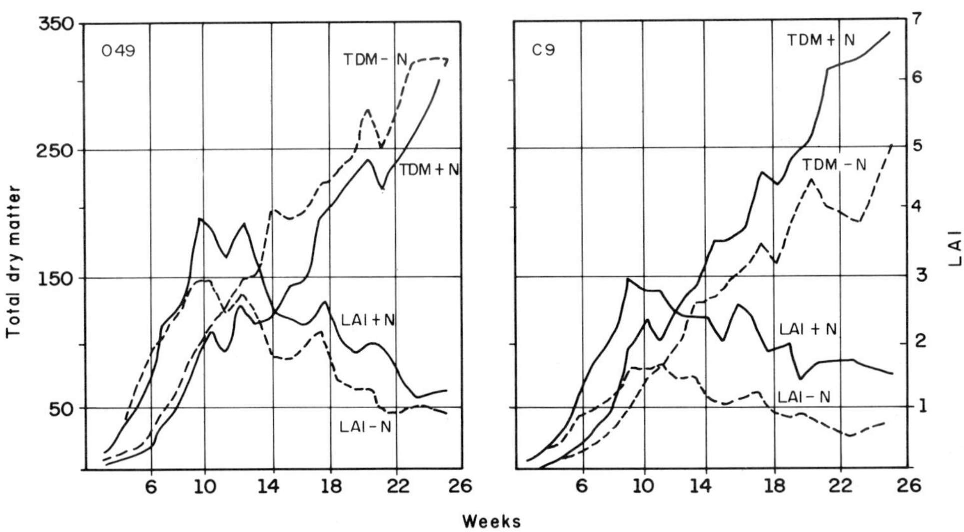

Fig. 5 Total dry matter (TDM) and leaf area indexes (LAI) of two sweet potato cultivars (049, C9) at two levels of nitrogen supply. After Haynes *et al.* (1967).

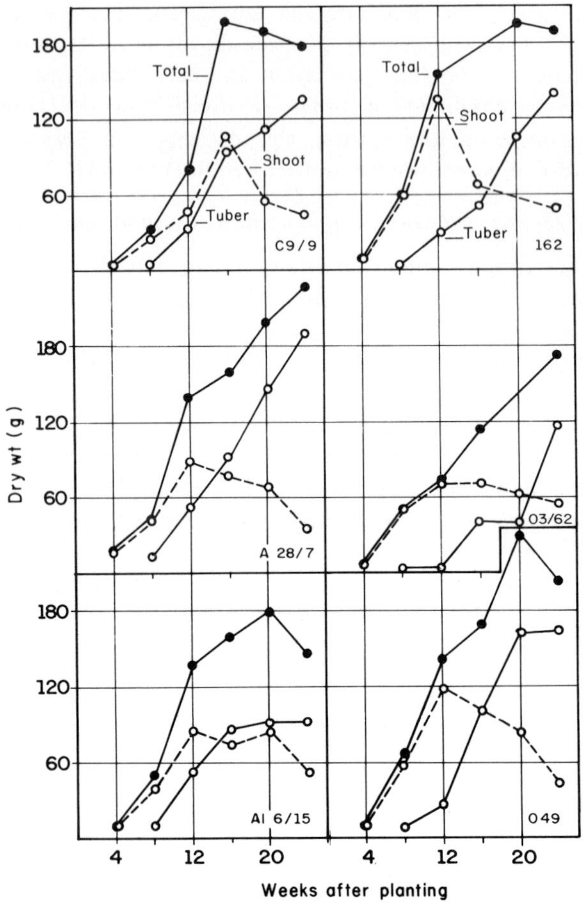

Fig. 6 Distribution of dry matter in six sweet potato cultivars. From Lowe and Wilson (1974).

Other environmental factors have similar differential effects on shoot and tuber development, e.g., short daylength, which stimulates tuberization but reduces shoot growth in *Solanum* potato. Accordingly, the growth pattern of *Solanum* potato can be changed from one with simultaneous rapid increase in LAI and tuber weight (Fig. 3) to a pattern in which rapid tuber growth follows attainment of maximum LAI (Fig. 4) by growth in different climates. Such changes illustrate that growth patterns of root crops can be manipulated through judicious use of the environment to maximize the genetic potential and variation (Fig. 6) of the species.

Maximum LAI varies quite considerably in root crops (Table XII) from 3.5 in *Dioscorea trifida* (Ferguson and Haynes, 1969) to approximately 4 in *Solanum* in intermediate climates (Radley, 1963), *Ipomoea* (Walter, 1966), *Xanthosoma* (Enyi, 1973), and *Dioscorea alata* (Chapman, 1965a) to 5.8 in *Solanum* cultivated tropically and 10.0 in *Dioscorea esculenta* (Ferguson and Haynes, 1969). Nitrogen supply is a major factor affecting LAI in root crop species, although daylength, light intensity, temperature, and water supply have important effects on LAI in *Solanum* potato.

The rate and duration of tuber bulking ultimately determine tuber yield, and there is evidence of considerable interspecific variation in these parameters (Figs. 3 and 4). Effects of environmental factors on these processes are dealt with in Section III, C on tuberization. However, it is here noted that a phase of constant, linear, and rapid rate of increase in tuber weight is the critical period for tuber yield determination in all root crops. During this period, it must be deduced either that the rate of assimilate supply from the leaf system is constant or rate limiting, or that the physical sink capacity in tubers develops at a constant rate or at a rate which is limiting to tuber development.

B. Effects of Components of Total Environment on Growth and Development

The total environment is made up of a number of individual components; growth and development depend on the net interaction between the genetic

TABLE XII
Maximum Recorded Leaf Area Indexes of Root Crop Species[a]

Crop	Maximum LAI
Solanum[b] (tropical)	5.8
Solanum[c] (temperate)	4.0
Ipomoea[d]	4.0
Xanthosoma[e]	4.0
D. alata[f]	4.0
D. esculenta[g]	10.0
D. trifida[g]	3.5

[a]From Wilson (1974).
[b]From Chapman (1967).
[c]From Radley (1963).
[d]From Walter (1966).
[e]From Enyi (1973).
[f]From Chapman (1965b).
[g]From Ferguson and Haynes (1969).

potential of the individual crop species and these components. The individual effects of components of the total environment on growth and development are, however, examined separately in order to explore the different responses of species, varieties, and indeed, individual organs to the environment. The components considered include (1) the aerial environment: light intensity, daylength, temperature, evapotranspiration; and (2) the soil environment: water, oxygen, soil physical conditions, and nutrient supply. Most available information on effects of environment on growth and development refers to *Solanum* potato, so that responses inferred for other root crops are often speculative rather than experimentally determined. In any case, the discussion of environmental effects on individual root crops is brief and comparative.

1. The Aerial Environment

a. Light Intensity. Bodlaender (1963) showed that increasing light intensities in the range of 3000–16,000 lux had the following effects on growth and development of *Solanum* potato: tuber initiation and early tuber growth were stimulated, maximum stem length was achieved earlier, and senescence and death advanced by higher light intensities. Total dry matter and distribution of dry matter to tubers were also increased at high light intensities, but final tuber weight was limited by earlier leaf senescence. Similar controlled experiments have not been conducted in tropical root crops, but it has often been observed that irrigated dry season sweet potato crops are higher yielding than wet season crops in which cloud cover reduced light intensity (Walter, 1966; Lowe, 1971; Lowe and Wilson, 1975a).

b. Daylength. Effects of daylength on growth and development are particularly important in *Solanum* potato cultivation. European potato varieties initiate tubers under any daylength conditions, but development is greatly accelerated and plants die earlier under short-day conditions. Thus, in the early stages of development, tuber weight is higher in short than in long days, but owing to larger tops and an extended cropping period long-day plants have higher final yields. Also, the efficiency of yield production is often greater (1.1 tons/acre/week) under long days (Fig. 4) than under short days (0.8 tons/acre/week) (Fig. 3). Tropical root crops are for the most part confined to short-day conditions of the tropics in which tuberization occurs readily. Evidence of the influence of short days in stimulating tuberization has been reported for cassava (Bolhuis, 1966), yams (Njoku, 1963), and *Colocasia* aroids (Tsukamoto and Inaba, 1961).

c. Temperature. Like daylength, effects of temperature on growth and development are more relevant for *Solanum* potato, since in the tropics seasonal variations in temperature are rather small. Thus, *Solanum* potato tubers must be stored at temperatures above 0°C to maintain viability and subsequent germination will take place only at temperatures above 8°C. At temperatures above

45°C, tubers and potato plants are damaged. The optimal average temperature for growth and development of the potato crop, including tuber setting, is 18°–24°C. This average may be achieved either by a narrow, 15°–25°C, or wide, 0°–40°C, temperature range. Potato production also occurs in the tropical lowlands in the temperature range of 22°–30°C. In general, high temperatures are favorable for stem growth, but unfavorable for leaf expansion and tuber production. The optimum temperature for leaf growth has been reported to be 12°–14°C, while that for maximum stem weight was 18°C (Bodlaender, 1963). Consequently, the ratio of the weight of leaves to stems decreases with increasing temperature. Borah and Milthorpe (1959) reported a temperature optimum of 20°C for tuber formation and showed that the number of tubers per plant was higher at lower temperatures. These effects of temperature are, however, modified considerably by light intensity, daylength, and water supply, e.g., Bodlaender (1963) reported that in greenhouse experiments, tuber weights were higher at 12°–14°C under low light intensity winter conditions but optimum temperature for high tuber weight was 18°–20°C under high light intensity summer conditions. Such temperature optima may be further increased under the higher light intensity conditions of tropical potato cultivation.

Diurnal temperature variation is also known to affect potato development. Thus, leaf growth and high tuber number are favored more by low night temperature than by low day temperature. Tuber yield is also decreased more by high night temperatures than by high daytime temperatures (Gregory, 1954; Bodlaender, 1958).

Soil temperature also has important effects on growth and development of potato. Thus, Gregory (1954) observed that high soil temperature increased stem length and top weight, especially at low air temperature (17°C day and 11°C night) and decreased tuber weight especially at high air temperature (30°C day–23°C night). The influence of soil temperature on growth of tubers was, therefore, greatest when air temperature was unfavorable for tuber growth.

Little work has been done on the effects of temperature on tropical root crops. However, Kim (1961) established that low nighttime temperatures increased tuber weight in sweet potato. Spence and Humphries (1972) also showed that tuber development in sweet potato was more rapid at 25°C than at 30°C, and no tubers were formed at 10°C and 15°C. Therefore, higher yields obtained in sweet potato dry season crops might also be explained by the lower temperatures prevalent in that season.

d. Evapotranspiration. Penman (1962, 1963) established a linear relationship between *Solanum* potato yield and an adjusted value of potential evapotranspiration [$E_T(A)$] (Fig. 7). Yields increased from less than 5 to more than 20 tons/acre as $E_T(A)$ increased from 4 to 15 inches, and the maximum possible yield response to irrigation was calculated at 2 tons/acre/inch. Similar studies on tropical root crops have not been reported.

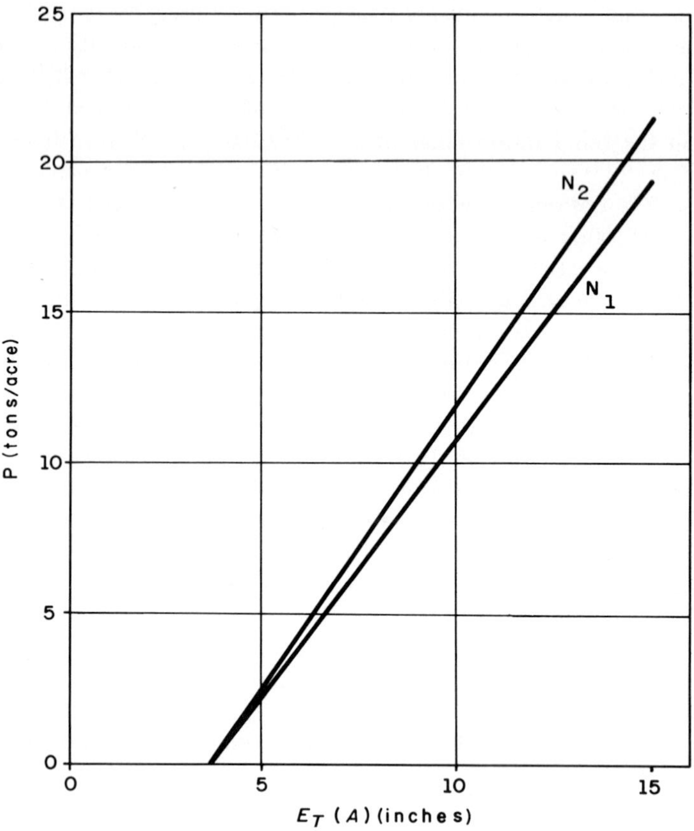

Fig. 7 Total tuber weight (P) in relation to adjusted potential transpiration $[E_T(A)]$ for potatoes at two nitrogen levels. From Penman (1963).

2. *The Soil Environment*

a. Water. Water supply is perhaps the most critical environmental factor affecting the growth and development of tropical root crops, since precipitation is subject to seasonal and diurnal fluctuations similar to the temperature fluctuations experienced in intermediate climates. Water supply is particularly important for the germination and establishment of all root crops because of the methods of vegetative propagation used in cultivation. For the further growth of the crops, water requirements are different. Thus, in the *Colocasia* aroids, higher yields are obtained under lowland paddy field conditions than in drier upland conditions (de la Pena and Plucknett, 1967). *Xanthosoma* aroids also have high water requirements (but not paddy field conditions) for highest yields. The yams require adequate supplies of water (115–300 cm) for growth and development, and there is a high correlation between precipitation, vine growth, and tuber yield. For optimum yields, adequate moisture between the fourteenth and twentieth weeks of growth is of great importance (Kay, 1973). The requirement for water in sweet potato is in the range of 50–100 cm/year, but the crop can tolerate considerable periods of drought, although yields are much reduced if water shortage occurs 50–60 days after planting, i.e., at the onset of tuber bulking. Under conditions of high rainfall, waterlogged soils inhibit sweet potato tuber initiation and tuber growth, and pencil roots (Wilson, 1970) rather than tubers are formed. Except at planting, cassava can withstand prolonged periods of drought and is a valuable crop in regions of low or uncertain rainfall (30–50 cm). Optimal water requirements are, however, in the region of 100–150 cm of well-distributed precipitation (Kay, 1973). Although the water requirement for potato is modest (50–75 cm) by comparison with tropical root crops, water supply needs to be evenly distributed (2.5 cm/week) because the crop is very intolerant of even short periods of drought, especially in the last 9 weeks of growth. Inadequate and/or irregular water supply lead not only to poor yields but also to malformed tubers (Kay, 1973).

b. Oxygen. The deleterious effects of oversupply of water, e.g., in waterlogged soils on tuberization in tropical root crops are due to restriction of oxygen supply shown to be critical for tuberization in sweet potato (Togari, 1950). Clearly, tuberization is not so restricted in *Colocasia* aroids grown under paddy field conditions. However, the oxygen requirement for tuber growth in this crop might be obtained from the aerial environment by transport in the aerenchyma tissue as in rice. Unlike *Colocasia* aroids, *Xanthosoma* aroids cannot withstand waterlogging.

c. Soil Physical Conditions and Nutrient Supply. In crops in which the yield organ, the tuber, must grow against the resistance of its solid soil environment, soil physical conditions are important requirements for high tuber yield. However, very little work has been done on interrelationships between physical properties of soils and tuberization. Gumbs and Ferguson (1975) measured the positive pressures which develop in the course of yam tuber growth in specially designed root chambers, and found a pressure increase of 0–1.008 bars over a 10-day period, at four months after planting. They also demonstrated that tuber growth as well as shoot growth were inhibited by increasing bulk density in the range of 1.1–1.6 g/ml applied separately to tubers and roots in separate compartments. On a field scale, it was shown that increasing soil compaction caused by different cultural practices reduced tuber yield. Highest yields (24.4 tons/ha) were recorded at a penetration resistance of 380 psi and lowest yields (21.9 tons/ha) at 465 psi. Generally, root crops give higher yields on light loams, but both tropical root crops and *Solanum* potato can be cultivated in heavy clays, with proper management.

Nutrient supply is the environmental condition that perhaps can be most precisely brought under control by cultural practice, e.g., fertilizer application. Although the fertilizer requirement of root crops is outside the scope of this chapter, some general responses of root crops to nutrient supply are worthy of note. Thus, the effect of high nitrogen supply in promoting shoot growth at the expense of rapid tuber growth (which is postponed) is of major importance for high productivity in the shortest possible crop period. Potassium supply, which is important for photosynthesis as well as for translocation (Fujise and Tsuno, 1967) and storage (Murata and Akazawa, 1968) of assimilate in sweet potato, is critical for high root crop yield. Recently, Ferguson and Haynes (1970) demonstrated N-depressive and K-responsive effects of yield in *Dioscorea esculenta,* and N-responsive and K-depressive effects on *Dioscorea alata* cultivar White Lisbon. Tsunoda (1965) also reported the occurrence of low leaf area N-responsive and high leaf area N-depressive sweet potato types, while Haynes and Wholey (1968) classified some 100 sweet potato varieties into N-responsive, N-indifferent, and N-depressive types, according to the effects of nitrogen fertilizer application on final tuber yield. De la Pena and Plucknett (1967) demonstrated differential responses to N, P, and K in lowland and upland cultivated taro.

C. Effects of Environment on Quantitative and Qualitative Processes Leading to Tuber Yield

Although many physiological and metabolic processes contribute to growth and development, the following qualitative (ontogenetic) and quantitative (dry matter production) processes are briefly reviewed as being important in the productivity of root crops: (1) qualitative processes–shoot ontogeny and tuber ontogeny; and (2) quantitative processes–photosynthesis and protein synthesis.

1. Qualitative Processes

a. Shoot Ontogeny. During shoot ontogeny of root crops, the following stages are recognized: (a) germination to produce a seedling; (b) period of juvenile growth, i.e., until leaves characteristic of the mature plant in size and shape are first produced; (c) period of growth of the mature plant to maximum leaf area; (d) leaf senescence.

Although there are considerable differences between root crop species and varieties in shoot ontogeny, e.g., the cataphyll production in yams and juvenile leaves of high anthocyanin content in cassava, the effects of the environment on these factors have not yet been critically examined.

b. Tuber Ontogeny. Several studies have been made of the external factors affecting tuberization. Here, the morphogenetic process of tuber initiation must be carefully distinguished from the tuber growth which occurs subsequent to tuber initiation. Thus, the only environmental factor categorically shown to have an absolute effect on tuber initiation is photoperiod in daylength-sensitive root crops, e.g., short-day tuberizing *Solanum* potato varieties. The reports that sweet potato roots do not tuberize in the presence of light (Fujise and Tsuno, 1967) are yet to be confirmed at the cellular level (Togari, 1950; Wilson and Lowe, 1973) where the change from longitudinal to the lateral growth characteristic of tuberization occurs. The cellular events leading to tuberization in root crops other than *Solanum* potato and sweet potato have not been as thoroughly studied. However, the anatomy of tuberization in cassava (Doku, 1970), yams (Martin and Ortiz, 1962), and *Xanthosoma* (E. Duncan and E. Thornhill, private communication, 1970) has been described. Effects of other environmental factors, e.g., light intensity, temperature, water supply, mineral nutrients, discussed earlier apparently operate through their influence on tuber growth rather than on tuber initiation.

It is generally accepted that effects of the environment on tuber initiation and the cellular component of tuber growth, i.e., increase in cell number and cell size (Wilson, 1967, 1970; Wilson and Lowe, 1973) are mediated through hormonal responses. Studies on application of externally applied hormones and endogenous changes in hormone contents of potentially tuberous structures and tuberizing organs (Wilson *et al.*, 1973) indicate the participation of several plant growth substances in tuber growth; these include auxins, gibberellins, cytokinins, ethylene, and abscisic acid. However, the exact hormonal nature of the tuber forming stimulus suggested by Madec (1963) has not yet been identified. Work by Kumar and Wareing (1973) in *Solanum andigena* suggests that unidentified growth inhibitors play an important role in tuberization. It has, however, been established that the inhibiting effects of light and low oxygen supply on sweet potato tuber growth and development are mediated through inhibition of meristematic activity in tubers (Togari, 1950; Ito, 1947). Effects of environmental conditions on *Solanum* potato tuber growth, e.g., the inhibiting effect of water shortage, also operate through decreased cell division and expansion.

2. *Quantitative Processes*

a. Protein Synthesis. The protein synthesis involved in cell division and expansion is an important aspect of growth of root crops. Wilson *et al.* (1973) referred to competition between the tuber sink and the alternative shoot sink in view of the exponential increase in cell number occurring both in tubers (Plaisted, 1957) and leaves (Wilson, 1964). It is interesting, therefore, that the effect of increased nitrogen supply on reduced tuber growth is to make the alternative shoot sink a stronger competitor for available carbohydrates during early growth. Wilson and Knox (1973) demonstrated that leaf nitrate reductase in sweet potato was an extremely active enzyme and the preeminence of shoot growth in the presence of high nitrogen supply might be due to the capacity of this adaptive enzyme (Beevers and Hageman, 1969) to direct carbohydrate to protein synthesis in leaves rather than to storage in tubers. Alternatively, the reduction of nitrate reductase activity would release carbohydrates for storage in tubers. The role of nitrate reduction in root crop metabolism needs further study.

b. Photosynthesis. The process of photosynthesis leads to the increase in total dry matter in the developing crop. Parameters such as crop growth rate (CGR) and relative growth rate (RGR) express the rate of increase in dry matter while net assimilation rate (NAR) expresses the efficiency of the leaf surface in accumulation of dry matter. Direct relationships have been established between NAR and yield in sugar beet (Watson, 1958) and potato (Milthorpe, 1967), between LAI and tuber bulking in cassava (Enyi, 1972), and between CGR and LAI in sweet potato (Tsuno, 1971) (Fig. 8), indicating close interrelationships between the parameters of dry matter accumulation. Since accumulation of dry matter depends on photosynthesis, it is instructive to examine quantitative interactions between the environment and photosynthesis. Lemon (1966) estimated that only 2.9% of the total incident radiation was used for dry matter production during the active growing period of corn, a crop with vertically displayed leaves, and efficient C_4 photosynthesis. Later, the net efficiency, corrected to absorbed

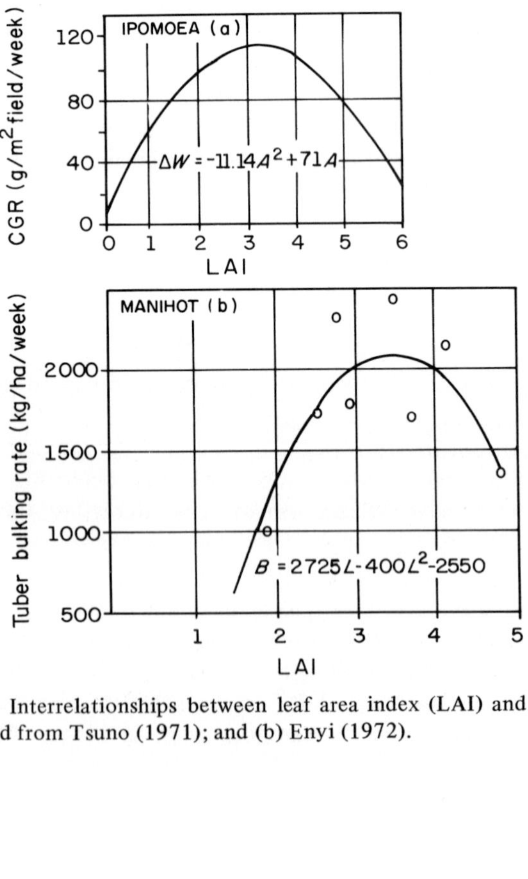

Fig. 8 Interrelationships between leaf area index (LAI) and growth rate: (a) adapted from Tsuno (1971); and (b) Enyi (1972).

visible radiation, was still only 7.3%. Loomis and Williams (1963) estimated theoretical dry matter production from incident radiation of 500 cal/cm^2/day at 71 g/m^2/day, or 633 lb/acre/day, or 710 kg/ha/day (Table XIII). Lemon (1963) indicated that only 230 lb/acre/day or 225 kg/ha/day can be achieved in red clover. A major problem in crop production, therefore, is increasing the efficiency of the conversion of radiant energy into dry matter (Tables IX–XI).

Quite apart from considerations of leaf display and the mechanism of photosynthesis, Wilson (1974) drew attention to the efficiency of light interception in root crops, assuming that optimal light interception occurred at LAI = 3. Thus, 50–60% of the crop growth cycle elapses (Table XIV and XV) before a LAI of 3 is achieved in yams, compared with 27–31% in *Xanthosoma* and *Ipomoea* species, and 40–46% in *Solanum* potato grown both in the tropics and intermediate climates. Rapid increase in leaf area during early growth would, therefore, increase the efficiency in light interception. On the other hand, high LAI's, e.g., > 10 in *Dioscorea esculenta,* reduce the effectiveness of photosynthesis due to mutual shading of leaves. However, Bonhomme and Chartier (1972) recorded values of gap frequency for sweet potato which suggested that for a canopy of LAI = 2.8, the foliage was underdispersed at sun elevations greater than 35° (Fig. 9).

Working with sweet potato Tsuno and Fujise (1965) obtained data which elucidated many of the principles of photosynthesis and dry matter accumulation in tropical root crops as follows:

(1) The maximum value of photosynthetic activity for mature sweet potato leaves was 20 mg CO_2/dm^2/hour and differences between varieties were small.

TABLE XIII
Estimation of Potential Daily Production by a Crop Surface Receiving 500 cal/cm² /day[a]

	Potential production
1. Total radiation/day	500 cal/cm²
2. Visible radiation, 400–700 μm	222 cal/cm²
3. Total quanta, 400–700 μm	4320 μE/cm²
Albedo loss	−360μE/cm²
Inactive absorption	−432 μE/cm²
4. Total available for photosynthesis	3528 μE/cm²
5. (CH_2O) produced	353 μmole/cm²
6. Respiration loss	−116 μmole/cm²
7. Net production of (CH_2O)	237 μmole/cm²
8. Net production in grams	71 g/m²
9. % Total visible radiation	6.7

[a]From Loomis and Williams (1963).

TABLE XIV
Time Scale for Leaf Area Index Development in Some Root Crops[a]

Crop	Time scale, LAI development (weeks)					Total time (weeks)
	LAI < 1	LAI > 1 < 3	LAI > 3	LAI < 3 > 1	LAI < 1	
Solanum[b] (tropical)	4.0	1.6	4.0	4.4		14.0
Solanum[c] (temperate)	6.5	3.5	9.0	1.5	1.5	22.0
Ipomoea[d]	4.5	3.5	5.5	12.5		26.0
Xanthosoma[e]	6.0	8.0	12.0	10.0	10.0	36.0
D. alata[f]	9.0	8.0	8.0	6.5	2.5	34.0
D. esculenta[g]	14.0	4.0	21.5	1.0		40.5
D. trifida[g]	17.0	10.0	6.5	4.5	2.5	40.5

[a]From Wilson (1974).
[b]From Chapman (1967).
[c]From Radley (1963).
[d]From Walter (1966).
[e]From Enyi (1973).
[f]From Chapman (1965b).
[g]From Ferguson and Haynes (1969).

TABLE XV
Time Scale for Leaf Area Index Development in Some Root Crops[a]

Crop	Time scale, LAI development (% total crop time)					
	LAI < 1	LAI > 1 < 3	LAI < 3	LAI > 3	LAI < 3 > 1	LAI < 1
Solanum[b] (tropical)	28.6	11.4	40.0	28.6	31.4	
Solanum[c] (temperate)	29.5	16.0	45.5	40.9	6.8	6.8
Ipomoea[d]	17.4	13.4	30.8	21.2	48.1	
Xanthosoma[e]	20.0	6.7	26.7	40.0	13.3	20.0
D. alata[f]	26.5	23.5	50.0	23.5	19.1	7.4
D. esculenta[g]	34.5	9.9	44.4	53.1	2.5	
D. trifida[g]	41.6	24.7	66.3	16.0	11.1	6.6

[a]From Wilson (1974).
[b]From Chapman (1967).
[c]From Radley (1963).
[d]From Walter (1966).
[e]From Enyi (1973).
[f]From Chapman (1965b).
[g]From Ferguson and Haynes (1969).

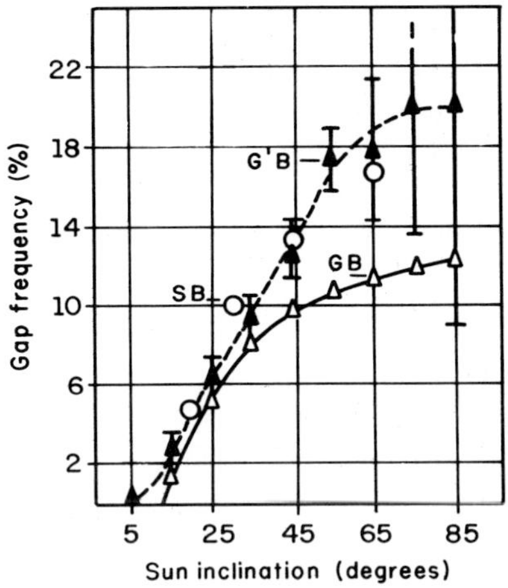

Fig. 9 Gap frequency as calculated from the exponential profile (GB), and sunflect frequency (SB) and gap frequency as measured by means of fish eye photographs (G′B) against sun elevation for a sweet potato canopy. (Standard deviations of gap frequency values are indicated on the figure.) From Bonhomme and Chartier (1972).

(2) Saturating light intensity for individual, intact, or detached leaves was 30 klux, but although light-photosynthesis curves based on diurnal data collected under field conditions showed a close relationship between light intensity and photosynthetic rates, light saturation was never achieved and was, therefore, far higher than that for the leaf. Maximum dry matter production per unit field area was 120 g/m^2/week and this was achieved at midcrop. Maximum values of NAR and leaf area index were 55 g/m^2/week and 4.3, respectively. Optimum LAI under solar radiation of 400 g cal/cm^2/day was 3.2.

(3) Photosynthesis was increased by a tenfold increase in CO_2 concentration, at a light intensity of 40 klux and 26.5°C.

(4) Photosynthesis of individual sweet potato leaves at saturating light intensity was unaffected by temperature in the range of 23°–33°C.

(5) Photosynthesis was closely related to the potassium content of leaves. Leaves containing more than 4% potassium showed high rates of photosynthesis provided nitrogen contents were above 2.2%. Photosynthetic rate, however, decreased with leaf age, irrespective of mineral nutrient contents.

(6) In a study of four sweet potato varieties, it was shown that rates of photosynthesis per plant were similar in high leaf area (leaf area types) and low leaf area types (net assimilation rate types) during the first half of their growth period. Thus, per plant photosynthesis could be classified into two types, i.e.,

high leaf areas with low photosynthesis per unit area of leaf, and low leaf areas with high photosynthesis per unit leaf area.

(7) This study also suggested that apart from a direct effect on photosynthesis, potassium, by increasing tuber development and hence, the transport of assimilates from leaves to tubers, might have had an indirect stimulatory effect on the rate of photosynthesis. Fujise and Tsuno (1967) also showed that photosynthesis was depressed and starch storage in leaves increased when tuber formation was inhibited by exposure of the sweet potato root system to light.

Later, Spence (1971) found a rate of photosynthesis of 15 mg CO_2/dm^2/hour in rooted sweet potato leaves and Spence and Humphries (1972) showed that increasing tuber development in rooted leaves either by water supply or temperature treatment to the rooting medium led to increased dry matter production per unit leaf area. Despite the similarity of photosynthetic rates in high and low leaf area varieties shown by Tsuno and Fujise, Haynes *et al.* (1967) drew attention to the potential advantage of N-induced high LAI, during the later stages of the sweet potato crop grown in the tropics, in providing a source of assimilate for rapid tuber bulking. However, reciprocal graft experiments with high and low yielding sweet potato varieties conducted by Wilson (1967) and Hozyo (1970) showed that high tuber yield was determined by the root stock irrespective of the leaf shape and yield characteristics of the scion used.

On balance, therefore, it appears that the quantitative yield limiting process in sweet potato is tuberization rather than photosynthesis, and hence the external factors which influence this process, e.g., water supply and oxygen, are likely to have considerable effect on yield. Milthorpe (1967), however, concluded that in *Solanum* potato the rate of tuber growth was controlled by the supply of photosynthate, which, in turn, was mainly controlled by the extent and duration of the leaf surface. Apparently higher yields in cassava root stocks grafted to *Manihot glaziovii* might indicate a similar yield limitation by the quantitative process of photosynthesis.

Both light and dark respiration might function to reduce the efficiency of photosynthesis. Thus, although it has been suggested that high leaf to stem ratios in sweet potato (Tsuno and Fujise, 1965; Austin and Aung, 1973) might lead to reduced dry matter production, the occurrence of peroxisomes and photorespiration in leaves might make leaf tissue an equally effective site of nonproductive dry matter utilization.

Despite the existence of environmental factors in tropical ecosystems which could function to reduce the potentially high productivity of root crops, e.g., extremes of water availability, a major factor responsible for the low productivity of tropical ecosystems is the relatively unimproved crop performance of tropical root crop species, e.g., in the patterns of growth and development (Figs. 3

and 4). However, record yields in existing cultivars of tropical root crop species indicate that productivities, comparable with those of cereals, can be achieved (Table XVI).

IV. SPECIES-DEPENDENT ECOPHYSIOLOGY

Previous sections considered responses of root crops to naturally occurring variables both in the soil and aerial environment. However, in modern agricultural practice, many of these variables are manipulated to increase crop productivity. Thus, in the soil environment, nutrient status is improved by fertilizer application and water supply regulated by irrigation and/or drainage. Variations in the aerial environment, e.g., temperature and daylength, can also be maximally utilized for increasing productivity by choice of planting date to realize crop growth in the most favorable season and use of varieties capable of completing their growth cycles in this season. An excellent example of such utilization of variation in daylength is the high-density cultivation of existing dwarf, determinate *Cajanus cajan* cultivars as a row crop under short-day conditions in Trinidad (West Indies) from October to January (Spence and Williams, 1972). The most dramatic recent advances in crop production have, however, been made through improved utilization of environmental resources, particularly light and nutrient supply for increased crop productivity, by development of new varieties responsive to high density cultivation at high levels of fertilization. Such successes include dwarf IRRI rice varieties and U.S. maize hybrids. Another significant recent success is the development of subtropical wheat varieties which have made countries such as India and Mexico almost self-sufficient in wheat flour within a few years. The transfer of *Solanum* potato

TABLE XVI
Maximum Yields in Selected Experiment Stations in the Tropics[a]

Crop	Tons per ha per harvest	Tons per ha per year	Cal/ha/day × 10^3
Rice	16.4	26.0	176
Wheat	3.9	11.7	110
Maize	5.5	20.0	200
Cassava	77.0	71.1	250
Sweet potato	41.0	65.2	180

[a]From de Vries *et al.* (1967).

cultivation from tropical to temperate latitudes through modifications of the species' response to temperature and daylength is an achievement of longer standing.

These examples illustrate approaches to the ecophysiology of crop production which depend on development of new varieties: (a) either to increase crop productivity through more efficient utilization of the environmental resources of a particular ecosystem, or (b) to transfer production of a particular species to a new ecosystem with different environmental characteristics, e.g., in temperature and daylength. Such an approach is here referred to as species-dependent ecophysiology. In this section, recent developments in species-dependent ecophysiology relevant to increasing the productivity of tropical root crops are discussed. However, problems of transfer of tropical root crops to new ecosystems are not here considered, despite earlier attempts to cultivate *Dioscorea batatas* in Europe in the nineteenth century during the *Solanum* potato famine, and more recent reintroduction of *Solanum* potato to tropical latitudes.

A. Crop Ideotypes

Donald (1968) called attention to the limitations of classical plant breeding methods which were based on (a) either defect elimination in breeding programs for disease resistance, or (b) selection of individual plants for high yield without reference to designated physiological or morphological characters which determined such high yield. Thus, although these methods have, in the past, led to realization of immediately desired objectives, i.e., increased yields and/or resistance to disease, the methodology employed did not generate any information on the plant characters associated with improved crop performance. A major disadvantage of this approach is that it does not allow for improvement in crop performance to the asymptotic limit of the species by systematically optimizing the determinants of yield (or resistance to disease) in new cultivars. The failure to achieve any major improvement in European potato productivity over the last 50 years is perhaps a good example of the inadequacy of such conventional breeding methods.

Donald (1968) proposed a third approach to improvement of productivity that specifically aims to achieve the objective of systematic improvement in crop productivity to the species asymptote through breeding for model plants or ideotypes. In this approach, a model is first designed from a number of model characters, shown to be determinants of high yield in a range of contrasting cultivars and a biological prototype later synthesized by appropriate breeding methods, using these cultivars. Accordingly, an ideotype was defined by Donald (1968) as "a biological model which is expected to perform or behave in a predictable manner within a defined environment and to yield a greater quantity or quality of useful produce than existing varieties when developed as a cultivar."

Design of an ideotype, therefore, presupposes detailed and accurate knowledge of the range of anatomical, morphological, physiological, and biochemical characters existing in varieties and/or cultivars within a species. In ideal circumstances, it also assumes the availability of information on the development and performance of cultivars and varieties with contrasting habits in different ecosystems. However, data on the performance of cultivars in the ecosystem for which the ideotype is being designed may be adequate. Examples of model characters which have been suggested for cereals are shown in Table XVII. Donald (1968) stated that the successful crop plant will be of low competitive ability relative to its mass and high efficiency relative to its environmental resources. Such an ideotype maximizes productivity by intensive high-density cultivation.

The design of an ideotype for tropical root crops is complicated by the diversity of habits in existing species (Table VII) ranging from plants with cormous stems in the edible aroids, through the trailing stem of sweet potato, to the upright, secondarily thickened, woody stem of cassava. The yield organ also varies considerably in origin, from lateral roots in cassava and sweet potato, through reduced stems in the edible aroids, to a rudimentary rhizome in the yams. In addition, apart from sweet potato, crop performance data for a wide

TABLE XVII
Some Model Characters for a Cereal Ideotype[a]

Morphological
Short, erect stem, resistant to lodging
Single culm
Short, erect, dark green leaves
Few small leaves
Physiological
Shade tolerance
Tolerance of high density planting at high rate of fertilization
High LAD[b] above flag leaf
Rapid translocation of sugars
High nitrate reductase activity
High interspecific competitive ability
Low intraspecific competitive ability
Reproductive
Early flowering and maturity
High harvest index
Large erect ears
Many florets per unit dry matter
High (nonlimiting) sink capacity of ears

[a]After Donald (1968).
[b]LAD, Leaf area duration.

range of contrasting cultivars are not available for the tropical root crops, and much of the data for sweet potato has been obtained for subtropical ecosystems in Japan. Thus, although it is not possible at this time to design with confidence either a single ideotype or range of ideotypes for tropical root crops, some of the principles that might be applied to the design of such an ideotype will be discussed.

B. Potential Productivity of Tropical Seasons

A characteristic feature of the tropical environment is the existence of suitable temperatures for growth throughout the year. However, in most climates, the year is divided into wet and dry seasons of varying lengths. In turn, the wet season is characterized not only by high precipitation, but also by longer daylengths (in northern latitudes) and lower light intensities due to high cloud cover. Conversely, in the dry season there are low precipitation, shorter daylengths, and higher light intensities. Thus, although water is not a limiting factor in crop growth in the wet season, soil oxygen may become limiting due to soil waterlogging, leading to limited nutrient uptake and slow growth. Heavy precipitation results in loss of soil nutrients (both applied and native) by leaching and surface movement. In addition, lowered light intensities might lead to reduced rates of photosynthesis, leaf density thickness, increased stem length, area of individual leaves, and consequently reduced availability of assimilates for storage. The prevalent, long days and high nighttime temperatures of the wet season might also result in postponement of tuber initiation and slower tuber growth, while high humidities might increase the incidence of pests and diseases. Alternatively, apart from availability of water, conditions of soil oxygen, and nutrient supply, light intensity, daylength, and night temperature are more favorable for growth in the dry season. As a result, in root crop species with short-enough growth cycles to be cultivated in either season, e.g., sweet potato, irrigated dry season crops usually produce higher yields (Walter, 1966; Lowe and Wilson, 1975a). It would seem, therefore, that the potential productivity of the dry season is higher than the wet season and in climates with scattered precipitation in the dry season, tropical crop productivity is probably maximal.

C. Alternative Model Characters for Tropical Root Crop Ideotypes

1. Ideotypes for Cultivation Systems

Perhaps the first alternative in the considerations of root crop ideotypes is whether there should be a single ideotype, two ideotypes for wet and dry seasons, or ideotypes for each species, which would maximize the specific climate adaptation of the species, e.g., paddy conditions in the *Colocasia* aroids, resistance to wind damage due to trailing habit in sweet potato, relative drought tolerance in cassava. Then, attention must also be given to whether in the context of current agricultural development in individual tropical countries, and the relative importance of tropical root crops in the agricultural sector, extensive or intensive systems of root crop production are desirable. This consideration would lead to a choice of characteristics for an aggressive ideotype with strong interspecific competitive ability, e.g., for suppression of weeds, perhaps at the expense of tolerance of high density cultivation. Choice of length of crop growth cycle for maximizing root crop productivity in the tropics is also relevant in this connection. It seems that the possible alternatives include (a) long crop growth cycles of up to 10 months in which crops are planted at the outset of the wet season and harvested in the following dry season, e.g., cassava, yams, edible aroids; (b) growth cycles of intermediate length of 5–6 months, which would allow for cultivation of successive wet and dry season crops in one year, e.g., sweet potato; (c) short growth cycles of 3–4 months, which would permit at least three successive crops in one year, e.g., *Solanum* potato or early sweet potato varieties.

2. *Stem Habit*

The ideal character of the stem is another important consideration in the design of root crop ideotypes. In root crops, the stem has the following functions: (i) support, exposure, and display of leaves; (ii) transport of assimilate from leaves to tubers; (iii) storage in the cormous stem of the *Colocasia* and *Xanthosoma* aroids, and the rhizomatous tuber of the yams; (iv) genesis and attachment of tubers in sweet potato and *Solanum* potato.

Thus, the stem is not involved in providing aerial support for the yield organ as it is in the cereals, and provision of such support, which is a central feature in the design of cereal ideotypes, is, therefore, of no importance in root crops. However, the functions of leaf display and transport of assimilate require opposing characteristics. Thus, reduction of the stem to the point where leaves present as a rosette, e.g., in the aroids, would reduce the transport pathway for assimilate from leaf to tuber thus increasing the efficiency of such transport, but could lead to mutual shading of leaves, poor display to light, and reduced photosynthesis. Alternatively, the relative advantage of the elongated stem for leaf display may be partially offset by increase in assimilate transport distances, as well as the diversion of considerable material to stem growth, e.g., cassava. The elongated trailing stems of sweet potato would appear to be disadvantageous for leaf display, but comparative measurements of gap frequency in sweet potato and maize by Bonhomme and Chartier (1972) indicated that light penetration in a sweet potato canopy (at LAI = 2.8) may be adequate. Haynes and Wholey (1968) and Wholey (1968) classified sweet potato cultivars into those with gathering as opposed to spreading habits, and the latter investigator attempted to quantify stem habit in terms of lateral branches per unit stem length. The character of the canopy in which gap frequency was measured was not defined. It would seem, however, that the long twining stems of yams, which in practice are supported either on stakes or trelliswork, offer neither an ecological nor an economic advantage, and the ideal yam plant would perhaps be better served by a short, semi-erect stem. In cassava, a single short stem rather than many long stems would seem to be an advantage for efficient utilization of environmental resources (Enyi, 1972). Model characters for stem habits in tropical root crops might therefore be tentatively suggested as follows: cassava–short, erect stem; yam–semi-erect stem; sweet potato–short, trailing stem with gathering habit; *Solanum* potato–short, erect stem. The relative merits of divided or entire leaves in all species must also be considered. But these suggestions should be as far as possible confirmed by experimentation before adoption, e.g., by reciprocal graft experiments (Wilson, 1967) or by physiologically induced changes in stem habit (Spence, 1970; Spence and Boerboom, 1975).

3. *Production and Distribution of Assimilate*

The total production of dry matter as well as the distribution to tubers of the portion of dry matter surplus to that necessary for maintenance of the individual plant in a crop community as a morphogenetic entity are ultimate determinants of yield. However, the characteristics of the ideal morphogenetic entity change

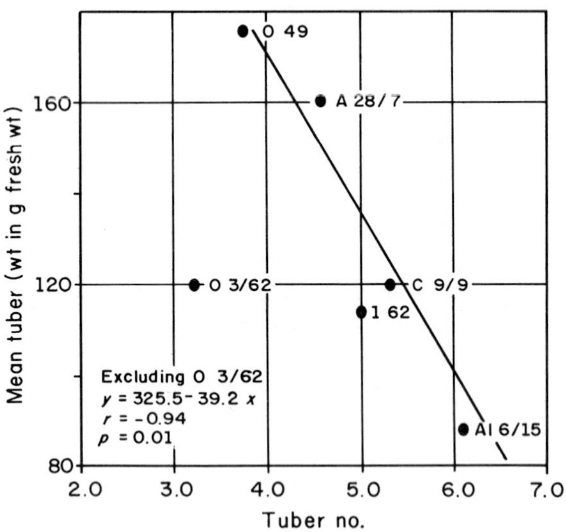

Fig. 10 Relationship between tuber number and tuber weight in six sweet potato cultivars. From Lowe and Wilson (1974).

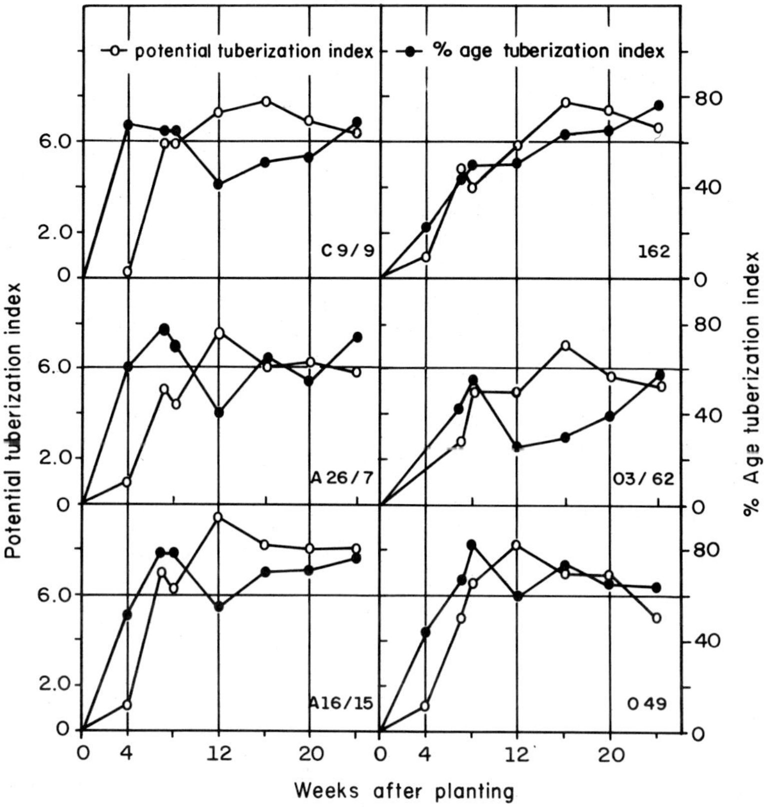

Fig. 11 Tuberization indexes in six sweet potato cultivars. From Lowe and Wilson (1974).

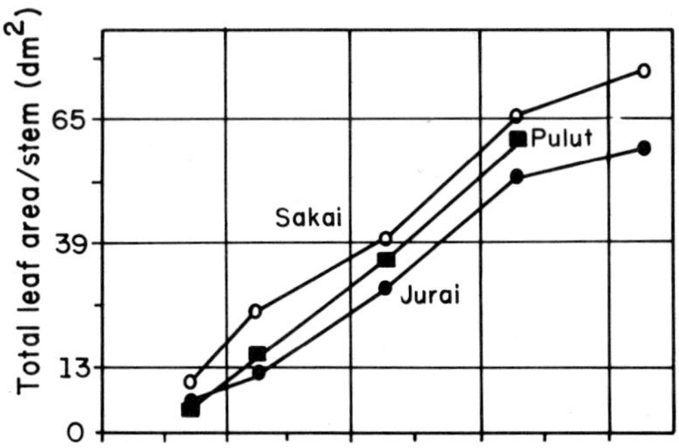

Fig. 12, Leaf area trends in shoots of three tapioca varieties. From Williams and Ghazali (1969).

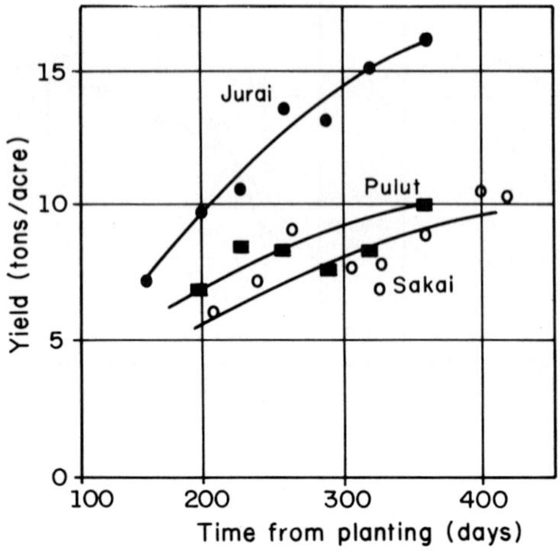

Fig. 13 Yields of three varieties of tapioca. From Williams and Ghazali (1969).

throughout ontogeny (Wilson, 1967) and harvest index (Donald, 1962), as the term implies, considers plant form and yield only at harvest.

Some model characters associated with assimilate production and distribution include (1) rapid, early leaf area development to LAI = 1 for effective light interception during early growth (Wilson, 1974); (2) development of leaf area to a maximum LAI of 3–4, and slow turnover of leaves after maximum LAI has been attained (Spence, 1970); (3) slow decline of leaf area during leaf senescence, since this period often coincides with maximum rates of tuber bulking (Haynes *et al.,* 1967); (4) early initiation and cellular development of tubers to create an active tuber sink for assimilate (Wilson, 1970; Wilson and Lowe, 1973; Lowe and Wilson, 1974, 1975a,b).

The examples of model characters considered are given only to illustrate the sort of plant characteristics that might be improved in order to systematically approach the yield asymptote of root crop species. Such characters must, however, be closely studied for each species, taking into account the environment for which the ideotype is being designed. Some indications of interrelationships between the characters mentioned in sweet potato and cassava cultivars are discussed by Wilson (1974) and Lowe and Wilson (1975a,b), and are shown in Figs. 10–13.

ACKNOWLEDGMENT

The assistance of Professor K. Caesar in providing information on crop performance and ecophysiology of *Solanum* potato is gratefully acknowledged.

REFERENCES

Akita, S., Yamamoto, F., Ono, M., Kushara, M., and Ikemoto, S. (1962). Studies on the small tuber sett method in sweet potato cultivation. *Bull. Chugoku Agric. Exp. Stn.* **8,** 75–128.

Austin, E. E., and Aung, L. H. (1973). Patterns of dry matter distribution during development of sweet potato (*Ipomoea batatas*). *J. Hortic. Sci.* **48,** 11–17.

Beevers, L., and Hageman, R. H. (1969). Nitrate reduction in higher plants. *Annu. Rev. Plant Physiol.* **20,** 495–522.

Bodlaender, K. B. A. (1958). *Jaarb. Inst. Biol. Scheik Onderzkd., Wageningen* **45,** 69–72.

Bodlaender, K. B. A. (1963). Influence of temperature, radiation and photoperiod on development and yield. *In* "Growth of the Potato" (J. D. Ivins and F. L. Milthorpe eds.), p. 199. Butterworth, London.

Bolhuis, G. G. (1966). Influence of length of illumination period on root formation in cassava (*Manihot utilissima* Pohl.). *Neth. J. Agric. Sci.* **14,** 251–254.

Bonhomme, R., and Chartier, P. (1972). The interpretation and automatic measurement of hemispherical photographs to obtain sunlit foliage area and gap frequency. *Isr. J. Agric. Res.* **22,** 53–61.

Borah, M. N., and Milthorpe, F. L. (1959). *Rep. Univ. Notts. Sch. Agric.*

Boyd, D. A., and Dyke, G. V. (1950). National Agricultural Advisory Service. *Q. Rev., Chem. Soc.* **3,** 47.

Bremner, P. M., and El Saeed, A. K. (1963). The significance of seed size and spacing. *In* "Growth of the Potato" (J. D. Ivins and F. L. Milthorpe, eds.), p. 267. Butterworth, London.

Chapman, T. (1965a). Some investigations into factors limiting yields of the White Lisbon yam (*Dioscorea alata* L.) under Trinidad conditions. *Trop. Agric.* Trinidad. **42,** 145–152.

Chapman, T. (1965b). Experiments with Irish potatoes (*Solanum tuberosum*) in Trinidad. *Trop. Agric. (Trinidad)* **42,** 189–194.

Chapman, T. (1967). Some experiments with potatoes (*Solanum tuberosum*) in Trinidad 1963–63. *Proc. Int. Symp. Trop. Root Crops, 1st, 1967* Vol. 1(iii), 171–179.

Coursey, D. G. (1967). "Yams." Longmans, Green, New York.

Coursey, D. G., and Haynes, P. H. (1970). Root crops and their potential as food in the tropics. *World Crops,* **22**(7), 1–5.

de la Pena, R. S., and Plucknett, D. L. (1967). The response of Taro (*Colocasia esculenta* [L.] Schott) to N, P and K fertilization under upland and lowland conditions in Hawaii. *Proc. Int. Symp. Trop. Root Crops, 1st, 1967* Vol. 1, pp. 70–85.

de Witt, C. T. (1967). Photosynthesis: Its relationship to overpopulation. *In* "Harvesting the Sun" (A. San Pietro, F. A. Greer, and T. J. Army, eds.). Academic Press, New York.

Doku, E. V. (1969). "Cassava in Ghana." Ghana University Press, Accra.

Donald, C. M. (1962). In search of yield. *Aust. Inst. Agric. Sci.* **28,** 171–178.

Donald, C. M. (1968). The breeding of crop ideotypes. *Euphytica* **17,** 385–403.

Enyi, B. A. C. (1972). Effect of shoot number and time of planting on growth, development and yield of cassava (*Manihot esculenta* Crantz). *J. Hortic. Sci.* **47,** 457–466.

Enyi, B. A. C. (1973). Growth, development and yield of some tropical root crops. *Proc. Int. Symp. Trop. Root Crops, 3rd* (in press).

Ferguson, T. U. (1973). The effect of sett characteristics and spacing on growth, development and yield of yams (*Dioscorea* spp.). Ph.D. Thesis, University of West Indies, St. Augustine, Trinidad (unpublished).

Ferguson, T. U., and Haynes, P. H. (1968). Annual Report. Faculty of Agriculture, University of West Indies, St. Augustine, Trinidad.
Ferguson, T. U., and Haynes, P. H. (1969). Half-yearly Report. Faculty of Agriculture, University of West Indies, St. Augustine, Trinidad.
Ferguson, T. U., and Haynes, P. H. (1970). The response of yams (*Dioscorea* spp.) to nitrogen, phosphorus, potassium and organic fertilizers. *Proc. Int. Symp. Root Tuber Crops, 2nd, 1970* Vol. 2, pp. 93–96.
Food and Agriculture Organization Production Yearbook, 1971.
Fujise, K. (1970), Sweet potato and its breeding efficacy in Japan. *Proc. Int. Symp. Trop. Root Crops, 2nd, 1970* Vol., 2, pp. 19–21.
Fujise, K., and Tsuno, Y. (1967). Effect of potassium on the dry matter production of sweet potato. *Proc. Int. Symp. Trop. Root Crops, 1st, 1967* Vol. 1, pp. 20–29.
Gaussen, A. (1955). Theories and classification of climates and microclimates. *Congr. Int. Bot., Rapp. Commun., 8th, 1954* Sect. 7 and 8, pp. 125–130.
Godfrey-Sam-Aggrey, W. (1974). Effects of cutting lengths of sweet potato yields in Sierra Leone. *Exp. Agric.* **10,** 33–38.
Gregory, L. E. (1954). Some factors controlling tuber formation in the potato plant. Ph.D. Thesis, University of California, Los Angeles.
Gumbs, F., and Ferguson, T. U. (1975). Root Crop Programme Report. Faculty of Agriculture, University of West Indies, St. Augustine, Trinidad.
Haynes, P. H., and Wholey, D. W. (1968). Annual Report. Faculty of Agriculture University of West Indies, St. Augustine, Trinidad.
Haynes, P. H., Spence, J. A., and Walter, C. J. (1967). The use of physiological studies in the agronomy of root crops. *Proc. Int. Symp. Trop. Root Crops, 1st, 1970* Vol. 1, pp. 1–17.
Hernandez, T. P., and Hernandez, T. (1967). Irrigation to increase sweet potato production. *Proc. Int. Symp. Trop Root Crops, 1st, 1967* Vol. 1, pp. 31–36.
Hozyo, Y. (1970). Growth and development of tuberous roots in sweet potato. *Proc. Int. Symp. Trop Root Tuber Crops, 2nd, 1970* Vol. 1, pp. 22–23.
Ito, K. (1947). A fundamental study on the formation of the sweet potato tuber. *J. Hortic. Assoc. Jpn.* Reports 1, 2, and 3.
Kay, D. E. (1973). "Crop and Product Digest 2, Root Crops." Tropical Products Institute, Foreign and Commonwealth Office, ODA, London.
Kim, Y. C. (1961). Effects of thermoperiodism on tuber formation in *Ipomoea batatas* under controlled conditions. *Plant Physiol.* **36,** 680–684.
Kumar, D., and Wareing, P. F. (1973). The physiology of tuberization in *Solanum adigena. Proc. Int. Symp. Trop. Root Crops, 3rd* (in press).
Lemon, E. R. (1963). Energy and water balance of plant communities. *In* "Environmental Control of Plant Growth" (T. L. Evans, ed.), pp. 55–78. Academic Press, New York.
Lemon, E. R. (1966). Energy conversion and water use efficiency. *In* "Plant Environment and Efficient Water Use" (W. H. Pierre *et al.,* eds.), pp. 28–48. Am. Soc. Agron., Madison, Wisconsin.
Loomis, R. S., and Williams, W. A. (1963). Maximum crop productivity: An estimate. *Crop Sci.* **3,** 67–72.
Lowe, S. B. (1971). Tuberisation in the sweet potato. M.Sc. Thesis, University of West Indies, St. Augustine, Trinidad (unpublished).
Lowe, S. B., and Wilson, L. A. (1974). Comparative analysis of tuber development in six sweet potato (*Ipomoea batatas* [L.] Lam) cultivars. I. Tuber initiation, tuber growth and partition of assimilate. *Ann. Bot. (London)* [N.S.] **38,** 307–317.
Lowe, S. B., and Wilson, L. A. (1975a). Yield and yield components of six sweet potato cultivars. I. Contribution of yield components to tuber yield. *Exp. Agric.* **11,** 39–48.

Lowe, S. B., and Wilson, L. A. (1975b). Yield and yield components of six sweet potato (*Ipomoea batatas*) cultivars. II. Variability of possible sources of variation. *Exp. Agric.* **11**, 49–58.

Madec, P. (1963). Tuber forming substances in the potato. *In* "Growth of the Potato" (J. D. Ivins and F. L. Milthorpe, eds.), p. 121. Butterworth, London.

Martin, F. W., and Ortiz, S. (1962). Origin and anatomy of tubers of *Dioscorea floribundta* and *D. speculiflora. Bot. Gaz. (Chicago)* **124**, 416–424.

McDavid, C. R. and Alanu, S. (1976). Promotion of flowering in Tannia (*Xanthosoma sagittifolium* by gibberellic acid. *Trop. Agric. Trinidad* **53**, 373–374.

Milthorpe, F. L. (1967). Some physiological principles determining yield of root crops. *Proc. Int. Symp. Trop. Root Crops, 1st, 1967* Vol. 1, pp. 1–19.

Murata, T., and Akazawa, T. (1968). Enzymic mechanism of starch synthesis in sweet potato roots. *Arch. Biochem. Biophys.* **126**, 893–897.

Njoku, A. (1963). The propagation of yams (*Dioscorea* spp.) by vine cuttings. *J. West Afr. Sci. Assoc.* **8**, 29–32.

Penman, H. L. (1963). Weather and water in the growth of potatoes. *In* "Growth of the Potato" (J. D. Ivins and F. L. Milthorpe, eds.), p. 191. Butterworth, London.

Plaisted, P. H. (1957). Growth of the potato tuber. *Plant Physiol.* **32**, 445–453.

Purseglove, J. W. (1968). "Tropical Crops–Dicotyledons," Vols. 1 and 2. Longmans, Green, London and Harlow.

Purseglove, J. W. (1972). "Tropical Crops–Monocotyledons." Vol. 1. Longmans, Green, London.

Radley, R. W. (1963). The effect of season on growth and development of the potato. *In* "Growth of the Potato" (J. D. Ivins and F. L. Milthorpe, eds.), p. 211. Butterworth, London.

Rankine, L. B., and Ferguson, T. U. (1974). "Agro-economic Evaluation of Yam Cultivation in the Commonwealth Caribbean," Root Crop Programme Report. Faculty of Agriculture, University of West Indies, St. Augustine, Trinidad.

Sadik, S. and Okerere, O. U. (1975). Flowering, pollen grain germination, fruiting, seed germination and seedling development of white yam *Dioscorea rotundata* Poir. *Ann. Bot.* **39**, 597–604.

Spence, J. A. (1970). Growth and development of tannia (*Xanthosoma* spp.). *Proc. Int. Symp. Trop. Root Tuber Crops, 2nd, 1970* Vol. 2, pp. 47–52.

Spence, J. A. (1971). Cultivation of detached sweet potato (*Ipomoea batatas* [L.] Lam) leaves with tuberous roots for photosynthetic studies. *Photosynthetica* **5**, 424–425.

Spence, J. A., and Boerboom, B. (1975). Root Crop Programme Report. Faculty of Agriculture, St. Augustine, Trinidad, West Indies.

Spence, J. A., and Humphries, E. C. (1972). Effect of moisture supply, root temperature and growth regulators on photosynthesis of isolated leaves of sweet potato (*Ipomoea batatas*). *Ann. Bot. (London)* [N.S.] **36**, 115–121.

Spence, J. A., and Williams, S. J. A. (1972). Use of photoperiod response to change plant design. *Crop Sci.* **12**, 121–122.

Togari, Y. (1950). A study of tuberous root formation in sweet potato. *Bull., Natl. Agric. Exp. Stn., Tokyo* **68**, 1–96.

Tsukamoto, Y., and Inaba, K. (1961). The effect of daylength upon the cormal formation in Taro (*Colocasia antiquorum*). *Mem. Res. Inst. Food Sci., Kyoto Univ.* **23**, 15–22.

Tsuno, Y. (1971). "Sweet Potato: Nutrient Physiology and Cultivation." Int. Potash Inst., Berne, Switzerland.

Tsuno, Y., and Fujise, K. (1965). Studies on dry matter production of sweet potato. *Bull. Natl. Inst. Agric. Sci., Ser. D* No. 13.

Tsunoda, S. (1965). "The Mineral Nutrition of the Rice Plant," Proc. Symp. IRRI, 1964. Johns Hopkins Press, Baltimore, Maryland.

van Royen, W., and Bengston, N. A. (1964). "Fundamentals of Economic Geography," 5th ed. Constable, London.

Walter, C. J. (1966). The effect of nitrogen on leaf area and tuber production in the sweet potato (*Ipomoea batatas* L. Lam). M.Sc. Thesis, Univeristy of West Indies Library, Trinidad (unpublished).

Walter, H., and Lieth, H. (1960). "Klima diagram Weltatlas Veb." Fischer, Jena.

Watson, D. J. (1958). The dependence of net assimilation rate on leaf area index. *Ann. Bot. (London)* [N.S] **22,** 37–45.

Watson, D. J. (1963). Some features of crop nutrition. *In* "Growth of the Potato" (J. D. Ivins and F. L. Milthorpe, eds.). Butterworth, London.

Wholey, D. W. (1968). Sweet potato (*Ipomoea batatas* L. Lam). A study of foliage characteristics and separation of types. D.T.A. Thesis, University of West Indies, St. Augustine, Trinidad.

Williams, D. B., and Cope, F. W. (1967). Notes on self-incompatibility in the genus *Ipomoea* L. *Proc. Int. Symp. Trop. Root Crops, 1st, 1967* Vol. 1, pp. 1–16.

Williams, C. N., and Ghazali, S. M. (1968). Growth and productivity of tapioca (*Manihot utilissima*). 1. Leaf characteristics and yield. *Exp. Agric.* **5,** 183–194.

Wilson, L. A. (1964). Growth of tomato and radish species in relation to zinc supply, light intensity and daylength. Ph.D. Thesis, University of Bristol, Library, Bristol, England.

Wilson, L. A. (1967). The use of rooted leaves and grafted plants for the study of carbohydrate metabolism in sweet potato (*Ipomoea batatas* L. Lam). *Proc. Int. Symp. Trop. Root Crops, 1st, 1967* Vol. 1, pp. 46–55.

Wilson, L. A. (1969). Alternative interpretations of a critical experiment in the physiological determinants of sweet potato yield. *Biol. J.* **3,** 29–38.

Wilson, L. A. (1970). The process of tuberisation in sweet potato (*Ipomoea batatas* [L.] Lam). *Proc. Int. Symp. Trop. Root Tuber Crops, 2nd, 1970* Vol. 2, pp. 24–26.

Wilson, L. A. (1974). Improvement and development of tropical root crops. *In* "Interaction of Agriculture with Food Science" (R. MacIntyre, ed.) p. 65. CIDR-033e.

Wilson, L. A., and Knox, D. (1973). Joint report on nitrogen utilization on sweet potato. University of Illinois/University of the West Indies Programme.

Wilson, L. A., and Lowe, S. B. (1973). The anatomy of the root system in West Indian sweet potato (*Ipomoea batatas* [L.] Lam) cultivars. *Ann. Bot. (London)* [N.S.] **37,** 633–643.

Wilson, L. A., Posthumus, A. C., Wareing, P. F., and Jolivet, E. (1973). The process of tuberisation in root crops. *Proc. Int. Symp. Trop. Root Crops, 3rd* (in press).

CHAPTER

8

Sweet Potato

S. K. HAHN

I. INTRODUCTION

Sweet potato (*Ipomoea batatas*) is indigenous to tropical America from which it was disseminated, first to tropical islands of the Pacific and northern New Zealand, and later to tropical Asia and Africa by Spanish and Portuguese explorers and/or traders after Columbus. Sweet potato is widely grown as one of the important staple food crops in many parts of the tropics and subtropics. In addition to its importance as human food, sweet potato provides animal feed and raw material for industrial purposes. It is now grown in areas reaching 40°N and 40°S latitudes and as high as 2000 m above sea level. Statistics on world production of sweet potato are obscure, but a rough approximation of about 15 million hectares and 130 million tons, respectively, can be made from the FAO 1973 Production Yearbook.

Sweet potato is grown over a wide range of environmental conditions, including land of low fertility and relatively low pH where it can still produce a considerable yield. Sweet potato has good drought tolerance. It requires relatively little attention and labor and its production costs are low compared with other crops. It has the highest solar energy fixing efficiency among the food crops, primarily because of its tremendous capacity to produce dry matter for a long period of time. A national average of about 20 tons fresh yield per hectare in 4 to 5 months has been recorded in several countries. This is equivalent to about 6 tons dry yield. As such it has high production potential and provides food at a time when the staple diet is in short supply. In some highly populated areas, traditional staple cereals have been replaced by the high calorie producing sweet potato.

Although sweet potato has been under cultivation for a long time and is economically important in the tropics and subtropics, little attention has been given to its improvement. There is, however, tremendous potential for sweet potato improvement as human and animal food as well as for industrial raw materials in the tropics and subtropics.

II. EFFECT OF ENVIRONMENT ON DRY MATTER PRODUCTION

Since the tuberous root comprises a high proportion of total dry weight, the increase in total dry matter has a direct connection with tuberous root yield. Productivity of the tuberous root is, therefore, primarily a function of dry matter accumulation.

Dry matter production is determined by many environmental factors, with mineral nutrients, radiation, temperature, and moisture most important. The relationships among the factors in relation to source and sink are diagrammatically shown in Fig. 1. Tsuno and Fujise (1965a) reported significant positive correlations of photosynthetic activity with potassium, nitrogen, and phosphorus content in the leaf blade (Fig. 2). Photosynthetic activity showed the highest positive correlation with potassium and the lowest with phosphorus, but a negative correlation with carbohydrate content in the leaf blade, which in turn showed the highest negative correlation with potassium and the lowest with phosphorus. There were also high positive correlations between nitrogen, phosphorus, and potassium. These correlations will be discussed relative to dry matter production in terms of nutrient absorption, photosynthetic activity, and partitioning of photosynthate.

Adventitious roots of sweet potato, formed at the node, were classified into four kinds including young root, fibrous root, hard root, and tuberous root (Togari and Fujise, 1962). The young root becomes either tuberous, hard, or fibrous depending on the activity of primary cambium and degree of lignification of stele cells which are in turn dependent on environmental factors even at

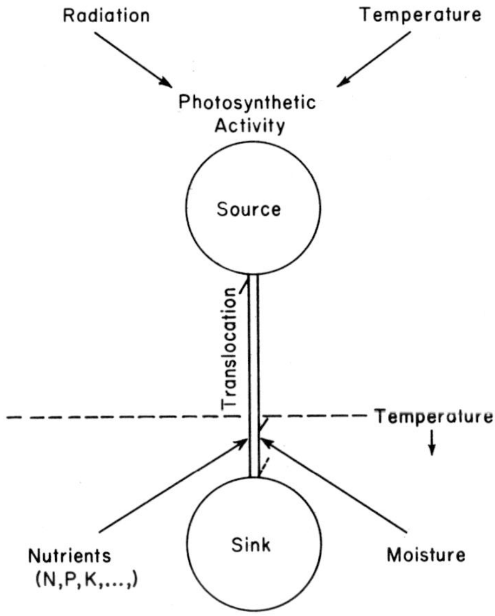

Fig. 1 Relationships between the most important environmental factors affecting dry matter production.

the early growth stages. The relationship is diagrammatically shown in Fig. 3 from which it can be seen that the young root develops into tuberous root when the primary cambium activity is great and the degree of lignification of stele cells is small, particularly under conditions of low temperature and high potassium supply.

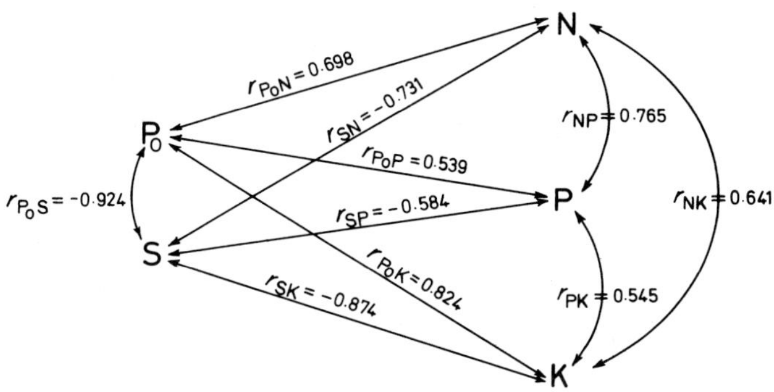

Fig. 2 Diagrammatic summary of Tsuno and Fujise's correlations of three major nutrients (N, P, K), photosynthetic rate (P_0), and starch content in the leaf blade (S).

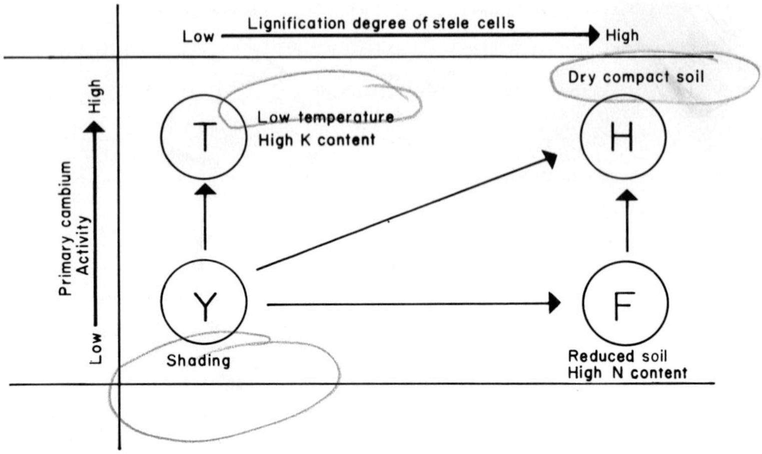

Fig. 3 Development of respective roots at early stage of growth in relation to cambium activity and lignification degree of stele cells. Y stands for young root, T for tuberous root, H for hard root, and F for fibrous root.

A. Mineral Nutrients

Nitrogen increases dry matter production by increasing leaf expansion of leaf area index and increases the distribution ratio of above-ground parts to underground parts (Tsuno and Fujise, 1964b). There is a linear relationship between nitrogen content and leaf area index. However, it is well established that the linear increase with leaf area index reduces the net assimilation rate in the natural population because of increased mutual shading of lower leaves (Fujise and Tsuno, 1962). Similarly, the leaf area duration increased with increased nitrogen absorption but net assimilation rate decreased.

At fixed levels of potassium, distribution of dry matter to roots was affected by nitrogen content in the leaf blade (Tsuno and Fujise, 1964b). When nitrogen content was high in the leaf blade, most of the photosynthates were utilized in top growth.

It is well known that among mineral nutrients, potassium affects tuberous root yield most. Potassium affects dry matter production by increasing net photosynthetic activity of a given leaf area (Fujise and Tsuno, 1962). The increased potassium caused no change in the leaf area index but increased net assimilation rate, which finally increased total dry matter production (Tsuno and Fujise, 1964a).

Increased potassium prolonged leaf area duration (Tsuno and Fujise, 1964b). When potassium was increased, excessive top growth was greatly suppressed, even under high nitrogen conditions, and higher tuberous root yield was accordingly produced.

When potassium was increased, photosynthetic activity was promoted by decreasing photosynthate in the blade (Fujise and Tsuno, 1962). Increased potassium accelerated the translocation of photosynthates and was effective in increasing the size of sink or tuberous root (Tsuno and Fujise, 1965a). Therefore potassium accelerated translocation of photosynthates into the tuberous root by increasing sink capacity, which resulted in low carbohydrate content in the blade and finally increased photosynthetic activity.

The relationship between photosynthetic rate and nitrogen percent varied with potassium level in the leaves (Fujise and Tsuno, 1962). When potassium levels were high (above 4%) in the leaves, the rate of photosynthesis was very high even at low nitrogen levels (2.2%), while the rate was low when potassium content was low (below 4%) even under high nitrogen levels (about 3%). Dry matter distribution ratio of above-ground parts over underground parts increased as nitrogent percent in the blade was increased at a fixed level of potassium. However, the ratio continuously decreased as potassium was used less for formation of above-ground parts but used primarily by underground parts, mainly tuberous roots. For high tuberous root yield it is therefore important to keep the ratio of potassium to nitrogen high (Tsuno and Fujise, 1965b).

Path coefficient analysis was applied using the correlation coefficients of Fig. 2 to evaluate direct and indirect effects of nitrogen and potassium on photosynthetic activity and starch content in the leaf blade (Fig. 4). Potassium showed higher positive direct effects on photosynthetic activity (0.640) and higher negative direct effect on starch content in the leaf blade (−0.688) than those for

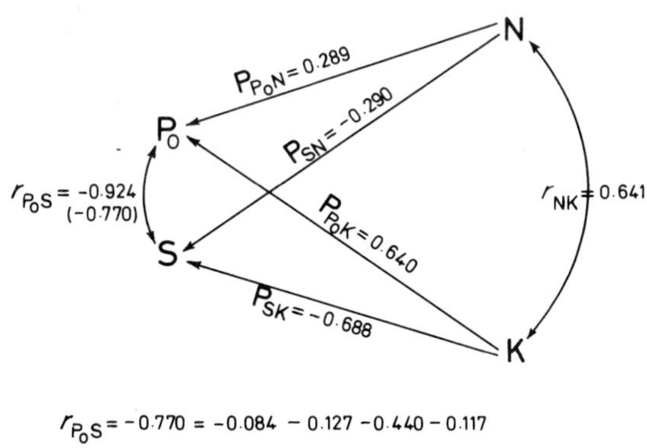

Fig. 4 Direct and indirect effects of nitrogen (N) and potassium (K) on photosynthetic rate (P_0) and starch content.

nitrogen (0.289 and −0.290, respectively). The indirect effect of potassium on photosynthetic activity by reducing starch content also was very high (−0.440).

The key factors for success of sweet potato production are (1) to maintain optimum nitrogen levels to limit excess growth of above-ground parts thus increasing light interception, and (2) to provide sufficient potassium to increase sink capacity and photosynthetic rate (Tsuno and Fujise, 1968).

B. Radiation, Temperature, and Moisture

Sweet potato grows well under conditions of high temperature, high light intensity, and considerable rainfall, providing there is good drainage during the growing period, but moderately dry weather is favorable to formation and development of tuberous roots.

Relatively low temperature combined with low light intensity and short photoperiod promotes development of tuberous roots to a greater extent than that of the vines. Photoperiod appears to be most important (Edmond and Ammerman, 1971).

The photosynthetic process depends primarily on external conditions, especially radiation and temperature (Spence and Humphries, 1972). At low temperatures, respiration rate decreases, thus reducing the energy needed for nutrient uptake and distribution (Tsuno and Fujise, 1964b). Dry matter production increased with increasing soil temperature from 20° to 30°C (Tsuno and Fujise, 1964a).

When rooted leaves were held at temperatures ranging from 10° to 35°C, tuberous root formation occurred maximally at 25°C. At 15° and 35°C few or no roots swelled to form tuberous roots and more fibrous roots developed, while at 10°C there were no tuberous roots and few fibrous roots (Spence and Humphries, 1972). However, this experiment was conducted at constant temperatures and when naturally fluctuating temperatures are considered, especially night temperature, the situation might be different.

Sweet potato maintains a more satisfactory agronomic balance between the growth of vine and yield of tuberous roots in well-drained sandy or silt loams (Johnson and Ware, 1948). Less tuberous root and total dry matter per unit area of lamina were produced in a wet regime (Spence and Humphries, 1972). There were no significant differences in yield from irrigating at moisture levels higher than 20% at any stage of growth (Jones, 1961).

Heavy clay results in growth of stems and leaves but not of tuberous roots which are poorly shaped. In dry compact soil, both cambial activity and the degree of lignification were high, giving young roots of a hard fibrous nature. Oxygen deficiency in the soil at an early growth stage increases the degree of lignification of stele cells and suppresses the activity of primary cambium, resulting in development of young roots into fibrous roots. Lack of oxygen and excess moisture in the soil have an adverse effect on fleshy root formation and enlargement.

The optimum pH of soil for sweet potato is 6.1–7.7 but it performs well even in soil with a relatively lower pH (4.2) (Kotama, 1962).

Under shaded conditions, the degree of lignification of stele cells is not high but cambial activity is lower, which delays differentiation and development of tuberous roots. The rate of photosynthesis is highest between 9 AM and 1 PM, being about 12 mg $CO_2/100\ cm^2$/hour, decreasing gradually to about 2 mg $CO_2/100\ cm^2$/hour at 5 PM (Fujise and Tsuno, 1962). Reduction of natural light intensity by 30–50% did not significantly affect dry matter production (Tsuno and Fujise, 1964a), suggesting source activity was not limiting. Light saturation of photosynthesis occurred at around 30 klux with no change in photosynthetic rate at higher intensities (Fujise and Tsuno, 1962).

Even though sweet potato has a high photosynthetic activity per unit of leaf area, mutual shading occurs because of a leaf arrangement which causes poor light penetration. Therefore, the net assimilation rate decreases with an increase of leaf area index in natural populations (Tsuno and Fujise, 1963). Consequently it is difficult to increase dry matter production, especially in terms of tuberous root yield, through increasing leaf area in the natural population. Therefore, the effectiveness of leaf area in terms of net assimilation rate will be mainly important for dry matter production. Since photosynthetic activity per unit of leaf area in the main contributes to the net assimilation rate, the most important consideration is to maintain net photosynthetic activity as high as possible for a long period of time (Fujise and Tsuno, 1962).

Therefore, the favorable external conditions for high yield are those which promote nutrient absorption, improve light interception in natural populations, increase net photosynthetic activity, and improve formation and enlargement of tuberous roots.

III. SOURCE–SINK RELATIONS

Tuberous root yield is a function of both sink capacity and source potential, and yield is reduced when either is limiting. Questions then arise as to which has more effect on yield and how the source and sink are related in terms of tuberous root yield.

In crop populations which have been improved for many years in a given environment, sink may be less limiting than source. In this instance, source activity may have a greater effect on yield. However, with sweet potato in the tropics, which has not been subjected to much improvement, the sink may be more limiting on tuberous yield than source activity.

There are some reports on source–sink relationships in sweet potato. Spence and Humphries (1972) reported that there was more dry matter per unit area in lamina and petiole when tuberous roots did not form. Photosynthetic activity was drastically reduced when tuberous root enlargement was restrained by exposing tuberous roots to light (Tsuno and Fujise, 1965a). They also reported that low tuber sink capacity was a limiting factor for photosynthetic activity. From the reciprocal graft between cultivated sweet potato *Ipomoea batatas* and its related species, *Ipomoea trifida,* Hozyo and Park (1971) reported that photosynthetic activity was higher when cultivated sweet potato with larger sink was used as a stock. Increased tuberous root growth was related to increased net assimilation rate (Spence and Humphries, 1972). Based on these results Hozyo and Park (1971) and Spence and Humphries (1972) support the view that the rate of photosynthesis depends on the demand of sink for photosynthates.

Photosynthetic activity was low when carbohydrate concentration was high in the leaf blade. Thus photosynthates in the blade must be translocated into a different sink such as the tuberous roots in order to maintain high photosynthetic activity (Tsuno and Fujise, 1965a). By feeding $^{14}CO_2$ into the fifth leaf of reciprocal grafts of *I. batatas* and *I. trifida,* Kato and Hozyo (1974) confirmed that photosynthates were translocated mainly into stock as sink capacity increased. Therefore, it can be concluded that sink capacity affects translocation of photosynthates, which reduces the carbohydrate content in the blade and in turn increases photosynthetic activity.

During 1974, a 4 × 4 diallel graft was made by the author using sweet potato varieties with different source potentials and sink capacities to investigate their relationships with special reference to tuberous root yield (Hahn, 1977). The variety with large sink capacity showed a high response of sink to good source. On the other hand, the variety with poor sink capacity showed a low response of sink to source even though a good source was provided. This suggests that sink is the primary factor in higher yield and source is secondary. Genotypes with large sink capacity should, therefore, be selected primarily in breeding procedures with subsequent improvement by incorporating a large source potential. Sink capacity was more important than source potential in determining tuberous root yield but only within certain limits. The combination of the variety with the highest sink capacity and the other variety with the highest source potential gave the highest value of dry tuberous root yield by harvest index.

IV. PHOTOSYNTHETIC EFFICIENCY

Sadik (1973) screened some 10,000 sweet potato genotypes in the "Moss chamber" for low CO_2 compensation concentration to test photosynthetic efficiency and obtained some plants which remained green and either had low

CO_2 compensation points or some other mechanism of survival at low CO_2. These were designated as having high photosynthetic efficiency. It was later reported (International Institute of Tropical Agriculture, 1973) that there were no consistent differences in the biomass and components of biomass yields between the high and low groups of photosynthetic efficiency.

During 1973, a 4 X 4 diallel cross of two clones each for high and low photosynthetic efficiency was made in order to study the genetic and physiological mechanisms of photosynthetic efficiency (International Institute of Tropical Agriculture, 1974). The cross, high to high photosynthetically efficient clones, did not produce average progenies of high photosynthetic efficiency. It appeared that photosynthetic efficiency was mainly due to nonadditive effects and may also have been influenced by other factors with which it might be inversely associated. When average harvest indexes and yield of the six families were plotted on a fresh weight basis against photosynthetic efficiency, it was observed that photosynthetic efficiency to a certain point might not be a limiting factor in high yield but that distribution of assimilates might be (Fig. 5).

It was pointed out earlier that net assimilation is most important in increasing dry matter production in natural populations. It was also pointed out that source potential must be improved to increase tuberous root yield when the sink is not limiting. Therefore, importance of sweet potato improvement for high photosynthetic efficiency is stressed for a further increase in yield.

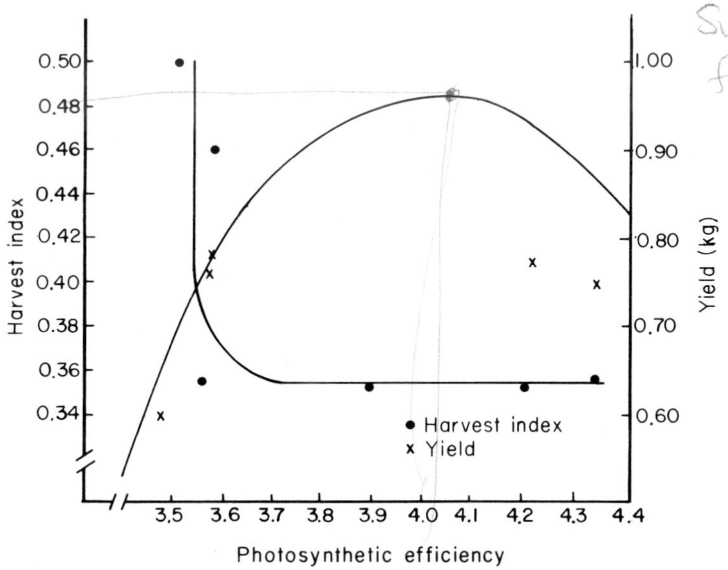

Fig. 5 Harvest index for fresh yield in relation to photosynthetic efficiency.

V. DROUGHT TOLERANCE

Sweet potato is resistant to drought (Kotama, 1962), but little has been reported on the mechanisms involved. The drought scores of a 4 × 4 diallel graft in the field during the dry season are summarized in Table I. The low drought tolerance of variety TIb 4 was due to the stock (root) and not to the scion (top) which in fact had the most drought-resistant top. The low tolerance of variety TIb 10 was associated with both root and top. The high tolerance of TIb 5 was due to the root and not to the top, while that of TIb 1499 was due to both root and top. When a drought-resistant scion was grafted to a drought-resistant stock, the combination produced a synergistic level of greater drought resistance in the combination of TIb 1499 as stock and TIb 4 as scion than that of either of the components. The synergistic mechanism of drought tolerance and higher production under drought stress in the graft combination was due to higher leaf water content, smaller size of stomata, better ability to reduce bottom leaves of TIb 4 under stress, and more and larger fibrous roots of TIS 1499. Because of its resistance to drought, sweet potato is widely grown and continuously expanding in drought-affected areas. Thus, to meet with this trend sweet potato needs to be further improved relative to drought resistance and the mechanisms should be further studied.

VI. GENOTYPE × ENVIRONMENT INTERACTION

Sweet potato is grown in a wide range of soil and climatic environments and cultural practices. This is primarily because sweet potato has the ability to adapt well to environment, and many different cultivars adapted to a given environ-

TABLE I
Root–Top Relation in Drought Tolerance in a 4 × 4 Diallel Graft[a]

Stock	Scion 1	Scion 2	Scion 3	Scion 4	Stock average
1 TIb 4	*1.80*	2.13	2.67	1.56	2.04
2 TIb 5	0.29	*0.45*	1.00	0.81	0.64
3 TIb 10	0.83	1.17	*1.50*	1.12	1.16
4 TIS 1499	0.25	0.81	0.83	*0.81*	0.68
Scion average	0.79	1.14	1.50	1.08	1.13

[a]The italic numbers indicate self-grafts. Lower figures indicate resistance to drought. Scoring was made on December 16.

ment have emerged through natural and artificial hydridization of mutation and natural and artificial selection for many years.

Environment, including different soils, climates, cultural practices, diseases and insects, varies with years or seasons and locations. The differential response of various genotypes of cultivars to specific environments is called genotype X environment interaction. Large interaction effects indicate less adaptation stability of genotypes to the range of environment. For stable and dependable sweet potato production over large areas every year and every season, cultivars with the least interaction effect are most essential.

From preliminary yield trials conducted in both wet and dry seasons, a high genotype X season interaction was observed for fresh tuberous root yield. The differences between seasons in temperature, soil moisture, radiation, etc., might result in a differential response of different genotypes. Other preliminary yield trials were conducted in a sandy, high rainfall area (2600 mm) and/or clay soil receiving relatively lower rainfall area (1800 mm). A high interaction effect between genotypes and locations was also noted in the trials. This suggests that high yielding varieties may not produce high tuberous root yield in different seasons and locations. Thus screening of germ plasm for genotypes with wide adaptation assumes high priority.

ACKNOWLEDGMENTS

The author is grateful to Dr. G. A. Schaefers, Dr. K. J. Treharne, and Dr. S. Sadik for reviewing the manuscript.

REFERENCES

Edmond, J. B., and Ammerman, G. R. (1971). "Sweet Potatoes: Production, Processing and Marketing." Avi Publ., Westport, Connecticut.

Fujise, K., and Tsuno, Y. (1962). Study on the dry matter production of sweet potato. I. Photosynthesis in the sweet potato with special reference to measuring of intact leaves under natural conditions. *Proc. Crop Sci. Soc. Jpn.* **13,** 145–149.

Hahn, S. K. (1977). A quantitative approach to source potentials and sink capacities among reciprocal grafts of sweet potato varieties. *Crop Sci.* **17**(3), in press.

Hozyo, Y., and Park, C. Y. (1971). Plant production in grafting plants between wild type and improved variety in *Ipomoea. Bull. Natl. Inst. Agric. Sci., Ser. D* No. 12.

International Institute of Tropical Agriculture. (1973). Annual Report of Root and Tuber Improvement Program for 1973, Ibadan, Nigeria.

International Institute of Tropical Agriculture. (1974). Annual Report of Root and Tuber Improvement Program for 1974, Ibadan, Nigeria.

Johnson, W. A., and Ware, L. M. (1948). Effects of rates of nitrogen on the relative yields of sweet potato vines and roots. *Proc. Am. Soc. Hortic. Sci.* **52,** 313–316.

Jones, S. T. (1961). Effects of irrigation at different levels of soil moisture on yield and evapotranspiration rate of sweet potatoes. *Proc. Am. Soc. Hortic. Sci.* 77, 458–462.

Kato, S., and Hozyo, Y. (1974). Translocation of ^{14}C-photosynthates in several growth stage of the grafts between improved varieties and wild type plants in *Ipomoea. Bull. Natl. Inst. Agric. Sci., Ser. D* No. 25.

Kotama, T. (1962). Culture of sweet potato. "Encyclopedia of Crops," Root Crop Sec. Yokendo Inc., Tokyo.

Sadik, S. (1973). Screening sweet potato for low CO_2 compensation point. (In preparation.)

Spence, J. A., and Humphries, E. C. (1972). Effect of moisture supply, root temperature, and growth regulators on photosynthesis of isolated root leaves in sweet potato (*Ipomoea batatas*). *Ann. Bot. (London)* [N.S.] **36**, 115–121.

Togari, Y., and Fujise, K. (1962). Growth of sweet potato. "Encyclopedia of Crops," Root Crop Sect. Yokendo Inc., Tokyo.

Tsuno, Y., and Fujise, K. (1963). Studies on the dry matter production of sweet potato. II. Aspect of dry matter production in the field. *Proc. Crop Sci. Soc. Jpn.* **231**, 285–288.

Tsuno, Y., and Fujise, K. (1964a). Studies on the dry matter production of sweet potato. III. The relation between the dry matter production and the absorption of mineral nutrients. *Proc. Crop Sci. Soc. Jpn.* **32**, 297–300.

Tsuno, Y., and Fujise, K. (1964b). Studies on the dry matter production of sweet potato. IV. The relation between the concentration of mineral nutrients in plant distribution ratio over dry matter produced. *Proc. Crop Sci. Soc. Jpn.* **32**, 301–305.

Tsuno, Y., and Fujise, K. (1965a). Studies on the dry matter production of sweet potato. VIII. The internal factors influencing photosynthetic activity of sweet potato leaf. *Proc. Crop Sci. Soc. Jpn.* **33**, 230–235.

Tsuno, Y., and Fujise, K. (1965b). Studies on the dry matter production of sweet potato. IX. The effect of potassium on the dry matter production of sweet potato. *Proc. Crop Sci. Soc. Jpn.* **33**, 236–241.

Tsuno, Y., and Fujise, K. (1968). Studies on the dry matter production of sweet potato. X. An explanation of the determination of dry matter percentage in tuber with the view point of nutrient conditions and dry matter production. *Proc. Crop Sci. Soc. Jpn.* **37**, 12–16.

CHAPTER

9

Coffee

MOACYR MAESTRI and RAIMUNDO SANTOS BARROS

I. INTRODUCTION

Various aspects of ecophysiology of the coffee plant (*Coffea arabica* L.) have been periodically reviewed (Sylvain, 1954; Alvim, 1958, 1973; Franco, 1965; Hernández Medina, 1965; Wormer, 1965a; Ramaiah and Gopal, 1969; Huxley, 1970; Barros and Maestri, 1972). As emphasized by Sylvain (1954) and Huxley (1970), information on physiology of the coffee tree is needed to establish the best management practices and to provide guidance in breeding programs.

This chapter discusses some aspects of photosynthesis, productivity, vegetative and reproductive growth, and their regulation by climatic factors. It also deals with the influence of these factors on distribution of Arabica coffee.

II. CLIMATIC FACTORS AFFECTING DISTRIBUTION OF COFFEE

Many crop plants have been very widely dispersed by man, often because high profits more than justified low yields in less-than-optimal environments (Loomis *et al.,* 1971). Crop plants generally show wide morphological pasticity in most diverse environments. These considerations make it difficult to define the specific ecological requirements of these plants. Some investigators believe, however, that ecological requirements are best studied in the area of origin of a species. Others believe it sometimes is more useful to study plant growth in areas to which species have become dispersed, especially when plants grow better there than in regions of their origin.

There appears to be no doubt that Arabica coffee originated in Ethiopian tropical forests at altitudes of 1600 to 2800 m at 6° to 9°N and 34° to 40°E. In these regions the mean temperatures are about 20°C, and rainfall is well distributed, varying from 1600 to more than 2000 mm, and there is a dry season of 3 to 4 months. In these areas the coffee tree occurs as part of a four-layer forest complex. The upper layer is approximately 30–40 m high and the dominant trees are species of *Albizzia, Celtis, Chrysophyllum, Clausenopsis,* and *Cordia.* The second layer, at 20 m, consists mainly of *Bersama, Bridelia, Croton, Ehretia, Ekbergia, Ficus, Morus, Polyscias, Pygeum,* and *Syzygium.* The main genera of the two shrubby layers are *Carissa, Coffea, Gymnosporia,* and *Sideroxylon* (Sylvain, 1955a). Hence, under original conditions, coffee grows permanently under shade, is not exposed to excessively high temperatures, and undergoes a relative dry period during the year.

Commercial coffee plantations are largely distributed from Cuba, 22°N (Gindel, 1962) to 26°S in Paraná, Brazil, although some plantations occur elsewhere (Nosti Nava, 1953). Beyond those limits, temperatures below 12°C for long periods inhibit growth and development of coffee (Alègre, 1959).

Several factors influence growth and yield of coffee. High temperatures inhibit growth, because above 24°C net photosynthesis begins to decrease and becomes negligible at 34°C (Nunes *et al.,* 1968). Perhaps this is one of the reasons coffee is not recommended for equatorial lowlands. Nevertheless, successful coffee plantations occur in or close to equatorial regions, as in Kenya, Tanzania, and Colombia. In these areas, altitudes above 1000 m ensure success of the crop (Alègre, 1959; Haarer, 1962, 1963). Coffee does not tolerate a wide range of temperatures. Mean temperatures below 16° and above 23°C are not suitable, and the optimum varies from 18° to 21°C (Alègre, 1959).

Coffee grows under a wide range of rainfall conditions. In some parts of Kenya annual rainfall is about 800 mm (Alègre, 1959), whereas in Costa Rica and in Mysore, India, it exceeds 2500 mm a year. Nevertheless, Alègre (1959) suggests that the optimum annual rainfall is 1200 to 1800 mm. Good seasonal distribution of rainfall and short dry periods appear to favor growth of coffee. The dry period seems to be important for root growth, maturation of branches formed during the preceding rainy season, flower initiation, and ripening of fruits (Haarer, 1962, 1963). Nevertheless, Wellman (1961) stated that in regions without a definite dry season, as in Costa Rica, coffee grows as well as elsewhere. However, in Costa Rica distribution of rainfall is not uniform and temperature effects during the months of lowest rainfall should be taken into account (McFarlane, 1949). On the other hand, intensive rainfall throughout the year often is responsible for scattered harvests and low yields. Lack of a dry period can also limit coffee cultivation in low tropical rainy regions.

Figure 1 shows the diverse climatic conditions of four coffee regions. The lower horizontal dotted line represents the minimum limit of mean temperatures recommended by Alègre (1959) for the best development of coffee; the vertical dotted line represents the minimum rainfall for the driest month. The upper horizontal line represents the temperature above which net photosynthesis begins to decline (Nunes *et al.*, 1968). In Ruiru (01°08′S), development of the coffee tree is controlled mainly by distribution of rainfall. In Turrialba (10°N), temperature and photoperiod appear to be most important, and in El Salvador (13°04′N), distribution of rainfall and photoperiod are controlling factors. In

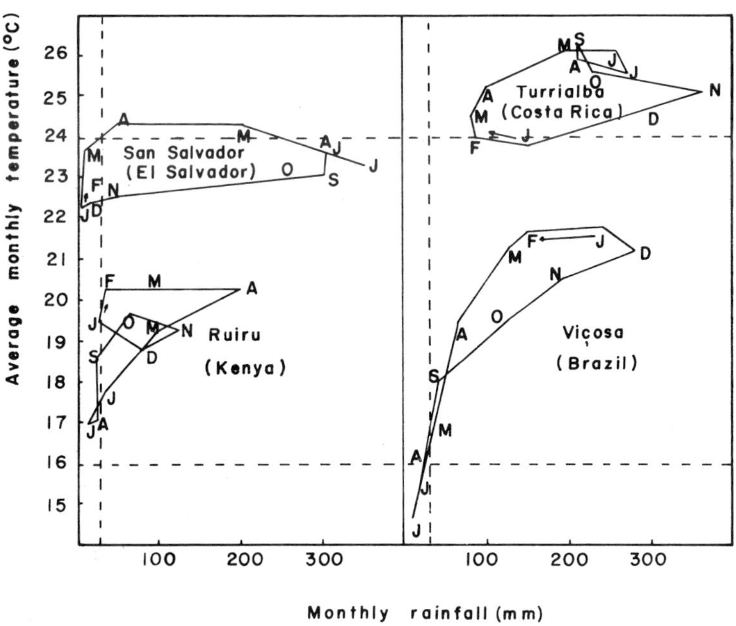

Fig. 1 Climatograms of four coffee regions in the world (Barros and Maestri, 1972).

Viçosa (20°45′S), temperature, rainfall, and photoperiod may be important factors. In El Salvador and Turrialba, mean temperatures are much higher than in Viçosa and Ruiru (Fig. 1). Curiously, in the first two localities coffee is raised under shade, and in the last two, in full sun. At temperatures above 24°C, the rate of net photosynthesis decreases. Hence, Barros and Maestri (1972) suggested that coffee should be shaded in regions of high average temperatures. Franco (1951, 1952) emphasized that overhead trees compete with coffee for available soil water.

III. PHOTOSYNTHESIS AND PRODUCTIVITY

Investigations on photosynthesis of coffee were summarized by Sylvain (1955b) and Cannell (1971c). The pioneer investigations of Nutman (1937a) stimulated much discussion and, as pointed out by Sylvain (1955b), often were incorrectly interpreted. Nutman (1937a) reported that the rate of photosynthesis of coffee in Tanzania varied directly with light intensity when it was low, but at high light intensity, the rate decreased. Nutman's conclusion that the assimilation rate was greater in moderate than in high light intensity, and that total daily assimilation was greater in the shade than in the sun, led to the erroneous interpretations cited by Sylvain (1955b). However, Nutman (1937c) was very cautious in extrapolating his results, obtained with individual leaves, to whole trees.

The maximum rate of net photosynthesis as reported by Nutman (1937a) was 4.5 mg CO_2 dm^{-2} h^{-1}. This ranks coffee among the least efficient plants (Zelitch, 1971). Perhaps the most interesting feature of Nutman's (1937a) data was a midday decrease in photosynthesis that could not be explained by leaf water balance or accumulation of assimilates. Nutman (1937b) concluded that the midday decrease in photosynthesis could be explained by stomatal closure induced by direct action of sunlight on stomata and not by leaf water relations. He concluded that stomatal aperature was directly correlated with radiation intensity up to 0.7 cal cm^{-2} min^{-1} and inversely correlated at intensities above 0.9 cal cm^{-2} min^{-1}. Alvim (1968) observed that stomatal opening increased gradually with light intensity up to about 20,000 lux and showed a sharp decrease at 60,000 and 90,000 lux.

The separate effects of light and temperature were not evaluated by Nutman (1937a). As pointed out by Cannell (1971c), the leaves of Nutman's plants must have reached at least 30°C during sunny intervals, when net photosynthesis was very low. This did not happen in the study of Alvim (1968). However, since leaf temperature was not measured by Alvim, the possibility of a heating effect should not be ruled out.

Nunes *et al.* (1968) found a maximum rate of photosynthesis about three times higher than the highest rate reported by Nutman (1937a). Nunes *et al.*, however, showed that maximum photosynthesis could be attained when internal CO_2 concentration was equal to zero (Figs. 2 and 3). This occurs only at temperatures below 24°C (Fig. 3), with light intensity higher than 0.15 cal cm^{-2} min^{-1}, which normally occurs in the field. Above 24°C, internal CO_2 concentration increases linearly at a rate of 20 ppm per °C. Maximum photosynthetic rates have lower values at temperatures above 24°C because of higher internal CO_2 concentrations and higher total diffusive resistance. This last point is discussed later. Nunes *et al.* (1968) suggested that the lower rates reported by Nutman (1937a) and Tió (1962) (Fig. 4) could be attributed to high leaf temperatures.

Ever since Nutman (1937a,b) emphasized the role of stomata in photosynthesis of coffee at high light intensities, attention was given to stomatal responses under different environmental conditions. Franco (1938), in São Paulo,

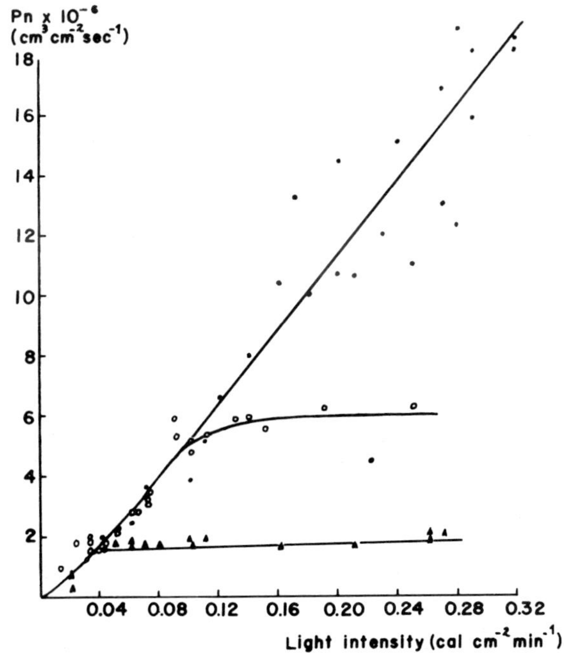

Fig. 2 The effect of light intensity under natural conditions on net photosynthesis of *Coffea arabica* at an external CO_2 concentration of 300 ppm and internal CO_2 concentration of 0 (filled circles), 70 (open-circles), and 180 (triangles) ppm (Nunes *et al.*, 1968).

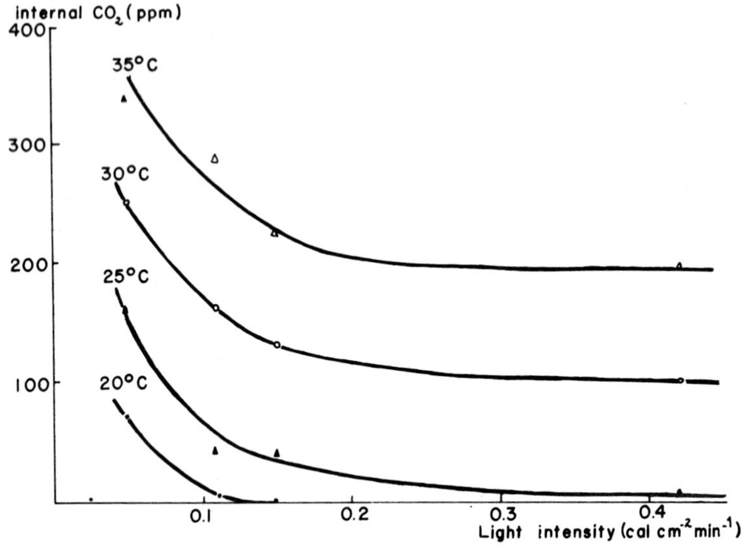

Fig. 3 The effect of artificial light and temperature on the internal CO_2 concentration of *Coffea arabica* (Nunes *et al.*, 1968).

Brazil, made virtually the same observations as Nutman (1937b), namely that in cloudy weather or shade, the stomata remained open throughout the day, whereas on clear days they closed near midday and remained closed for the rest of the day. Alvim and Havis (1954), in Turrialba, Costa Rica, obtained data similar to those of Nutman and Franco, but they attributed stomatal closure during periods of high light intensity to leaf dehydration. Similar conclusions were reached by Maestri and Vieira (1958) in Brazil. They found that, during the dry season, stomata were almost completely closed during the day in both shaded and unshaded coffee trees. During the rainy season, stomata were fully open on cloudy days. On sunny days they began closing early in the morning and were almost completely closed by late afternoon. Shade had a minor effect on diurnal changes in stomatal aperture, suggesting that water stress might be the controlling factor. During the dry season, shade trees may compete with coffee for soil moisture (Franco, 1952), and this may be one cause of failure of coffee cultivation under trees in Brazil and Kenya. Nutman (1937b) had previously shown that drought induced early stomatal closure, but his observations were limited.

Thoday (1938) suggested that insolation of coffee leaves might affect stomatal aperture by lowering the epidermal water content even though the leaf remained turgid. This was considered unlikely by Nutman (1938). Since loss of turgidity, as Nutman (1937b) contended and Bierhuizen *et al.* (1969) demon-

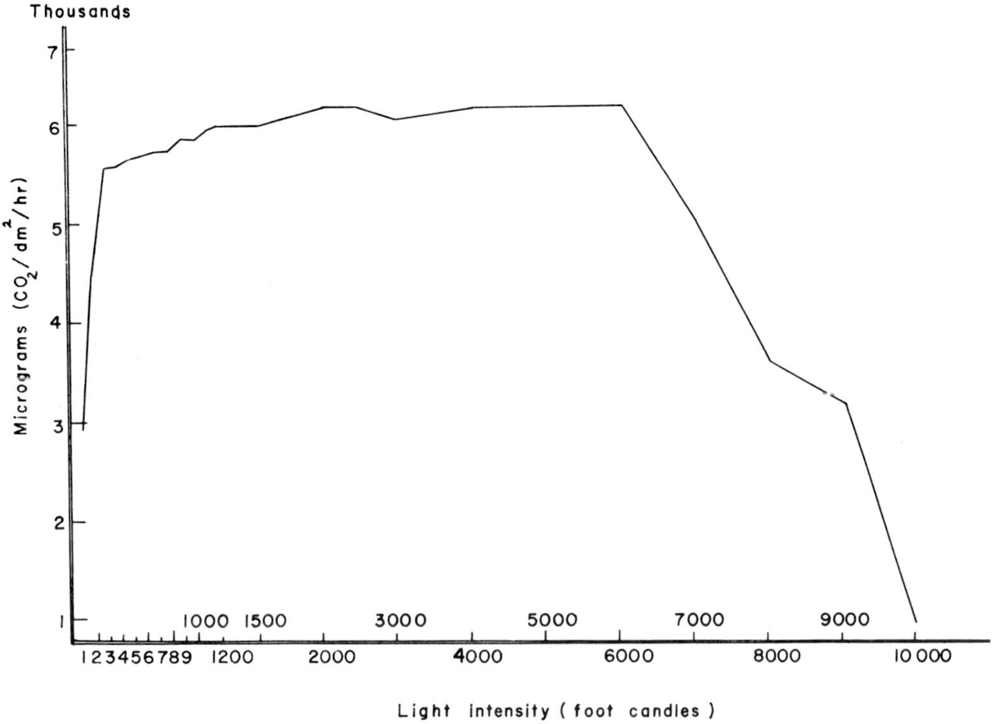

Fig. 4 Light saturation curve for Arabica coffee var. Bourbon (Tió, 1962).

strated, was not the direct cause of stomatal closure in coffee under high light intensities, other explanations were sought.

Heath and Orchard (1957) found that the CO_2 content in intercellular spaces of coffee leaves increased logarithmically up to 30°C and then more rapidly to 35°C. They suggested that the midday closure of stomata observed by Nutman (1973b) could be a temperature effect acting on intercellular space CO_2 content. Nunes *et al.* (1968) found that for each degree of temperature above 24°C there was an increase of 20 ppm in internal CO_2 concentration. This increase was accompanied by stomatal closure (Bierhuizen *et al.*, 1969). Wormer (1965b) also found that temperature was inversely correlated with stomatal opening.

The CO_2 compensation point for coffee is not a fixed value. Working with detached leaves in continuous light of 8000 lux at 23°C, Jones and Mansfield (1970) observed an endogenous rhythm in compensation point. This rhythm was susceptible to phase shift and had a period of slightly more than 24 hours. They considered that measurements of the CO_2 compensation point were most meaningful when determined near the middle of the normal photoperiod. This corresponded to the lowest diurnal values which did not vary greatly from test to test.

Few studies of productivity using growth analysis techniques have been carried out with coffee trees in the field. Huerta S. and Alvim (1962) reported that leaf area index (LAI) of 2- and 4-year-old trees of two cultivars under strip cultivation in Chinchiná, Colombia, varied from 3.6 to 11. Cannell (1971b) studied seasonal production and distribution of dry matter in 3.5- to 5-year-old trees in Kenya. During the hot, dry season (January–February), net assimilation rate (E_A) of deblossomed trees was very low (0.09 g dm^{-2} $week^{-1}$) and then increased in the rainy season (0.13–0.14 g dm^{-2} $week^{-1}$). During the cool, dry season (July–September), E_A was as high as during the rainy season (0.13 g dm^{-2} $week^{-1}$). The low values in the hot, dry season were attributed to high day temperatures as well as water stress. E_A of fruiting trees was up to 0.19 g dm^{-2} $week^{-1}$, when expressed on a leaf area basis, and up to 0.16 g dm^{-2} $week^{-1}$ when total green surface area was included in the calculation. This was in contrast to 0.13 g dm^{-2} $week^{-1}$ in deblossomed trees. Significant correlations were found between E_A of individual trees and the number of fruits they produced.

Leaf area index of deblossomed trees, at a density of 6.5 m^2 per tree, was 0.8, 1.1, 1.4, and 2.2 for the cool–dry, short-rain, hot–dry, and long-rain periods, respectively. These values are much lower than those reported by Huerta S. and Alvim (1962). Apart from planting density, this difference may be due to the pruning system used in Kenya (Cannell, 1971b). The rate of dry matter production was, following the order given above, 0.45, 0.76, 0.53, and 1.36×10^4 kg ha^{-1} $year^{-1}$, a pattern similar to that for growth of branches in Kenya (Fig. 5). Considering that the plants covered only one-third of the ground surface, and using the last value mentioned, productivity of coffee in a closed canopy compared favorably to that of other tropical plants (Cannell, 1971b). Huerta S. and Alvim (1962) reached a similar conclusion. Assuming E_A to be 8 mg dm^{-2} day^{-1} and LAI equal to 10, they calculated productivity to be about 30 tons ha^{-1} $year^{-1}$. This classifies coffee among the most productive plants. The assumption of closed canopy or high leaf area indexes seems unrealistic, however, because growth parameters change at high plant densities.

A few determinations of net assimilation rate of seedlings of young coffee plants have been carried out under different climatic conditions (Alvim, 1953, 1960; Huerta S., 1954; Castillo Z., 1961; Huerta S. and Alvim, 1962; Magalhães, 1964; Hollies, 1967a; Huxley, 1967; Silveira and Maestri, 1973). E_A of coffee ranged from 0.11 to 0.32 g dm^{-2} $week^{-1}$. Shade decreased E_A in Costa Rica (Alvim, 1953; Huerta S., 1954), Peru (Alvim, 1960), and Colombia (Castillo Z., 1961), but not in Viçosa, Brazil, although in the latter case the data did not show a definite trend (Silveira and Maestri, 1973). In Campinas, Brazil, Magalhães (1964) observed an increase in E_A as leaf area was reduced by cutting off part of each leaf, an affected attributed to better light distribution, implying

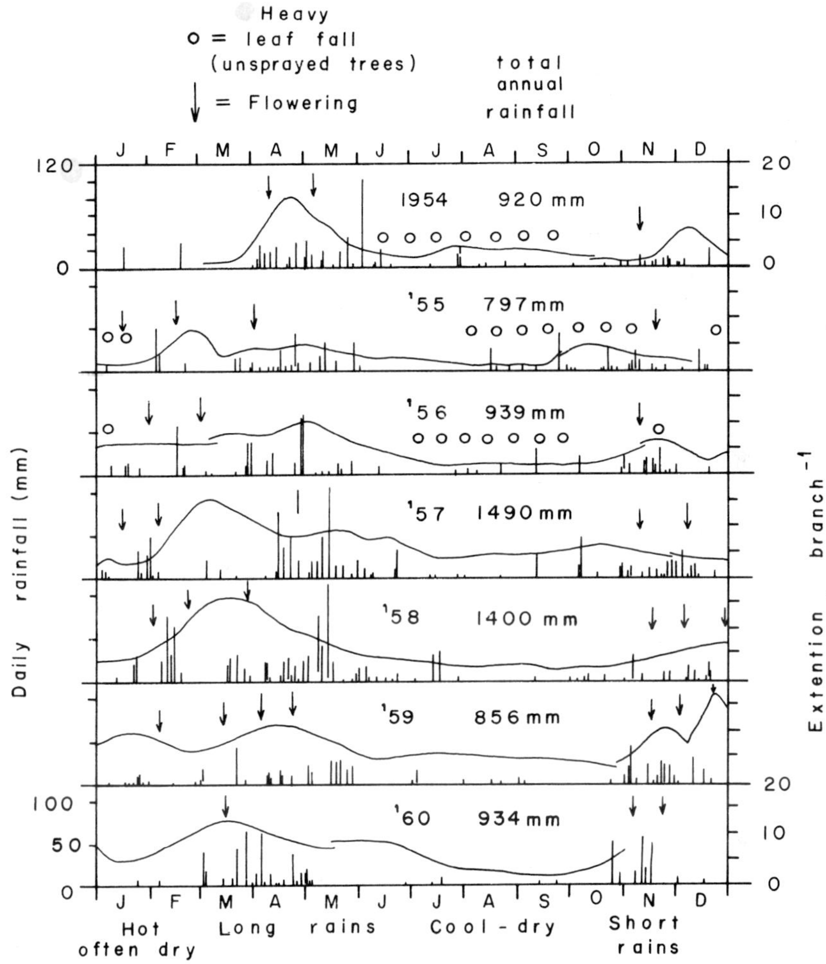

Fig. 5 Changes in the rate of extension of branches of Arabica coffee at the Coffee Research Station, Ruiru, redrawn after Rayner (1956–1959). The times of flowering and, for 1954–1956, the times of leaf fall are also shown. Note that growth is always rapid at the start of the long rains and always slow during the cool dry season (Cannell, 1971c).

therefore, that shade is detrimental to photosynthesis. Age of plants, size of containers (Huerta S. and Alvim, 1962), soil fertility (Huerta S. 1954), and spacing (Huerta S. and Alvim, 1962) may account for some but not all the differences reported in the literature.

The above values are considered low, when compared to 0.2–0.5 g dm^{-2} $week^{-1}$ for temperate woody plants (see Sylvain, 1955b; Alvim, 1960; Kozlowski and Keller, 1966), but are well within the range of 0.09–0.34 g dm^{-2} $week^{-1}$ found for other tropical perennials (Cannell, 1971b).

Differences in net photosynthesis between species and varieties of *Coffea* were reported by Nunes *et al.* (1969). They also showed that light saturation occurred at approximately 0.11 cal cm^{-2} min^{-1}, whereas net photosynthesis decreased with increasing temperatures above 20°C, but the high levels of photosynthesis reported earlier (Nunes *et al.*, 1968; Bierhuizen *et al.*, 1969) were not found. Their data suggested the possibility of selecting coffee for higher potential dry matter production in breeding programs.

Under field conditions, coffee has a fairly low rate of net photosynthesis, as a result of high CO_2 compensation point and high diffusive resistance of leaves. Leaf temperature and soil water availability appear to be the main factors involved. The physiological mechanisms underlying their action are, however, still to be determined.

Decker and Tió (1959) showed that when coffee leaves were alternately illuminated and darkened, initial rates of intake or evolution of CO_2 were momentarily high. They suggested that most of the photosynthate was immediately cancelled by photorespiration and, consequently, only a minor part remained as net gain or dry-weight increment. The relative importance of respiration and/or photorespiration still awaits evaluation, but it does appear to be important, decreasing net photosynthesis either directly by CO_2 loss or indirectly by increased overall resistance (Nunes *et al.*, 1968).

Dark fixation of CO_2 respired was suspected by Nunes *et al.* (1968). If this should be confirmed, some observations on photosynthesis of coffee plants and rhythmic autonomous stomatal movement may be explained (Nunes *et al.*, 1968).

IV. VEGETATIVE GROWTH

A. Leaf Initiation, Expansion, and Abscission

Leaf production of coffee occurs throughout the year, but the rate varies with changes in climatic factors (Cannell, 1971a; Barros and Maestri, 1972). In Ruiru, Kenya, leaf growth occurs most rapidly during the wet season, with larger leaves produced than during the dry season (Cannell, 1971a). In Viçosa, Brazil, more leaves are produced on lateral branches in the hot, rainy season (October–March) than in the dry, cool season (April–September). The rate of leaf expansion, as well as final size of leaves follow a similar trend. Leaves that appeared in October reached an average area of 55 cm^2, with a mean growth rate of 9.2 cm^2 $week^{-1}$; in January 27 cm^2 and 4.5 cm^2 $week^{-1}$; in February 41 cm^2 and 6.8 cm^2 $week^{-1}$; and in June 9 cm^2 and 0.9 cm^2 $week^{-1}$ (Barros and Maestri, 1974a).

A similar trend is shown in Fig. 6. Since there were no significant differences between irrigated and nonirrigated plants (Fig. 7), and both final leaf area and rate of leaf growth began to decrease in April when temperatures were still favorable, Barros and Maestri (1974a) attributed the effect mainly to shorter photoperiods (Figs. 6 and 7). The lower values for leaves which appeared in January were explained by high temperatures and solar radiation. Growth of leaves was completed in about 6 weeks, except for those appearing in June, which grew for approximately 10 weeks (Barros and Maestri, 1972).

In South India, maximum leaf area, which was due to both leaf expansion and production of new leaves, was recorded in August–October. Thereafter, leaf area decreased because of leaf fall of prematurely and naturally yellowed leaves (Vasudeva, 1967).

Gindel (1962) concluded that radiation and temperature were the main factors controlling leaf development of coffee in Israel. Even on the same plant, leaf area was low for trees exposed to prolonged and strong light. Leaves that appeared at the beginning and end of the growing season (May–November) were

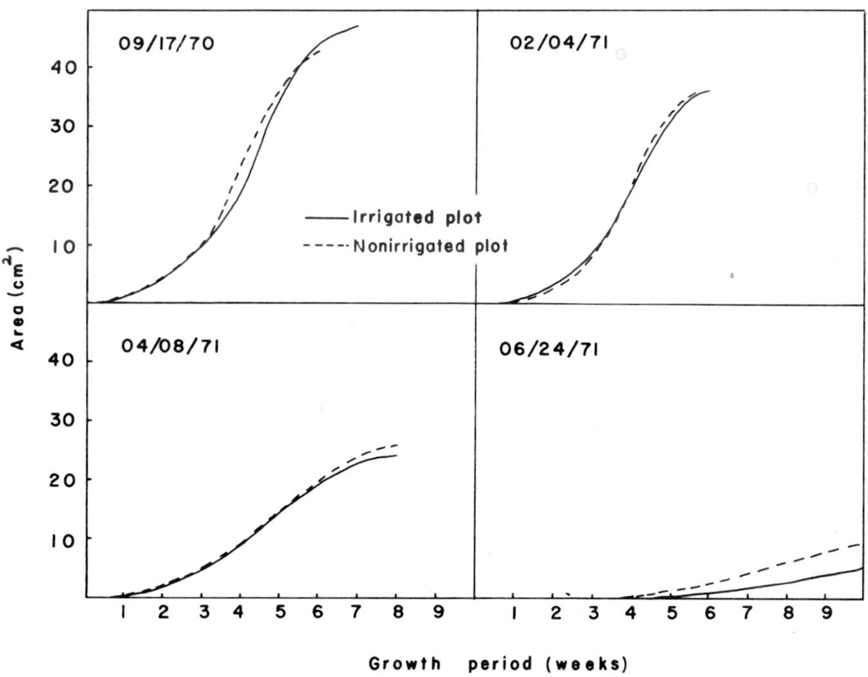

Fig. 6 Growth curves of coffee leaves in various periods of the year. Leaves appeared at dates on upper left in each rectangle. Note the similar growth between leaves in irrigated and nonirrigated plots (Barros and Maestri, 1972).

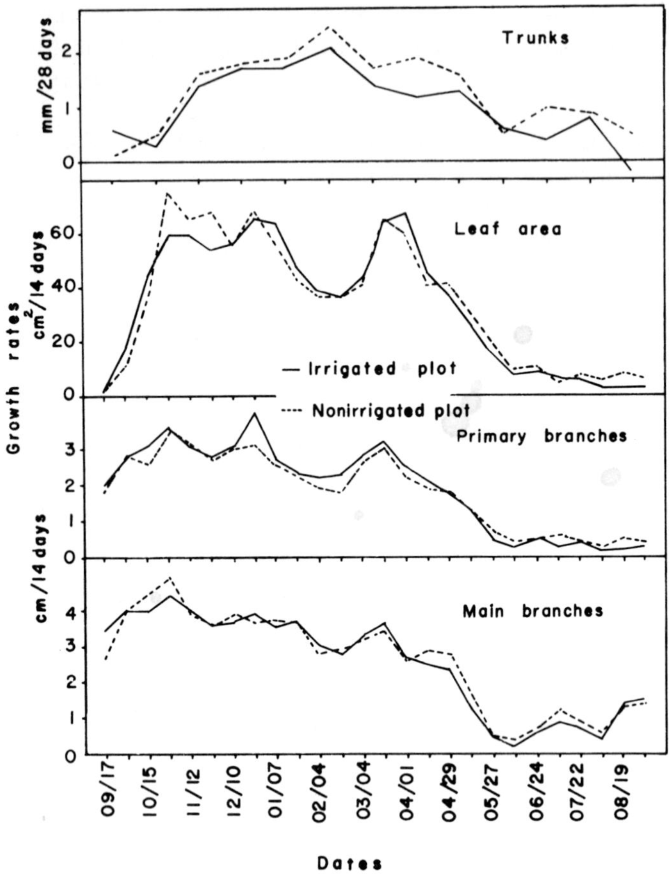

Fig. 7 Rates of absolute growth of various parts of coffee plant in Viçosa, Minas Gerais, Brazil, from September 1970 through August 1971. Average per plot of 12 main branches, 24 primary branches, leaf area on primary branches, and of 12 trunk bases (Barros and Maestri, 1974a).

the smallest; those appearing in June or July were the largest. The area of leaves produced in August was lower than that of leaves produced in preceding and succeeding months, and this was attributed to very high temperatures in August. Hollies (1967b), in Lyamungu, Tanzania, reported that leaf area increased progressively with artificial shade up to the maximum level tested (52%) and also varied with the season of leaf formation.

Coffee seedlings in Viçosa, Brazil had most leaves and a greater total leaf area under approximately 50% shade, but the area of individual leaves increased with shade up to 75% (Maestri and Gomes, 1961; Silveira and Maestri, 1973).

Specific leaf area also increased linearly with shade over the range studied (0–75% shade) (Silveira and Maestri, 1973). Similar results were obtained by Castillo Z. (1961) in Chinchiná, Colombia. He found that the number of leaves per plant was significantly reduced under 75% shade, but total leaf area per plant was higher under 50% shade. In Turrialba, Costa Rica, coffee seedlings reacted differently, with the number of leaves per plant decreasing as shading was increased, but total leaf area per plant was unaffected by shading treatments (Alvim, 1953; Huerta S., 1954). Apparently, the area of individual leaves increased with the level of shade, thus compensating for reduction in leaf number. Similar results were obtained in Peru by Alvim (1960).

It should be remembered that in shading experiments much more than the light intensity is changed. Soil, leaf and air temperatures, soil moisture, air humidity, and consequently water status of leaves may be altered at the same time. Hence, there are real difficulties in ascribing the effects of shading to light intensity changes alone.

Nunes *et al.* (1968) concluded that temperature was more important than light in controlling leaf growth. The best growth occurred at a day/night temperature of 24°/20°C. Growth declined somewhat at a lower temperature (20°/12°C), and was only half as great at the highest temperatures used (32°/25°C). Light intensity had no significant effect. These results are in good agreement with photosynthetic responses.

Alvim (1958) found that temperature of coffee leaves sometimes was as much as 20°C above air temperature. At Ruiru, Kenya, Cannell (1971c) measured temperatures near 40°C in leaves exposed to full sun. These observations, together with the findings of Nunes *et al.* (1968), support the observations of Gindel (1962) and Barros and Maestri (1974a) that inhibition of leaf growth occurs during the hot months.

In greenhouse experiments total leaf area of coffee seedlings was greatly increased by nitrogen fertilizers when soil moisture content was high. When soil moisture content was low, the effect was less marked, especially when sodium nitrate was the fertilizer. Plants responded better to urea than to nitrate fertilizers (Bravo C. and Fernandez, 1964).

In tropical climates, leaf abscission had been associated with distribution of rainfall and daylength (Alvim, 1964). Since coffee is an evergreen and grows more or less continuously, it does not normally lose all of its leaves at a specific time of year. Individual leaves have a life span of a few months to more than a year (Vasudeva, 1967).

As may be seen in Fig. 5, coffee loses its leaves during droughts (Camargo and Telles, 1953; Alègre, 1959; Haarer, 1963; Gopal and Ramaiah, 1971). Haarer (1962) recommended that subsoils should have a certain water holding capacity throughout the year. In Campinas, Brazil, coffee trees had an average leaf area of approximately 32 m^2 during the rainy season and only 12 m^2 in the dry season (Franco and Inforzato, 1950). Vasudeva (1967), on the other hand, attributed premature leaf fall of coffee in South India, during the postmonsoon dry period (October onward), to high temperature and light intensity.

Leaf shedding undoubtedly conserves water during critical periods. According to Levitt's (1972) terminology, coffee is drought resistant through a drought avoidance mechanism. In addition, stomatal closure during drought (Maestri and Vieira, 1958; Bravo C. and Fernandez, 1964) helps to conserve water (Bierhuizen *et al.,* 1969).

The life span of coffee leaves is greatly influenced by drought and high temperatures, which decrease leaf carbohydrate levels (Nunes *et al.,* 1968; Bierhuizen *et al.,* 1969). Because fruits are strong sinks for carbohydrates (Cannell, 1970, 1971a; Patel, 1970), the detrimental effects of drought and high temperature may be greatest during the fruiting or postharvest periods. Defoliation and dieback diseases often have been associated with, or ascribed to, low carbohydrate reserves (Nutman, 1933; Thorold, 1945; Robinson and Bull, 1961; Burdekin and Baker, 1964; Janardhan *et al.,* 1971; Clowes, 1973).

B. Shoot Growth

As pointed out by Went (1957), the coffee plant has a remarkable system of vegetative buds. The apical bud develops into an orthotropic shoot, which bears in its leaf axils, at first one and then (beginning at the eighth to tenth node) two types of buds (Carvalho *et al.,* 1950). The primary axillary buds are dormant and usually develop only if the stem is decapitated or the plant is continuously exposed to temperatures above 30°C (Went, 1957). The larger bud, which appears a little later during seedling development, is located above the primary dormant bud, and gives rise to a plagiotropic shoot (primary lateral). The leaf axils of plagiotropic shoots have buds which may develop into plagiotropic shoots or inflorescences (Carvalho *et al.,* 1950). A plagiotropic shoot almost never reverts to an orthotropic shoot and vice versa (Carvalho and Antunes Filho, 1952). Growth of lateral branches of coffee, similarly to leaves, exhibits a seasonal fluctuation that is related to climatic factors.

In the area of Turrialba, Costa Rica, Boss (1958) observed minimum shoot growth from August to the first half of January, when temperatures were low and days were short. Rapid growth began in January and showed a double peak (March and July) (Fig. 8). The basic periodicity pattern was not altered by nitrate fertilization, rainfall, soil moisture, flowering, or fruiting. However, the rate of growth was higher in defruited or immature (nonflowering) trees, but no correlation was found between growth rate and rainfall or soil moisture. This is reasonable because the soils of Costa Rica have high water holding capacity (Franco, 1951). High rainfall in Costa Rica may be negatively correlated with growth (McFarlane, 1949). No apparent correlations were found between

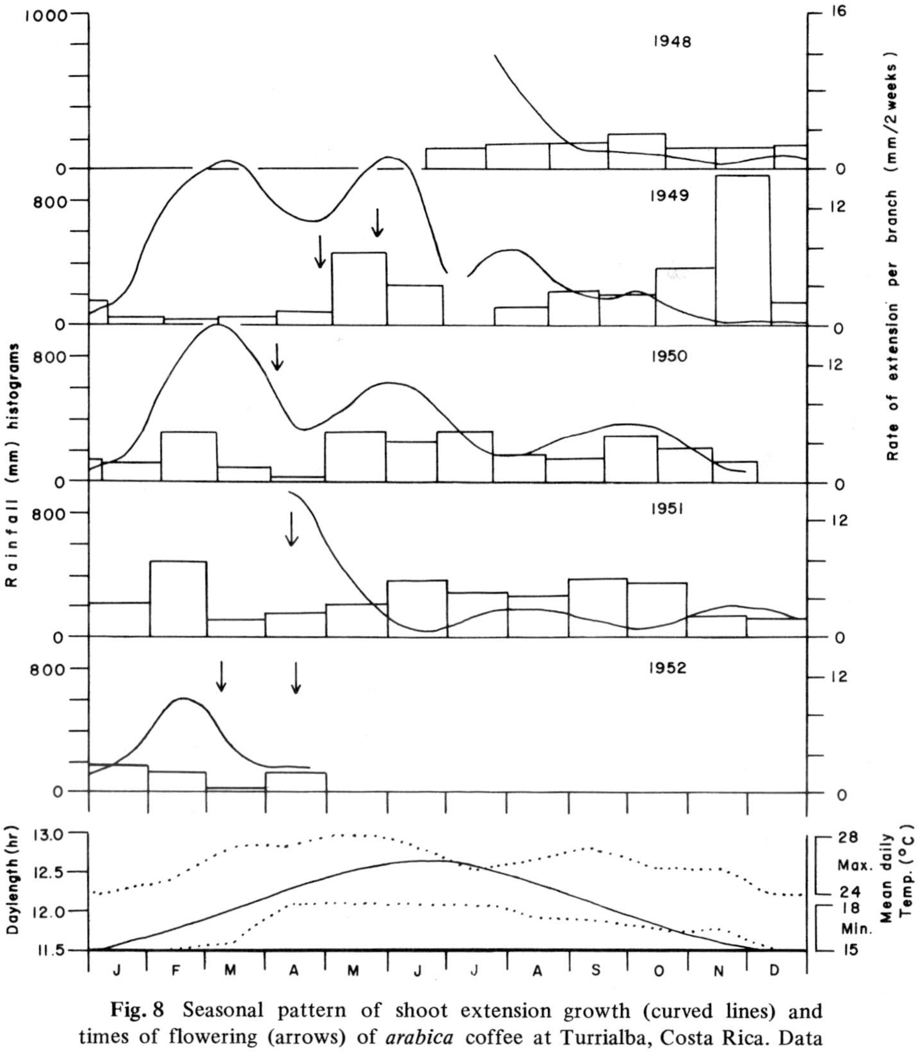

Fig. 8 Seasonal pattern of shoot extension growth (curved lines) and times of flowering (arrows) of *arabica* coffee at Turrialba, Costa Rica. Data from McFarlane, 1948–1949 (1949), Boss, 1949–1950 (1952), and Newton, 1951–1952 (1952) (Cannell, 1972).

growth rate and temperature, especially considering the narrow range of variation in the Turrialba area. A decrease in growth rate in April–May was attributed to low rainfall in the preceding period (McFarlane, 1949).

Newton (1952) did not detect a correlation between shoot growth and temperature or precipitation. He noted that in three years (1949–1951), in the Turrialba region, the maximum rate of growth occurred when daylength was increasing (McFarlane, 1949; Boss, 1958; Newton, 1952). Cannell (1972) compared the Turrialba studies of McFarlane, Boss, and Newton (Fig. 8) and concluded that it was unlikely that coffee trees responded to an annual fluctuation in daylength of only 1 hour and 10 minutes.

Most growth of coffee trees in Santa Tecla, El Salvador occurred during the rainy season, although the growth cycle began in January, more than three months before the onset of rains (Reeves and Vilanova, 1948). Maximum growth rates occurred in May and the first half of June and a rapid decline occurred thereafter, when precipitation was at a maximum. A secondary growth peak occurred in July, followed by a gradual decrease from August through December. Although this decline was correlated with decreasing rainfall, it could not be entirely explained by soil moisture stress. In fact, the soils of Santa Tacla still have large amounts of available water even after 4 months of scanty rainfall (Franco, 1951).

In Kenya (Fig. 5) and Tanzania (Wakefield, 1933), where two wet and two dry seasons occur, growth is correlated with distribution of rainfall. Shoots grow slowly during the cool, dry season (Rayner, 1946).

Although coffee shoots grow more or less continuously in Central Colombia, two seasons of rapid shoot growth are recognized (March–May and August–September). Both of these are also flowering seasons (Suárez de Castro and Rodriguez G., 1956). Least growth occurs during a two-month rainy period during November and December. A relationship between maximum and minimum growth and sum of mean hourly temperature has been reported (Suárez de Castro and Rodriguez G., 1956). Similar observations were made by Castillo Z. (1957). A significantly high correlation was found only with high temperatures above 20°C, recorded 3–4 weeks before the growth measurements.

In Viçosa, Brazil, growth of main and lateral shoots is rapid during the rainy, warm season (September–March) and slow during the dry, cool season (March–September) (Barros and Maestri, 1972, 1974a). A slight decrease in growth rate occurs in January–February, probably because temperatures are the highest of the year. Since regular irrigation during the dry season did not influence either the pattern or rate of growth (Fig. 6), Barros and Maestri (1974a) concluded that moisture supply had an effect only during the season of rapid growth and that the lowered growth rate was induced by decreasing daylength.

In the coffee growing areas of Rhodesia, which have a climate similar to that of Viçosa, Brazil, shoot growth occurs primarily during the rainy season (Clowes and Wilson, 1974).

In South India, after a period of inactivity during the dry, cool season (November–March), shoot growth of coffee begins with an increase in temperature, even without rain (Mayne, 1944; Rayner, 1946). The rate of growth is highest during the rainy, hot season (March–June). Growth almost stops during much of the southwest monsoon season (June–July), and resumes by the end of the season (August). Growth is then continuous to the end of the northeast monsoon season (September–December). A decrease in growth by the end of May and beginning of June has been attributed to the competition of reproductive growth (Mayne, 1944), leaching of nitrates (high rainfall) (Rayner, 1946), and excess soil moisture.

The brief review given here indicates that several factors are implicated in explaining growth periodicity of coffee in various regions. These include drought, temperature, photoperiod, excess soil water, leaching of nitrates by high rainfall, and reproductive growth (Sylvain, 1958).

Some investigators associate the decrease in growth after heavy rains with nitrogen deficiency (Saunders and Wakefield, 1932; Trench and Beckley, 1935; Rayner, 1946; McFarlane, 1949). According to Boss (1958), soil nitrate content fluctuates markedly but does not appear to inhibit growth at any time in Turrialba, Costa Rica. He found considerable leaching of nitrates by heavy rains in November and December. Although nitrate content is negatively correlated with rainfall, nitrate-fertilized trees do not show significantly increased growth or altered growth pattern. By comparison, Montoya *et al.* (1961) found that high levels of urea nitrogen stimulate shoot growth of coffee in Turrialba. Possibly urea remains longer than nitrate in the root zone.

Periodicity in growth of coffee has been correlated with drought in several regions, including Ruiru, Kenya (Rayner, 1946), Tanzania (Saunders and Wakefield, 1932; Wakefield, 1933), Santa Tecla, El Salvador (Reeves and Vilanova, 1948), Rhodesia (Clowes and Wilson, 1974), and Viçosa, Brazil (Barros and Maestri, 1974a). However, in some cases, the dry seasons coincide with periods of low temperatures and short days (Mayne, 1944; Rayner, 1946; Barros and Maestri, 1974a; Clowes and Wilson, 1974), making it difficult to identify a single growth controlling factor. Drought may not be the primary factor regulating the growth rhythm of coffee in Viçosa, Brazil, because irrigation did not alter the pattern or rate of growth (Figs. 6 and 7) (Barros and Maestri, 1972, 1974a). In Kenya (Fig. 5) and Colombia, where daylength changes little throughout the year (Trojer, 1968; Cannell, 1972), rainfall and temperature appear to have an important effect on growth of coffee. However, the observations of Rayner (1946) that growth of irrigated trees continued to decrease during an extended drought in Kenya and by Suárez de Castro and Rodriguez G. (1956), in Colombia, that least growth occurred during a rainy period of two months, suggest that temperature may be the major growth controlling factor. In fact, in Colombia most growth occurs during the 3 months of highest sums of mean temperature, and least growth occurs during the 2 months of lowest temperature sums (Suárez de Castro and Rodriguez G., 1956). Castillo Z. (1957), however, found that growth was significantly correlated only with night temperatures above 20°C, recorded 3–4 weeks before the growth measurements were taken.

Coffee is very sensitive to climatic and soil conditions that induce high leaf temperatures (Bierhuizen *et al.*, 1969). Barros and Maestri (1974a) attributed a temporary decline in growth of lateral shoots in January–February (Fig. 7) to high temperatures. Similar responses were observed in irrigated trees.

The effect of root temperature on shoot growth has received little attention. Franco (1958) found that 26°/20°C was the best day/night root temperature combination for height growth of seedlings and dry matter accumulation in roots and shoots. Either continuous or alternating lower and higher temperatures had an inhibitory or even lethal (above 33°C) effect.

Boss (1958) called attention to the fact that flowering and vegetative growth of coffee generally occur at the time of year when daylength begins to increase on both sides of the equator. Very near the equator coffee shows periodic growth and flowering.

Barros and Maestri (1974a) suggested that daylength was involved in the decrease in growth rate which starts in the middle of March in Viçosa, Brazil (Fig. 7). They noted that least growth occurs in June–August when daylength is about 11 hours and maximum growth when daylength is greater than 12 hours. They also suggested that the low temperatures of June–August may prepare the plant for growth, which starts in September. Piringer and Borthwick (1955) had previously reported that lateral shoots of plants exposed to long days tended to be longer than those on short days. Cannell (1972) questioned the controlling role of daylength on seasonal growth and flowering of coffee. However, he added that changes in temperature and rainfall may condition responses of coffee trees to changing photoperiods. A conditioning effect of low temperature was also postulated by Barros and Maestri (1972).

Relations between vegetative and reproductive growth are not fully understood. In South India rapid expanion of fruits alternated with rapid vegetative growth (Mayne, 1944). McFarlane (1949) attributed the vegetative growth rhythm of coffee in the Turrialba area to setting and expansion of fruits. However, Boss (1958) showed that removal of fruits did not change growth periodicity, even though the growth rate was greater in defruited trees. According to McFarlane (1949), W. H. Cowgill found in Guatemala that fastest shoot growth occurs at the same time that fruits grow rapidly. This also is true in Viçosa, Brazil (R. S. Barros and M. Maestri, unpublished observation).

The relative size of the shoot tips, fruits, and trunk–root system as sinks for assimilates varied seasonally in Kenya (Cannell and Huxley, 1969). Expanding fruits are the major sink for photosynthate and hence limit the amounts mobilized by vegetative tissues (Cannell, 1971a). An increase in net assimilation rate of fruiting trees more than compensates for the smaller leaf area (Cannell, 1971a). Although growth of vegetative tissues is reduced by the presence of fruits (Cannell, 1971b), the seasonal pattern of growth is not altered (Cannell and Huxley, 1969; Cannell, 1971b).

Flowering occurs just before or simultaneously with period(s) of maximum growth of lateral shoots in several regions (Wakefield, 1933; Dean, 1939; Rayner, 1946; Suárez de Castro and Rodriguez G., 1956; Clowes and Wilson, 1974) (Fig. 5). The effect of flowering on vegetative growth has not been adequately studied.

C. Stem and Root Growth

The morphology and distribution of the root system of coffee in different soil types have been studied by several investigators. References for early studies can be found in Sáiz del Rio *et al.* (1961) and Bull (1963). However, periodicity of root growth, as affected by climatic and edaphic factors, has received little attention.

Saunders and Wakefield (1932) and Wakefield (1933) noted that surface rootlets of coffee in Tanzania showed two periods of rapid growth (May–July and January–March). As these periods coincided with dry periods, growth periodicity was ascribed to accumulation and production of nitrates in the soil. However, no evidence was given to support this contention.

In Ruiru, Kenya, the trunk–root system was a relatively important sink during the dry seasons, importing photosynthate from nonfruiting and even sometimes from fruiting branches (Cannell and Huxley, 1969; Cannel, 1971a).

In Tanzania (Bull, 1963), irrigation brought about a reduction of tap root penetration and development of the primary and secondary roots in the deeper layers of the soil profile.

Franco (1958) studied growth of roots and tops of coffee seedlings under controlled conditions at several root temperatures (continuous or alternate). Best growth, as measured by dry matter accumulation, was found with a combination of 26°/20°C day/night temperatures. Top/root ratios were lower under the below "normal" (26°/20°C) treatment, and higher at temperatures above "normal." Within the range of 13° to 33°C, low temperatures appeared to have an indirect effect on root growth, and high temperatures a more direct effect. Franco (1974) and Magalhães (1974) partly supported this view. Phosphorus uptake was inhibited at high temperature (Franco, 1974), even if exposure was for short periods of up to 2 hours a day at a somewhat higher temperature (45°C). Plant response was explained as a direct detrimental effect on root activity (Magalhães, 1974).

Gindel (1962) reported seasonal periodicity of cambial growth of coffee in Israel. In Viçosa, Brazil, cambial growth began in October and continued to June; no measurable growth was detected from July through September. Higher growth rates occurred from the end of December to mid March, simultaneously with a temporary inhibition of growth of lateral shoots (Barros and Maestri, 1974a,b) (Figs. 7 and 9). These observations are in accord with those of Cannell (1971b) who established the following order of sink strength: leaves > branches > root system > trunk.

Both high and low temperatures cause injury to coffee stems. Even before visible injury occurs, extreme temperatures may affect growth of both roots and tops by impeding translocation, as in other species (Crafts and Crisp, 1971). Plants grown under controlled conditions with 30°/30°C or 30°/23°C day/night temperatures developed small tumors at the base of the stem (Franco, 1958). Lesions at the stem collar, which appear during the summer in Brazil in young coffee plants in the field, can be induced by localized treatments with high temperatures of 45°C or above (Franco, 1961). In Brazil, soil surface temperatures reached 45°–51°C in the summer (Franco, 1961).

"Strangulation" of the lower part of the stem that appears in young coffee plants during very cold winters in southern Brazil was reproduced experimentally by low temperatures (Drummond and Costa, 1952; Franco, 1960; Rossetti *et al.*, 1960). Temperatures as high as 11°C caused such injury (Rossetti *et al.*, 1960).

It would be interesting to know how extreme temperatures affect phloem translocation in coffee seedlings and what the implications of these effects are on need for shade for best growth of young plants, as found in some places.

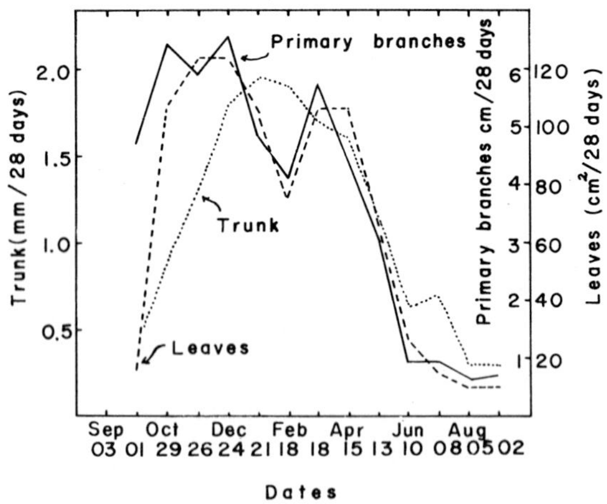

Fig. 9 Rates of growth of coffee trunk (average of 24 trunks) as compared with growth of lateral branches and leaf area on these branches in Viçosa, Minas Gerais, Brazil. This figure was drawn for a better understanding of Fig. 7 (Barros and Maestri, 1974b).

V. REPRODUCTIVE GROWTH

A. Flowering

Flowering in coffee has been reviewed by Alvim (1973) and only a few additional comments will be made.

Cannell (1972) investigated photoperiodic responses of mature coffee trees in Kenya by extending the "natural" short photoperiod (12 hours) to over 15 hours, or by interrupting the night by 15 minutes or 3 hours. In one out of three cases the treated plants appeared to retain the short-day stimulus for floral induction for at least 6 months (the natural photoperiod is always inductive in Kenya). Only thereafter did the plants tend to respond to long days, i.e., they initiated flowers more slowly and extended their branches faster than untreated trees. However, the response was much weaker than that reported for seedlings (Piringer and Borthwick, 1955). Hence, Cannell (1972) suggested that seasonal changes in growth and floral development of adult trees in most coffee growing areas of the world are regulated by factors other than photoperiod. He also suggested that in Turrialba, for example, flowering could be associated with a preceding relatively cool period as shown in Fig. 8. However, since Turrialba has a very narrow temperature range (Fig. 8), the temperature effect on flowering is difficult to understand. It would be interesting to study photoperiodic sensitivity of coffee trees at higher latitudes, where the seasonal changes in photoperiod are greater. Cannell (1972) suggested that responses of trees to changing daylength may be conditioned by fluctuations in temperature and rainfall. Wormer and Gituanja (1970a) had previously noted that the rate of floral differentiation in Kenya was highest in the cool months and also proceeded without interruption for a considerable time at Marianne where temperatures are lower than in Ruiru. Temperature might, therefore, be an important factor in floral differentiation. Incidentally, in the areas east of Rift Valley, Kenya, two main growth flushes and flowering periods occur, coinciding with the first half of each of the two wet seasons. Floral initiation begins some time before the terminal bud of lateral branches enters a period of quiescence (Wormer and Gituanja, 1970a,b).

Browning (1973) concluded that release of flower bud dormancy is controlled by hormones. After dormancy was broken by rain or irrigation, gibberellin content of buds increased rapidly, whereas it was not altered in the xylem sap. Bud gibberellins began to decrease as soon as rapid growth was under way. Absolute levels of abscisic acid did not change until four days before anthesis, and they increased thereafter. Browning (1973) suggested that resumption of growth leading to anthesis may be regulated by liberation of free gibberellins from a bound form in the buds. In view of the role played by water stress on release of bud dormancy, Browning acknowledged that a second xylem-transported stimulus, possibly a cytokinin, might also be involved.

After release of dormancy of flower buds by rain, the starch content of the corolla increased until 9 days before anthesis. Thereafter it decreased, slowly at first and then sharply beginning on the fourth day before anthesis. Sugars increased steadily until blooming (Croope *et al.,* 1970). Starch grains almost disappeared from the open corolla (Maestri *et al.,* 1970). It may be that starch in the corolla is associated with changes in gibberellin level, as reported by Browning (1973). Apparently bud growth involves two distinct metabolic phases, one characterized by starch accumulation and the other by starch hydrolysis (Croope *et al.,* 1970).

Frederico and Maestri (1970) and Gopal and Vasudeva (1973) studied growth of flower buds in Minas Gerais (Brazil) and South India. Flower buds were larger and heavier in South India than in Brazil. Greater availability of soil moisture and higher air temperatures in South India, as well as differences in varieties, may explain these discrepancies.

B. Fruiting

Growth curves of coffee fruits show a double sigmoid shape (Leon and Fournier, 1962; Gindel, 1962; Wormer, 1964; Ramaiah and Vasudeva, 1969; Cannell, 1971d). Figure 10 shows fruit growth curves of five cultivars of coffee. At first growth is negligible. This is followed by a period of rapid growth until the green berry approaches its final size. An extended period of arrested growth follows and lasts until the fruit begins to ripen. Finally growth is resumed and the berry enlarges rapidly. Wormer (1964) and Cannell (1971d) added another period, during which dry weight of the fruit increased regularly while fresh weight increased little. They divided fruit growth into five stages: (1) "pin-head" stage, (2) rapid swelling stage, (3) stage of suspended and slow growth, (4) endosperm filling stage, and (5) ripe stage.

The rate of respiration of the pericarp and seed of developing fruits decreases greatly with age, even though oxygen uptake per fruit does not change. The respiration rate of the pericarp, during and after ripening, was higher than before ripening. Fruits in the pin-head stage have high respiration rates and should not be regarded as dormant (Cannell, 1971a).

In Kenya, final size of the bean depended greatly on rainfall 4–6 months before picking (10–17 weeks after blooming) (Cannell, 1971a). This period corresponded to the stage of rapid expansion of the berry after which the endocarp hardened (Leon and Fournier, 1962). Thereafter the seeds could not increase in size, because of the mechanical constraint imposed by the hard endocarp (Leon and Fournier, 1962). The size attained by the locules at that time determined the potential size of the bean (Wormer, 1966). In Kenya, the

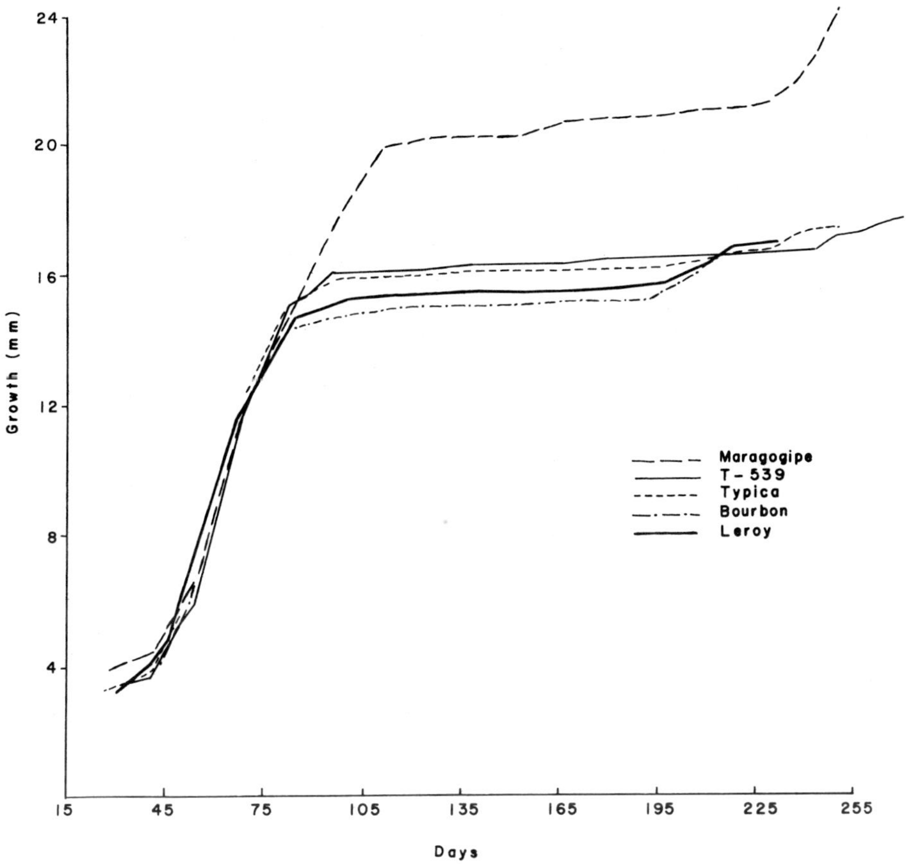

Fig. 10 Growth curves of coffee fruit length in five varieties (Leon and Fournier, 1962).

only cultural practices that increased bean size were irrigation and mulching, both of which improved the water balance of trees (Cannell, 1971d).

Cannell and Huxley (1969) showed that CO_2 labeled assimilates move preferentially into fruits without modifying the overall seasonal pattern of translocation. Cannell (1971a) measured dry-weight changes in defoliated and ring-barked shoots, and concluded that carbohydrates are imported by heavily fruiting branches at the time of rapid fruit expansion. This excess carbohydrate supply enables the branches to produce more leaves and retain more fruits than similar branches that are ring-barked at the base. Small lateral shoots also contribute largely to fruit growth, especially when the fruiting branches have many fruits and there are only few lateral shoots (Cannell, 1970). Whereas growing fruits are strong sinks and monopolize available carbohydrates, they also stimulate photosynthesis (Cannell, 1971b). Furthermore green berries have stomata (Dedecca, 1958) and may synthesize some carbohydrates. However, the photosynthetic contribution of developing fruits to their own development is small (Cannell, 1971b).

Several investigators noted a relation between carbohydrate reserves and fruit growth (Cooil and Nakayama, 1953; Wormer and Egabole, 1965; Patel, 1970; Janardhan *et al.,* 1971). Fruits continued to grow even after complete withdrawal of starch from the wood, suggesting that the growing crop might first mobilize carbohydrates from the wood and then from the leaves (Wormer and Egabole, 1965). However, Patel (1970) and Janardhan *et al.* (1971) found that developing fruits withdraw starch from the wood and leaves concurrently.

The starch content of leaves appears to be a good index of carbohydrate economy of trees. Cooil and Nakayama (1953) observed that yield in the current year is small if carbohydrate reserves are low during flowering and early fruit growth. Patel (1970) found that crop yield correlates better with starch content of leaves than that of wood.

Physiological disorders, such as "overbearing dieback" in Kenya (Wormer, 1965a), "Lyamungu dieback" in Tanzania (Burdekin and Baker, 1964), and "debility growth" in Tanzania (Robinson and Bull, 1961), have been attributed to low starch reserves associated with heavy cropping and other conditions leading to carbohydrate deficiency.

Fruit shedding in coffee usually occurs 8 to 12 weeks after flowering, when the berries are in the first month of rapid expansion (Montoya and Sylvain, 1962; Huxley and Ismail, 1969). Fruits may also be shed later if the carbohydrate supplies are exceptionally low (Cannell, 1971a). However, Gopal's (1971) observation that 2,4-D or 2,4,5-T sprays reduce fruit drop suggests that hormonal controls are also involved.

VI. CONCLUSIONS

Basic information on physiology of coffee is essential for increasing crop yield (Huxley, 1970). Little is known and much more research is necessary.

Although coffee is classified as a C_3 plant and appears to exhibit photorespiration (Decker and Tió, 1959; Heath and Orchard, 1957; Nunes *et al.,* 1968, 1969), the origin of the CO_2 which increases in the leaf mesophyll in response to high temperatures is in doubt. Under field conditions, high temperatures coincide with high light intensities (Barros and Maestri, 1972), thus complicating the problem.

High temperature, high light intensities, and dry periods reduce shoot growth of the coffee tree in favor of the trunk–root system (Cannell, 1971a; Cannell and Huxley, 1969; Barros and Maestri, 1974b). The relative strengths of metabolic sinks appear to be controlled by hormones. However, little specific information is available on levels or kinds of hormones involved and their effects in response to climatic changes.

Much research is needed on flower initiation and release of flower buds from dormancy. Coffee apparently is a short-day plant (Franco, 1940; Piringer and Borthwick, 1955; Went, 1957). The work of Cannell (1972) with adult plants under field conditions has modified somewhat this point of view. Such work should be repeated with adult plants under controlled environmental conditions. A very important question to be answered is why a drought period is necessary for release of flower bud dormancy (Alvim, 1958; Browning *et al.*, 1970) when inhibitor levels increase with the advancement of the dry period and the level of promoters (Browning, 1973) in buds increases only after the first rains. Research is also needed on the physiology of biennial bearing and fruit set after pollination. Answers to these and many other questions would contribute enormously to plant breeding and management programs so as to improve coffee yields.

REFERENCES

Alègre, C. (1959). Climats et caféiers d'Arabie. *Agron. Trop.* **14**, 23–58.

Alvim, P. de T. (1953). Algunos estudios sobre la fisiologia del cafeto. *Suelo Tico* **7**, 58–62.

Alvim, P. de T. (1958). Recent advances in our knowledge of coffee trees. I. Physiology. *Coffee Tea Ind. Flavor Field* **81**, 17–18, 20, 22, and 24–25.

Alvim, P. de T. (1960). Physiology of growth and flowering in coffee. *Coffee* (*Turrialba*) **2**, 57–62.

Alvim, P. de T. (1964). Tree growth periodicity in tropical climates. *In* "The Formation of Wood in Forest Trees" (M. H. Zimmermann, ed.), pp. 479–495. Academic Press, New York.

Alvim, P. de T. (1968). Some factors to be considered in diagnosing moisture deficiency by stomatal opening. *Cienc. Cult (São Paulo)* **20**, 726–732.

Alvim, P. de T. (1973). Factors affecting flowering of coffee. *In* "Genes, Enzymes and Population" (A. M. Srb, ed.), Vol. 2, pp. 193–202. Plenum, New York.

Alvim, P. de T., and Havis, J. R. (1954). An improved infiltration series for studying stomatal opening as illustrated with coffee. *Plant Physiol.* **29**, 97–98.

Barros, R. S., and Maestri, M. (1972). Periodicidade do crescimento em café. *Rev. Ceres* **19**, 424–448.

Barros, R. S., and Maestri, M. (1974a). Influência dos fatores climáticos sobre a periodicidade de crescimento vegetativo do café (*Coffea arabica* L.). *Rev. Ceres* **21**, 268–279.

Barros, R. S., and Maestri, M. (1974b). Ritmo de crescimento do tronco do café. *Turrialba* **24**, 127–131.

Bierhuizen, J. F., Nunes, M. A., and Ploegman, C. (1969). Studies on the productivity of coffee. II. Effect of soil moisture on photosynthesis and transpiration of *Coffea arabica*. *Acta Bot. Neerl.* **18**, 367–374.

Boss, M. L. (1951). Some external and internal factors related to the growth cycle of coffee. Int.-Am. Inst. Agric. Sci., Turrialba (unpublished Master's thesis).

Boss, M. L. (1958). Some environmental factors related to the growth cycle of *Coffea arabica* L. *Proc. Fla. State Hort. Soc.* **71**, 327–332.

Bravo C., M., and Fernandez, C. E. (1964). Respuesta de plantas jóvenes de café a la aplicación de tres niveles de humedad en el suelo y dos fertilizantes nitrogenados. *Turrialba* **14**, 15–23.

Browning, G. (1973). Flower bud dormancy in *Coffea arabica* L. I. Studies of gibberellin in flower buds and xylem sap and of abscisic acid in flower buds in relation to dormancy release. *J. Hortic. Sci.* **48**, 29–41.

Browning, G., Hoad, G. V., and Gaskin, P. (1970). Identification of abscisic acid in flower buds of *Coffea arabica* L. *Planta* **94**, 213–219.

Bull, R. A. (1963). Studies on the effect of mulch and irrigation on root and stem development in *Coffea arabica* L. 1. Changes in the root system induced by mulching and irrigation. *Turrialba* **13**, 96–115.

Burdekin, D. A., and Baker, R. M. (1964). "Lyamungu dieback" of Arabica coffee in Tanganyika. II. Relation of starch reserves to Lyamungu dieback. *Ann. Appl. Biol.* **54**, 107–113.

Camargo, R. de, and Telles, A. de A., Jr. (1953). Botânica do cafeeiro. *In* "O Café no Brasil, sua Aclimatação e Industrialização" Vol. 1, Ser. Estud, Bras. No. 8, pp. 38–88. Serv. Inf. Agric. Minist. Agric., Rio de Janeiro, Brasil.

Cannell, M. G. R. (1970). The contribution of carbohydrates from vegetative laterals to the growth of fruits on the bearing branches of *Coffea arabica*. *Turrialba* **20**, 15–19.

Cannell, M. G. R. (1971a). Effects of fruiting, defoliation and ring-barking on the accumulation and distribution of dry matter in branches of *Coffea arabica* L. in Kenya. *Exp. Agric.* **7**, 63–74.

Cannell, M. G. R. (1971b). Production and distribution of dry matter in trees of *Coffea arabica* L. in Kenya as affected by seasonal climatic differences and the presence of fruits. *Ann. Appl. Biol.* **67**, 99–120.

Cannell, M. G. R. (1971c). Seasonal patterns of growth and development of Arabica coffee in Kenya. Part III. Changes in the photosynthetic capacity of the trees. *Kenya Coffee* **36**, 68–74.

Cannell, M. G. R. (1971d). Seasonal patterns of growth and development of Arabica coffee in Kenya. Part IV. Effects of seasonal differences in rainfall on bean size. *Kenya Coffee* **36**, 176–180.

Cannell, M. G. R. (1971e). Changes in the respiration and growth rates of developing fruits of *Coffea arabica* L. *J. Hortic. Sci.* **46**, 263–272.

Cannell, M. G. R. (1972). Photoperiodic response of mature trees of Arabica coffee. *Turrialba* **22**, 198–206.

Cannell, M. G. R., and Huxley, P. A. (1969). Seasonal differences in the pattern of assimilate movement in branches of *Coffea arabica* L. *Ann. Appl. Biol.* **64**, 345–357.

Carvalho, A., and Antunes Filho, H. (1952). Novas observações sobre o dimorfismo dos ramos em *Coffea arabica* L. *Bragantia* **12**, 81–84.

Carvalho, A., Krug, C. A., and Mendes, J. E. T. (1950). O dimorfismo dos ramos em *Coffea arabica* L. *Bragantia* **10**, 151–159.

Castillo Z., J. (1957). Observaciones sobre la relación del crecimiento del cafeto y temperatura, en condiciones de campo. *Cenicafé* **8**, 305–313.

Castillo Z., J. (1961). Ensayo de analisis del crecimiento en café. *Cenicafé* **12**, 1–16.

Clowes, M. St. J. (1973). Physiological dieback of coffee. *Rhod. Agric. J.* **70**, 85–86 and 89.

Clowes, M. St. J., and Wilson, J. H. (1974). Physiological factors influencing irrigation management of coffee in Rhodesia. *Rhod. Agric. J.* **71**, 53–55.

Cooil, B. J., and Nakayama, H. (1953).. Carbohydrate balance as major factor affecting yield of the coffee tree. *Hawaii Agric. Exp. Stn., Prog. Notes* **91,** 1–16.

Crafts, A. S., and Crisp, C. E. (1971). "Phloem Transport in Plants," pp. 370–376. Freeman, San Francisco, California.

Croope, S. M. S., Frederico, D., Maestri, M., and Sant'Anna, R. (1970). Variações no conteúdo de amido e açúcares nas corolas de café (*Coffea arabica* L.) durante a floração. *Rev. Ceres* **17,** 217–226.

Dean, L. A. (1939). Relationship between rainfall and coffee yields in the Kona district, Hawaii. *J. Agric. Res.* **59,** 217–222.

Decker, J. P., and Tió, M. A. (1959). Photosynthetic surges in coffee seedlings. *J. Agric. Univ. P. R.* **43,** 50–55.

Dedecca, D. M. (1958). Recent advances in our knowledge of coffee tree. 4. Anatomy. *Coffee Tea Ind. Flavor Field* **81,** 46–50.

Drumond, D. A., and Costa, A. C., Jr. (1952). Ocorrência de uma destruição da casca do cafeeiro em Minas Gerais. *Reun. Lat. Am. Fitogenet. Fitoparasitol. Bol., 2nd,* p. 67.

Franco, C. M. (1938). Sobre a fisiologia dos estomas do cafeeeiro *Coffea arabica* L. *An. Reun. Sul-Am. Bot., 1st, Rio de Janeiro* Vol. 3, pp. 293–302.

Franco, C. M. (1940). Fotoperiodismo em cafeeiro (*C. arabica* L.). *Rev. Inst. Café* **15,** 1586–1592.

Franco, C. M. (1951). A água no solo e o sombreamento dos cafezais na América Central. *Bragantia* **11,** 100–119.

Franco, C. M. (1952). A água no solo e o sombreamento dos cafezais em São Paulo. *Bul. Superintend. Serv. Café* **27,** 10–19.

Franco, C. M. (1958). Influence of temperature on growth of coffee plant. *IBEC Res. Inst.* [Bull.] **16,** 1–21.

Franco, C. M. (1960). Estrangulamento do caule do cafeeiro, causado pelo frio. *Bragantia* **19,** 515–521.

Franco, C. M. (1961). Lesão do colo do cafeeiro causada pelo calor. *Bragantia* **20,** 645–652.

Franco, C. M. (1965). Fisiologia do Cafeeiro. *In* "Cultura e Adubação do Cafeeiro," 2nd ed., pp. 63–80. Inst. Bras. Potassa, São Paulo.

Franco, C. M. (1974). Efeito da temperatura do solo e suas variações sobre o crescimento do cafeeiro. *Simp. Lat.-am. Fisiol. Veg., 5th, 1974* Abstracts, p. 54.

Franco, C. M., and Inforzato, R. (1950). Quantidade de água transpirada pelo cafeeiro cultivado ao sol. *Bragantia* **10,** 247–257.

Frederio, D., and Maestri, M. (1970). Ciclo de crescimento dos botões florais do café (*Coffea arabica* L.). *Rev. Ceres* **17,** 171–181.

Gindel, I. (1962). Ecological behavior of the coffee plant under semi-arid conditions. *Coffee (Turrialba)* **4,** 49–63.

Gopal, N. H. (1971). Preliminary studies on the control of fruit drop in Arabica Coffee. *Indian Coffee* **35,** 413–417.

Gopal, N. H., and Ramaiah, P. K. (1971). Studies on wilting and die-back of Arabica coffee plants. *Indian Coffee* **35,** 249–252 and 257.

Gopal, N. H., and Vasudeva, H. (1973). Physiological studies on flowering in Arabica coffee under South Indian conditions. I. Growth of flower buds and flowering. *Turrialba* **23,** 146–153.

Haarer, A. E. (1962). "Modern Coffee Production." Leonard Hill, London.

Haarer, A. E. (1963). Best environment for coffee. *Indian Coffee* **27,** 289–291.

Heath, O. V. S., and Orchard, B. (1957). Midday closure of stomata. Temperature effects on the minimum intercellular space carbon dioxide concentration "Γ". *Nature (London)* **180,** 180–182.

Hernández Medina, E. (1965). Consideraciones fisiologicas en torno al desarrollo y fructificación del cafeto. *Rev. Agric. P. R.* **44,** 8–14.

Hollies, M. A. (1967a). Chronic leaf fall in Arabica coffee in Tanzania. *East Afr. Agric. For. J.* **32,** 404–410.

Hollies, M. A. (1967b). Effects of shade on the structure and chlorophyl content of Arabica coffee leaves. *Exp. Agric.* **3,** 183–190.

Huerta S., A. (1954). La influencia de la intensidad de luz en la eficiencia assimilatoria y el crecimiento del cafeto.Inst. Interam. Cienc. Agric., Turrialba (unpublished Master's thesis).

Huerta S., A., and Alvim, P. de T. (1962). Indice de área foliar y su influencia en la capacidad fotossintetica del cafeto. *Cenicafé* **13,** 75–84.

Huxley, P. A. (1967). The effects of artificial shading on some growth characteristics of Arabica and Robusta coffee seedlings. I. The effects of shading on dry weight, leaf area and derived growth date. *J. Appl. Ecol.* **4,** 291–308.

Huxley, P. A. (1970). Some aspects of the physiology of Arabica coffee—the central problem and the need for a synthesis. *In* "Physiology of Tree Crops" (L. C. Luckwill and C. V. Cutting, eds.), pp. 255–268. Academic Press, New York.

Huxley, P. A., and Ismail, S. A. H. (1969). Floral atrophy and fruit set in Arabica coffee in Kenya. *Turrialba* **19,** 345–354.

Janardhan, R. V., Gopal, N. H., and Ramayiah, P. K. (1971). Carbohydrate reserves in relation to vegetative growth, flower and fruit formation, and crop levels in Arabica Coffee. *Indian Coffee* **35,** 145–148.

Jones, M. B., and Mansfield, T. A. (1970). A circadian rhythm in the level of the carbon dioxide compensation point in *Bryophyllum* and *Coffee. J. Exp. Bot.* **21,** 159–163.

Kozlowski, T. T., and Keller, T. (1966). Food relations of woody plants. *Bot. Rev.* **32,** 293–382.

Leon, J., and Fournier, L. (1962). Crecimiento y desarrollo del fruto de *Coffea arabica* L. *Turrialba* **12,** 65–74.

Levitt, J. (1972). "Responses of Plants to Environmental Stresses." Academic Press, New York.

Loomis, R. S., Williams, W. A., and Hall, A. E. (1971). Agricultural productivity. *Annu. Rev. Plant Physiol.* **22,** 431–468.

McFarlane, W. L. (1949). Some factors affecting growth and yield of coffee. Int. Am. Inst. Agric. Sci., Turrialba (unpublished Master's thesis).

Maestri, M., and Gomes, F. R. (1961). Crescimento de mudas de café (*Coffea arabica* L. var. Bourbon) sob diferentes níveis de luz. *Rev. Ceres* **11,** 265–271.

Maestri, M., and Vieira, C. (1958). Movimento de estômatos em café, sob condições naturais. *Rev. Ceres* **10,** 324–331.

Maestri, M.; Shimoya, C., Croope, S. M. S., and Frederico, D. (1970). Observações sobre a distribuição de grãos de amido e modificações anatômicas na carola de café durante a floração. *Rev. Ceres* **17,** 227–234.

Magalhães, A. C. (1964). Efeito da redução da superficie foliar sobre o desenvolvimento de cafeeiros. *Bragantia* **23,** 337–342.

Magalhães, A. C. (1974). Efeito da temperature das raízes sobre os processos de translocação em cafeeiros. *Simp. Lat.-am. Fisiol. Veg., 5th, 1974* Abstracts, p. 36.

Mayne, W. W. (1944). The growth and bearing habits of *Coffea arabica* L. under South India conditions. *Plant. Chron.* **39,** 284–286.

Montoya, L. A., and Sylvain, P. G. (1962). Aplicacion de soluciones de azúcar en aspersiones foliares, para prevenir la caída del grano de café. *Turrialba* **12,** 100–101.

Montoya, L. A., Sylvain, P. G., and Umaña, R. (1961). Effect of light intensity and nitrogen fertilization upon growth differentiation balance in *Coffea arabica* L. *Coffee (Turrialba)* **3**, 97–108.
Newton, O. A., Jr. (1952). A preliminary study of the growth and flower habits of *Coffea arabica* L. Int.-Am. Inst. Agric. Sci., Turrialba (unpublished report).
Nosti Nava, J. (1953). "Cacao, Café y Té." Salvat, Barcelona.
Nunes, M. A., Bierhuizen, J. F., and Ploegman, C. (1968). Studies on the productivity of coffee. I. Effect of light, temperature and CO_2 concentration on photosynthesis of *Coffea arabica. Acta Bot. Neerl.* **17**, 93–102.
Nunes, M. A., Bierhuizen, J. F., and Ploegman, C. (1969). Studies on productivity of coffee. III. Differences in photosynthesis between four varieties of coffee. *Acta Bot. Neerl.* **18**, 420–424.
Nutman, F. J. (1933). The root system of *Coffea arabica* L. I. Root systems in typical soils of British East Africa. *Exp. Agric.* **1**, 271–284.
Nutman, F. J. (1937a). Studies on the physiology of *Coffea arabica*. I. Photosynthesis of coffee leaves under natural conditions. *Ann Bot. (London)* [N.S.] **1**, 353–367.
Nutman, F. J. (1937b). Studies on the physiology of *Coffea arabica*. II. Stomatal movements in relation to photosynthesis under natural conditions. *Ann. Bot. (London)* [N.S.] **1**, 681–693.
Nutman, F. J. (1937c). Bearing of recent physiological research on the shade problem in Arabica coffee cultivation. *East. Afr. Agric. J.* **2**, 366–370.
Nutman, F. J. (1938). Stomatal movement and epidermal water content. *Nature (London)* **141**, 608.
Patel, R. Z. (1970). A note on the seasonal variations in starch content of different parts of Arabica coffee trees. *East Afr. Agric. For. J.* **36**, 1–4.
Piringer, A. A., and Borthwick, H. A. (1955). Photoperiodic responses of coffee. *Turrialba* **5**, 72–77.
Ramaiah, P. K., and Gopal, N. H. (1969). Effects of temperature and light on Arabica coffee. *Indian Coffee* **33**, 254–259 and 263.
Ramaiah, P. K., and Vasudeva, N. (1969). Observations on the growth of coffee berries in South India. *Turrialba* **19**, 455–464.
Rayner, R. W. (1946). Growth and bearing habits of *Coffea arabica* L. in Kenya and Southern India. *East Afr. Agric. J.* **11**, 251–255.
Reeves, R. G., and Vilanova, T. (1948). Estudio preliminar acerca del crecimiento periódico de los cafetos. *Café de El Salvador* **18**, 1085–1092.
Robinson, J. B. D., and Bull, R. A. (1961). Debility growth symptoms in Arabica coffee. *Kenya Coffee* **26**, 251–256.
Rossetti, V., Pinheiro, E. P., and Abrahão, J. (1960). Estrangulamento da haste do cafeeiro e reprodução experimental de um dos tipos dessa doença. *Arq. Inst. Biol.*, São Paulo **27**, 179–186.
Sáiz Del Rio, J. F., Fernández, L. E., and Bellavita, O. (1961). Distribution of absorbing capacity of coffee roots determined by radioactive tracers. *Proc. Am. Soc. Hortic. Sci.* **77**, 240–244.
Saunders, F. R., and Wakefield, A. J. (1932). "Coffee Cultivation with Special Reference to Northern Province Tanganyika Territory," Pam. No. 7. Department of Agriculture, Tanganyika.
Silveira, A. J. da, and Maestri, M. (1973). Crescimento de mudas de café. (*Coffee arabica* L. cv. 'Bourbon') sob quatro níveis de luz, em Viçosa, Minas Gerais. *Rev. Ceres* **20**, 354–369.
Suárez de Castro, F., and Rodriguez G., A. (1956). "Relaciones entre el crecimiento del cafeto y algunos factores climáticos." (Bol. Tec. No. 16). Campaña de Defensa y

Restauración de Suelos, Federación Nacional de Cafeteros de Colombia, Chinchiná.
Sylvain, P. G. (1954). Long-range objectives in studies of the physiology of coffee. *Turrialba* **4,** 13–22.
Sylvain, P. G. (1955a). Some observations on *Coffea arabica* in Ethiopia. *Turrialba* **5,** 37–53.
Sylvain, P. G. (1955b). "The Photosynthesis of *Coffea arabica* L." Basic Inf. Mat. No. 1 (mimeo.). Int.-Am. Inst. Agric. Sci., Turrialba.
Sylvain, P. G. (1958). "El ciclo de crecimiento de *Coffea arabica.*" (mimeo.) Inst. Int.-Am. Cienc. Agric., Turrialba.
Thoday, D. (1938). Stomatal movement and epidermal water-content. *Nature (London)* **141,** 164.
Thorold, C. A. (1954). Elgon dieback disease in coffee. *East Afr. Agric. J.* **10,** 198–206.
Tió, M. A. (1962). Effect of light intensity on the rate of apparent photosynthesis in coffee leaves. *J. Agric. Univ. P. R.* **46,** 159–166.
Trench, A. D., and Beckley, V. A. (1935). Observations on coffee in Kenya. I. Chlorosis and die-back in coffee. *Emp. J. Exp. Agric.* **3,** 203–209.
Trojer, H. (1968). The phenological equator for coffee planting in Colombia. *In* "Agroclimatological Methods," Proc. Read. Symp., pp. 107–113. UNESCO, Paris.
Vasudeva, N. (1967). Preliminary studies on the growth of coffee leaves. *Indian Coffee* **31,** 5–6.
Wakefield, A. J. (1933). "Arabica coffee. Periods of Growth and Seasonal Measures." Pam. No. 9. Department of Agriculture, Tanganyika Territory.
Wellman, F. L. (1961). "Coffee. Botany, Cultivation and Utilization." Leonard Hill, London.
Went, F. W. (1957). "Experimental Control of Plant Growth," pp. 164–168. Chronica Botanica, Waltham, Massachusetts.
Wormer, T. M. (1964). The growth of the coffee berry. *Ann. Bot. (London)* [N.S.] **28,** 47–55.
Wormer, T. M. (1965a). Some physiological problems of coffee cultivation in Kenya. *Café (Lima, Peru)* **6,** 1–20.
Wormer, T. M. (1965b). The effect of soil moisture, nitrogen fertilization and some meteorological factors on stomatal aperture of *Coffea arabica* L. *Ann. Bot. (London)* [N.S.] **29,** 523–539.
Wormer, T. M. (1966). Shape of bean in *Coffea arabica* L. in Kenya. *Turrialba* **16,** 221–236.
Wormer, T. M., and Egabole, H. E. (1965). Visual scoring of starch in *Coffea arabica* L. II. Starch in bearing and non-bearing branches. *Exp. Agric.* **1,** 41–54.
Wormer, T. M., and Gituanja, J. (1970a). Floral initiation and flowering of *Coffea arabica* L. in Kenya. *Exp. Agric.* **6,** 157–170.
Wormer, T. M., and Gituanja, J. (1970b). Seasonal patterns of growth and development of Arabica coffee in Kenya. Part II. Flower initiation and differentiation in coffee East of the Rift Valley. *Kenya Coffee* **35,** 270–277.
Zelitch, I. (1971). "Photosynthesis, Photorespiration, and Plant Productivity," pp. 269–272. Academic Press, New York.

CHAPTER

10

Cacao

Paulo de T. Alvim

I. INTRODUCTION

In characterizing appropriate environmental regimes for growth of cacao (*Theobroma cacao* L.), some information on climate and soils of representative producing areas, as well as the origin and distribution of the species, is reviewed briefly. This is followed by a discussion of the physiological determinants of yield and effects of environmental factors on vegetative and reproductive growth of cacao.

II. ORIGIN AND DISTRIBUTION

Cacao is a native species of the rain forest regions of tropical America. Its origin is believed to be the headwaters of the Amazon basin, where *Theobroma* and related genera are found in greater proportion than elsewhere (Cuatrecasas, 1964). From its center of origin, the species spread mostly in two directions, resulting in the two primary groups of cultivars known as "criollo" and "forastero." The former originated from dispersion across the Andes toward the lowlands of Venezuela, Colombia, and Ecuador, and northward to Central America and Mexico. The latter resulted from dispersion down the Amazon valley toward northern Brazil and the Guianas. Over 95% of the current world production of cacao is of forastero origin and comes mainly from the West African countries and Brazil.

As a typical crop of the humid tropics cacao is commercially grown within a fairly narrow belt about 15°N and 15°S from the equator. Exceptionally, some small producing areas are found in subtropical latitudes as, for instance, in the lowlands of the state of São Paulo, Brazil, at 23°–25°S.

As a rule, cacao is grown at relatively low altitudes, usually below 300 m. However, in places near the equator, plantings are sometimes found at fairly high elevations, as for instance in some regions of Colombia (1000–1200 m), in the Chama valley of Venezuela (900 m), and in Uganda (1100–1300 m).

Important developments in cacao production began about a century ago, mainly in Ecuador and in the state of Bahia, Brazil (where cacao was introduced from the Amazon region in 1746). Following introduction to the West African countries, world production of cacao increased very rapidly from about 100,000 tons at the beginning of the 20th century to the 1.4–1.5 million tons presently produced. About 85% of this amount is produced by the following countries: Ghana (25%), Brazil (17%), Nigeria (15%), Ivory Coast (13%), Camaroon (7%), Ecuador (4%), Dominican Republic (2%), and Papua and New Guinea (2%). The remaining 15% comes from about 30 different countries.

III. CLIMATE

In defining a suitable climate for cacao, reference is usually made to temperature and rainfall, generally considered to be the most critical climatic factors for growth. In some areas wind is troublesome and deserves very careful attention. Solar radiation and relative humidity obviously affect many physiological processes of the plant, but are not generally thought to play such an important role as the previously mentioned factors in defining ecological limitations for the crop.

It is worth pointing out that the traditional practice of growing cacao under the shade of larger trees greatly modifies the microclimate inside the plantation, particularly with regard to solar radiation, air turbulence, and relative humidity. For this reason it is convenient to discuss the problem of shading together with a review of climate, although it should be recognized that soil factors, mineral nutrition, and disease and insect incidence must also receive careful attention when studying the effect of shade on cacao yield.

A. Temperature

Mean monthly temperatures of representative cacao growing areas are shown in Fig. 1. The annual means vary from 22.4° to 26.7°C and the monthly averages vary from 18.8°C in the coldest areas (Pariquera Açu) to 27.9°C in the warmest regions (Manaus).

Ernholm (1948) concluded that the lower limits for successful growth of cacao are a mean monthly minimum temperature of 15°C and an absolute minimum of 10°C. These extreme temperatures are frequently mentioned in the literature as factors which define the limits of altitude and latitude for the crop. There is, however, no experimental evidence indicating that cacao will not tolerate such temperature limits. In fact, in the small areas planted with cacao in the State of São Paulo, Brazil, for example, where the mean annual temperature

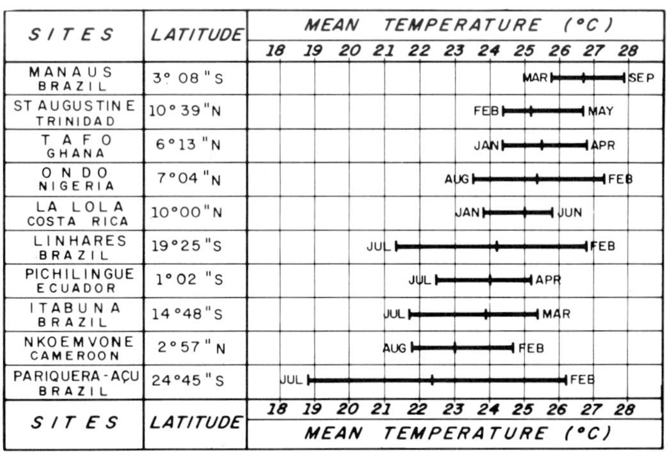

Fig. 1 Amplitude and mean annual temperature of different cacao growing areas. The indicated ranges show the months with the lowest and highest mean monthly temperature as well as the annual mean (middle mark).

is about 22.5°C, the absolute minimum often drops to 4°–6°C and the mean monthly minimum during the coldest month reaches about 10°C. As with most plants, actual damage by low temperature appears to occur only when freezing takes place, as I had the opportunity to observe in an experimental cacao plot in Ribeirão Preto, State of São Paulo, in 1965, following a temperature drop to –3°C. Leaf scorch and drying out of all flowers were the main symptoms observed. The plants, however, were not irreversibly damaged by frost, and recovery took place after about two months. The main effect of relatively low temperatures appears to be reflected in the rate of vegetative growth (Alvim *et al.*, 1969; Sale, 1969b), fruit development, and to a lesser extent, flowering intensity (Sale, 1969a). The influence on fruit development is clearly observed in places where seasonal differences in temperature occur. In Bahia, Brazil, for example, the interval between fertilizing the flower and harvesting the mature fruits varies in accordance with the prevailing temperature during the pod growing period. Fruits of the "temporão" crop, which develop during the warmer months (harvested between April to August), mature in 140–175 days, whereas fruits of the "safra" crop (harvested between September and January) need 167–205 days to mature. For the Catongo cultivar in Bahia, the following empirical formula has been used in calculating the mean number of days between fruit setting and harvesting:

$$N = 2500/(T-9)$$

where N is the number of days to maturity, and T is daily mean temperature after the date of pollination.

Figure 2 gives the estimated number of days from pollination to harvest for fruits set at different periods of the year calculated on the basis of the above equation, using the mean temperature values for Uruçuca, Bahia.

Temperature also affects chemical and physical characteristics of cacao butter. Relatively low mean temperatures during early stages of fruit development result in a high proportion of unsaturated fatty acids in seeds and consequently low melting points of cacao butter (Berbert and Alvim, 1972; Berbert, 1976).

B. Rainfall

Annual rainfall in cacao growing regions generally exceeds annual water loss by evapotranspiration. As a rule, the annual rainfall varies from 1400 to 2000 mm in most regions. When rainfall exceeds 2500–3000 mm, yield is reduced because of soil waterlogging part of the year and/or high incidence of disease, especially black pod caused by *Phytophthora palmivora.* In places where rainfall is less than 1200 mm cacao can only be grown successfully when irrigation is used (as in the small valleys of northern Venezuela where the rainfall is 700–800

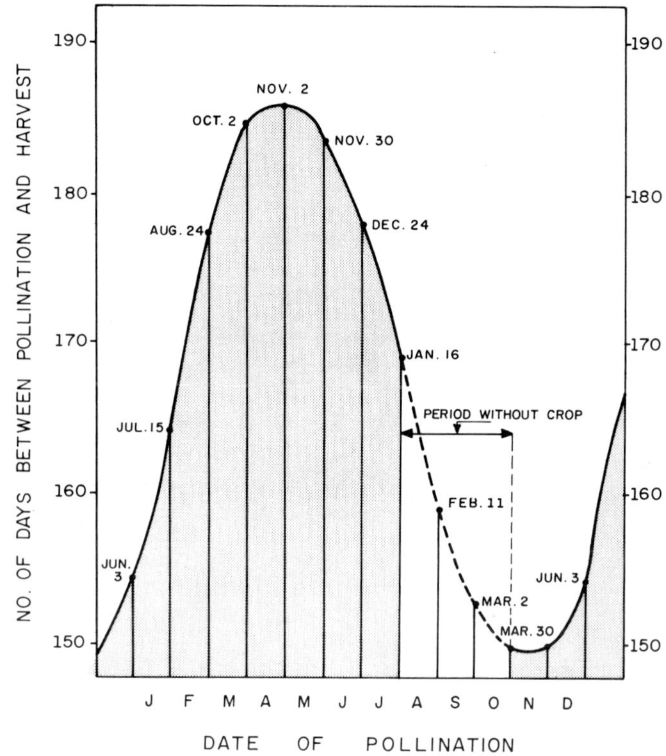

Fig. 2 Theoretical periods between fruit setting (pollination) and harvest, expressed in days, as a function of date of pollination. Dates on the curve were calculated on the assumption that pollination occurred on the first day of each month. In Bahia, as a rule, there is no crop between mid January and mid April. (From Alvim *et al.*, 1972.)

mm), or in places where the ground water table is relatively high (as appears to be the case in some alluvial soils of the state of Espirito Santo, Brazil where the rainfall is 1000–1100 mm).

Rainfall distribution varies greatly from region to region and is the main factor responsible for differences in cropping patterns. As seen in Fig. 3, most cacao producing areas have between 3 and 5 months during which water loss exceeds average precipitation; others, such as Itabuna and Uruçuca, have practically no dry season. In the West African countries, where the dry season is relatively long, about 80% of the crop is harvested within a relatively short time, usually 2–3 months. In Bahia, Brazil, harvesting is more evenly distributed through the year, usually extending over a period of 8–10 months, but varying

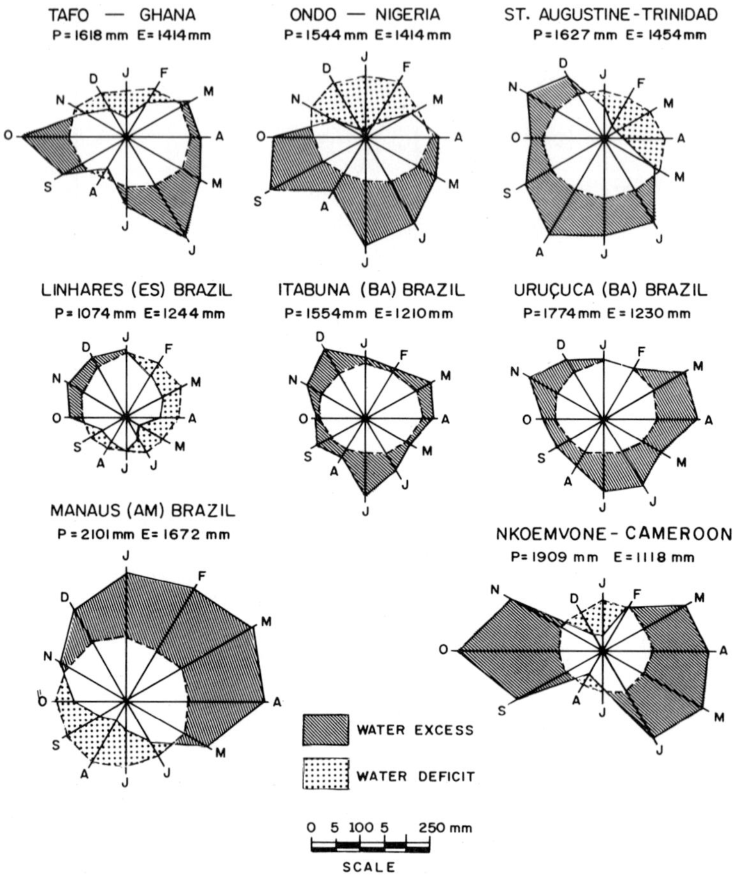

Fig. 3 Monthly values (mm) for rainfall (solid line) and potential evapotranspiration (dotted line) in different cacao producing areas, indicating periods of moisture deficit and excess.

from year to year following changes in the rainfall pattern. The effect of rainfall distribution on cacao is discussed further when physiological responses to climatic factors are considered.

C. Wind

Cacao is very wind sensitive and, in areas exposed to frequent breezes, cannot be grown without wind breaks. Wind breaks are commonly used, for example, in Grenada and other cacao producing islands of the West Indies, particularly in places relatively near the coast where strong winds are frequent. In these areas the trees most frequently recommended as wind breaks are mango (*Mangifera indica*) and Malacca apple (*Syzygium malaccensis*). The main effect of excessive wind is to cause defoliation or premature leaf fall, presumably because of excessive water loss and/or mechanical injury. In Brazil frequent defoliation by wind occurs in Linhares, Espirito Santo, particularly when cacao is grown without or with little shade, but in the nearby valley of Colatina which is protected by mountains, cacao has been under cultivation without shade for several years and defoliation does not occur. Annual rainfall is about the same in both places (1200–1300 mm), but in Colatina the mean wind speed is 1 m/sec compared with 4 m/sec in Linhares.

In some parts of Ghana and western Nigeria cacao yield is reduced by dry "harmattan" winds, which blow from the Sahara Desert between December and March. The duration and intensity of this wind vary from place to place, and are important factors in determining cacao productivity.

D. Shading

In most cacao areas it is virtually impossible to start a cacao field without shading the young plants during the first 2–3 years. This fact has been interpreted as an indication that cacao behaves as a true shade-loving species in its early stages of development. It has been demonstrated, however, that when even mild moisture stress is avoided by continuous watering, even very young seedlings can thrive in the absence of shade (Cunningham and Burridge, 1960). In places with very low light intensity or low potential evapotranspiration, as in some regions of Ecuador, or in areas exceptionally well protected against wind, as in the Colatina valley of Espirito Santo, Brazil, it has been possible to establish cacao plantation without shading the young plants. These, however, are exceptional cases.

Some recent studies in Bahia (Alvim *et al.,* 1974) have shown that the beneficial effect of early shading is due not only to reduced exposure to solar radiation but also to reduced air movement around the young plants, with both factors contributing to prevent plant moisture stress. This was demonstrated in an experiment with seedlings about 4 months old transplanted to an unshaded area and submitted to three treatments: (a) control plants fully exposed to sunlight and without any protection against wind, (b) lateral protection against wind provided by vertical clear plastic sheets, and (c) the same as "b" but using black plastic sheets. Although severe defoliation occurred in all plants during the first 2 months, presumably due to absence of overhead shade, control plants were virtually decimated, whereas plants protected by the vertically oriented plastic sheets refoliated well. The black plastic sheets were more effective than clear ones in preventing leaf fall (Fig. 4).

After passing the juvenile stage, or when the leaf canopy is sufficiently

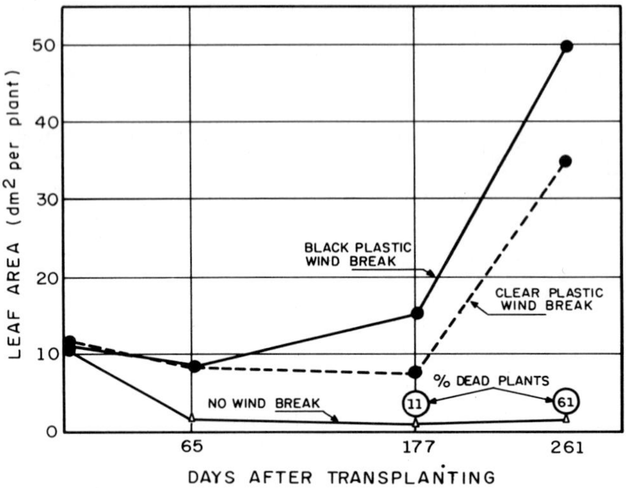

Fig. 4 Effect of black and clear plastic "wind breaks" on leaf area and percentage of dead seedlings, following transplanting to an unshaded site. (From P. de T. Alvim *et al.,* 1974.)

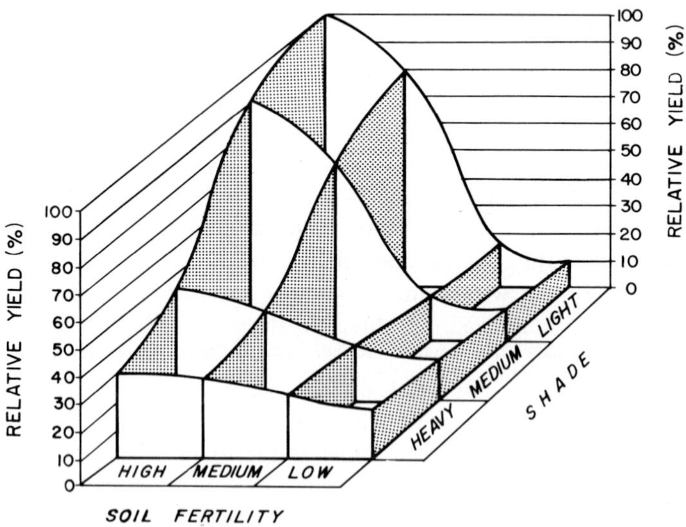

Fig. 5 Interacting effects of soil fertility and shade on yield of cacao pods. (From Alvim, 1972.)

developed to provide some self-shading, cacao production is usually higher with little or no shade than when plants are shaded (Murray, 1957; Alvim, 1960a; Cunningham and Arnold, 1962). Because of the yield increase, which is an indication of higher photosynthetic activity, the nutrient requirement of cacao also increases when overhead shade is not used (Evans and Murray, 1953; Murray 1957). For this reason, fertilizers usually are necessary in unshaded plantings, except in unusual situations where the soil has a very high natural fertility. The theoretical relationship between degree of shade and fertilizer requirement is shown in Fig. 5. In the traditional cacao farms of Bahia there are approximately 80 shade trees per hectare (Alvim and Pereira, 1972). In this area shade thinning combined with fertilizer application has proved very effective in increasing yield.

An average increase in yield of over 100% has been obtained for the last 10 years on several sites in Bahia following shade removal and fertilizer application (Cabala *et al.*, 1970). Because the experimental plots were relatively small (50 × 50 m), shade removal did not eliminate lateral shading and wind protection provided by tall trees in neighboring areas, and should therefore be regarded as "partial" rather than "complete" deshading. Because of the results of this experiment, as well as many demonstration plots, over 150,000 ha of cacao in Bahia have been partially deshaded and treated with fertilizers during the 5-year period from 1968–1972. Such treatments have been major factors in the increase in total production of Brazil in recent years. Shade thinning as recommended in Bahia is accomplished by injections of arboricides with the objective of reducing the number of shade trees to 25–30 trees/ha.

Figure 6 summarizes the results of a shade and manurial trial in Ghana over a 17-year period (Ahenkorah *et al.*, 1974). Shade removal and fertilizer application increased yield about 200%, but after about 10 years the unshaded trees showed signs of deterioration which was attributed to a number of factors such as high loss of exchangeable bases, depletion of reserves caused by higher yield, and high incidence of insects pests and parasitic epiphytes (*Tapinnathus bangwensis*).

Another factor contributing to the general deterioration of unshaded plants and apparently overlooked by Ahenkorah *et al.* (1974) was the fact that several acres of forest around the experimental plots were cleared in 1964/1965 (E. J. A. Asomaning, personal communication). This may have caused some important microclimatic changes, especially in wind velocity. As mentioned earlier, in places well protected against wind, such as the Colatina valley, Espirito Santo, cacao has been grown without overhead shade for several years with no signs of dieback or yield decline. This of course does not invalidate the conclusion of Ahenkorah *et al.* (1974) for the conditions in Ghana, where both heavy overhead shade and lack of shade were unsuitable for maintaining high yield of cacao for more than 8–10 years.

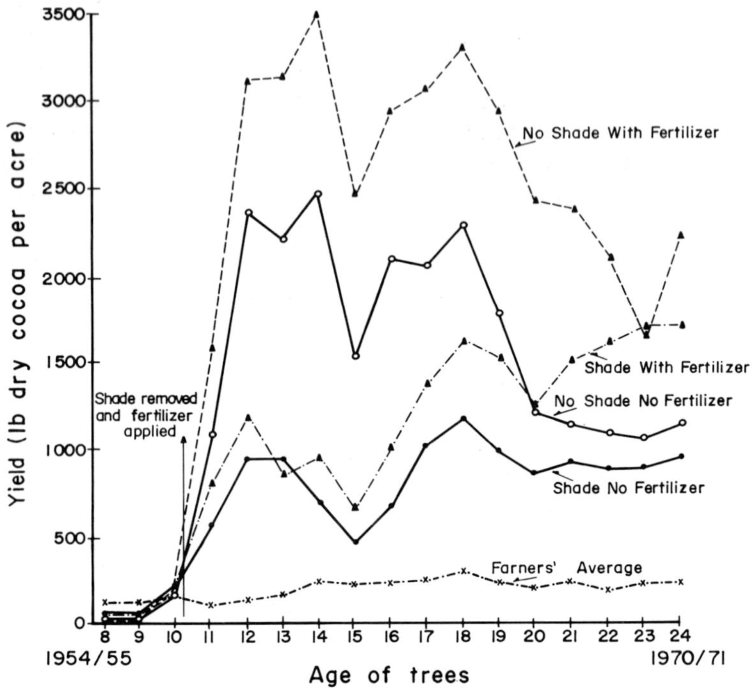

Fig. 6 Effects of shade and fertilizer on yield of 10-year-old cacao trees during the subsequent fourteen years. (From Ahenkorah *et al.*, 1974.)

The above findings are in agreement with practical observations by cacao farmers in many countries, showing that increase in yield following complete removal of shade usually does not last very long and is followed by rapid decline of plantations, with many plants showing severe defoliation and dieback after the third or fourth year. This decline is accelerated by drought and in most areas is associated also with intensive insect attack, particularly by thrips. It is commonly believed that fertilizer application, efficient insect control, and irrigation could prevent deterioration of fully exposed plantations. However, this has not yet been demonstrated in extensive plantings in the traditional cacao producing regions.

From the above considerations, it may be concluded that the advantage of shade in cacao cultivation is not to provide a low light intensity considered optimal for growth and yield; its main function is to counteract unfavorable ecological factors, such as low soil fertility, wind damage, high transpiration leading to moisture stress, and increased insect attack. If all such unfavorable factors can be controlled by methods other than by use of shade, the highest production of cacao is obtained without shade. From a practical point of view, however, moderate shading appears to be the safest and the most economical method of counteracting unfavorable conditions in most regions, especially in areas where soils and water supply are not entirely satisfactory. Excessive shade is a common problem in many regions and should obviously be avoided. Besides decreasing yield by lowering photosynthetic production, excessive shade increases the incidence of black pod disease caused by *Phytophthora palmivora* (Cabala *et al.*, 1970).

Future research should attempt to define the minimum amount of shade for counteracting unfavorable conditions without excessively hampering cacao production. The use of economically important trees for shading and for wind breaks should also be investigated.

IV. SOILS

As shown by Hardy (1960) and Smyth (1966) cacao can be grown in many different soils, provided they have satisfactory physical and chemical properties. Such properties, especially the physical ones, must be considered in connection with climate, particularly the volume and distribution of rainfall.

A. Physical Properties

Ideal cacao soils must have good moisture retention, and be well drained and aerated for unrestricted root growth. The thickness of the layer of the soil which provides such conditions, termed "root room" or "physiological depth" by Hardy (1958, 1960), is generally considered to be the main factor which differentiates "good" from "bad" soils. Based on studies carried out mainly in Trinidad, Hardy (1960) concluded that the physiological depth of soil should be at least 150 cm. This probably applies also to soils of West Africa and also where a long dry season prevails, as in Trinidad. However, in Bahia, where rainfall is well distributed, a root room layer thicker than 1.2 m is seldom found and successful plantations have been established in soils only 60–70 cm deep (Silva, 1969). Distribution of roots in cacao soils rated as "good" and "bad" in Bahia is shown in Fig. 7 (Cadima and Alvim, 1973).

The best cacao soils usually are formed by aggregated clay or loamy sand, the proportions of the different particles usually being 30–40% clay, 50% sand, and 10–20% silt-sized particles (Smyth, 1966). Very heavy soils impede root growth because of reduced soil aeration. On the other hand, although sandy soils allow root penetration, they usually have low moisture holding capacity and are recommended for cacao only in areas with high and well-distributed rainfall. A

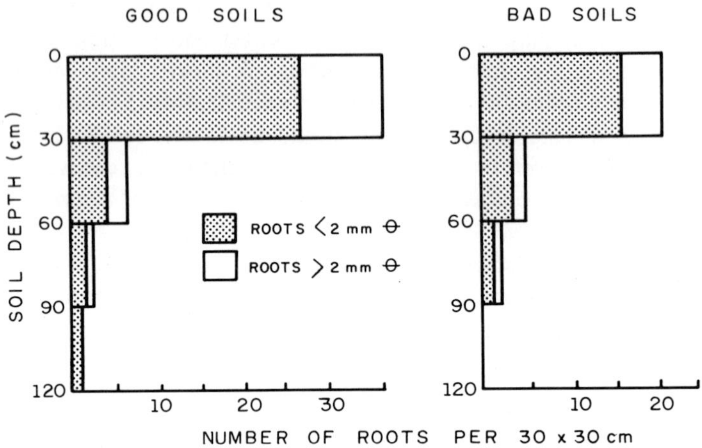

Fig. 7 Root development in "good" and "bad" soils in Bahia, Brazil, as expressed by root counts in 30 X 30 cm squares at different depths of a profile near the plant (mean values for sites). (From Cadima and Alvim, 1973.)

study by Cadima and Alvim (1973) on good and bad cacao soils in Bahia showed that yield was negatively correlated with percent of clay in the C horizon, with an average of 41.5% in the good soils and 54.2% in the bad ones.

Soil consistency or compactness also is important. Preference should always be given to loose, crumbly, soft, and plastic soils, and not hard and compact ones which hinder root penetration (Hardy, 1960). This is particularly important in the surface soil layers, since about 80% of the absorbing roots of cacao are usually found in the top 20 cm of soil.

Soil consistency can be assessed by measuring the shear strength using a "Torvane" vane shear. In Bahia, Brazil, good soils showed a shear strength of only about half that of bad ones, the mean values being 0.19 and 0.36 kg/cm^2, respecitvely (Cadima and Alvim, 1973).

B. Chemical Properties

Cacao is not as tolerant to low soil fertility as some other tropical crops, such as oil palm and rubber (Alvim *et al.*, 1969). However when cacao is grown under shade or as a mixed crop, reasonably productive cacao plantings can be established in relatively poor soil, provided suitable management practices are used. As mentioned previously, when cacao is grown with little shade and in relatively poor soil, yield is greatly improved by fertilizers.

In Bahia, the physical and chemical characteristics of cacao soils have been quantified (Table I) from data obtained from many soil analyses, information on

TABLE I
Characterization of Low, Medium, and High Fertility Brazilian Cacao Soils in Top 20 cm of the Profile

Parameters	Relative fertility		
	High	Medium	Low
1. pH (in water, 1:2.5)	7.5–6.0	6.0–5.0	<5.0
2. % Organic matter (moist combustion)	>3.5	3.5–2.5	<2.5
3. P (ppm) (Mehlich's method)	>15	6–15	<5
4. K (mEq/100 g) (Mechlich's method)	>0.31	0.30–0.11	<0.11
5. Ca + Mg (mEq/100 g) (KCl extraction)	12–6	6–3	<3
6. % Al saturation (KCl extraction)	0–10	10–25	>25

cacao production in many areas, and extensive fertilizer experiments with field and pot cultures.

To illustrate how soil fertility varies from place to place, nine soil types of Brazil are illustrated in Fig. 8 using the graphic method of Alvim and Cabala (1974).

Experience indicates that in soil of relatively low fertility, such as "Vermelhão," "Colônia," and "Itaituba," cacao plantations cannot be established without applications of lime and chemical fertilizers, especially phosphorus. However, other crops, with lower fertility requirements or high tolerance to acidity and relatively high concentration of aluminum, can be cultivated on these soils. In fact rubber has been cultivated with some success on poor soils such as "Colônia" with addition of only small amounts of lime.

V. PHYSIOLOGICAL DETERMINANTS OF YIELD

The main physiological determinants of yield of a fruit crop are (a) the size of the leaf canopy, usually expressed by the ratio of leaf surface to ground surface or the leaf area index (LAI), (b) the rate of photosynthesis per unit leaf area, and (c) the partitioning of photosynthates between the fruits and other parts of the plant. A high yielding crop should have a fast growing leaf canopy, high net photosynthetic rate, and an inherent ability to channel a relatively high proportion of its photosynthates into the fruits. In the past, research on the physiolog-

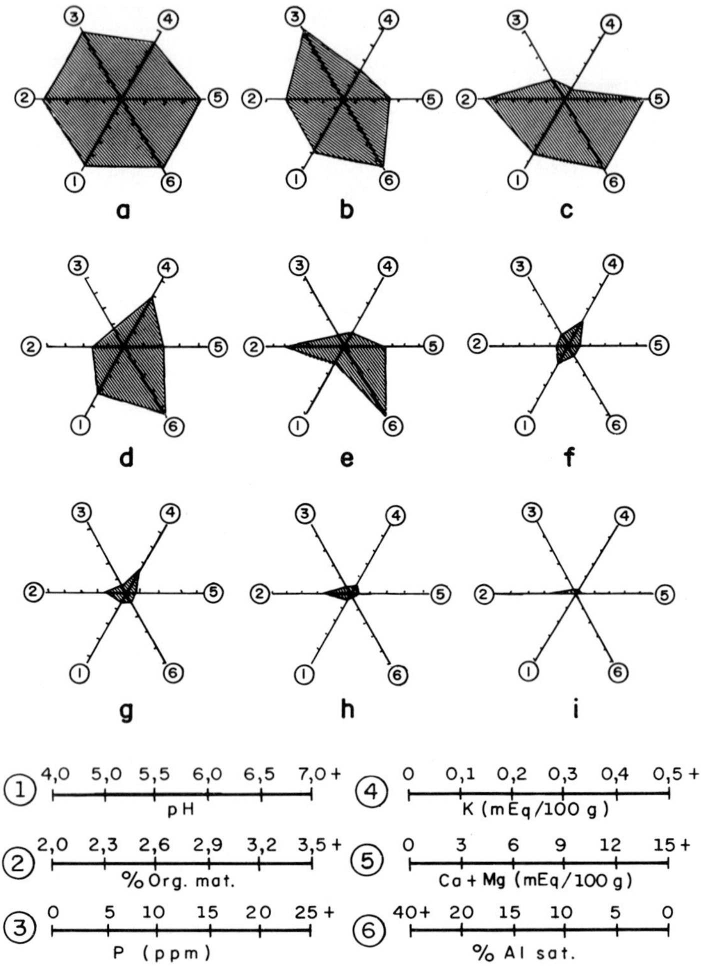

Fig. 8 Graphic representation of some representative soils in rain forest regions of Brazil showing different levels of fertility: (a) "CEPEC" (Bahia), (b) Hydromorphic (Bahia), (c) "Rio Branco" (Bahia), (d) "Ouro Preto" (Rondonia), (e) "Itabuna" (Bahia), (f) "Paraiso" (Rondonia), (g) "Vermelhão" (Rondonia), (h) "Colonia" (Bahia), (i) "Itaituba" (Pará). (From Alvim and Cabala, 1974.)

ical determinants of yield in cacao has not received enough attention. The following is a brief account of some relevant studies on leaf area development, photosynthetic production, and fruit bearing capacity of cacao.

A. Canopy Architecture and Leaf Area Index

No research has been conducted under field conditions on the relation of canopy development to yield of cacao, but some studies with seedlings seem to indicate that genetic differences in leaf growth rate are associated with differences in yield of some cacao cultivars. Alvim and Grangier (1966) reported, for instance, that the cultivars "Catongo" and hybrid "Catongo X Comum" showed a higher leaf area growth rate, and are known to be more productive, than the unselected "Comum." No significant difference in net assimilation rate was found between these cultivars, but Comum showed a lower relative growth rate than the others.

Using hybrid cacao plants about 8 years old, which showed a relatively high yield (about 2000 kg/ha), research was carried out in Bahia on canopy structure, leaf area index, leaf angle, and light interception. Some representative results obtained in 3 X 3 m and 2 X 2 m plantings with practically no overhead shade are shown in Fig. 9 (P. de T. Alvim, unpublished observation). Light interception was measured with two radiometers provided with filters for measuring photosynthetically active radiation. One radiometer was used for readings above the canopy or outside the plantation and the other was utilized for simultaneous readings at different heights within the canopy. A portable metal tower was used at selected places for measurements of light interception at successive horizontal planes 25 cm apart throught the canopy, and for leaf sampling for measurements

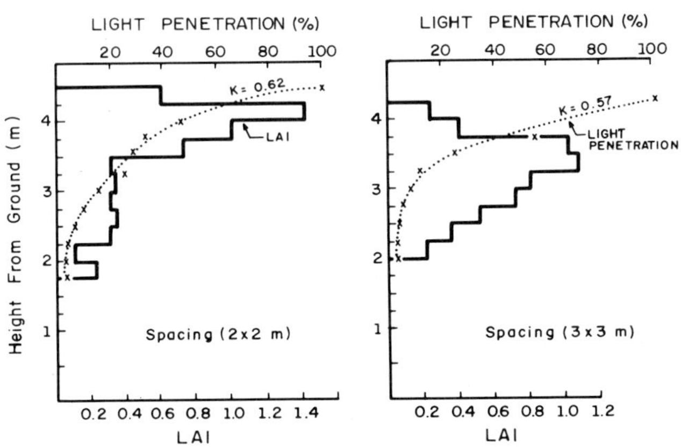

Fig. 9 Leaf area and light penetration through the canopy of 8-year-old cacao trees planted at 2 X 2 and 3 X 3 m spacing.

of leaf area index of the various canopy layers. Leaf angles were also measured with a clynometer.

With the results obtained on light interception at different heights an attempt was made to calculate the extinction coefficient of cacao using the formula proposed by Saeki (1960) based on Beer's law:

$$I = I_0 e^{-KL}$$

where I is the light intensity under the canopy, I_0 is the light intensity at the top of the canopy, K is the light extinction coefficient, and L is the leaf area index. Replicated measurements in six different sites gave a mean value for K of 0.62, varying from 0.57 to 0.67. As shown in Fig. 9, higher values of K were obtained in 2 X 2 m than in 3 X 3 m spacing, but further research is needed to confirm these results. In the 2 X 2 m sites K was not uniform throughout the canopy but decreased in the upper layers. Presumably such variations in K values were due to differences in leaf angles, which were found to vary from 42.6° ± 18° above 3.5 m to 31.5° ± 15° between 2.5 and 3.5 m, and 26.2° ± 3° below 2.5 m. It is well known that the extinction coefficient increases as the leaves come closer to a horizontal orientation.

Light measurements under the canopy of mature cacao trees usually show an interception which varies from 90 to 97%. Assuming a mean K value of 0.62, this would indicate a LAI varying from 3.7 to 5.7 which is an agreement with some measurements of LAI reported for Bahia (Alvim, 1967).

B. Photosynthetic Production

Studies on photosynthesis in cacao have not been carried out with mature plants under field conditions. Photosynthetic (primary) production of cacao plantations can be roughly esimated by extrapolating results obtained from growth analysis and CO_2 absorption studies with seedlings under various light intensities, and taking into account the canopy structure and light profile of mature plants in the field.

Cacao has a relatively low net assimilation rate (NAR) in comparison with other crops (Goodall, 1950; Murray, 1953; Alvim and Grangier, 1966). On the average, NAR varies from 5 to 20 mg dm^{-2} day^{-1} as against 30 to over 100 mg dm^{-2} day^{-1} for other cultivated plants (Watson, 1952). Lemée (1955) found net photosynthesis of cacao to vary from 7 to 27 mg CO_2 dm^{-2} day^{-1}. This is less than 30% and may be as low as 10% of the rate reported for maize, potato, apple, and other crops (Spector, 1956).

Lemée (1955) also observed that photosynthesis of cacao increased from 7 to 22 mg CO_2 dm^{-2} day^{-1} as light intensity was increased from 2 to 25% of full sunlight, but further increases in light intensity caused very little increase in photosynthesis.

The effect of light intensity on photosynthetic production has also been studied by growth analysis (Murray, 1953). Results obtained with 3-month-old seedlings grown under different light intensities (as controlled by shading with different types of plastic screening) are shown in Fig. 10 (P. de T. Alvim, unpublished observation). NAR increased from a mean value of 4.5 to 15.2 mg dm^{-2} day^{-1} when light intensity increased from 5 to 60% of full daylight and showed no further increases at higher light intensities. At 100% exposure NAR appeared to be slightly lower (mean value of 13.6 mg dm^{-2} day^{-1}) but this was not statistically significant. Lemée (1955) also observed reduction in photosynthesis of individual leaves when measurements were made on very bright days, or when solar radiation intensity was above 0.5–0.75 cal cm^{-2} min^{-1}. In leaves developed in the shade, the depressing effect of direct sunlight occurred at lower radiation intensity.

These studies indicate that some degree of shading appears to be beneficial to photosynthesis in individual cacao leaves or to young seedlings having all leaves exposed to sunlight. Obviously this conclusion does not apply to whole plants or to plantations in which many leaves are partially to heavily shaded. As previously mentioned, experiments on effects of shading on mature cacao showed that fruit production increased in proportion to increased illumination up to maximum light intensity.

One of the features that differentiates cacao from typical sun-loving species is its high degree of tolerance of heavy shade. This tolerance is indicated not only by data in Fig. 10 but also by studies on light compensation point (Alvim, 1960a, 1967). The fact that cacao can be grown at close spacing and with a relatively high LAI is further evidence of its shade tolerance.

On the basis of the above mentioned results it is not possible to estimate with precision the mean photosynthetic rate of cacao leaves in mature plants under

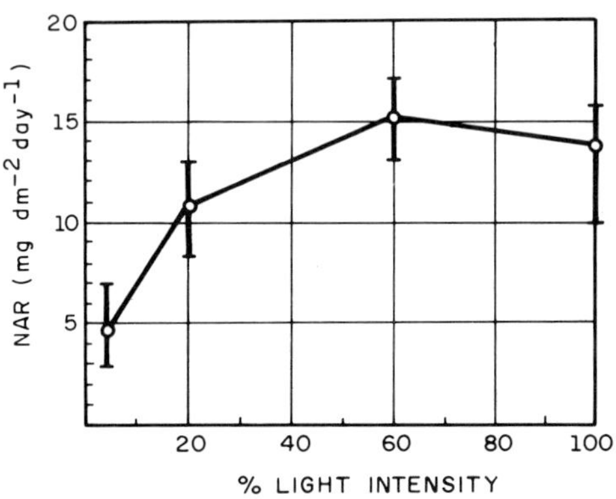

Fig. 10 Effect of light intensity on net assimilation rate (NAR) of 3-month-old cacao seedlings. Vertical lines indicate range of variation among replications.

field conditions. However, from data on canopy structure and light profiles, together with data on NAR responses to light intensity, it seems reasonable to estimate a mean NAR within the range of 7–10 mg dm^{-2} day^{-1} for the whole canopy. Extrapolating these theoretical figures to a cacao plantation showing a mean LAI of 5 would lead to an estimated rate of dry matter production varying from 3.5 to 5.0 mg dm^{-2} day^{-1}, which is equivalent to a primary production of 12.8–18.2 ton ha^{-1} $year^{-1}$.

C. Fruit Bearing Capacity

Cacao yield varies widely from place to place, depending on cultural practices and on the planting material used. In most producing areas where traditional methods are used the mean yield varies from 300 to 500 kg of dry beans ha^{-1} $year^{-1}$. With high-yield cultivars and improved cultural practices, or in places with exceptionally fertile soils, it is possible to obtain a crop of 2000–3000 kg ha^{-1} $year^{-1}$. Very likely, the record yield of about 3700 kg ha^{-1} $year^{-1}$ reported for some unshaded and fertilized experimental plots (Ahenkorah *et al.*, 1974) is close to the maximum potential yield of the presently known cultivars.

The dry weight of beans comprises about one-third of the dry weight of a cacao pod (Alvim, 1967). Hence, a high yield 2000–3000 kg ha^{-1} $year^{-1}$ would correspond to a total fruit dry weight of 6000–9000 kg ha^{-1} $year^{-1}$. This is equivalent to about 50% of our estimate for the annual photosynthetic (primary) production, assuming a LAI of 5 and a NAR varying from 7 to 10 mg dm^{-2} day^{-1}.

When studying the fruit bearing capacity of cacao, attention should be given to the physiological disorder known as cherelle wilt characterized by yellowing and subsequent shriveling and browning of young fruits. From a physiological point of view, this disease is probably analogous to fruit shedding common in many fruit trees, such as citrus, avocado, mango, apple, etc., except that in cacao most pods do not drop but remain shriveled on the plant. Any treatment that increases fruit yield also increases the number of fruits which wilt or shed. This indicates that cherelle wilt is controlled, in the same way as fruit shedding of other species, by an internal correlative mechanism.

Two types of internal competition appear to be involved in induction of cherelle wilt: (a) competition between growing pods and (b) competition between new leaf flushes and the young fruits. Also, any condition that reduces availability of photosynthates to the growing pods, either by lowering photosynthesis or inhibiting translocation of photosynthetic products, also increases the incidence of cherelle wilt. This explains why the malady becomes more severe during extended periods of moisture stress or excessive rainfall (Alvim, 1954). On the other hand, wilt is reduced under conditions which favor high photosynthetic production as, for example, when cacao is grown without shade. Asomaning *et al.* (1971) concluded that the beneficial effect of reducing shade on cacao yield was basically the result of greatly reduced incidence of cherelle wilt. They also observed that increases in yield following fertilizer application reflected increased pod setting rather than reduced cherelle wilting. Losses by wilting were also reduced by adding fertilizers but to a much lesser extent than by shade removal, "suggesting that minerals are not as critical as certain photosynthetic products, probably carbohydrates." Those conclusions are in agreement with an earlier hypothesis of Alvim (1954, 1967) that the main cause of cherelle wilt is a carbohydrate strain brought about by internal competition or by inhibition of photosynthesis or translocation of carbohydrates.

Evidence that competition between growing pods plays a decisive role in cherelle wilt was first presented by Humphries (1943) who demonstrated that the young fruits tended to wilt at progressively earlier stages as the fruiting season advanced. It is well established that the fruits are subject to wilting only during the first 70–80 days of their life cycle. According to Nichols (1964, 1965) this age limit is related to some anatomical changes, with the beginning of the nonwilting phase apparently occurring soon after the vascular bundles become fully differentiated. Nichols also observed that during the wilting phase growth involved cell division as well as enlargement, whereas during the nonwilting phase only cell enlargement occurred. Alvim (1967) reported that this second phase was also characterized by a very high growth rate. Dry weight increment was from 500 to 700 mg day^{-1} at the beginning of the phase to over 2 g day^{-1} at the period of maximum growth (14–16 weeks after fruit set). Pods in the nonwilting phase are therefore much stronger sinks than those in the wilting phase.

Several investigators (Pound, 1933; Voelcker, 1938; Humphries, 1943; Alvim, 1954) have shown that cherelle wilt is also closely correlated with vegetative growth of the plant and is particularly severe during or soon after a period of intensive leaf flushing. This indicates that the sink strength of growing leaves is also higher than that of young pods.

Figure 11 gives monthly data on fruit setting and cherelle wilting in an experimental plot of 12 clonal plants in Bahia over a 2-year period. As also observed in Ghana (Asomaning *et al.*, 1971) cherelle wilt incidence followed a similar seasonal pattern of pod setting, with a time lag of 1–2 months. Wilting accounted for loss of about 75% of the fruits which set. Asomaning *et al.* (1971) also reported losses of 71 to 86% because of wilting.

The fruit bearing capacity of the tree, and consequently its final yield, exhibit a seasonal pattern which appears to be greatly affected by seasonal changes in the ratio pod set:wilted cherelles. Thus the absence of a crop in Bahia during January to March appears to result not only from reduced flowering during the

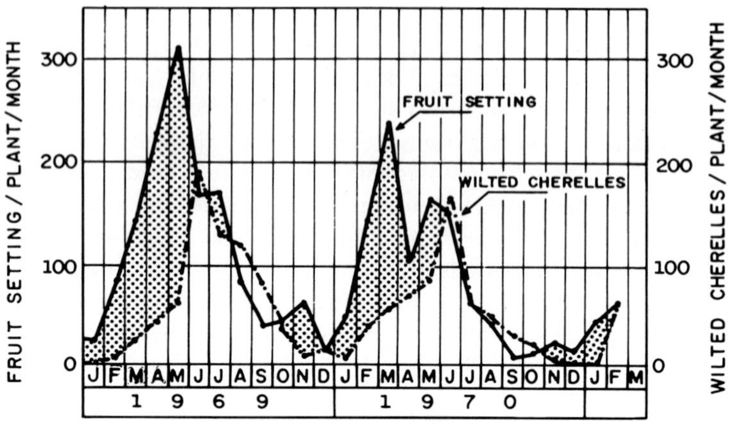

Fig. 11 Seasonal variation in fruit setting and cherelle wilt at monthly interval in Juçari, Bahia, from January 1969 to February 1971. Mean values for 12 clonal plants. (From Alvim *et al.*, 1972.)

months of July to September, as previously suggested (Alvim, 1967), but also—and perhaps chiefly—because cherelle wilt incidence is higher than fruit set during the months of August to mid October (Fig. 11).

The seasonal pattern of crop harvesting also appears to be related to changes in vegetative growth. Figure 12 shows the monthly numbers of harvested pods and developing pods in an experimental area in Bahia. On the basis of a previous study on changes in pod dry weight as related to fruit age (Alvim, 1967), and using the final number of harvested pods as a reference, the total dry weight of the hanging pods was estimated. Wilted cherelles were not included in this estimate. These data showed that the period of maximum fruit load on a dry weight basis did not coincide with the period of maximum number of pods per tree. As a rule, the latter occurred from March to July with the peak in May–June, whereas the former came about three months later, with the peak in August–September. The period of minimum fruit load occurred between December and February.

As will be seen later, cacao plants in Bahia usually are in complete vegetative rest between May and September when fruit load increases and they show periodic growth flushes between October and April when the fruit load decreases. Competition between vegetative growth (flushing, root, and cambial growth) and young fruits could be an important factor determining these changes in fruit load during the year. Hence, when external conditions are favorable for growth (October–April) the strength of the vegetative sink increases and fruit-bearing capacity decreases; when the environment does not favor vegetative growth (May–September), the vegetative sink decreases and fruit load

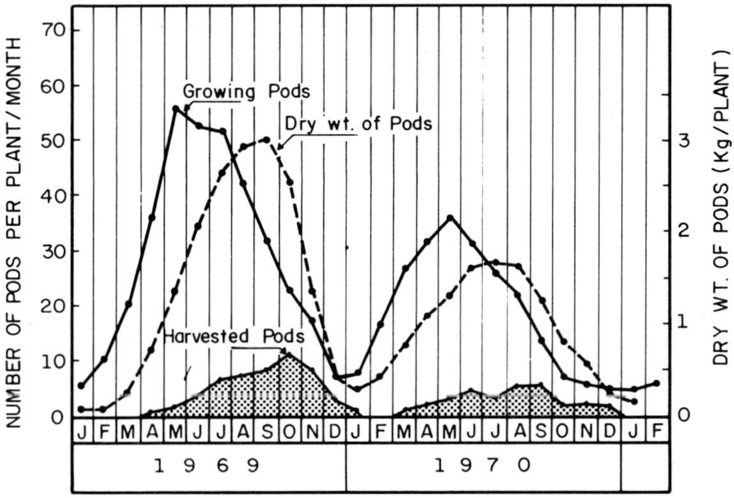

Fig. 12 Seasonal variation in pod numbers and pod dry weight for 12 clonal plants in Juçari, Bahia, from January 1969 to February 1971. (From Alvim *et al.*, 1972.)

increases. This hypothesis assumes that changes in fruit load during the year are affected by environmental effects on vegetative growth, and this in turn affects fruiting. If this assumption is correct, a period of vegetative rest should be of benefit for increasing fruit-bearing capacity, possibly by preventing or reducing cherelle wilt.

VI. ENVIRONMENTAL EFFECTS ON GROWTH AND FLOWERING

Cacao is an intermittently growing evergreen which exhibits periods of leaf flushes alternating with periods of vegetative rest. In most places, adult plants show two or three intensive flushes of shoot growth during the year and an equal number of less pronounced ones, with only a few branches renewing their leaves. As a rule there is a "rest period" of about 3–4 months during which flushing is negligible or absent. Flowering also shows much variation during the year, but in most regions a few flowers are present throughout the year.

It is well established that juvenile cacao plants flush at a more or less fixed interval of approximately two months and usually do not show any synchronism among themselves with regard to their growth periodicity (Greathouse *et al.*, 1971). This suggests that flushing of young plants is under the control of an endogenous mechanism. According to Borchert (1973), such an endogenous mechanism could be related to a feedback interaction between two potentially continuous processes, such as shoot-and-root growth, if the slower process is rate limiting for the faster one. In other words, the rhythmic behavior is a consequence of feedback mechanisms needed for maintaining a constant shoot:root ratio.

Whatever the nature of the internal mechanism associated with rhythmic growth and flowering of cacao, flushing and flowering periodicity in mature plants follow a "seasonal" pattern which varies according to the region, but within a region usually shows little change from year to year. This suggests that flushing and flowering of mature cacao are under the control of some seasonal change in environmental conditions. If an endogenous mechanism exists in mature plants, this appears to be commonly overshadowed by the effect of exogenous factors.

A. Flushing Rhythm

Early theories about the effect of environmental factors on growth rhythms of cacao were reviewed by Alvim (1967). More recent studies carried out in Bahia (Alvim *et al.,* 1969) support a theory first suggested by McDonald (1933) which emphasizes the importance of soil moisture and/or air relative humidity on flushing. On the basis of field studies in Brazil, the following hypothesis was proposed to explain the relationship between moisture availability and flushing (Alvim *et al.,* 1969): (a) decreased rainfall (or increased moisture stress) induced leaf abscission, (b) leaf abscission breaks bud dormancy because of reduced production of an inhibitor originating in leaves, (c) flushing occurs soon after leaf abscission or, under conditions of severe moisture stress, after the onset of rains.

Flushing and leaf fall are closely related phenomena (Alvim, 1967; Boyer, 1974). Intensive leaf fall invariably occurs simultaneously with, or just before or after, an intensive flush or shoot growth. Ordinarily, the main flush in Bahia occurs in September–October and is followed by two or three minor flushes between November and April, with flushing in February–March usually less important. Maximum leaf fall as a rule takes place soon after the intensive growth flush of September–October. From May to mid September the plants remain in a steady state, with practically no flushing and very limited leaf-fall taking place.

A recent experiment with 3-year-old cacao trees grown in large pots (250 liters) and submitted to different water regimes illustrated the relationship between soil moisture status, variation in stem diameter, leaf shedding, and flushing (Alvim and Alvim, 1975). Soil moisture was monitored by means of Bouyoucos gypsum blocks placed at depths of 6, 12, 18, and 24 inches. Trunk diameter was measured with a highly sensitive liquid displacement dendrometer (Alvim, 1975). A decrease in soil moisture caused stem shrinkage as well as chlorosis, especially in older leaves, but very few leaves were shed during wilting. However, soon after the soil was rewatered, there was an abrupt increase in both stem swelling and leaf shedding (Fig. 13). The increase in leaf fall appeared to result from rupture at the abscission layers following the rapid swelling of the shoot bark. Seven to eight days after rewatering, many terminal buds started swelling, initiating a new flushing cycle.

Sale (1970a) and Boyer (1973) reported that each time potted cacao plants were watered following drought vigorous flushing occurred, beginning about 10 days after watering. Many of the flushes were from axillary buds as well as from terminal buds. Frequently irrigated plants flushed less vigorously than droughted ones and only at the terminal buds, with peaks of flushing occurring every 5–6 weeks. In spite of less intensive flushing, the total leaf area was higher in the frequently watered plants than for droughted ones because of reduced leaf shedding or longer leaf life of the former. Sale (1968) also demonstrated under controlled conditions that flushing was not induced by an increase in temperature above 83°F, as previously proposed by Humphries (1944) and Greenwood and Posnette (1950), or by increase in temperature range (thermoperiodicity) as suggested in earlier papers by Alvim (1957). Sale (1968) found, however, that

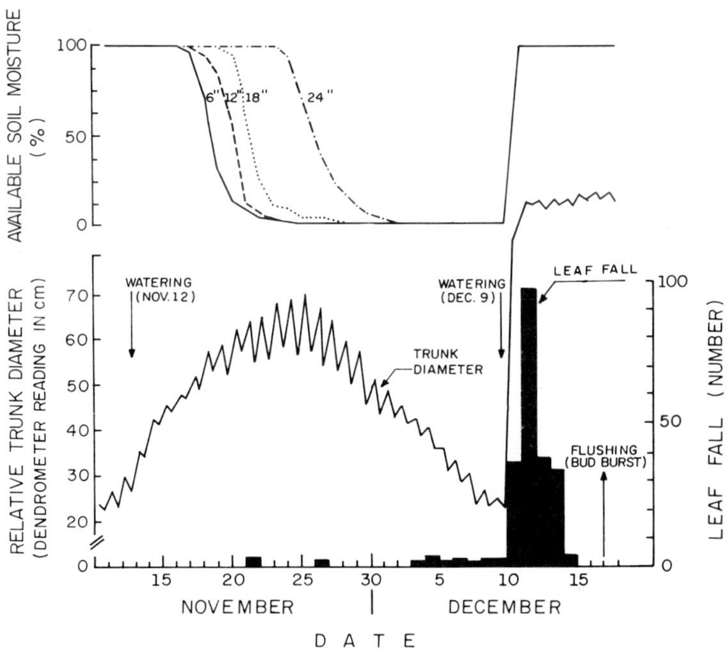

Fig. 13 Effect of soil moisture on leaf fall, flushing, and changes in trunk diameter. (From Alvim and Alvim, 1975.)

the intervals between flushes were reduced when temperature was raised, with plants growing at 74°F flushing approximately every 95 days and plants growing at 80° and 86°F flushing every 36 to 20 days, respectively.

Figure 14 shows data on flushing and leaf shedding of 12-year-old plants in Juçari, Bahia, from July 1969 to November 1971, together with data on rainfall, mean temperature, and cumulative soil moisture deficit and excess. For the sake of simplicity, data on bud break and number of newly formed leaves are presented for two-week intervals.

Cumulative soil moisture deficits were estimated following the method of Smith (1966), assuming a soil depth (root space) of 45 cm, an apparent specific gravity of the soil of 1.4, a field capacity of 32%, and a wilting percentage of 16%. These values give a K value of 0.01, and a water retention capacity of 100

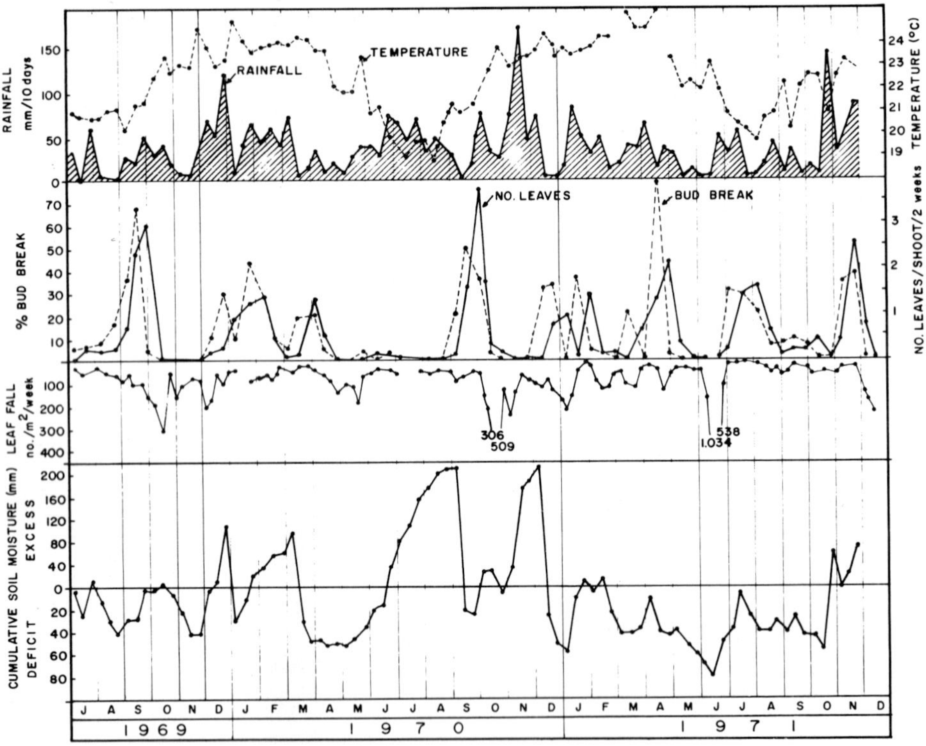

Fig. 14 Seasonal variations in leaf flushing of cacao in Juçari, Bahia, from July 1969 to November 1971. Data are expressed as percent of bud break and number of newly formed leaves, together with data on leaf fall m^{-2} $week^{-1}$, temperature, and rainfall. (From Alvim *et al.*, 1972.)

mm. The cumulative soil moisture excess was derived from the balance between rainfall and potential evapotranspiration. Data on solar radiation, relative humidity, and vapor pressure deficit were also analyzed in this study but are not included in Fig. 12 as they did not show any correlation with growth rhythms.

From a previous study (Alvim *et al.*, 1969), two hypotheses were suggested to explain the correlation between leaf fall and leaf flushing: (a) the concentration of an inhibitor responsible for bud dormancy and originating in the leaves is reduced as a consequence of leaf abscission, thus triggering bud expansion, (b) flushing imposes a strain for nutrients and/or hormones, which accelerates senescence and abscission of old leaves. The first hypothesis is supported by the fact that flushing can be induced during any time of year by heavy pruning or by various degrees of defoliation. Studies in Bahia have shown that leaf removal caused intensive flushing even during a time when plants normally are in a state of rest (e.g., in June).

As shown in Fig. 14 a drought occurred during June/July, normally the wettest period of the year in Bahia, causing off-season leaf shedding which was immediately followed by intensive leaf flushing in July–August. This was the first time in many years that flushing had been observed in Bahia at this time of year.

In support of the second hypothesis involving competition between young and older leaves, on the other hand, leaf-fall in October 1969, October 1970, and December 1971 occurred 2–3 weeks after flushing (Fig. 14). This does not necessarily invalidate the first hypothesis, since the abscission layer is obviously formed before actual leaf shedding occurs and could therefore precede the onset of flushing.

A growth promoter moving from the roots to the shoots through the phloem also appears to play in important role in control of flushing. Indirect evidence for this is the following (Alvim *et al.*, 1969): (a) when a complete ring of bark is removed from the trunk or from one of the main branches, thereby preventing upward phloem transport, flushing is inhibited in the whole tree or in the ringed branch, respectively; (b) when all but one of the main branches are ring barked, the ungirdled branch is stimulated to flush off-season, whereas all ringed branched do not flush, suggesting that the growth promoter originating in the roots moved in higher concentrations to the ungirdled branch; (c) flushing is also inhibited when some lateral roots are pruned.

When data on flushing were examined with reference to cumulative soil moisture deficit and excess (Fig. 14) and to a lesser extent with water supply, periods of increased bud break and subsequent leaf expansion were almost always preceded by a period of soil moisture depletion. In most cases the beginning of a flushing cycle took place soon after the first rainy days following a relatively dry period. Two apparent exceptions were observed, in September and December, 1970, following relatively long periods of continuous heavy rains which caused severe waterlogging. It seems likely that in these cases the plants were under "physiological drought" as a result of poor soil aeration.

After the vegetative buds start to open an adequate moisture supply is essential for fast leaf expansion. This explains why a vigorous flushing cycle is always associated with well-distributed rainfall following a period of moisture stress. In April 1971, for instance, bud break was very intensive, but actual leaf expansion was less pronounced, apparently because limited rainfall from mid April to June. Again, in August and September 1971 some buds were induced to open but many new flushes aborted, seemingly as a result of soil moisture deficit.

The importance of moisture stress in breaking dormancy has been demonstrated for coffee flower buds (Alvim, 1960b). However anthesis could take place only after irrigation or rain. A similar type of response to a relatively dry period followed by a wet one has been observed in citrus, in connection with flowering and flushing cycles (Cassin, 1969), and probably occurs commonly in tropical trees. The term "hydroperiodicity" has been suggested to refer to this relationship between growth and flowering pattern and the exogenous conditions affecting the internal moisture status of the plant (Alvim, 1964).

As mentioned previously, flushing very seldom occurs in Bahia during the months from May to August, which correspond to the period of relatively frequent rainfall, reduced solar radiation, and low temperature. In 1971 intensive leaf fall followed by vigorous flushing occurred for the first time during the month of July, and this was associated with an unusually severe drought which occurred in June (Fig. 14). Temperature and solar radiation showed no apparent relationship to this off-season flush. The data indicated that the absence of flushing normally observed from May to August in Bahia cannot be attributed to low temperature or low solar radiation as previously proposed (Alvim, 1967), but to the normal absence of moisture stress during that period.

In Sales's experiments (1970a) frequently watered plants did not flush at all after they reached the age of about 2 years, whereas plants of the same age but previously submitted to soil moisture stress exhibited vigorous flushing. Sale carried out his experiments with plants that were 12–30 months old and it seems likely that the "juvenility effect" overshadowed responses to the water regime during the earlier part of his experiments, i.e., when plants were younger. He agreed with the previous suggestion of Alvim *et al.* (1969) that water stress caused the buds to develop to a stage at which they were ready to grow, and appeared to be a "necessary requirement for the plants to continue to either initiate or expand new leaves." Sale recognized that this conclusion did not invalidate the suggestion of Cheesman (1948) "that the absence of a dry season may be wholly favourable to cacao—for under most field conditions it is probable that soil water would be depleted occasionally to at least 50%, and thus provide any growth check that might be necessary for continued flushing."

Cacao growing without shade flushes much more intensively than does shaded cacao. This probably is because of the higher internal moisture stress in unshaded plants (Alvim *et al.*, 1972).

Moisture stress increases the abscisic acid (ABA) content of cacao leaves (R. Alvim *et al.*, 1974). The increase in ABA is known to induce formation of abscission layer as well as to impose bud dormancy (Smith, 1967; El-Antably *et al.*, 1967). R. Alvim *et al.* (1974) also demonstrated that defoliation, which is known to induce flushing, not only caused a decrease in ABA content, but also an increase in the cytokinin activity of cacao buds.

Based on these findings, an analogy was made between the processes leading to an interflush period in cacao and the gradual decrease of extension growth during the preparative stages for dormancy in temperate species (R. Alvim *et al.*, 1974). In both cases, an inhibitor–promoter balance appears to be involved. It may be that, with an increasing leaf area, more and more ABA is translocated from the leaves to the growing points and larger amounts of root cytokinins are diverted to the leaves. The inhibitor–promoter ratio in the shoot tips would consequently be gradually increased. A situation may be reached in which the leaf area attains a critical value, at which the action of ABA totally overcomes the effects of cytokinins and other promoters. For growth to be resumed following this stage of total correlative inhibition, the critical leaf area would have to be somewhat reduced. Perhaps at this stage a dry spell, which appears to be obligatory for renewal of growth in some species, exerts its role. A critical number of water-stressed leaves could increase ABA levels to the point of inducing formation of abscission layers. These layers would act as barriers against the diversion of root cytokinins to the leaves and of leaf ABA to the shoot tips, thus triggering growth flushes.

In a juvenile plant, which flushes episodically throughout the year, the leaf area might not be adequate to maintain inhibition of the terminal bud through an adverse ABA/cytokinin balance, when large supplies of cytokinins are available from a rapidly developing root system. This model resembles the one suggested by Borchert (1973) to explain endogenous rhythms in young plants, except that the stimulus for the rhythmic behavior would come from changes in hormone balance resulting from changes in the shoot:root ratio. In mature cacao trees, which are more likely to develop internal water deficits due to a higher resistance to water transport, the change in the ABA/cytokinin balance necessary for flushing would result from moisture stress induced by changes in climatic conditions. Hence, the well-documented synchronization of flushing patterns among mature plants in the field appears to result from an exogenous, hydroperiodic stimulus leading to changes in internal moisture status, and hence in the ABA/cytokinin balance.

B. Flowering

The age at which cacao starts producing flowers in the field varies greatly with genetic origin and cultural practices. Under good management, most of the new hybrids now being planted in Brazil start flowering at an age of about 18 months after transplanting to the field. An extreme case of genetically controlled premature flowering was observed by Soria (1961) on some seedlings of pure criollo cacao from Mexico, which began to produce normal flowers at the age of three months. Cotyledons still had not fallen from these plants. Plants growing in nutrient solution under greenhouse conditions in Bahia usually start producing flowers at the age of about 12 months from seed. Unselected amelonado plants under the traditional cultural practices used by farmers in Bahia require from 3 to 4 years, on the average, to begin flowering.

As previously reported (Alvim, 1967), the flowering cycle of mature cacao plants in Bahia follows a clearly established seasonal pattern, with very few flowers appearing in July–October. In young plants flowering is reduced but not completely arrested during the same period. The difference between mature and young plants with regard to flowering has recently been confirmed in a study comparing the flowering cycle of the cultivar Catongo aged 5, 12, and 23 years growing in neighboring areas, in Uruçuca, Bahia. To estimate flowering intensity, the soil surface under each tree was covered with a layer of sand and the number of unfertilized flowers falling on the sand was counted at intervals of two or three days (Fig. 15). Seasonal variability in flowering increased in direct proportion to plant age. The 23-year-old plants produced about ten times as many flowers as the 5-year-old plants during the peak of flowering (February–April) and about ten times less during the low flowering period (August–September). The data for the 12-year-old plants fell between the other two age groups.

It may be possible that the microclimate around the plant rather than plant age was responsible for the difference in flowering and cropping patterns between mature and young plants. Because of their lower leaf area index, young plants receive more solar energy per unit leaf area than older plants, and this might be an important factor in the difference in their response to variations in climate. Another possibility is that older plants, because of their larger proportion of nonphotosynthetic tissues (roots, stems, etc.) which act as a sink for photosynthates, are more readily depleted of flower-forming substances than younger plants.

A chemical stimulus originating in the leaves and moving downward through the inner bark (phloem) appears to play an important role in the mechanism of flowering. This can be demonstrated by removing a ring of bark from the trunk of mature cacao trees. Studies in Brazil have shown that, following such stem girdling during the normal blossoming season, flowering was very profuse on the

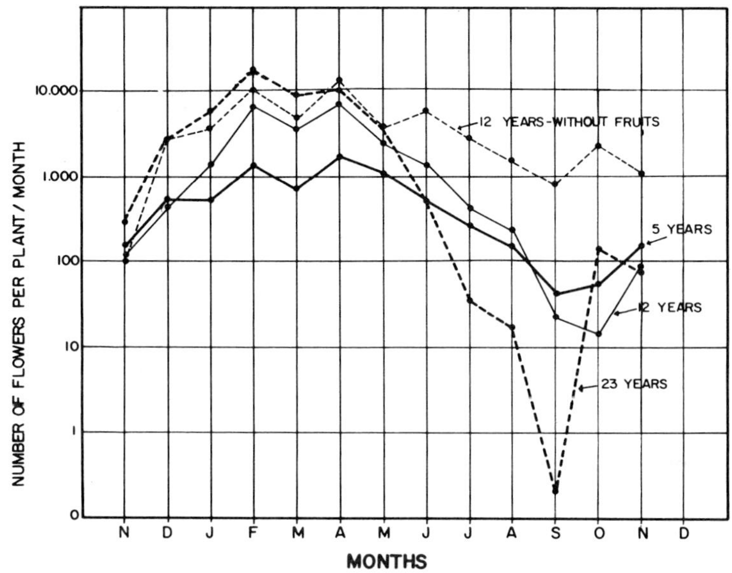

Fig. 15 Seasonal variations in flower production in Catongo trees aged 5, 12, and 23 years, in Uruçuca, Bahia, from November 1970 to December 1971 (average for eight trees in each group). Data for a sterile 12-year-old tree which produced only one pod during the period of the study are included to illustrate the effect of fruit sink on flowering. (From Alvim *et al.,* 1972.)

portion of the trunk above the ring and was completely arrested below the ring. This response occurred within 2–3 weeks following treatment. Accumulation of photosynthates and/or flowering hormones is apparently responsible for the enhancement of flowering above the stem girdle. Similar responses to bark ringing has been reported for citrus (Furr *et al.,* 1947) and other species (Kozlowski, 1971).

The fruit load of the plant also affects flowering intensity. This was observed during the course of our study on flowering in plants at different ages. One of the 12-year-old Catongo plants included by chance in the experiment was practically sterile, as it produced only one pod. The flowering cycle of this plant is also included in Fig. 15 (12 years, without fruit). It not only blossomed much more heavily than the normal bearing 12-year-old plants, but showed a relatively slight decrease in flowering from July to November when other plants, irrespective of age, showed much lower flowering intensity. Together with our observations on the effect of ringing, these results seem to indicate that fruits compete with flowers for a substance or substances (which could very well be carbohydrates) whose concentration is related to flowering intensity. Competition between fruits and flowers is known to play an important role in determining flowering intensity of other tree crops, such as citrus (Lenz, 1969) and many forest trees (Kozlowski, 1971). Flowering cycles of cacao were studied in Juçari, Bahia, for a period of 38 months (January, 1968 to February, 1971), using 12 clonal plants which were about 8 years old at the beginning of the study. The average number of flowers/day/plant is given in Fig. 16 for 10-day intervals. Data on cumulative soil moisture deficit and excess, calculated by the method mentioned under growth rhythms and mean temperature data, are also presented. Other climatological data, such as solar radiation, relative humidity, and vapor pressure deficit, did not show any obvious relationship with the flowering data and will not be included in this discussion.

Intensive flowering usually occurred from November to June; least flowering from August to October. Previous studies at another station (Uruçuca) with older trees (Alvim, 1967) showed minimum flowering from June to September. It is not known if this discrepancy is due to differences in climate, age of trees, varietal differences, or methods of study. However, data given in Figs. 15 and 16

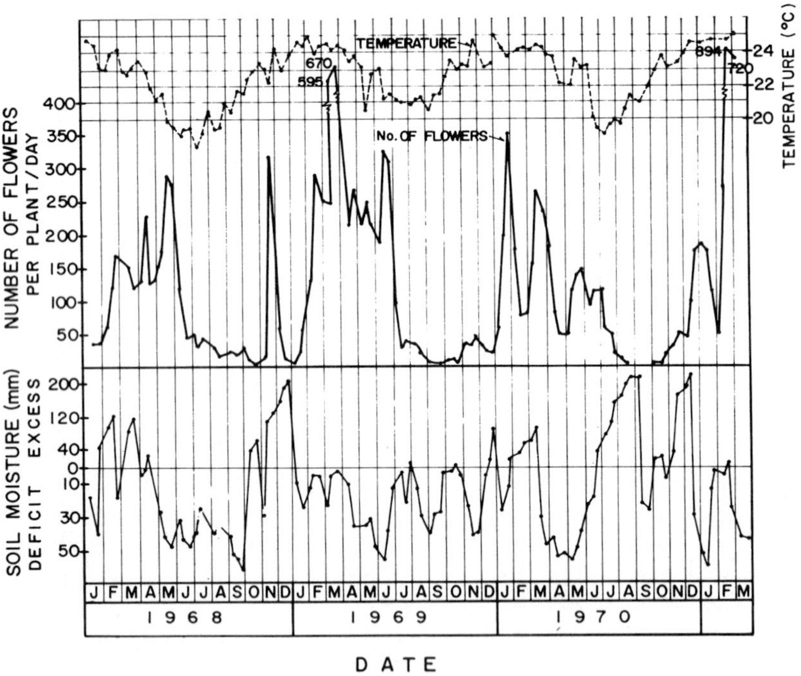

Fig. 16 Seasonal variation in intensity of cacao as expressed in number of flowers per plant per day (mean figure for 12 clonal plants) in Juçari, together with data on temperature and cumulative soil moisture deficit and excess, at 10 days intervals from January 1968 to March 1971. (From Alvim *et al.*, 1972.)

are considered to be more accurate than those previously reported, and probably are more reliable.

Previous studies on flowering of cacao in Bahia (Alvim, 1967) suggested that a decrease in flowering from June to September was the result of low temperature. Because the temperature in Bahia starts to decrease in May when flowering is still quite intensive, it was postulated that the effect of low temperature was indirect in the sense that it first affected vegetative growth, which in turn affected flowering, this accounting for the time lag between supposed cause and effect. This theory needs to be revised in view of recent findings. As shown in Fig. 14, temperature was not related to shoot growth rhythms. Intensive bud break and leaf expansion could take place even during time when temperatures were relatively low (e.g., in July 1971). Sale (1969a) also did not find support for an "indirect" effect of temperature on flowering. He confirmed that low temperature decreased flowering, but the effect seemed to be direct. It now appears that reduced flowering in Bahia from June–July to September–October is caused by internal competition between fruits and flowers, with the effect of low temperature apparently being less important. As shown in Fig. 12 the period of maximum fruit load in Bahia, as expressed in total dry weight of hanging pods, coincides with the period of minimum flowering.

The flowering season for plants in our studies occurred from November to June and the beginning of this season coincided each year with the time when plants had the lowest fruit load (Fig. 12). This lends support to the emphasis on the importance of internal competition between fruits and flowers. Our data suggest that as long as this competition exists flowering shows practically no response to changes in environmental conditions. Only after the plants reach the period of minimum fruit load, which means decreased internal competition between fruits and flowers, does blossoming respond to environmental conditions, and exhibit wide fluctuation from time to time.

It is well established that cacao flowering is inhibited during periods of moisture stress (Alvim, 1964; Sale, 1970a) and becomes particularly profuse after the first heavy rains followed by extended dry period (Alvim, 1967). Comparisons of the data on cumulative soil moisture deficit and excess with the number of flowers per plant during the flowering season (Fig. 16) show that period of increased flowering are usually preceded by a relatively dry period followed by a wet period and that decreased flowering is associated either with a dry spell (e.g., April 1970) or excessive soil moisture (e.g., December 1968, December 1969, February 1970, and November 1970). In some instances, such as in December 1970 and January 1971, increased flowering appeared to result from decreased soil moisture following a period of moisture excess.

Sale (1970a) obtained profuse flowering in potted cacao plants whenever the soil was watered following a period of moisture stress, and concluded that flower initiation had been enhanced during the dry period and only flower growth was inhibited by moisture deficiency. In coffee, the flower buds remain in a state of complete rest for as long as the plants are frequently watered and kept under high atmospheric humidity, a "dry shock" apparently being necessary to break the resting period (Alvim, 1960b). As pointed out by Sale (1970a) this does not appear to be the case with cacao, because the flower buds do not go through a rest period imposed by low water stress and some flowers always occur even under conditions of frequent watering or high humidity. Sale (1970b) also demonstrated, under controlled conditions, that flowering can be induced by transferring plants from a low (50–60%) or medium (70–80%) relative humidity to a high one (90–95%).

It is worth pointing out that flowering of citrus in tropical areas was controlled by alternations between dry and rainy periods and that, in irrigated citrus orchards, it is possible to induce out-of-season flowering by irrigation following a drought (Cassin, 1969).

From the foregoing discussion it appears that the major factors controlling flowering of cacao are, internally, competition between fruits and flowers, and externally, an alternation between dry and wet periods. Hence, hydroperiodicity of the environment appears to be the most important exogenous stimulus for inducing opening of vegetative buds as well as for enhancing flowering.

REFERENCES

Ahenkorah, Y., Akrofi, E. S., and Adri, A. K. (1974). The end of the first cocoa shade and manurial experiment at the Cocoa Research Institute of Ghana. *J. Hortic. Sci.* **49**, 43–51.

Alvim, P. de T. (1954). Studies on the cause of cherelle wilt of cacao. *Turrialba* **4**, 72–78.

Alvim, P. de T. (1957), Fatores que controlam o lançamento do cacaueiro. *Conf. Interam. Cacau, 6th, 1956* pp. 117–125.

Alvim, P. de T. (1960a). El problema del sombreamiento del cacao bajo el punto de vista fisiológico. *Conf. Interam. Cacau, 7th, 1958* pp. 294–303.

Alvim, P. de T. (1960b). Moisture stress as a requirement for flowering of coffee. *Science* **132**, 354.

Alvim, P. de T. (1964). Tree growth periodicity in tropical climates. *In* "The Formation of Wood in Forest Trees" (M. H. Zimmermann, ed.), pp. 479–495. Academic Press, New York.

Alvim, P. de T. (1967). Eco-physiology of the cacao tree. *Conf. Int. Rech. Agron. Cacaoyeres, 1965* pp. 23–25.

Alvim, P. de T. (1972). Sombra e adubação na cultura do cacaueiro. *Cacau Atual. (Brasil)* **9**(1), 2–5.

Alvim, P. de T. (1975). A new dendrometer for monitoring cambium activity and changes in the internal water status of plants. *Turrialba* **25**, 445–447.

Alvim, P. de T., and Alvim, R. (1975). Hidroperiodicidade dos fluxos foliares do cacaueiro. *Inf. Tec. CEPEC, 1975* pp. 42–48.

Alvim, P. de T., and Cabala Rosand, F. P. (1974). Um novo sistema de representação gráfica da fertilidade dos solos para cacau. *Cacau Atual. (Brasil)* **11**(2), 2–6.

Alvim, P. de T., and Grangier, A., Jr. (1966). Estudio sobre el analysis crecimiento de plántulas de cacao de diferente origen genética. *Cacao* **11**, 1–3.

Alvim, P. de T., and Pereira, C. P. (1972). Sombra e espaçmento nas plantações de cacau da Bahia. *Cacau Atual. (Brasil)* 9(3), 2–3.

Alvim, P. de T., Machado, A. D., and Grangier, A., Jr. (1969). Alguns estudos sobre as relações de água, solo e crescimento do cacaueiro. *Mem. Conf. Int. Pesquis. Cacau, 2nd, 1967* pp. 316–326.

Alvim, P. de T., Machado, A. D., and Vello, F. (1972). Physiological responses of cacao to environmental factors. *Int. Cacao Res. Conf., 4th, 1972* pp. 210–225.

Alvim, P. de T., Alvim, R., and Smith, G. E. (1974). Ação de quebraventos sobre o crescimento de mudas de cacau. *Inf. Tec. CEPEC, 1974* pp. 29–30.

Alvim, R., Alvim, P. de T., Lorenzini, R., and Saunders, P. F. (1974). The possible role of abscisic acid and cytokinins in growth rhythms of *Theobroma cacao* L. *Rev. Theobroma* **4**(3), 3–12.

Asomaning, E. J. A., Kwakwa, R. S., and Hutcheon, W. V. (1971). Physiological studies on an Amazon shade and fertilizer trial at the Cocoa Research Institute, Ghana. *J. Agric. Sci.* **4,** Part 1, 47–64.

Berbert, P. R. (1976). Influência das condições climáticas na composição química e características físicas da manteiga do cacau. *Rev. Theobroma* **6**, 67–76.

Berbert, P. R., and Alvim, P. de T. (1972). Fatores que afetam o índice de iodo da manteiga de cacau do Brasil. *Rev. Theobroma* **2**(3), 3–16.

Borchert, R. (1973). Simulation of rhythmic tree growth under constant conditions. *Physiol. Plant.* **29,** 173–180.

Boyer, J. (1973). Etude particulière des facteurs hydriques de la croissance des cacaoyers. *Cafe, Cacao, The* **15**, 189–201.

Boyer, J. (1974). Etude écophysiologique du développement de cacaoyers cultivés au Cameroun. *Cafe, Cacao, The* **18**, 3–30.

Cabala Rosand, F. P., de Miranda, E. R., and do Prado, E. P. (1970). Efeitos da remoção de sombra e da aplicação de fertilizantes sobre a produção do cacaueiro na Bahia. *Turrialba* **15**, 1–10.

Cadima Z., A., and Alvim, P. de T. (1973). Algunos factores del suelo asociados con la productividad del cacaotero en Bahia, Brasil. *Rev. Theobroma* **3**(2), 13–26.

Cassin, J. (1969). The influence of climate upon the blooming of citrus in tropical areas. *Proc. Int. Citrus Symp. 1st, 1968* Vol. 1, Sect 1, pp. 315–323.

Cheesman, E. E. (1948). "Report on the Potentialities for the Cultivation of Cocoa in Malaya, Sarawak, and N. Borneo." HM Stationery Office, London.

Cuatrecasas, J. (1964). "Cacao and its Allies: A Taxonomic Revision of the Genus Theobroma." Smithsonian Institute, Washington, D.C.

Cunningham, R. K., and Arnold, P. W. (1962). The shade and fertilizer requirements of cacao (*Theobroma cacao*) in Ghana. *J. Sci. Food Agric.* **13**, 213–221.

Cunningham, R. K., and Burridge, J. C. (1960). The growth of cacao (*Theobroma cacao*) with and without shade. *Ann Bot. (London)* [N.S.] **24,** 458–462.

El-Antably, H. M. M., Wareing, P. F., and Hillman, J. (1967). Some physiological responses to D. L. Abscisin (Dormin). *Planta* **73,** 74–90.

Ernholm, I. (1948). "Cacao Production in South America, Historical Development and Present Geographical Distribution." Gothemburg, Sweden.

Evans, H., and Murray, D. B. (1953). A shade and fertilizer experiment on young cacao. *Annu. Rep. Cacao Res., 1945–1951, Imp. Coll. Trop. Agric., St. Augustine, Trinidad* pp. 67–76.

Furr, J. R., Cooper, W. C., and Reece, P. C. (1947). An investigation of flower formation in adult and juvenile citrus trees. *Am. J. Bot.*, **34**, 1–4.

Goodall, D. W. (1950). Growth analysis of cacao seedlings. *Ann. Bot. (London)* [N.S.] **14**, 291–294.

Greathouse, D. C., Laetsch, W. M., and Phinney, B. O. (1971). The shoot growth rhythm of a tropical tree, *Theobroma cacao. Am. J. Bot.* **58**, 281–286.

Greenwood, M., and Posnette, A. F. (1950). The growth flushes of cocoa. *J. Hortic. Sci.* **25**, 164–174.

Hardy, F. (1958). Cacao soils. *Soil Crop Sci. Soc. Fla., Proc.* **18**, 75–87.

Hardy, F. (1960). "Cacao Manual." Inter-American Institute of Agricultural Sciences, Turrialba, Costa Rica.

Humphries, E. C. (1943). Studies in the physiology of *Theobroma cacao,* with special reference to cherelle wilt. *Annu. Rep. Cacao Res., Imp. Coll. Trop. Agric., St. Augustine, Trinidad* pp. 23–27.

Humphries, E. C. (1944). A consideration of the factors controlling the opening of buds in the cacao tree (*Theobroma cacao* L.). *Ann. Bot. (London)* **8**, 259–267.

Kozlowski, T. T. (1971). "Growth and Development of Trees," Vol. 2. Academic Press, New York.

Lemée, G. (1955). Influence de l'alimentation en eau et de l'ombrage sur l'économie hydrique et la photosynthèse du cacaoyer. *Agron. Trop. (Paris)* **10**, 592–603.

Lenz, F. (1969). Effects of day length and temperature on the vegetative and reproductive growth of Washington navel orange. *Proc. Int. Citrus Symp., 1st, 1968* Vol. 1, Sect. 1, pp. 333–337.

McDonald, J. A. (1933). An environment study of the cacao tree. *Annu. Rep. Cacao Res., 2nd, 1932, Imp. Coll. Trop. Agric., St. Augustine, Trinidad,* pp. 29–38.

Murray, D. B. (1953). A note on the growth analysis of cacao. *Annu. Rep. Cacao Res., 1945–1951, Imp. Coll. Trop. Agric., St. Augustine, Trinidad* pp. 77–78.

Murray, D. B. (1957). The use of shade for cacao. *Conf. Interam. Cacau, 6th, 1956.* pp. 111–116.

Nichols, R. (1964). Studies of fruit development of cacao (*Theobroma cacao*) in relation to cherelle wilt. I. Development of the pericarp. *Ann. Bot. (London)* [N.S.] **28**, 619–625.

Nichols, R. (1965). Studies of fruit development of cacao (*Theobroma cacao*) in relation to cherelle wilt. III. Effects of fruit thinning. *Ann. Bot. (London)* [N.S.] **29**, 197–203.

Pound, P. J. (1933). Studies of fruitfulness in cacao. X. Physiological effects of application of nitrogen, potassium, and phosphorus on the cacao tree. *Annu. Rep. Cacao Res., 2nd, 1932, Imp. Coll. Trop. Agric., St. Augustine, Trinidad* pp. 37–40.

Saeki, T. (1960). Interrelationships between leaf amount, light distribution, and total photosynthesis in a plant comunity. *Bot. Mag.* **73**, 404–408.

Sale, P. J. M. (1968). Flushing and leaf growth of cacao under controlled temperature conditions. *J. Agric. Sci.* **43**, 475–489.

Sale, P. J. M. (1969a). Flowering of cacao under controlled temperature conditions. *J. Hortic. Sci.* **44**, 163–173.

Sale, P. J. M. (1969b). Extension growth of cacao under controlled temperature conditions. *J. Hortic. Sci.* **44**, 189–193.

Sale, P. J. M. (1970a). Growth, flowering and fruiting of cacao under controlled soil moisture conditions. *J. Hortic. Sci.* **45**, 99–118.

Sale, P. J. M. (1970b). Growth and flowering of cacao under controlled atmospheric relative humidities. *J. Hortic. Sci.* **45**, 129–132.

Silva, L. F. (1969) Solos bons para cacau. *Cacau Atual. (Brasil)* **6**(2), 28–31.

Smith, G. W. (1966). The relation between rainfall, soil, water and yield of copra on a coconut estate in Trinidad. *J. Appl. Ecol.* **3**, 117–125.

Smith, O. E. (1967). Abscission physiology of abscisic acid. *In* "Biochemistry and Physiology of Plant Growth Substances" (F. Wightman and G. Setterfield, eds.), pp. 1547–1560. Runge Press, Ottawa.

Smyth, A. J. (1966). "The Selection of Soil for Cocoa," FAO Soils Bull. No. 5. FAO, Rome.

Soria, V., J. (1961). A note on cacao plants which flowered at the age of three months. *Cacao* **6**, 11–12.

Spector, W. S. (1956). "Handbook of Biological Data." Saunders, Philadelphia Pennsylvania.

Voelcker, O. J. (1938). A note on the behavior or cacao flowers after pollination and on the incidence of cherelle wilt. *Annu. Rep. Cacao Res., Imp. Coll. Trop. Agric., St. Augustine, Trinidad* Vol. 7, pp. 5–8.

Watson, D. G. (1952). The physiological basis of variation in yield. *Adv. Agron.* **4**, 101–145.

CHAPTER

11

Rubber

VICENTE H. F. MORAES

Natural rubber holds and will continue to hold a leading position among the commodities of the world. It is relatively a newcomer among the necessities of our advancing civilization, yet its absence would change the conveniences of modern life to drudgery (Seibert, 1947).

I. INTRODUCTION

Rubber (*Hevea brasiliensis* Muell. Arg.) seedlings were first brought to Ceylon in 1876 via the Royal Botanical Garden at Kew, England. The plants were grown from seeds collected in Brazil by Sir Henry Wickham. The introduction of rubber plants to Ceylon was a very decisive step in increasing the importance of rubber as a crop. If domestication had been first attempted in the Amazon, the results would have been discouraging because of the occurrence of South American leaf blight, an endemic disease under Amazon forest conditions. In Ceylon, isolated from the threat of leaf blight, the introduced rubber seedlings grew vigorously.

A few decades elapsed before a strong interest developed in cultivation of rubber. Among the factors that eventually stimulated such an interest were the occurrence of *Hemileia* rust in coffee, deficiency of more fertile soils, and increasing rubber prices (Dijkman, 1951). As soon as the potential of growing rubber was realized, its cultivation spread rapidly through the old Dutch and British Indies.

Hevea brasiliensis probably was the first tropical species to receive the attention of research centers specializing in a single crop. Most natural rubber since then has been produced in the Far East, with considerable recent progress in West Africa. Even though tropical Latin America must cope with the leaf blight disease, there are prospects of overcoming it.

The origin of research data is obviously proportional to the density of the geographical distribution of the crop. It is, however, unfortunate that native populations of the Amazonian region were neglected for such a long time. Early rapid success of the first introductions with a probable restricted genetic potential could explain this.

According to Mann (1940, cited in Seibert, 1947), it seemed fairly clear, however, that all of the rubber produced in Malayia originally came from 27 seedlings of the original Wickham collection that went to Singapore in 1826. Dijkman (1951) stated that 1900 seedlings were sent in 1876 to Ceylon; some to Malaysia, and only two to the Culturtuin Buitenzorg, West Java. Other less important and later introductions are reported by Dijkman (1951).

Wickham collected seeds near Boim, a small village on the river Tapajos, where wild rubber trees occur on the upland, unflooded soils. Seibert (1947) believes the morphological evidence indicates that the wild, upland rubber tree is a widely distributed ecotype derived through introgression of *Hevea guianensis* var. *lutea* into *Hevea brasiliensis. Hevea brasiliensis* occurs on periodically flooded, alluvial soils along the rivers. *Hevea guianensis* var. *lutea* is found only on upland, well-drained soils.

The genus *Hevea* is a natural taxonomic entity and it is easily shown that a given tree belongs to this genus. However, the differences between species are many and complex. The lack of genetic barriers has led to frequent hybridization and this may account for the known ecological plasticity of the cultivated rubber tree. Plasticity may even be increased through breeding programs. This is suggested by the diversity of natural environments inhabited by different species. Hence, *Hevea camporum* and *Hevea nitida* var. *toxicodendroides* are found in plant communities of xerophytic physiognomy on very sandy soils. *Hevea spruceana, H. paludosa,* and *H. microphylla* occur in swampy or permanently inundated areas, and *H. benthamiana,* in the alluvial periodically flooded soils, similar to the lowland *Hevea brasiliensis.*

Except for information on cultivated *Hevea brisiliensis,* almost nothing is known about the ecophysiology of *Hevea* species or even whether there is a significant physiological difference between the upland and lowland *Hevea brasiliensis.* However, brief references are made to other important species which might be applicable in the case of the upland *Hevea brisiliensis.*

II. SOIL CONDITIONS AND ROOT GROWTH

Research at the Rubber Research Institute of Malaysia (1958) with clonal seedlings growing on light soils has provided a detailed description of rooting habits of rubber. Trees up to three years old had surface laterals concentrated close to the trees. At or about four years after planting, the feeder roots of adjacent rows had met. In 5- to 7-year-old trees, the density of roots in the center of the interrow was greater because of intermingling of roots from adjacent rows. In mature trees (15 years or older), root density across the interrows was fairly even and not significantly related to the distance from the trees.

Three-year-old trees had taproots 1.5 m long; lateral roots were 6 to 9 m long. In 7- and 8-year-old trees, taproots were 2.4 m long and some laterals were more than 9 m long. In the deep fertile basaltic soils of Vietnam and Cambodia, taproots up to 10 m deep have been excavated in old stands (G. Losson, personal communication, 1975). An interesting feature was the profuse development of lateral roots all along the taproot. This is not in accord with the widespread belief that the root system consists of only a deep taproot and the lateral and feeder roots, down to a 20 cm depth.

The preceding description applies to clonal seedlings in light soils without hard, ironstone concretions or pans. Planting techniques and soil conditions may change this pattern. For example, old budded stumps with taproots sectioned on mature tissue of thick suberized bark cannot regenerate strong taproots. Poor soil aeration, due to impeded drainage or compaction of heavy soils, also inhibits root growth.

In shallow alluvial soils, the taproots of rubber trees are confined to not more than 30–40 cm of the A horizon. In such soils more vigorous growth of laterals compensates for the decreased growth of taproots. This is the case for both cultivated and native rubber trees. When planted in deep, well-aerated, upland soils, seedlings of the lowland *Hevea brasiliensis* produce long, well-developed taproots.

An unexpected root characteristic was found in *Hevea marajoensis,* a new species recently discovered by J. M. Pires (personal communication, 1975). This species has a xylopode (an underground organ, common to plants of savannas). Each plant has many shoots derived from buds in the xylopode.

Alvim *et al.* (1965) concluded that roots of 3-month-old *Hevea seedlings* were more efficient in absorbing nutrients than cacao roots of the same age. Respiration of *Hevea* roots was about twice as high as that of cacao roots. The shoot-to-root ratio of rubber was lower than in cacao, which has twice the leaf area of rubber.

Endotrophic mycorrhizae were found in *Hevea*, but it could not be demonstrated that the mycorrhizae were able to assist *Hevea* roots in the uptake of nutrients (Waistie, 1965).

A. Physical Characteristics of Soils

Because of the extensive root system, as well as the high oxygen demand associated with high respiration rates, the better soils for *Hevea* must be very deep and well drained, with good moisture retaining capacity. Sandy soils should not have less than 20% clay in the surface horizons, and more at lower depths. Heavy soils must not lose their structure in order to be kept well aerated.

Shallow-rooted trees growing in soils with a hard pan may be uprooted by strong winds. Chan and Pushparajah (1972) cited the Selangor soil series of Malaysia, where a low rubber yield was associated with the presence of a hard pan. Physical factors of soils, especially soil depth and soil drainage, frequently are overlooked with respect to their adverse effects on yield.

B. Chemical Characteristics of Soils

Because of its extensive root system and high nutrient absorbing capacity, *Hevea* can tolerate very acid and infertile soils. It has been found growing in soils with a pH as low as 3.8 (Vollema, 1941). Reasonable economic returns cannot be expected from rubber trees on poor soils, because both establishment and exploitation of *Hevea* are very expensive. The highest yields have been recorded in the fertile basalt-derived soils of Vietnam and Cambodia. This also applies to naturally occurring rubber trees. The highest yields were reported for trees on fertile basalt-derived soils in the territory of Rondonia (Brazil) along the Jaru River (Moraes, 1963).

Early theories on mineral nutrition of rubber trees considered the nutrient loss in the crop to be very low since rubber itself is a hydrocarbon. However, it has been shown more recently that addition of fertilizers is necessary to sustain high yields (Pushparajah *et al.*, 1972; Sivanadyan *et al.*, 1972). These observations will change the traditional view that application of fertilizer should be restricted to young plants.

III. CLIMATIC CONDITIONS

A. Rainfall and Water Balance

Only indirect references to macroclimatic conditions in natural and cultivated areas of production are found in the literature (Dijkman, 1951; Sudhevea, 1971; Bastos and Diniz, 1975).

Table I gives some examples of annual rainfall and Thorntwaite water balances according to Sudhevea (1972). The higher water deficiency recorded in this table is 232 mm, but Laikhê in Cambodia had an annual water deficit of 350 mm (Institut des Recherches sur la Caoutchouc au Cambodge, 1965). However, very high yields have been recorded under these conditions, because of deep root penetration in the fertile basaltic soil. High yields resulted not only because of the drought avoiding adaptation of the plants, but also because of refoliation in a drier atmosphere, unfavorable to fungal attacks. Rainfall and other climatic variables found in the natural habitats of *Hevea brasiliensis* are of the same order of magnitude in areas where rubber is cultivated (Bastos and Diniz, 1975).

B. Temperature

The cultivation of *Hevea brasiliensis* is concentrated between latitudes of 10°N and 10°S of the equator. It grows well at latitudes above 20°S, in the State of São Paulo, Brazil, where it displays a better resistance to frost than coffee (*Coffea arabica* L.). Low temperatures at high altitudes retard growth. At altitudes higher than 200 m there was a 6-month delay in reaching tappable trunk size for each 100 m increase in altitude (Dijkman, 1951). This corresponds to approximately

TABLE I
Water Balance Data from Various Areas Where Rubber Is Cultivated[a,b]

Localities	Rainfall	Evapotranspiration		Water deficiency	Water excess
Djakarta	1797	1540	1308	232	489
Siantar	2705	1665	1665	0	1040
Yangamki	1840	1333	1329	4	511
Malaca	2190	1665	1655	10	535
Singapore	2282	1716	1715	1	567
Medan	1939	1579	1560	11	371
Kuala Lumpur	2499	1709	1705	4	794

[a]From Sudhevea (1972).
[b]All values in millimeters.

a 0.5°C decrease per 100 m of altitude. Based on comparisons of growth rates in Campinas and São Paulo (Brazil), an annual mean temperature of 20°C was considered the lower limit of thermal adaptation of rubber (Sudhevea, 1971).

The apparent effect of low temperature is shown in Fig. 1. This effect, if reproducible, can only be attributed to decrease in rubber regeneration at the bark drainage areas after tapping, since latex flow is greater at lower temperatures and associated lower transpiration rates.

There is little to say with respect to other important environmental variables due to the lack of published quantitative data. Some references will, however, be made to other topics. No quantitative relationship has been shown between light intensity and photosynthesis in *Hevea brasiliensis.* The species is known to be light demanding, however, and regeneration depends on natural opening or clearings in forests. This also is true for other tropical species of the primary forest such as *Goupia glabra* and *Bertholetia excelsa* (Schulz, 1960). However, unlike *Goupia, Hevea brasiliensis* produces a large number of young seedlings (Moraes and Muller, 1976).

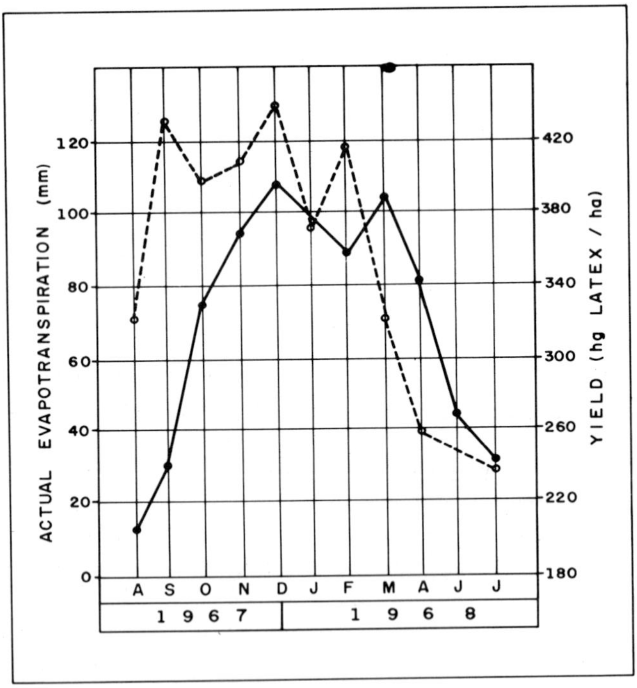

Fig. 1 Actual evapotranspiration and rubber yield, Colina, São Paulo, Brazil. (From Sudhevea, 1971.)

IV. SHOOT GROWTH

A. Seasonal Periodicity

Shoot growth of *Hevea* is characterized by periodic rapid elongation, alternating with periods of inactivity. In young, unbranched trees, new growth flushes occur successively during the year, with leaves of each flush separated into layers. According to Hallé and Martin (1968), flushing coincides with periods of cambial activity and a positive correlation exists between flushing frequency and the number of latex vessel rings in the stem.

Adult trees exhibit prominent annual flushing periodicity. *Heavea brasiliensis, Hevea benthamiana,* and *Heva microphylla* shed their leaves almost completely before the new growth flushes begin. The degree of flushing depends on environmental conditions and individual or clonal variation in the completeness of "wintering." *Hevea spruceana, H. pauciflora, H. camporum, H. rigidifolia, H. nitida,* and *H. Guianensis* retain leaves of the previous year until after the new annual flushes appear and even after the new leaves mature. "Wintering" species go through a dormant, leafless period of two to three weeks, followed sequentially by appearance of inflorescences and then rapid growth of new vegetative flushes. Senescence and leaf abscission begin with onset of the dry period.

Species with persistent leaves produce conspicuous short shoots which appear as terminal or lateral spurs during the maturation of the growth flush. These short shoots, which often are only about 1 cm long, are covered with caducous scales and have as many as 100 very compact nodes. In old portions of branches these short shoots alternate with long shoots. Wintering species, such as *Hevea brasiliensis,* have small or inconspicious short shoots. Naturally or hand-pollinated hybrids show that conspicuous short shoots and the persistence of leaves are dominant characters.

In addition to the main annual period of refoliation, growth flushes also occur sporadically with a frequency and intensity that vary among individuals and clones. Such sporadic flushing seems to be characteristic in clones with a dense canopy and also a late "wintering" habit. Clonal differences in the "wintering" habit are only evident in regions with a gradual decrease in rainfall. In Thailand, Cambodia, and Vietnam, all clones start wintering at almost the same time, and the trees remain leafless for a long time (Boychou, 1962). Differences in the nature of the dry season are shown by the pronounced dry season of Bogambo (ex Belgian Congo) and the less well-defined dry period of Yangambi (Evers *et al.,* 1960). Wintering tends to occur faster and is more complete in clones with light foliage than in those with dense foliage. In the Köppen Afi climate near Belem (Para, Brazil) there was a difference of two and a half months between senescence, leaf fall, and refoliation of some early and late wintering oriental clones planted on alluvial soil (Moraes, 1972). Clones PB 86 and PB 186 were early wintering and AV 1301 and AV 1279 wintered later (Moraes, 1972). Clone PB 86 was also reported to be an early wintering clone in Malaysia and the Ivory Coast (Martin, 1969).

Even near Belem, where there is no definite dry season, late wintering clones grow better than early wintering ones, since the former are affected less by South American Leaf Blight because they refoliate during months of low rainfall. In Malaysia where leaf fall occurs during a short main dry season, an early wintering habit is necessary for avoiding attacks by *Oidium* and *Phytophthora,* because refoliation occurs during the drier period (Ho *et al.*, 1960).

B. Leaf Senescence and Abscission

The triggering mechanism of leaf senescence and abscission is uncertain. It is known, however, that hormonal control of protein synthesis and general metabolic activity is important in regulating leaf longevity (Chua, 1970). The balance between growth promoters and inhibitors is upset by a decrease in the level of free auxins in leaves between three and five months before leaves are shed. No significant change in kinetin or kinetin-like compounds and abscisic acid or other inhibitors was detected during the entire life span of leaves. Young leaves from both juvenile and mature trees contain more auxin than mature leaves. The decrease in free auxin during October, November, and December is followed by an increase (Fig. 2). Such an increase is probably due to release of bound auxin in senescing tissues. This is followed by a sharp drop in auxin level at the time of leaf fall in February and March. Abscission layers are initiated three to four weeks before there is any reduction in chlorophyll content.

Chua (1970) did not find any additional evidence for formation of kinetin-like compounds by roots of cuttings or rooted petioles. These roots prevented leaves from senescing, as found for other plants, with the effect attributed to the action of kinetin supplied by roots to leaves (Chibnall, 1954; Mothes and Engelbrecht, 1963).

Chua (1970) suggested that *Hevea* roots may act by supplying peptides to maintain a high rate of protein synthesis in leaves or by supplying tryptophan acid thereby maintaining a high level of auxin in leaves (Bollard, 1956, 1957; Morgan and Reith, 1954; van Raalte and Helder, 1969; Medvedev, 1968). These effects would be especially important in young trees and could account for their lack of periodicity in leaf shedding. In young trees production of new leaves exceeds leaf losses. This enables a higher relative production rate of auxin for the leaf system of adult trees. The possibility that the root factor is a gibberellin was not investigated, but much evidence shows that roots produce gibberellin (Butcher, 1963; Carr *et al.*, 1964; Phillips and Jones, 1964; Skene, 1967; Kende and Sitton, 1967; Rien and Burrows, 1968; Jones and Larcy, 1968).

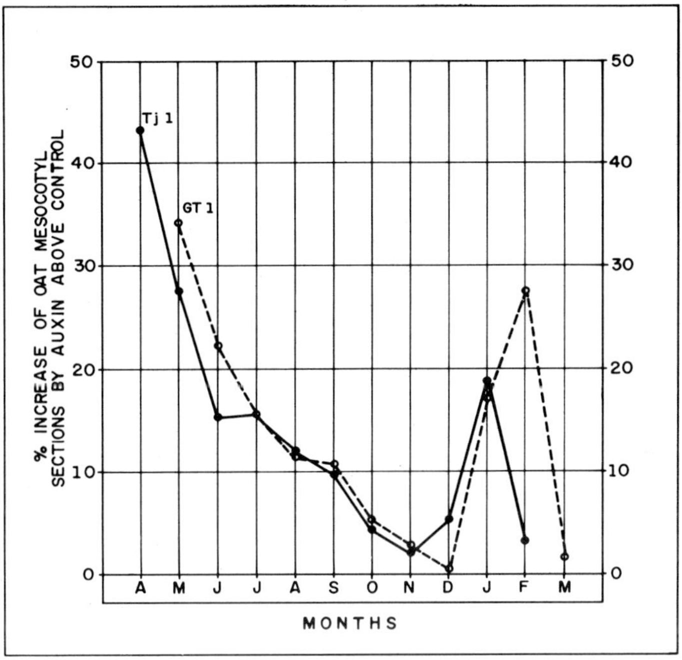

Fig. 2 Annual variation of auxin levels in leaves (Chua, 1970).

Chua's (1970) finding that a decrease in leaf auxin was a factor in senescence is consistent with the view that the early or late "wintering" habit of different clones is controlled by the crown and is slightly influenced by rootstock origin. Such independence of crown behavior is evident in the crown-budded rubber trees of Belterra (Para, Brazil near Santarem).

Waterlogged soils, which may alter the hormone supply from roots to shoots (Reid *et al.,* 1969), apparently do not influence wintering time. Oriental clones such as AV 1301 and AV 1279 senesced and shed their leaves at the same time when growing on upland, well-drained soils in Belem or on lowland soils with impeded drainage (Moraes, 1972). However, hormones supplied by roots must play a role in expression of juvenility by seedlings. This probably is due to an acropetally decreasing hormone gradient which is maintained in stems only up to a height of about 90 cm (McIndoe, 1958). On the other hand, the capacity of petioles of juvenile leaves to root and the induced capacity for rooting of mature tissue by leaves of budded juvenile tissue indicate that leaf factors also are involved in juvenility (Gregory, 1951). Other evidence of hormonal involvement was provided by Chua (1970). He found that application of IAA to leaf blades at the time of auxin decrease delayed senescence and leaf fall for three months. Leaf senescence was accelerated, however, when IAA was applied at other times.

The extremely low levels of auxin found before leaf fall did not resolve the question of what external environmental factors control leaf senescence or whether it might be governed solely by endogenous rhythms. It may be that the mature leaves are very sensitive to short periods of water stress, due to high transpiration rates on sunny days during the rainy season. Abrupt changes from wet to dry periods might hasten the rest of the process and this could explain the uniformity of wintering of different clones in areas of well-defined dry seasons. More research is needed to clarify these aspects. Studies of species with more persistent leaves as well as further studies of *Hevea brasiliensis* in contrasting climatic conditions may provide important new information. It would also be interesting to determine whether water stress increases ABA in leaves of rubber as it does in other species (Wright, 1969; Wright and Hiron, 1969; Mizaki *et al.*, 1972; Most, 1971; Loneys and Kridemann, 1973) or decreases kinetin (Itai and Vaadia, 1971).

V. FLOWERING

Inflorescences arise from the upper scales of terminal short shoots from the axils of the lower scale leaves of the new long shoots, and from axils of new flush leaves as in *Hevea brasiliensis, H. guianensis, H. nitida, H. benthamiana,* and *H. camporum. Hevea brasiliensis* and the other species with nonpersistent leaves come into flower during the wintering period, but may also flower again in areas where an additional shorter dry period occurs (Evers *et al.*, 1960). However, species with persistent leaves also flower during dry seasons. Alvim (1972) suggested that flower buds of coffee are maintained in a dormant state by a relatively high concentration of ABA supplied by the leaves. Internal moisture stress associated with lower moisture or high transpiration rates reduces the ABA concentration of flower buds, possibly because of reduced translocation from leaves. After rehydration, flower buds, now in a state of quiescence or imposed dormancy, begin to grow. As mentioned, the terminal or less frequently lateral short shoots of *Hevea* bear dormant flower buds in the axils of the upper scale. The flowering mechanism in *Hevea* might be similar to that suggested by Alvim for coffee.

In species which shed their leaves before flowering, the source of ABA is eliminated. Leaf water stress may restrict the supply of ABA in species with persistent leaves or in the sporadic flowering periods of the leaf-bearing *Hevea brasiliensis.* This may be the result of additional dry seasons during the year. Flower buds along the newly developing long shoots are differentiated when leaves are too young to act as an efficient source of ABA. Inflorescences may than arise from the axils of lower scales or true leaves of young shoots. Early induction of flowering in young plants is also preceded by senescence and leaf fall. After a period of dormancy, new shoots appear with the inflorescences (Chua, 1970; Camacho and Jimenez, 1963).

VI. LATEX FLOW

This section briefly summarizes the external and internal factors controlling flow of latex. For a comprehensive discussion of anatomy of latex vessels, tapping methods, and factors controlling latex flow, the reader is referred to the work of Buttery and Boatman (1976).

Hevea species have articulated latex vessels occurring in alternate concentric rings around the cambial layer where they originate. Turgor pressures as high as 10–14 atmospheres in the laticiferous phloem were observed before sunrise (Buttery and Boatman, 1964, 1966; Southorn, 1968b). Tapping causes a sudden drop of turgor pressure near the cut. Latex is then expelled by elastic contraction of the latex vessels, followed by a gradual increase in the area from where the latex is drained (Paardekooper and Samorson, 1969).

Diurnal variation in latex yield is highly correlated with variations in atmospheric water pressure. As shown in Figs. 3 and 4, increased transpiration rate decreases phloem turgor pressure by increasing xylem tension.

In Cambodia, Ninane (1967) found a yield decrease to 84% of the 6 AM yield of clone PR 107 by tapping at 2 PM. In Malaysia Parrdekooper and

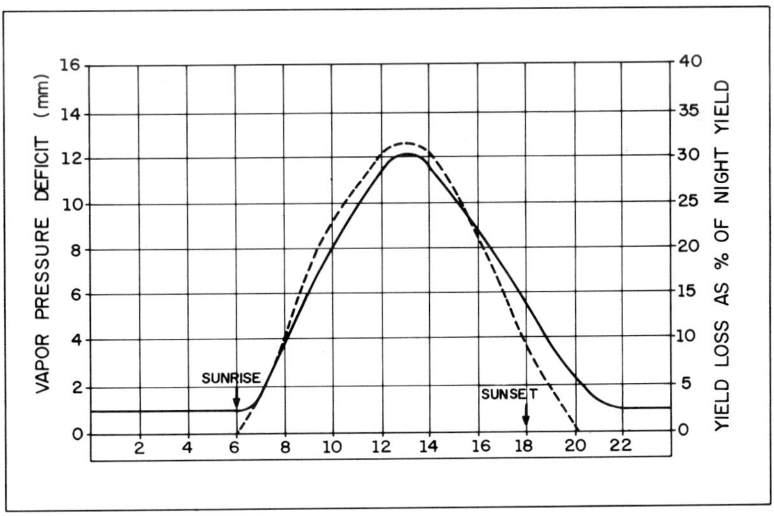

Fig. 3 Average diurnal variation in water vapor pressure deficit in the air (——) and % loss in yield (– – –). Adapted from Paardekooper and Sookmark (1969).

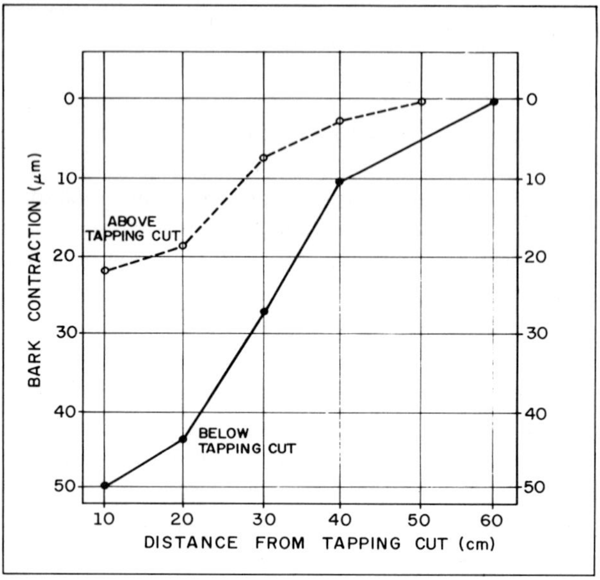

Fig. 4 Bark contraction due to tapping (Institut des Recherches du le Caoutchoue au Cambodge, 1965).

Sookmark (1969) found a 30% decrease in yield until 1 PM and maximum yield between 8 PM and 7 AM. Since the early periods of tapping wild rubber trees in the Amazon valley, tappers learned to begin work early in the morning or even at night. Tapping at night is also reported in Thailand (Paardekooper and Sookmark, 1969).

Xylem tension, caused by transpiration, is not instantaneously transmitted down the stem. Higher turgor pressures in phloem were found in the lower stem than in the upper stem. Pressure gradients up to a ratio of 6 atm/10 m down the stem were found during the day. During the night the corresponding ratio was 1 atm/10 m (Buttery and Boatman, 1966). These differences accounted for the greater influence of atmospheric conditions on latex flow in upper portions of the stem as reported by Schweitzer (1936).

Boatman (1966) made successive tapping cuts before each latex flow was completely stopped by sectioning only a few millimeters of the cut surface and obtaining new successive flows of latex during the day. This experiment eliminated a widely held view that latex stopped flowing because of equalization of phloem pressure with atmospheric pressure. Also, reduction of latex vessel lumens near the cut, as suggested by Pyke (1941), Gooding (1952), and Boatman (1966), did not prove to be a significant hindrance to the latex flow. Southorn (1967) found bark contraction of only 60 μm immediately below the cut surface. He estimated that a latex vessel 20 μm in diameter contracted only 3.5–6 μm. Bark contraction, due to tapping, was also measured in Cambodia (Institut des Recherches sur la Caoutchoue au Cambodge, 1965) with similar results (Fig. 4).

Southorn (1968a) demonstrated with electron microscopy that latex vessels had internal plugs of latex coagulum. Rubber caps formed in the cut surface and plugged some of the vessel ends. Lutoid particles in latex, unlike rubber particles, contain hydrolytic enzymes which are enclosed by a protein/lipid membrane as rubber particles, thus causing gradual plugging of latex vessels near the cut. This is probably due to a high osmotic gradient near the cut which causes rupture of the lutoid membranes (Southorn and Edwin, 1968). Another possibility could be a sudden change in the bioelectrical field near the cut, due to wounding which might depolarize the membranes of lutoids which normally have a positive internal and a negative external charge (Lim *et al.*, 1969).

A "Latex plugging index as a clonal characteristic" has been formulated by Milford *et al.* (1969) and has been correlated with responsiveness to yield stimulants which are thought to prolong latex flow through a protecting stabilizing action on lutoid membranes. Obstruction by lutoid damage is individual for each latex vessel being plugged. This successively leads to a gradual decrease in rate of latex flow. Spillage of the internal content of lutoids during the first phase of elastic contraction does not cause plugging because there is still enough internal pressure to expel the plugs.

VII. YIELD

A. Primary Productivity

In Malaysia, Templeton (1968) studied growth of *H. brasiliensis* trees up to seven years after bud grafting. Clone RRIM 501 trees, between 5.25 and 6.34 years after bud grafting, with an ideal, completely closed canopy (without missing trees due to thinning), could produce 35.5 tons ha^{-1} $year^{-1}$ of dry matter. A relatively high photosynthetic efficiency of 2.8% was also reported. A caloric content of rubber 2.5 times that of equal dry weight of vegetative organs was considered in calculating photosynthetic efficiency. Leaf area index (LAI) was established at 5.8 with parallel variations in primary production. Shorrocks (1965) found that primary production of *H. brasiliensis* increased from 1 to 14 tons ha^{-1} $year^{-1}$ from the first to the fifth year and, by the time the canopy closed, 24 tons ha^{-2} $year^{-2}$ of dry matter were obtained. Maximum LAI reached a value of 6.3 by the sixth year. It may be expected that growth analysis of *Hevea pauciflora* (Fig. 6) will show even higher primary production, because of its dense foliage and very vigorous growth. If so, this should be an additional reason to choose this species for crown budding, in areas where South American leaf blight is an endemic disease.

B. Partitioning of Assimilates

Templeton (1969a,b) showed that rubber clones varied widely in partition coefficients, with close correlation between yield and partition. In clone RRIM 501, the amount of rubber extracted was 11.1% of the total dry weight increment per year in comparison to only 3.0% in RRIM 612. The caloric content of rubber must be taken into account for a judicious calculation of the partition coefficient. For the synthesis of 1 mole of isoprene, 5 moles of ATP are needed.

Combining the highest primary production and the highest partition coefficient found in different clones, Templeton (1969b) estimated a peak production of 4 to 6 tons ha^{-1} $year^{-1}$ of dry rubber from the fourth to the sixth year of tapping. Record yields of farmers approach only half of this estimate and even the highest experimental yields are much lower. Nevertheless, Templeton's estimates provide a challenge to improve yield especially through research in breeding, pest and disease control, physiology, and crop management.

REFERENCES

Alvim, P. de T. (1972). Factors affecting flowering of coffee. *Symp. Fundam. Approaches Plant Anim. Improve.*, orig., pp. 1–17.

Alvim, P. de T., Grangier, A., and Santana, M. B. M. (1965). Estudo comparativo de crescimento em plantulas de cacaueiro e seringueira. *Inf. Anual CEPEC* pp. 23–25.

Bastos, T. X., and Diniz, T. O. A. S. (1975). Clima típico de seringueria. EMBRAPA, Representacão Estaduál do Para. *Bol. Tec.* (in press).

Boatman, S. G. (1966). Preliminary physiological studies on the promotion of latex flow by plant growth regulations. *J. Rubber Res. Inst. Malays.* **19,** 243–245.

Bollard, E. G. (1956). Nitrogen components in plant xylem Sap. *Nature (London)* **178,** 154–155.

Bollard, E. G. (1957). The translocation of organic nitrogen in the xylem. *Aust. J. Biol. Sci.* **10,** 38–41.

Boychou, J. G. (1962). "Manual du Planteur d'Hévéa." Societé parisiènne d'imprimérie, Paris.

Butcher, D. N. (1963). The presence of gibberellins in excised tomato roots. *J. Exp. Bot.* **19,** 272–280.

Buttery, B. R., and Boatman, S. G. (1964). Turgor pressures in phloem: Measurements on *Hevea* latex. *Science* **145,** 285.

Buttery, B. R., and Boatman, S. G. (1966). Manometric measurements of turgor pressures in laticiferous phloem tissues. *J. Exp. Bot.* **17,** 283–296.

Buttery, B. R., and Boatman, S. G. (1976). Water deficits and flow of latex. *In* "Water

Deficits and Plant Growth" (T. T. Kozlowski, ed.), Vol. 4, pp. 233–289. Academic Press, New York.

Camacho, E. V., and Jimenez, E. S. (1963). Resultados preliminares de una proba de inducción de floracion prematura en árboles jóvens. *Turrialba* **13,** 186–188.

Carr, D. S., Reid, D. M., and Skene, K. G. M. (1964). The supply of gibberellins from the root to the shoot. *Planta* **63,** 382–392.

Chan, H. Y., and Pushparajah, E. (1972). Productivity potentials of *Hevea* on West Malaysian soils. A preliminary assessment. *Proc. Rubber Res. Inst. Malays. Plant. Conf., 1972* pp. 97–125.

Chibnall, A. C. (1954). Protein metabolism in rooted runner bean leaves. *New Phytol.* **53,** 31–37.

Chua, S. E. (1970). The physiology of foliage senescence and abscission in *Hevea brasiliensis* Muell. Arg. Ph.D. Thesis, University of Singapore.

Dijkman, M. J. (1951). "*Hevea*: Thirty Years of Research in the Far-East." Univ. of Miami Press, Coral Gables, Florida.

Evers, E., Verbeke, R., and Marteus, C. (1960). Relations entre le climat, la phénologie et la production de l'Hevea. *INEAC, Ser. Sci.* No. 84, pp. 1–71.

Gooding, E. G. B. (1952). Studies in the physiology of latex. II. Latex flow in tapping *Hevea brasiliensis*; associated changes in trunk diameter and latex concentration. *New Phytol.* **51,** 11–15.

Gregory, L. E. (1951). Una nota sobre el enraizamiento de clones de *Hevea. Turrialba* **1,** 201–203.

Hallé, F., and Martin, R. (1968). Étude de la croissance rhythmique chez l'Hevea (Hevea brasiliensis Muell. Arg. Euphonbia-cées-crotonidées). *Adansonia* [N.S.] **8**(h), 475–503.

Ho, C. Y., Ng, A. P., and Subramanian, S. (1960). Choice of clones. *Plant. Bull. Rubber Res. Inst. Malays.* **104,** 226–240.

Institut des Researches sur le Caoutchouc an Cambodge. (1965). "Rapport Annuel." I.R.C.C.

Itai, S S., and Vaadia, Y. (1971). Cytokinin activity in water-stressed shoots. *Plant Physiol.* **47,** 87–90.

Jones, O. P., and Larcy, H. J. (1968). Gibberellin like substances in the transpiration stream of apple and pear trees. *J. Exp. Bot.* **19,** 526–531.

Kende, H., and Sitton, D. (1967). The physiological significance of kinetin and gibberellin like root hormones. *Ann. N. Y. Acad. Sci.* **144,** 235–243.

Lim, C. M., Southorn, N. A., Gomez, J. B., and Yip, E. (1969). Electrophysiological phenomena in *Hevea brasiliensis. J. Rubber Res. Inst. Malays.* **21,** 524–542.

Loneys, B. R., and Kriedmann, P. E. (1973). Rapid changes in abscisic acid like inhibitors following alterations in vine leaf water potential. *Physiol. Plant.* **28,** 476–479.

McIndoe, K. G. (1958). The development of clonal rootstocks in *Hevea. Rubber Res. Inst. Ceylon, Q. Circ.* **34,** 1–19.

Mann, C. E. T. (1940). Improvement in the quality of rubber planting material. *Planter* pp. 332–342.

Martin, R. (1969). Effect of tapping rest at different seasons. *J. Rubber Res. Inst. Malays.* **21,** 353–359.

Medvedev, Z. A. (1968). "Protein Biosynthesis and Problems of Heredity, Development and Ageing." Plenum, New York.

Milford, G. F. J., Paardekooper, E. C., and Yee, H. C. (1969). Latex vessel plugging: its importance to yield and clonal behaviour. *J. Rubber Res. Inst. Malays.* **20,** 274–282.

Mizraki, Y., Blumenfeld, A., and Richmond, A. E. (1972). Abscisic acid and transpiration in relation to osmotic root stress. *Plant Physiol.* **46,** 169–171.

Moraes, V. H. F. (1963). Selecão em seringais nativos. *Circ. Inst. Pesquis. Exp. Agropecu. Norte (Braz.)* 7, 1–31.

Moraes, V. H. F. (1972). Queda e renovacão de folhagem em clones orientais à margem do rio Guamá. *An. Semin. Nac. Seringueira, 1st, Brazil, 1972* pp. 65–67.

Moraes, V. H. F., and Muller, M. W. (1976). Resposta à aplicacão de Ethrel em seringal nativo de várzea do estuário amazônico. *Fac. Clienc. Agrar. Para., Bol. Tec.* (in press).

Morgan, C., and Reith, W. S. (1954). The composition and quantitative relations of protein and related fractions in developing root cells. *J. Exp. Bot.* **5,** 119–125.

Most, B. H. (1971). Abscisic acid in mature apical tissue of sugar cane and in leaves subjected to drought. *Planta* **101,** 67–75.

Mothes, K., and Engelbrecht, L. (1963). On the activity of a kinetin-like root factor. *Life Sci.* **11,** 852–857.

Ninane, F. (1967). "Relation entre les facteurs écologiques et les variations journalières dans la physiologie et les rendments de l'*Hevea brasiliensis* Opusc," No. 12/67. Techniques de l'Institue des Recherches sur le Caoutchoue au Cambodge.

Paardekooper, E. C., and Samorson, S. (1969). Clonal variations in latex flow pattern. *J. Rubber Res. Inst. Malays.* **21,** 264–273.

Paardekooper, E. C., and Sookmark, S. (1969). Diurnal variations in latex yield and dry rubber content and relation to the saturation deficit of the air. *J. Rubber Res. Inst. Malays.* **21,** 341–347.

Phillips, I. D. J., and Jones, R. L. (1969). Gibberellin like activity in bleeding sap of root systems of *Helianthus annus* detected by a new dwarf pea epicotyl assay and other methods. *Planta* **63,** 269–278.

Pushparajah, E., Siranadyan, K., P'ng, T. C., and Ng, E. K. (1972). Nutritional requirements of *Hevea brasiliensis* in relation to stimulation. *Proc. Rubber Res. Inst. Malays. Plant. Conf., 1971* pp. 189–200.

Pyke, E. E. (1941). Trunk diameter of trees of *Hevea brasiliensis:* Experiments with a new dendrometer. *Nature (London)* **148,** 51–56.

Reid, D. M., and Burrows, W. Y. (1968). Cytokinin and gibberellin like activity in spring sap of trees. *Experientia* **24,** 189–190.

Reid, D. M., Crozier, A., and Harvey, B. M. R. (1969). The effects of flooding on the export of gibberellins from root to the shoot. *Planta* **89,** 376–379.

Rubber Research Institute of Malaysia. (1958). Rooting habit. *Plant. Bull. Rubber Res. Inst. Malays.* **39,** 120–128.

Schulz, J. P. (1960). Ecological studies on rain forest of northern Suriname. *Meded. Landbouwhogesch. Wageningen* **163,** 216–244.

Schweitzer, J. (1936). Over den invloed van den wartelbast op de latex vloei van den stam van *Hevea brasiliensis. Arch. Rubber Cultiv.* **20,** 48–60.

Seibert, R. J. (1947). A study of *Hevea* (with its economic aspects) in the Republic of Peru. *Mo. Bot. Gard.* **34,** 261–352.

Shorrocks, V. M. (1965). Mineral nutrition, growth, and nutrient cycle of *Hevea brasiliensis.* I. Growth and nutrition content. *J. Rubber Res. Inst. Malays.* **19,** 32–47.

Sivanadyan, K., P'ng, T. C., and Pushparajah, E. (1972). Nutrition of *Hevea brasiliensis* in relation to Ethrel stimulation. *Proc. Rubber Res. Inst. Malays. Plant. Conf., 1972* pp. 83–96.

Skene, K. G. M. (1967). Gibberellin-like substances in root exudates of *Vitis vinifera. Planta* **74,** 250–262.

Southorn, W. A. (1967). Local changes in bark dimensions of *Hevea brisiliensis* very close to the tapping cut. *J. Rubber Res. Inst. Malays.* **20,** 36–43.

Southorn, W. A. (1968a). Latex flow studies. I. Electron microscopy of *Hevea brasiliensis* in the region of the tapping cut. *J. Rubber Res. Inst. Malays.* **20,** 176–186.

Southorn, W. A. (1968b). Physiology of *Hevea* (latex flow). *Plant. Bull. Rubber Res. Inst. Malays.* **99,** 181–199.

Southorn, W. A., and Edwin, E. E. (1968). Latex flow studies. II. Influence of lutoids in the stability and flow of *Hevea* latex. *J. Rubber Res. Inst. Malays.* **20,** 187–200.

Sudhevea. (1971). Viabilidade para a cultura de seringueira no Brasil. *In* "Plano Nacional da Borracha. Aneso X. Climas." Superintendencia da Borracha, Rio de Janeiro, Brazil.

Templeton, J. K. (1968). Growth studies in *Hevea brasiliensis.* I. Growth analysis up to seven years after budgrafting. *J. Rubber Res. Inst. Malays.* **20,** 136–146.

Templeton, J. K. (1969a). Where lies the yield sammit for *Hevea. Plant. Bull. Rubber Res. Inst. Malays.* **104,** 220–226.

Templeton, J. K. (1969b). Partition of assimilates. *J. Rubber Res. Inst. Malays.* **21,** 259–263.

Vollema, J. S. (1941). Over de invloed van der waterstof ionen concentratie op de groci von rubber. *Arch. Rubber Cultiv.* **26,** 257–269.

Waistie, R. L. (1965). The occurrence of an endogenous type of endotrophic mycorrhiza in *Hevea brasiliensis. Trans. Br. Mycol. Soc.* **48,** 167–178.

Wright, S. T. C. (1969). An increase in the "inhibitor" content of detached wheat leaves following a period of wilting. *Planta* **86,** 10–20.

Wright, S. T. C., and Hiron, R. W. P. (1969). (+)–Abscisic acid, the growth inhibitor induced in detached wheat leaves by a period of wilting. *Nature (London)* **224,** 719–720.

van Raalte, M. H., and Helder, J. J. (1969). The bleeding sap of plants. *Endeavour* **28,** 35–39.

CHAPTER

12

Tea

R. FORDHAM

I. INTRODUCTION

This chapter reviews the origin, distribution, and growth control of tea [*Camellia sinensis* (L.) O. Kuntze]. Earlier reviews of environmental requirements of tea include those of Eden (1965), Harler (1966), Purseglove (1968), and Carr (1972a).

The origin of tea is thought to be near the source of the Irrawaddy River in Southeast Asia, or possibly further north (Kingdom-Ward, 1950), from where it spread into southeastern China, Indochina, and Assam. Its natural habitat was therefore in areas of monsoon climate although, under cultivation, it is now found under a wide range of climates from the Mediterranean to hot, humid tropics. Its distribution in tropical and subtropical regions throughout the world is probably more dependent on the supply of labor for plucking than on specific environmental requirements. Eden (1965) gives a concise account of the development of the tea industry and its spread through Asia, to Africa, and South America. Southeast Asia remains the major source of tea with India, China, and Sri Lanka producing about two-thirds of the world's supply. Since tea was first commercially introduced to Malawi at the end of the last century, cultivation has spread to many other African countries of which the principal producers are now Kenya, Uganda, Malawi, and Mozambique. In South America small plantings of tea are to be found in Argentina, Brazil, and Peru.

II. BOTANY

Commercial tea consists of a heterogenous mixture of hybrids derived from two main types (Wight, 1959). The China form, *C. sinensis* (L.) var. *sinensis,* has relatively small, erect, hard, dark-green leaves, 3–6 cm long with a dull (matt) surface, whereas the Assam form, *C. sinensis* (L.) var. *assamica,* has supple, light-green, drooping leaves, 15–20 cm long with a glossy surface. The China form grows into small trees and is more resistant to cold conditions than the Assam type, which grows faster to produce fairly tall trees when unplucked.

In view of the marked variability within the species it is not surprising that responses to environmental conditions vary with different populations. Thus, there exists the possibility of selecting varieties to suit a wide range of ecological conditions. Information on the anatomical characteristics of the species has been given by Haan (1939).

III. PLANT WATER STATUS

A plant's physiological responses to water deficiency depend upon the development of moisture deficits *within* plant tissues rather than on soil moisture conditions. An internal deficit develops when the loss of water from leaves exceeds the uptake by roots and depends on atmospheric, soil, and plant factors. In recent years there has been an increasing interest in the plant water relations of tea stimulated largely by the development of irrigation projects in areas which experience marked dry seasons (see, for example, Carr and Carr, 1971).

A. Atmospheric Factors

Transpiration is an evaporative process dependent on solar radiation, temperature, humidity, and wind speed. A number of investigators have developed empirical relationships between one or more of these parameters and evaporation, but the method of Penman (1948), depending as it does on sound physical principles, is applicable to a wide range of environmental conditions. Using the Penman estimate of potential evaporation from an open water surface as a basis, Laycock (1964) calculated that mature tea growing in Malawi transpired at 0.85 of the potential rate. Subsequent studies in the same area (Willatt, 1971b, 1973) indicated that, during periods of readily available soil moisture, the ratio actual to potential evapotranspiration was about 0.9 with a reduction during periods of drought. Both Laycock and Willatt estimated transpiration rates from changes in soil water. In Kenya, Dagg (1970) employing a large hydraulic lysimeter obtained a ratio of actual to potential evapotranspiration of 0.9. This was later revised to 0.8 (Wangati and Blackie, 1971), a value supported by catchment studies.

While it is evident that under conditions of high soil moisture and low plant water stress transpiration depends largely on atmospheric conditions, it should be borne in mind that the crop microclimate may be modified by agronomic practices such as the use of shade and shelter (Ripley, 1967). Any effects of conventional irrigation on temperature and humidity conditions above the crop are, however, likely to be short lived (Fordham, 1970), although the frequent misting technique employed by some investigators in the U.S.S.R. (e.g., Petinov, 1961; Lebedev, 1961) may have a longer lasting effect.

B. Soil Factors

The total amount of water available for transpiration depends on soil textural characteristics and pore space distribution within the rooting profile. As this water becomes increasingly depleted a stage is reached at which water uptake can no longer keep up with potential transpiration rates. It must be appreciated that this stage is likely to be reached earlier under high potential rates than low ones.

Willatt (1971b, 1973) studied the reduction in transpiration induced by soil moisture depletion in Malawi and found that the ratio of actual to potential transpiration began to decrease when soil moisture deficits reached between 20 and 25 cm. This corresponded to a reduction of approximately 40% in the total available water content.

C. Plant Factors

Under conditions of plentiful soil moisture, and once a crop canopy has closed and is fully intercepting solar radiation, then further increases in leaf area will not significantly affect total transpiration. However, during postplanting and postpruning stages there will be a reduction in transpiration proportional to the amount of soil exposed (see, for example, Dagg, 1970; Willatt, 1971b).

Under situations of moisture stress, varieties may respond differently as a consequence of differences in the response of their stomata. Highly sensitive plants may, for example, respond by closing their stomata under conditions of mild stress and thus conserve water. In view of the dual role of stomata in affecting both water loss and carbon dioxide uptake, this topic is dealt with in more detail later.

Rooting depth is a major determinant of water availability. A general account of root systems of tea has been given by Kerfoot (1962), and more quantitative data have been published by Eden (1940) for Ceylon, Thomas (1944) for Uganda, Voroncov (1956) for Russia, and Barua and Dutta (1961) for India. In general, the roots are concentrated in the surface layer (e.g., Fordham, 1972) but roots of tea can, where soil conditions permit, be very deep. In Malawi, roots have been found at depths of over 5.5 m (Laycock and Wood, 1963), although in areas where high water tables or impermeable soil layers may occur (for example, Assam), rooting can be very superficial (Barua and Dutta, 1961).

Under conditions of a drying soil, water movement is negligible and roots must obtain water by extending into moister soil. Root growth in tea is not continuous, and Harada *et al.* (1957) observed alternation of shoot and root growth of young tea plants growing in glass fronted boxes. Similar findings under field conditions in Tanzania were reported by Carr (1971) and in Malawi, by Fordham (1972). Fordham also observed cessation of root elongation following pruning.

D. Measurement of Shoot Water Status

Data on the water status of tea have been collected by two methods, the relative water content technique of Barrs and Weatherley (1962) and the pressure chamber method described by Scholander *et al.* (1965).

Although leaf relative water content changes in response to diurnal changes in evaporative demand, this method does not appear to be sufficiently sensitive or convenient for routine use in field studies. Daytime values of between 95 and 85% have been recorded for mature tea (Fordham, 1970).

The pressure chamber, however, has proved to be a very practical method for following changes in shoot water status. In general, water status decreases (i.e., becomes more negative) during the early part of the day and then levels off until a subsequent recovery during the late afternoon (Fig. 1). Carr (1972b) in Tanzania found the midday shoot water status to decrease with increasing soil moisture deficit, with values of about −5 bars at field capacity to −15 bars at

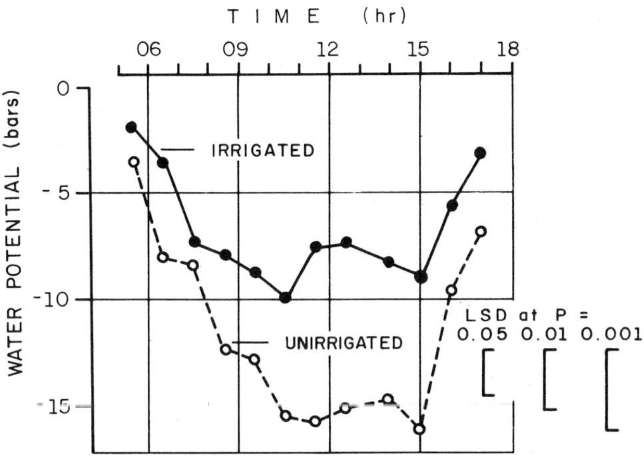

Fig. 1 Diurnal changes in leaf water potential on irrigated and unirrigated tea (after Fordham, 1970).

potential soil moisture deficits of 32 cm (equivalent to approximately 60% depletion of soil moisture). In Malawi, during a period of high potential evaporation, Fordham and Wilson (1969) recorded values of about −8 bars for irrigated tea compared with about −15 bars for unirrigated plants. Green (1971) and Williams (1971a) found that, in the absence of significant soil moisture deficits, the water status of tea shoots was closely correlated with vapor pressure deficits, although some clones of tea appeared to be more sensitive than others. Williams also found the more exposed leaves near the plucking table to be more stressed than leaves lower in the canopy.

IV. STOMATAL MOVEMENTS

Stomata play an important role in controlling exchange of water vapor by transpiration and carbon dioxide by photosynthesis. In tea they occur in the lower epidermis only, at an average density of about 190 per mm^2 (Fordham, 1971a). They are small, with a typical pore size of about $3 \times 11\ \mu m$ (G. R. Squire, personal communication, 1974), although variation in both density and size have been observed in different clones (Fordham, 1970; Carr, 1972b). Their movements have been followed in tea by viscous flow porometry (Fordham, 1968), diffusion porometry (Green, 1971), infiltration (Carr, 1971; Fordham, 1971a), and from measurement of leaf impressions (G. R. Squire, personal communication, 1974).

Daytime stomatal closure is induced both by poor water supply from dry soil and by conditions of high evaporation (Fig. 2). Fordham (1971a) found that

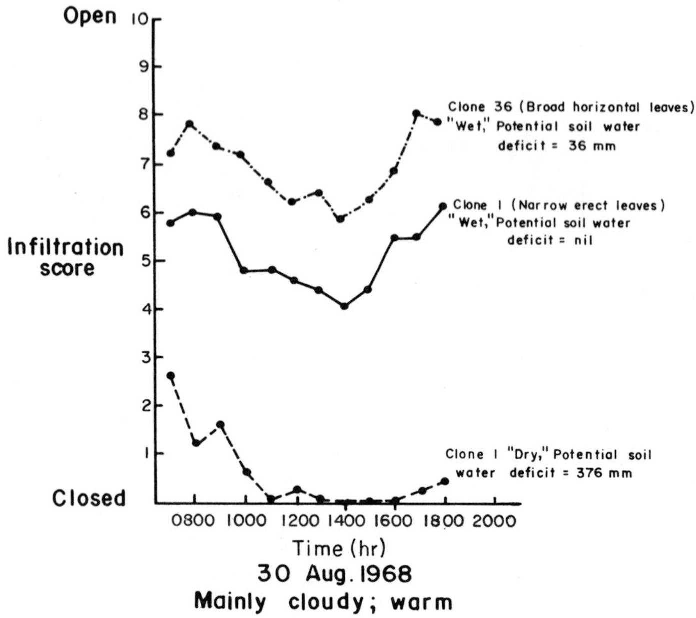

Fig. 2 Diurnal changes in stomatal aperture of unshaded tea (after Carr, 1970).

even in soil near field capacity, solar radiation, temperature, and vapor pressure deficit had a depressing effect on stomatal opening. Carr (1972b), studying the effects of shelter breaks, found that while wind induced stomatal closure this could lead to a conservation of water, with the result that, during periods of drought, exposed plants were stressed less than sheltered ones.

Stomatal movements are mediated by turgor changes and Carr (1972b) recorded a linear correlation between stomatal aperture, as recorded by the infiltration technique, and leaf water status, recorded by the pressure chamber technique. He also showed that the stomata of some clones were more sensitive than others to changes in water status. There is evidence to suggest that water stress may have a residual effect on stomatal opening, since Fordham (1971a) observed a delay in the opening response of stomata with rewatering following drought. Accumulation of abscisic acid (ABA) during periods of moisture stress could inhibit stomatal opening but this has not yet been demonstrated in tea.

G. R. Squire (personal communication, 1974) also observed an inhibitory effect of cold night temperatures (below 8°–10°C) on stomatal opening on the following day. The mechanism for the inhibition is not known but Squire has suggested that it may be a consequence of a reduction in the rate of removal of assimilates from the leaf.

V. PHOTOSYNTHESIS

The basis of productivity of all green plants is of course the photosynthetic process. This has been examined in tea at various levels of complexity ranging from laboratory experiments on excised leaves to field studies on complete canopies of mature bushes.

A. Single Leaf Studies

Photosynthetic capacity develops gradually in young expanding tea leaves and does not reach its maximum until the leaf reaches half its final size (Barua, 1960). The light response curves for leaves of different varieties differ significantly (Barua, 1964, 1970) and maximum rates of net photosynthesis appear to be of the order of 15–20 mg CO_2 dm^{-2} hr^{-1} (Fig. 3). As with other crops, shade leaves have significantly lower maximal rates and are about 10 mg CO_2 dm^{-2} hr^{-1}. Barua's measurements were made under controlled environment conditions at an ambient temperature of 25°C. Hadfield (1968) examined the effects of leaf temperature on photosynthetic rates and found that net photosynthesis declined sharply at leaf temperatures above 35°C and between 39°–42°C there was no net photosynthesis. Respiration continued up to about 48°C, and above this temperature leaf tissues were permanently damaged.

Studies under field conditions, using the $^{14}CO_2$ technique, indicate light

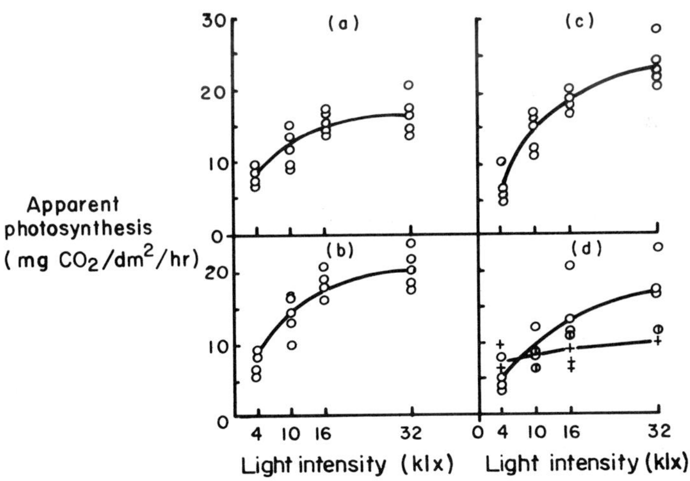

Fig. 3 Rate of apparent photosynthesis of mature leaves from four different sources. The four sources are (a) progeny of clone 19/29/13; (b) progeny of clone 14/5/35; (c) stock 203; (d) Indochina: o—o sun leaves; + — + shade leaves (after Barua, 1970).

saturation at an irradiance of 250–300 W m^{-2} (G. R. Squire, personal communication, 1974). Squire's data show reasonably high rates during the Malawi cold season at a time when shoot growth is minimal. This emphasizes that current assimilates need not always appear as new shoot growth but may be directed to other "sinks" within the plant.

B. Whole Plant Studies

The extension of single leaf studies to whole plant photosynthesis in tea has not yet been carried out in detail. Micrometeorological techniques for determining carbon dioxide flux (e.g., Slatyer and McIlroy, 1961) are likely to be applicable to extended areas of unshaded tea where they would provide valuable data on effects of changing environmental conditions on canopy photosynthetic rates. An alternative method of obtaining information on photosynthetic production of whole plants is by calculation from a knowledge of both the light response of individual leaves and light interception by the crop (Monteith, 1972). Green (1971) indicated that for tea, about half the incident solar radiation was intercepted within the plucking table and that the foliage in this region could be self-supporting. This conclusion is not in agreement with the findings of Sanderson and Sivaplan (1966) who showed, at least for isolated shoots of young tea plants, that carbon assimilated by mature leaves was translocated out of these leaves both into developing shoot tips and down the main stem, presumably to the roots. Alternations in shoot and root growth (Fordham, 1972) indicated that the pattern of translocation may vary with environmental conditions.

Barua (1970) measured the net assimilation rates (NAR) of young plants of different tea cultivars growing under varying degrees of shade. In all cases maximal rates were obtained under conditions of full sunlight, although absolute rates differed among varieties, ranging from 0.12 to 0.17 gm dm^{-2} $week^{-1}$. Data on NAR cannot be taken as a direct indication of photosynthetic efficiency since increases in leaf area will lead to increased mutual shading of leaves and a consequent decrease in net assimilation. However, the plants used in Barua's experiments were small and did not form part of a continuous canopy. Under these circumstances it is probable that the rates observed are evidence for differences in photosynthesis between varieties. The fact that NAR did not correlate with harvested yields of plucked shoots from older plants of the same varieties may be explained by differences in light interception or partitioning of carbohydrates within the plant.

VI. SHOOT GROWTH

Shoot growth, a major determinant of yield, has been examined in several studies. As in many other tropical crops such as mango and cacao, tea exhibits periods of intensive leaf growth alternating with periods of inactivity. A vegetative shoot that is allowed to grow uninterrupted unfolds leaves in a normal manner until leaf production ceases with the appearance of a small apical bud—the so-called "banjhi" (= sterile) bud. Following a period of apparent dormancy, described as the banjhi period, growth is resumed with further leaf unfolding, known as the flush period, after which the cycle is repeated. A detailed account of the phenomenon, as recorded on seed-bearing (i.e., unpruned) trees in Sri Lanka, has been given by Bond (1942, 1945) and briefly summarized by Eden (1965), Goodchild (1968) extended studies of the shoot growth cycle to the regrowth of tea shoots following pruning, and Wight (1955) and Wight and Barua (1955) examined the occurrence of the cycle on plucked tea.

It is evident from the literature that shoots can be either periodic or aperiodic in the manner in which they produce leaves, and young tea in Malawi has shown both types of growth on different bushes that, in other respects, appeared to be similar (Fordham, 1972). Individual branches on the same bush also can behave differently. Apparently the occurrence of the flushing cycle is not a simple or direct response to environmental changes, although these are likely to influence the endogenous factors that control the cycle. Bond (1945) suggested that the phenomenon might be controlled by competition for nutrients, whereas Kulasegaram (1969) favored a mechanism involving growth hormones.

Yield data from many tea growing areas show that tea production occurs in a number of peaks and troughs and most observers have inferred that these fluctuations are the result of the same type of flushing cycle that occurs on unplucked trees. However, Fordham (1970) called attention to the fact that flushing on unpruned trees occurs at intervals of approximately 3 months, whereas the marked peaks in tea yields are at intervals of about 6 weeks. This is the typical time required for an axillary bud to grow sufficiently large to be harvested. When a tea shoot, usually consisting of two young leaves and a bud, is plucked apical dominance is lost and regrowth then occurs from axillary buds. Fordham postulated that adverse weather conditions could temporarily inhibit outgrowth of these buds until the onset of more favorable conditions when the buds would all start to elongate at the same time. Within the limits imposed by variations in growth rates, the shoots thus attain harvestable size at approximately the same time, and their removal then leads to a further population of axillary buds. This inevitably produces a period of negligible yield during which shoots are elongating prior to attaining the required size for the next plucking. On reaching this stage, they are harvested and a second peak in production occurs.

It should be emphasized that the period between peaks is therefore one of active growth and not dormancy. Because of variations in growth rates and consequent differences in starting times for individual shoots, yields will not be in clearly defined peaks and troughs indefinitely, but will take on the form of a damped oscillation, until shoot growth is again synchronized—either as a result of adverse growing conditions or by deliberate manipulative plucking. The effects of weather are therefore twofold—adverse conditions followed by conditions more suited for growth result in a "flush" and the frequency of subsequent flushes depends on conditions prevailing during growth of the next generation of shoots.

Little work has been done on quantifying the environmental factors leading to inhibition of axillary bud growth but it is likely that both low and high temperatures and plant water status are implicated.

Fordham (1970) developed a model of tea growth which could simulate observed yield patterns under growing conditions in Malawi (Fig. 4). He estimated that the time for a shoot to reach a harvestable size was about 9.5 weeks toward the end of the dry season, reducing to approximately 6.5 weeks following the onset of rains. Subsequent work in Malawi (Williams, 1971b) examined growth of shoots in more detail and the model has since been modified to include these observations (Palmer-Jones, 1974; Fordham and Palmer-Jones, 1977). Although there is a need to alter certain aspects of the original model, the basic tenets of Fordham's hypothesis remain valid. Further development would almost certainly need to take into account the carbohydrate balance of tea plants.

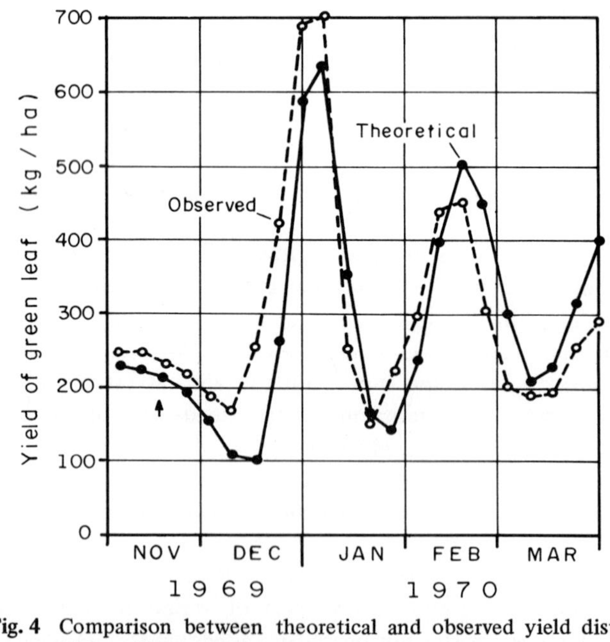

Fig. 4 Comparison between theoretical and observed yield distribution. Time at which axillary buds are assumed to have started growing synchronously is indicated by arrow (after Fordham, 1970).

Shoot growth rates have been studied in Tanzania by Carr (1970) and in Malawi by Fordham (1970) and Green (1971). Of the environmental variables only temperature appeared to be significantly correlated with shoot growth. Fordham found a small but significant correlation between air temperature (expressed as degree hours within the limits of 10° and 30°C) and leaf expansion. He also found a significant reduction in the leaf expansion rate at temperatures below 14.5°C. Green (1971) emphasized the role of shoot tissue temperature in preference to air temperature in controlling the rate of shoot growth and found that shoot extension was linearly related to leaf temperature in the range of 20°–35°C.

Data relating shoot elongation to soil temperatures recorded under grass on meteorological stations are difficult to interpret. Soil temperatures may be integrating the effects of solar radiation and air temperatures, or they may be having a more direct effect on plant growth. In Tanzania shoot growth stopped when soil temperatures (0.3 m beneath a grassed surface) fell below 17°–18°C and did not resume until soil temperature reached almost 20°C (Carr, 1970). Above this value growth rate was linearly related to soil temperature. In Malawi, Fordham (1970) obtained similar results with increases in leaf expansion up to 25°C but with a decline in growth rate at higher temperatures. However, soil temperature fluctuations under tea are likely to be less than those under grass (Willatt, 1969).

Experiments in Malawi with controlled environment chambers over tea in the field confirmed the inhibitory effects of low temperature on shoot growth (Herd and Squire, 1976). Work in Assam has implicated short days in shoot growth (Barua, 1969) and interrupting long nights with light appeared to influence shoot growth during winter months. Lebedev (1961) stated that very short days did not shorten the development time and growth of shoots in Russia. Evidence from Malawi is also conflicting (Fordham, 1970; Williams, 1971b) and it is probable that there is an interaction between day-length and temperature similar to that described by Lockhart and Bonner (1957) in the related species, *Camellia japonica.*

Of all the factors influencing shoot growth, plant water status appears to be the most clear cut and it is possible that the inhibition of growth reported at high temperatures is simply a reflection of reduced turgor. Factors leading to internal moisture stress have been described in an earlier section. There is a wealth of data to show that improved water status can increase shoot growth rates (e.g., Ali-Zade, 1950; Carr, 1970, 1974; Fordham, 1970) and the effect of irrigation on young plants with small root systems is particularly pronounced (Willatt, 1970, 1971a; Fordham, 1971b). Unfortunately most of the effects of water on shoot growth have been observed in irrigation studies and although it has been separately demonstrated that irrigation increases shoot growth and improves shoot water status, no concomitant data on water status and growth are available.

VII. FLAVOR

A detailed consideration of tea flavor is beyond the scope of this chapter, although it is generally acknowledged that, in addition to varietal and processing aspects of flavor development, tea flavors are improved by climatic conditions conducive to slow growth. Wickremasinghe (1974) stated that tea flavor in Sri Lanka is outstanding during seasons of cool (20°C) dry, windy, bright cloudless days followed by cold (6°–10°C) clear nights.

VIII. YIELD OF PLUCKED SHOOTS

Perhaps, due to the abundance of both yield and meteorological data, there is always a temptation to use statistical techniques to "determine" weather effects on production. It should, however, be obvious from the preceding sections that it is unlikely that there is a simple causal relation between any single meteorological factor and the complex of physiological responses leading to economic yield. Such correlations are likely to be relevant only to the regions for which they are derived and extrapolation to other areas should be carried out with extreme caution. Different environmental variables have been emphasized by workers in different countries and they are here considered separately.

A. Solar Radiation

Tea is traditionally interplanted with shade trees but in several areas it has been established that, like many other tropical crops, tea produces its highest yields in full sunlight, provided that nutrient supply is nonlimiting (Murray and Nichols, 1966). For example, McCulloch *et al.* (1965) in Kenya and Shaxon (1968) in Malawi showed that shade reduces tea crops in these areas. Shaxson (1969) subsequently derived a relation between hours of sun and yield. It must be appreciated that shade trees will not only influence light conditions but also change the entire microclimate surrounding tea bushes (Ripley, 1967). The complex nature of the light climate beneath shade trees of different species has recently been described by Hadfield (1974a). The actual interception of this light is further complicated by variations in leaf orientation of different cultivars (Hadfield, 1974b).

B. Rainfall

Laycock (1958) did not find any correlation between annual tea yields in Malawi and either annual or monthly rainfall, although he was able to fit a multiple regression equation of yield on rainfall when the latter was grouped into separate seasons. He later derived a highly significant linear relation between a weather parameter, based on both rainfall and evaporation, and annual yields (Laycock, 1964). Willatt (1971b), in the same area, established significant correlations between soil water deficits and subsequent tea yields. The relationship applied to a time of year (September to November) when other environmental factors, such as low temperatures, were unlikely to be limiting. The importance of rainfall is also emphasized in Hanna's (1969) studies in Uganda in which he used meteorological data to calculate the soil water balance of selected tea estates. Hanna concluded that high growth rates, as indicated by yield records, depended on readily available water in the surface layers of the soil. Multiple correlation techniques have also been applied to yield data for Assam (Sen *et al.*, 1966) where both rainfall and temperature were implicated. Clearly the distribution of rainfall over time is of paramount importance, although the absolute quantities needed to produce maximum growth depend on the atmospheric, soil, and plant factors discussed earlier.

C. Temperature

In general, yields of tea are adversely affected if mean monthly temperatures are lower than 13°C or higher than 29°C (Eden, 1965) and an optimum temperature of 22°C has been postulated by Lebedev (1961). Hasan *et al.* (1965) found that for satisfactory tea production in Pakistan, mean monthly temperatures should be in the range 18°–29°C and that best growth was achieved when the differences between maximum and minimum temperatures were least. The beneficial effects of increasing temperature during the cold season in Assam are evident in the analyses of Sen *et al.* (1966). In Malawi, Green (1971) reported a tendency for poor yield years to be associated with a high number of days in which temperatures fell below 21°C. Green also found that poor yield years were associated with maximum temperatures above 36°C. In considering the effects of temperature on growth it should be remembered that leaf temperatures of unshaded plants are likely to be many degrees higher than ambient temperature (Hadfield, 1968; Green, 1971; Fordham, 1973).

IX. CONCLUSIONS

In view of the complex interactions between environmental variables and physiological processes it is impossible to define a single set of optimum conditions for crop growth. From a practical standpoint, natural environments are undergoing continual change and all agricultural practices must be a compromise between the available site and crop requirements. The literature does, however, help to define climatic factors which might lower production or adversely affect crop distribution. Tea shoots must be harvested regularly and processed immediately and thus crop distribution has important bearings on both labor requirements and factory size. In some instances the literature indicates ways in which the effects of an adverse environment might be ameliorated.

Although tea may grow satisfactorily under a fairly wide temperature range, reduced yields can be expected when minimum temperatures fall below about 15°C. High temperatures also adversely affect growth, although the effect is likely to be influenced by humidity since high saturation deficits may also inhibit growth. Temperatures in excess of 35°C are likely to be particularly detrimental. Temperature effects can be moderated to some extent by the use of shade, although this should be used with discretion if production is not be be inhibited by reduced light.

Given suitable temperature conditions, rainfall distribution is probably the most important factor influencing growth. Ideally monthly rainfall should be sufficient to replace potential evaporation but in suitable soils, root development of mature bushes should enable growth to continue during dry periods, providing the atmosphere does not become too dry at the same time. In areas having dry weather, pruning could be timed to conserve water.

Periods of adverse weather may be succeeded by synchronized shoot growth and this is particularly likely to be so when heavy rain follows a severe drought. In such instances the resultant peaks in production may be suppressed by manipulative plucking.

One of the most interesting findings throughout the literature is the difference in response of different cultivars and it is clear that clones could be selected to suit a wide range of climates. In areas with seasonal variation in, for example, rainfall and temperature, it might be advantageous to select various clones for production at different times of the year in order to even out factory intakes.

REFERENCES

Ali-Zade, M. A. (1950). The effects of irrigation on the growth of young tea shoots. *Dokl. Akad. Nauk SSSR* **73,** 1057–1059.

Barrs, H. D., and Weatherley, P. E. (1962). A re-examination of the relative turgidity technique for estimating water deficits in leaves. *Aust. J. Biol. Sci.* **15,** 413–428.

Barua, D. N. (1960). Effects of age and carbon dioxide concentration on assimilation by detached leaves of tea and sunflower. *J. Agric. Sci.* **55,** 413–421.

Barua, D. N. (1964). Effect of light intensity on assimilation characteristics of detached tea leaves. *J. Agric. Sci.* **63,** 265–271.

Barua, D. N. (1969). Seasonal dormancy in tea (*Camellia sinensis,* L.). *Nature (London)* **224,** 514.

Barua, D. N. (1970). Light as a factor in metabolism of the tea plant (*Camellia sinensis,* L.). *In* "Physiology of Tree Crops" (L. C. Luckwill and C. V. Cutting, eds.), pp. 307–322. Academic Press, New York.

Barua, D. N., and Dutta, K. N. (1961). Root growth of cultivated tea in the presence of shade trees and nitrogenous manures. *Emp. J. Exp. Agric.* **29,** 287–298.

Bond, T. E. T. (1942). Studies in the vegetative growth and anatomy of the tea plant with special reference to the phloem. 1. The flush shoot. *Ann. Bot. (London)* [N.S.] **6,** 607–629.

Bond, T. E. T. (1945). Studies in the vegetative growth and anatomy of the tea plant with special reference to the phloem. 2. Further analysis of flushing behaviour. *Ann. Bot. (London)* [N.S.] **9,** 183–216.

Carr, M. K. V. (1970). The role of water in the growth of the tea crop. *In* "Physiology of Tree Crops" (L. C. Luckwill and C. V. Cutting, eds.), pp. 287–305. Academic Press, New York.

Carr, M. K. V. (1971). An assessment of some of the results of tea/soil/water studies in Southern Tanzania. *In* "Water and Tea Plant" (M. K. V. Carr and S. Carr, eds.), p. 21–43. Tea Res. Inst. East Africa, Kericho, Kenya.

Carr, M. K. V. (1972a). The climatic requirements of the tea plant: A review. *Exp. Agric.* **8,** 1–14.

Carr, M. K. V. (1972b). The internal water status of the tea plant (*Camellia sinensis*): Some results illustrating the use of the pressure chamber technique. *Agric. Meteorol.* **9,** 447–460.

Carr, M. K. V. (1974). Irrigating seedling tea in Southern Tanzania: Effects on total yields, distribution of yield and water use. *J. Agric. Sci.* **83,** 363–378.

Carr, M. K. V., and Carr, S., eds. (1971). "Water and the Tea Plant." Tea Res. Inst. East Africa, Kericho, Kenya.

Dagg, M. (1970). A study of the water use of tea in East Africa using a hydraulic lysimeter. *Agric. Meteorol.* **7,** 303–320.

Eden, T. (1940). Studies in the yield of tea. 4. The effect of cultivation and weeds on crop growth. *Emp. J. Exp. Agric.* **8,** 269–279.

Eden, T. (1965). "Tea," 2nd ed. Longmans, Green, New York.

Fordham, R. (1968). A pressure drop porometer and its application to irrigation research on the tea crop. *Spec. Comm. Appl. Meteorol., 4th 1968,* Nairobi, Kenya, 4 pp. (cyclostyled).

Fordham, R. (1970). Factors affecting tea yields in Malawi. Ph.D. Thesis University of Bristol (abridged version available in *Annu. Rep. Tea Res. Found. Cent. Afr., 1969/70, Mulanje, Malawi* pp. 71–130).

Fordham, R. (1971a). Stomatal physiology and the water relations of the tea bush. *In* "Water and Tea Plant" (M. K. V. Carr and S. Carr, eds.), pp. 89–100. Tea Res. Inst. East Africa, Kericho, Kenya.

Fordham, R. (1971b). Effects of soil moisture on growth of young tea in Malawi. *Exp. Agric.* **7,** 171–176.

Fordham, R. (1972). Observations on the growth of roots and shoots of tea (*Camellia sinensis,* L.) in Southern Malawi. *J. Hortic. Sci.* **47,** 221–229.

Fordham, R. (1973). The micro-meteorology of an extended area of tea before and after rain. *Agric. Meteorol.* **11,** 99–105.

Fordham, R., and Palmer-Jones, R. W. (1977). Simulation of intraseasonal yield fluctuations of tea in Malawi. *Exp. Agric.* **13,** 33–42.

Fordham, R., and Wilson, J. H. (1969). Diurnal changes in leaf water status of tea as measured by the pressure chamber. *Tea Res. Found. Cent. Afr. Q. Newsl.* **16,** 8–9.

Goodchild, N. A. (1968). Growth of tea shoots following pruning. *Ann. Bot. (London)* [N.S.] **32,** 567–572.

Green, R. M. (1971). Weather and the seasonal growth of tea in Malawi. M. Phil. Thesis, University of Nottingham.

Haan, de I. (1939). De anatomische bouw van de theeplant. *Arch. Teecult.* **13,** 318–343.

Hadfield, W. (1968). Leaf temperature, leaf pose and productivity of the tea bush. *Nature (London)* **219,** 282–284.

Hadfield, W. (1974a). Shade in north-east Indian tea plantations. 1. The shade pattern. *J. Appl. Ecol.* **11,** 151–178.

Hadfield, W. (1974b). Shade in north-east Indian tea plantations. 2. Foliar illumination and canopy characteristics. *J. Appl. Ecol.* **11,** 179–199.

Hanna, L. W. (1969). Climatic effects on tea yields in Uganda. *Spec. Comm. Soil Fertil., 11th, 1969,* Kampala, Uganda, 10 pp. (cyclostyled).

Harada, S., Watanabe, A., and Mitsui, M. (1957). Observation of the tea root growth with root-boxes. *Chagyo Gijutsu Kenkyu* **16,** 1–4.

Harler, C. R. (1966). "Tea Growing." Oxford Univ. Press, London and New York.

Hasan, K. A., Chaudhury, S. H., and Halim, M. A. (1965). Effect of climatic factors on the yield of tea. *Tea J. Pakistan* **3,** 4–13.

Herd, E. M., and Squire, G. R. (1976). Observations on the winter dormancy of tea in Malawi. *J. hortic. Sci.* **51,** 267–279.

Kerfoot, O. (1962). Tea root systems. *World Crops* **14,** 140–143.

Kingdon-Ward, F. (1950). Does wild tea exist? *Nature (London)* **165,** 297–299.

Kulasegaram, S. (1969). Studies on the dormancy of tea shoots. 1. Hormonal stimulation of the growth of dormant buds. *Tea Q.* **40,** 31–46.

Laycock, D. H. (1958). An empirical relationship between rainfall and tea yields in Nyasaland. *Spec. Meet. Appl. Meteorol., 1st, 1958,* Nairobi, Kenya, 4 pp. (cyclostyled).

Laycock, D. H. (1964). An empirical correlation between weather and yearly tea yields in Malawi. *Trop. Agric. (Trinidad)* **41,** 277–291.

Laycock, D. H., and Wood, R. A. (1963). Some observations on soil moisture use under tea in Nyasaland. *Trop. Agric. (Trinidad)* **40,** 35–48.

Lebedev, G. V. (1961). "The Tea Bush Under Irrigation." Izd. Akad. Nauk SSSR, Moscow (in Russian).

Lockhart, J. A., and Bonner, J. (1957). Effects of gibberellic acid on the photoperiod-controlled growth of woody plants. *Plant Physiol.* **32,** 492–494.

McCulloch, J. S. G., Pereira, H. C., Kerfoot, O., and Goodchild, N. A. (1965). Effect of shade trees on tea yields. *Agric. Meteorol.* **2,** 385–399.

Monteith, J. L. (1972). Solar radiation and productivity in tropical ecosystems. *J. Appl. Ecol.* **9,** 747–766.

Murray, D. B., and Nichols, R. C. (1966). Light, shade and growth in some tropical crops. *In* "Light as an Ecological Factor" (R. Bainbridge, G. C. Evans, and O. Rackham, eds.), pp. 249–263. Blackwell, Oxford.

Palmer-Jones, R. W. (1974). "Production and Marketing of Tea in Malawi," Dev. Study No. 15. University of Reading, England.

Penman, H. L. (1948). Natural evaporation from open water, bare soil and grass. *Proc. R. Soc. London, Ser. A* **193,** 120–145.

Petinov, N. S. (1961). Physiological basis for the high effectiveness of intermittent sprinkler irrigation in tea plantations. *Izv. Akad. Nauk SSSR* **5,** 717–728.

Purseglove, J. W. (1968). "Tropical Crops, Dicotyledons," Vol. 2. Longmans, Green, New York.

Ripley, E. A. (1967). Effects of shade and shelter on the microclimate of tea and their significance. *East Afr. Agric. For. J.* **33**, 67–80.

Sanderson, G. W., and Sivaplan, K. (1966). Translocation of photosynthetically assimilated carbon in tea plants *Tea Q.* **37**, 140–153.

Scholander, P. F., Hammel, H. T., Bradstreet, E. D., and Hemmingsen, E. A. (1965). Sap pressure in vascular plants. *Science* **148**, 339–346.

Sen, A. R., Biswas, A. K., and Sanyal, D. K. (1966). The influence of climatic factors on the yield of tea in the Assam Valley. *J. Appl. Meteorol.* **5**, 789–800.

Shaxson, T. F. (1968). Some long term effects of shade on yield. *Tea Res. Found. Cent. Afr. Q. Newsl.* **12**, 11–12.

Shaxson, T. F. (1969). Weather/yield relationships and the irrigation of tea in Malawi. *Congr. Int. Comm. Irrig. Drain, 4th, 1969,* Mexico City, 10 pp. (cyclostyled).

Slatyer, R. O., and McIlroy, I. C. (1961). "Practical Microclimatology." CSIRO, Australia, for UNESCO.

Thomas, A. S. (1944). Observations on the root systems of Robusta coffee and other tropical crops in Uganda. *Emp. J. Exp. Agric.* **12**, 191–206.

Voroncov, V. V. (1956). The dying off of the feeder root system in the tea plant. *Dokl. Vses. Akad. Skh. Nauk* **21**, 22–24.

Wangati, F. J., and Blackie, J. R. (1971). A comparison between lysimeter and catchment water balance estimates of evapotranspiration from tea crop. *In* "Water and the Tea Plant" (M. K. V. Carr and S. Carr, eds.), pp. 9–20. Tea Res. Inst. East Africa, Kericho, Kenya.

Wickremasinghe, R. L. (1974). The mechanism of operation of climatic factors in the biogenesis of the flavour. *Phytochemistry* **13**, 2057–2063.

Wight, W. (1955). Dormancy in relation to plucking and pruning of the tea plant. *Phytomorphology* **5**, 1–11.

Wight, W. (1959). Nomenclature and classification of the tea plant. *Nature (London)* **183**, 1726–1728.

Wight, W., and Barua, D. N. (1955). The nature of dormancy in the tea plant. *J. Exp. Bot.* **6**, 1–5.

Willatt, S. T. (1969). Soil physics. *Annu. Rep. Tea Res. Found. Cent. Afr., Mulanje, Malawi, 1968/9* pp. 85–94.

Willatt, S. T. (1970). A comparative study of the development of young tea under irrigation. 1. Establishment in the field. *Trop. Agric. (Trinidad)* **47**, 243–249.

Willatt, S. T. (1971a). A comparative study of the development of a young tea under irrigation. 2. Continued growth in the field. *Trop. Agric. (Trinidad)* **48**, 271–277.

Willatt, S. T. (1971b). Model of soil water use by tea. *Agric. Meteorol.* **8**, 341–351.

Willatt, S. T. (1973). Moisture use by irrigated tea in Southern Malawi. *In* "Ecological Studies Analysis and Synthesis" (A. Hadas *et al.*, eds.), Vol. 4, pp. 331–338. Springer–Verlag, Berlin and New York.

Williams, E. N. D. (1971a). Investigations into certain aspects of water stress in tea. *In* "Water and the Tea Plant" (M. K. V. Carr and S. Carr, eds.), pp. 79–87. Tea Res. Inst. East Africa, Kericho, Kenya.

Williams, E. N. D. (1971b). Physiology report. *Annu. Rep. Tea Res. Found Cent. Afr., Mulanje, Malawi, 1970/71* pp. 105–131.

CHAPTER

13

Oil Palm

JAN-DIRK FERWERDA

I. INTRODUCTION

In recent years, several monographs and textbooks on cultivation of the African oil palm (*Elaeis guineensis* Jacq.) have been published (Anonymous,

1966; Ferwerda, 1962, 1971; Hartley, 1967; Purseglove, 1972; Surre and Ziller, 1963; Turner and Gilbanks, 1974; Wastie and Earp, 1973; Williams and Hsu, 1970). Specific information on the ecophysiology of the oil palm may be found in Prévot (1961), Hartley (1970), and Corley (1973). This chapter discusses growth and development of the oil palm as affected by the environment. Emphasis is given to some controversial views following Hartley's and Corley's papers and to recent developments. Unpublished results of some investigations carried out on ecophysiology of oil palm at the Agricultural University of Wageningen are included.

A. Botany

The African oil palm, *Elaeis guineensis* Jacq., belongs to the family of the Arecaceae (formerly called Palmae). Its nearest relative is the American oil palm, commonly known as *Corozo oleifera* (HBK) Bailey or *Elaeis melanococca* Gaertn., but according to Wessels Boer (1965), it is more correctly *Elaeis oleifera* with which it can be readily crossed. It has a columnar well-developed, unbranched stem, 25–75 cm in diameter, which continues to increase in height as long as the palm lives. This may be more than 127 years, the age of two planted palms in the botanical garden of Bogor, Indonesia, but the life of plantation palms is usually terminated for economic reasons when the stems have attained a height of 10–11 m at an age of 25–35 years. The root system of adult palms consists of some 8000–10,000 adventitious primary roots, 4–10 mm in diameter, emerging mainly from the subterranean basal bole of the stem. Most of these roots grow more or less horizontally between 20 and 60 cm below the soil surface; individual roots may attain a length of 15–20 m. Some roots grow vertically downward. In deep, well-drained soils of good structure they commonly attain depths of 3 m and occasionally have been observed as deep as 9 m. Secondary roots, 2–4 mm in diameter and up to 150 cm long, emerge from the primary roots. Slightly more than half of them are ascending and almost reach the surface of the soil. Tertiary roots, 1–2 mm in diameter and 10–15 cm long, grow almost horizontally from the secondaries. They are most abundant on the secondary roots near the soil surface. Numerous quaternary roots, 0.5 mm in diameter and up to 2 cm long, are produced on the tertiaries. The tertiary and quaternary roots near the soil surface form a thick mat in the top 10 cm of the soil.

Under favorable conditions the crown of adult palms consists of 40–50 opened, green parapinnate leaves, produced at a rate of approximately two per month. Hence the functional life of an opened leaf is about two years. The number of juvenile leaves in various stages of development between the apical meristem and the youngest opened leaf is approximately the same, so that some four years elapse between the initiation of a leaf at the growing point and its death. Individual leaves are 5–7 m long and are composed of a rachis with 100–160 pairs of linear leaflets and a petiole with spines.

The length of the individual leaflets in the middle of the rachis is about 100 cm; they are shorter near the top and much reduced in size near the petiole. On planted palms the bases of dead leaves remain attached to the stem until the palm is 15–17 years old; on semiwild and wild palms much longer. The influence of the age of the palm on the area of its individual leaves and on the leaf area index (LAI) of a plantation is illustrated in Fig. 1 (Williams and Hsu, 1970). The LAI is an important criterion for the condition of oil palm plantations. It is closely related to dry matter production (Corley *et al.,* 1973), as may be seen in Fig. 2. From this it is apparent that the optimum leaf area for total dry matter production is not the same as that for production of palm oil. The amount of dry matter used per palm for production of vegetative tissues is fairly constant. When the requirements for vegetative growth are satisfied, the excess dry matter is used for production of fruit bunches.

The palm is monoecious. The inflorescence, which is either male or female and only rarely hermaphroditic, is a spadix. One inflorescence is formed in the axil of each leaf shortly after its differentiation from the stem apex. The sex of an inflorescence on an adult palm is determined about nine months after its initiation, but another 24 months elapse before it is ready to flower. Not all inflorescences reach this stage, however. During the period of rapid elongation, 5–6 months before anthesis, some of them abort. The number of individual flowers per inflorescence varies between cultivars, and increases with the age of the palm. The individual flowers of a female inflorescence usually open within three days and remain receptive for three to four days. The flowers of the male inflorescences shed their pollen within five days. Although self-pollination is not

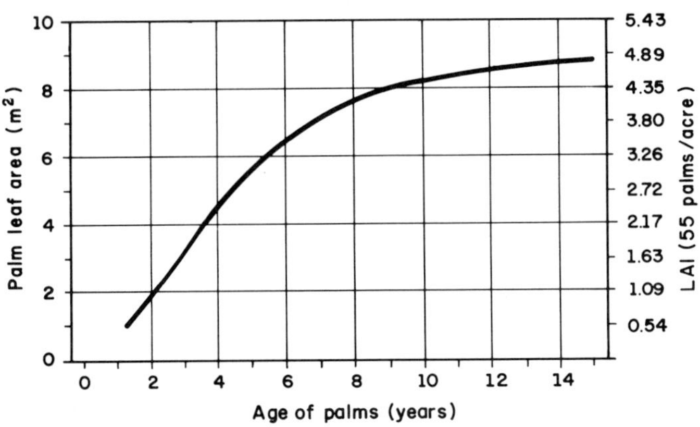

Fig. 1 Leaf area per palm and leaf area index (LAI) of palms of different ages from southern Malaysia (Williams and Hsu, 1970).

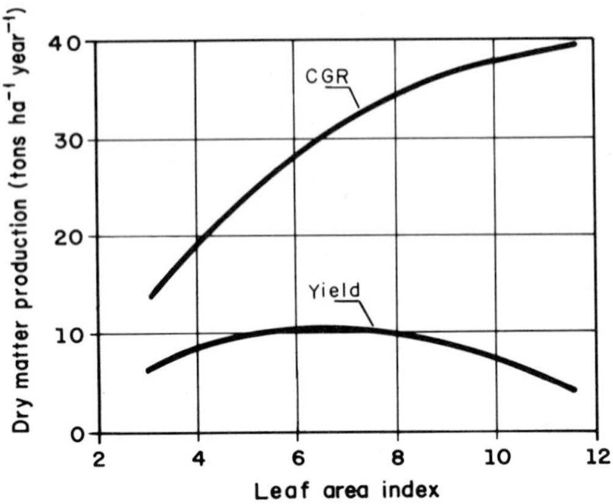

Fig. 2 Relationship between crop growth rate (CGR), yield, and leaf area index (Corley, 1973).

excluded, cross-pollination is the rule. The fruit is a sessile drupe with a fleshy mesocarp and a hard endocarp surrounding usually one, occasionally two, and rarely, three seeds. The mesocarp of ripe fruits contains 45–50% of an orange–red colored edible oil, containing approximately equal parts of saturated and unsaturated fatty acids. The ripe seed, commonly called kernel, contains 48–52% of an almost colorless oil, predominantly composed of saturated fatty acids. Fruit bunches develop from fertilized inflorescences in 5–6 months.

B. Propagation

Up to the present time oil palms have been propagated exclusively by seed. In plantation practice a "seed" is the proper seed surrounded by the hard endocarp. Methods of vegetative propagation are unknown, but the possibilities of propagation by means of tissue culture are being investigated (Jones, 1974; Staritsky, 1970). When seed is planted in moist sand, germination is irregular and may take 6–12 months. Various techniques to improve the rate of germination by means of heat treatments have been developed, both in Africa and in the Far East. At present only two methods are commonly in use, viz., the wet and the dry heat germination. The first technique consists of exposing previously soaked "seeds" placed in closed polythene bags to 38°–40°C in incubators. The seeds, which have germinated, are removed, and planted directly in the prenursery or nursery. The second technique consists of heating seeds containing about 18% moisture in the same way. No germination occurs during this process, but following this treatment during 30–60 days it is ready for germination at ambient temperature after bringing the moisture content up to about 22%. At present the classic prenursery beds have been largely replaced by polythene bags about 15 × 20 cm in size when laid flat.

Seedlings may be raised during some 4 months in these polythene bags before transferring them to a field nursery or to a polythene bag nursery. The latter consists of heavy quality polythene bags 30–38 × 45–60 cm in size, filled with good soil. Plants raised in polythene bags are easier to handle, but they cannot be kept much longer in these bags than scheduled, and their watering requires more attention than that of plants raised in field nurseries.

C. Field Planting

When the palms are 12–14 months old from the germinated seed stage, they are ready for planting out in the field. Field nursery palms require some leaf pruning with a view of reducing transpiration and are lifted with a ball of earth, which is protected with a gunny sack or polythene sheet. Polybag nursery palms require little special care. The spacing in the field is usually triangular and about 9 × 9 m, corresponding with 143 palms per net hectare planted. It is good plantation practice to keep the land permanently protected by means of creeping soil cover crops such as *Pueraria phaseoloides, Calopogonium mucunoides,* and *Centrosema pubescens.* A vegetable mulch applied around the young palm after planting out may help to keep the soil locally moist and cool during prolonged dry spells.

D. Field Maintenance

Field maintenance includes circle and path weeding and control of the cover crop. After planting, a circle with a radius of about 100 cm around each palm is kept weeded, which should be increased as the crown expands with a view to preventing the stemless palm from being overgrown by the cover crop of weeds. Later weeded circles and paths are required to facilitate harvesting and disease control.

Pruning of young palms, which produce too little to be harvested, is limited to periodic removal of dead leaves and young inflorescences just after their emergence. Pruning of producing palms may include removal of some green leaves to facilitate harvesting, but it has been known for many years that this should be restricted as much as possible (Turner, 1974). Although there is no evidence that stem and crown epiphytes reduce bunch yield, it is common practice to remove them periodically to facilitate harvesting.

E. Fertilizing

Fertilizing has become standard practice in oil palm nurseries and field plantings and may include adding both major nutrients and trace elements. There has been considerable progress during the last 30 years in diagnosis of nutritional deficiences by means of external symptoms, leaf analysis and, to a lesser extent, soil analysis. Although initial expectations about the reliability of leaf analysis as a diagnostic tool proved to be unfounded, analyses of leaves and other parts of the palm have undoubtedly contributed to a much better understanding of its nutritional requirements. Labeled plant nutrients have been successfully used to study placement and type of fertilizers.

F. Plant Protection

Control of pests and diseases in nurseries and field plantings by means of chemical pesticides is common practice in most plantations. Little progress has been made up to the present time in development of effective biological control methods. Mammalian pests, especially pigs and rats, are more important in oil palm cultivation than in most other crops. The former can be controlled by fences and hunting, but the latter are rather difficult to control without the use of poison baits.

G. Harvesting

Economic harvesting may commence during the second half of the third year after field planting. From young, almost stemless palms the fruit bunches may be removed by means of a chisel, without removing any leaves. For rough-stemmed palms, which can be climbed easily without a ladder, a cutlass or a harvesting axe is the usual tool. At present, smooth-stemmed palms are often harvested with a hooked knife mounted on a pole, but the traditional way of climbing the palm by means of a sling and cutting the fruit bunch with a harvesting axe or cutlass is still in use in Africa.

H. Breeding

The object of oil palm breeding is to increase the yields of oil per hectare and per annum and to improve the quality of the oil. Emphasis is usually given to the yield of palm oil, although the contributions of the kernels to the economy of a plantation are by no means negligible. The oil yield per hectare is the product of the number of palms per hectare $\times$ the number of fruit bunches per palm $\times$ the mean bunch weight $\times$ the fruit/bunch ratio $\times$ the mesocarp/fruit ratio $\times$ oil/mesocarp ratio. The interrelationships and heredity of these yield components have been the subject of extensive investigations during recent years (Hardon *et al.*, 1973). Another important breeding criterion is the annual height increase of the stems, which determines the length of the economic life of a plantation to a large extent. Finally, breeding for resistance against diseases, such as the wilting diseases caused by *Fusarium oxysporum* and *Ganoderma* spp., is receiving more and more attention.

II. GEOGRAPHY

Most geographic factors affect growth and development of plants through changes in other ecological factors, such as solar and terrestrial radiation, heat, water, atmospheric gases, and biotic factors. These are briefly reviewed and discussed in some detail only if their effect is more or less direct.

The longitude of a habitat is of no ecophysiological significance. The latitude, however, is very important because it affects the annual course of radiation and temperature. The distance and direction from the coasts of seas and large lakes work through the climate of the habitat. A review of the habitats of semiwild palm groves in Africa was given by Zeven (1967). Outside of Africa extensive semiwild palm groves exist only in Brazil, where they form a narrow belt along the Atlantic coast between 13°S and 14°S. Smaller areas may be found further north in the states of Sergipe, Alagoas, Pernambuco, and Para (Hartley, 1967). Other small areas and groups of scattered palms may be found in several other regions of the humid American tropics. It is usually difficult to trace their origin, but it seems likely that they were introduced as seed by African slave labor. Commercial oil palm plantations and experimental plantings have been established in many countries of the humid tropics between about 16°N and 10°S. In Africa and Asia commercial planting commenced before World War I; in America during and after World War II.

A. Erosion

Erosion by water removes the soil from between the tertiary and quaternary feeding roots near the soil surface, in particular in the weeded circle. Exposed roots dry up and die, so that the absorbing capacity of the root system is reduced. Although no experimental evidence is available, it is often quite obvious that palms growing under these conditions undergo water deficits and nutritional deficiencies. On the other hand, the dense mat of tertiary and quaternary roots near the soil surface, the cover crops, and the almost closed canopy of the crowns provide very effective protection against soil erosion. For this reason soil conservation works are not considered necessary. In Malaysia slopes up to 5° and 10° usually require some form of conservation if the soils have a tendency to erode (Williams and Hsu, 1970). Erosion caused by wind does not occur in the humid tropics and is therefore of no importance for oil palm cultivation.

B. Topography

The major topographic elements affecting plant growth and development are relief, slope angle, slope direction, and altitude of the land. They all work through the local climate of the habitat. In addition the relief works through the drainage of the land, the slope angle through the ratio between runoff and infiltration of water, and the altitude through many biotic factors. These are discussed later. Most commercial plantings have been made on land with slopes up to about 20°, mainly for technical and economic reasons (Williams and Hsu, 1970), and only limited ecophysiological considerations.

III. WEATHER AND CLIMATE

The effects of wether and climate will be discussed by analyzing the ėffects of the major elements, including radiation, temperature, water, and air. Most commercial plantings have been established in regions where the rainfall is in excess of evapotranspiration during at least nine months of the year, viz., in regions with climates classified as Af and Am by Köppen, or climates of the Equatorial Zone according to Walter (1973).

A. Radiation

Only solar radiation is considered under this heading. There is no information on the influence of cosmic radiation, while terrestrial radiation works mainly through soil and air temperature.

1. Wavelength

Field palms are normally grown under full sunlight, but prenursery palms and nursery palms are sometimes shaded, while young palms in replantings and in semiwild groves are usually shaded. Shading may be provided by screens made from palm leaves, wooden slats, Saran cloth, or living trees. Screens reflecting part of the radiation are not likely to affect its composition to a large extent. Light transmitted by living leaves, however, is relatively poor in the red and blue wavelengths. This, in addition to the effect of a lower intensity of solar radiation, may explain the increased size of whole plants and leaves of palms grown under the canopy of living leaves in comparison with palms of comparable morphogenetic age grown under full sunlight. From replanting experiments in Zaire (Ferwerda, 1955) it may be deduced that replanted palms growing under the canopy of an old plantation have longer leaves than palms with the same cumulative leaf production growing in full sunlight. This is also shown by the annual height increase of rough-stemmed palms growing under higher smooth-stemmed palms in a Nigerian semiwild palm grove (Zeven, 1967). The mean annual height increase of the rough-stemmed palms was 106 cm; that of the smooth palms only 30 cm.

2. *Light Intensity*

Photosynthesis is known to be quantitatively related to the intensity of the photosynthetically active part of the radiation, which corresponds approximately to visible light. When the sky is clear, light intensity at the equator varies between monthly minima of 1410 J cm^{-2} day^{-1} in June and December to about 1540 J cm^{-2} day^{-1} in March and September. At 10°N, however, there is already a distinct minimum of 1218 J cm^{-2} day^{-1} in December and a period of intensities over 1500 J cm^{-2} day^{-1} from March to September. When the sky is cloudy, these intensities are reduced to only 20% of the intensities on clear days, corresponding with a reduction in potential photosynthesis of about 50% (De Wit, 1965). It is not surprising, therefore, that dry matter production of oil palm seedlings during the first 13 weeks after germination is already related to light intensity. When 50% of natural daylight was intercepted by bamboo slats, reduction of dry matter production amounted to 24% for whole plants, 21% for aerial parts, and 33% for roots (Ferwerda, 1955).

Sparnaaij *et al.* (1963) found positive correlation between the number of effective sunshine hours per annum and yield of fruit bunches. They defined "effective" sunshine as total sunshine received during periods of moisture sufficiency plus a fraction of the sunshine received during periods of moisture stress, decreasing with the duration of moisture stress. Despite a somewhat arbitrary estimation of the influence of water deficit on the effect of sunshine, yields per palm increased by about 5.7 kg per increase of 100 hr of effective sunshine. Extrapolation of these figures to Malaysian conditions with values of 2400 hr or more of effective sunshine would give mean annual yields of about 125 kg per palm, or about 18 tons per hectare. These yields are obtained on comparable sandy soils in Malaysia, but yields on coastal clays usually are much higher (Williams and Hsu, 1970). It is evident, however, that positive or negative deviations from the calculated mean yield of some 25 kg per palm are within the fiducial limits. It would seem, therefore, that the influence of moisture stress is underestimated and/or that the potential yield of the Nigerian planting material is less than that of the Malaysian plants. The results obtained with genetically identical planting material at La Mé, Ivory Coast, and San Alberto, Colombia seem to confirm the beneficial effect of more sunshine in the absence of water stress (Table I).

Data obtained at La Arenosa estate near Turbo in Colombia, however, seem to cast serious doubt on this explanation (Coldesa, 1974). Despite an annual number of sunshine hours, which is even lower than that at Benin, and a dry season lasting an average of 75 days, yields of the young palms were suggestive of potential yields comparable with those obtained at San Alberto and in Malaysia (Table I). Perhaps the true explanation of the high correlations found

TABLE I
Fruit Bunches per Hectare and per Annum in Various Regions

Plantation	Sunshine hours	Planting material	tons/ha	Age
La Mé, Ivory Coast	1781	D × P La Mé	14–15	(adult)
San Alberto, Colombia	2070		25–27	(adult)
Turbo, Colombia	1470	D × P Malaysia	16–19	(5th–6th year)
		D × P Malaysia	15–18	(4th–5th year)
		D × P Surinam	14–21	(4th–6th year)
		D × P La Mé	13	(3rd year)
		D × P Coldesa (Turbo)	13	(3rd year)

by Sparnaaij *et al.* (1963) between effective sunshine and yield is that their estimate of the effective sunshine during the dry season is in fact an estimate of the degree of water stress suring that period.

3. *Periodicity*

Annual variations in duration of daily solar radiation are sufficiently large within the range of the oil palm to affect growth and development of many annual tropical plants, such as rice, cowpeas, and sesame. At about 16°N, the latitude of a commercial plantation in Honduras, the duration of the shortest day between sunrise and sunset is 11 hr 10 min and that of the longest day 13 hr 05 min. The duration of civil twilight at this latitude varies only between 21 and 24 min. The bunch yield of this plantation has a very irregular distribution throughout the year. There is almost no yield from January to May as compared with some 90% during June to December, with a high peak during the period of September to November (Hartley, 1967). It is unlikely, however, that these differences are related to variations in daylength. Young stemless palms submitted to constant daylengths of $10\frac{1}{2}$, $11\frac{1}{2}$, $12\frac{1}{2}$, and $13\frac{1}{2}$ hr, respectively, but receiving the same quantity of photosynthetically active light, showed no differences in leaf production after 28 weeks (J. D. Ferwerda, unpublished data, 1965). In view of the size of the palms, the experiment was continued with four palms per treatment until appearance of the first inflorescences. Then the palms were dissected to determine the presence of immature inflorescences (Table II).

Despite the small number of palms, the differences in production of new leaves between individual palms at the end of the experiment varied only between 15 and 17. The number of inflorescences in the axils of the 15 youngest leaves, including the spear leaf and five immature leaves preceding the spear leaf,

TABLE II
Influence of Constant Daylengths on Young Stemless Palms[a]

Daylength	New leaves/palm (9–28th week)	New leaves/palm at conclusion	Inflorescences in axil of leaves −5 to +15 (20)		
			Absent	Present	Aborted
10 hr 30 min	8.37	16.0	2.0	16.0	2.0
11 hr 30 min	8.53	16.8	3.1	13.8	3.1
12 hr 30 min	8.50	16.0	0.0	16.5	3.5
13 hr 30 min	8.33	15.8	3.8	14.2	2.0

[a]Mean initial leaf number 19. Leaves at the conclusion of the experiment numbered from the spearleaf = +1.

however, varied between only 2 well-developed + 9 aborted inflorescences and 20 well-developed + 0 aborted inflorescences, but these large differences were not related to the treatments.

B. Temperature

The influence of temperature on growth and development of the oil palm cannot be easily studied because of its size. Most information on temperature requirements has been deduced from the geographic distribution of wild, semi-wild, and planted palms. Experimental work has been carried out on the temperature requirements for germination, regulation of soil temperature in nurseries (Hartley, 1967), and temperature requirements for growth of seedlings and young stemless palms. Four aspects of temperature will be considered, viz., mean temperature, its annual variation, its daily variation, and the influence of extreme temperatures.

1. Mean Temperature

The annual mean temperature within the geographic distribution of commerical plantings is between 24° and 28°C, but fruiting semiwild palm groves occur near the equator at least up to 1300 m and, according to older observations even higher (Zeven, 1967). This means that the lowest annual mean temperature supporting palms may be close to 20°C. Growth of young seedlings is arrested at 15°C (Henry, 1957). J. D. Ferwerda and J. Ehrencron (unpublished data, 1975) studied growth in a Phytotron of young stemless palms with initially about nine leaves at different temperatures (Table III). The rate of leaf production showed an almost linear increase with mean temperature in the range 12°–22°C. The increase in mean dry weight of single

TABLE III
Growth of Young Stemless Palms at Different Temperatures in a Phytotron[a]

Temperature (°C)			Leaves after 4 months			
Light (12 hr 15 min)	Dark (11 hr 45 min)	Mean	Number	%	Weight (g)	%
32	22	27	6.5	100	19.7	100
27	17	22	6.0	92	17.1	87
22	12	17	3.6	55	12.3	62
17	7	12	0.5	8	1.5	8

[a]Light (52, 500 lux at plant level) provided 12 hr 15 min per day.

leaves showed a similar relation. Higher temperatures had much less effect on growth. It seems likely that annual leaf production of adult palms increased with the annual mean temperature as well, but direct comparisons of genetically identical planting materials are not available. Although the highest bunch yields are obtained in regions with annual mean temperatures between 25° and 27°C, the distribution of the semiwild palm groves suggests that the oil palm is well adapted to much lower temperatures.

2. *Annual Variation*

The difference between the highest and the lowest monthly mean temperatures within the geographic distribution of the commercial oil palm growing regions of the world vary between only 1.1°C for two sites mentioned by Hartley (1967) in Malaysia, 3.8°C for a plantation at about 16°N in Honduras and 5.8°C for the semiwild groves at 12°S–14°S in Bahia, Brazil. The oil palm belt of equatorial West Africa is intermediate with temperature differences just over 3°C. The highest yielding plantations are found in regions with the smallest annual variation of monthly mean temperatures. In view of the relation between leaf production rate and mean temperature it seems unlikely that monthly variations in yield are related to temperature in regions with the largest annual variation in monthly mean temperature. The lowest temperatures will hardly slow down the growth rate and therefore not increase the time between appearance of consecutive inflorescences. Low temperatures may increase abortion of inflorescences before anthesis, however, and slow down the ripening of fruit bunches. High temperatures may have the opposite effect. This may account for the very irregular distribution of bunch yield in Honduras, which has a very pronounced peak in September, October, and November and almost no yield in January–April (Hartley, 1967). The highest production of the Bahia groves, on the contrary, seems to occur between November and March.

3. *Daily Variation*

The annual mean of the daily variation in temperature in the oil palm regions of the world listed by Hartley (1967) varies between 4.8° and 11.2°C, but 50% is in the narrow range of 8°–10°C. The highest yielding plantations are found in regions with an annual mean daily variation of temperature within this range. Although the monthly means of daily variation of temperature tend to be somewhat higher during the driest periods of the year, the annual variation of these monthly means is surprisingly small. As the driest periods usually correspond to the coldest periods, the adverse effects of low temperatures may be accentuated.

4. *Extreme Temperatures*

Very little is known about the effect of extreme high or low temperatures on growth and development of the oil palm. The absolute maximum in the oil palm regions of the world is about 38°C, the absolute minimum 8°C, but daily maxima exceeding 34°C and daily minima below 14°C are exceptions (Ferwerda, 1962). Brief exposures to temperatures exceeding 38°C and below 8°C are not lethal, however. Palms are occasionally exposed to higher temperatures during short periods in our greenhouses on clear and hot summer days without visual damage. Although palms exposed to day temperatures of 17°C and night temperatures of 7°C for 4 months in a Phytotron practically stopped growing and became very chlorotic, they were still able to resume growth under normal temperature conditions.

C. Water

Water requirements of the oil palm have been studied by several investigators. The results have been somewhat contradictory and inconclusive, to some extent at least because of incorrect interpretations of the data (Hartley, 1967; Surre and Ziller, 1963).

1. *Humidity of the Atmosphere*

All information available on the relation between growth and development of the oil palm and atmospheric humidity appears to be circumstantial. Wormer and Ochs (1959) demonstrated a parallelism between stomatal aperture and relative humidity changes during the day, but relative humidity is closely related to other climatic elements showing daily fluctuations, such as air temperature and soil moisture content. The same conclusion may be drawn from the observations of Rees (1961) on the relationship between stomatal opening, air temperature, and soil moisture. In one of the general textbooks (Surre and Ziller, 1963) a relative humidity of at least 75% is considered favorable for growth and development without other evidence than that these humidities obtain in the principal zones of production. The well-known beneficial effects of watering nurseries during prolonged rainless periods, and the spectacular effect of irrigation on producing palms under similar conditions (Maillard *et al.*, 1974), suggest that atmospheric humidity is of secondary importance, as these localized applications of water are unlikely to have much effect on air humidity.

2. *Soil Water*

The influence of soil water on growth and development has been the subject of several studies during the last 20 years. Curiously the fundamental investigation by Ringoet (1952) on transpiration and water balance of cacao, coffee, and oil palm is not even mentioned in the bibliographies of the two leading textbooks (Surre and Ziller, 1963; Hartley, 1967). Ringoet (1952) found a distinct effect of soil moisture content on the transpiration coefficient (Table IV). The range of the soil moisture content studied corresponded with the annual range occurring at Yamgambi (Zaire) in a local Oxisol. Comparison of the mean transpiration coefficients of oil palm and coffee with those of other plants cultivated or growing in the same region gave the following results (Table V). Oil palm had a relatively low transpiration coefficient (Table V). In addition, the large range of transpiration coefficients of the oil palm as compared with that of the other plants reveals its high adaptability to changes in soil moisture, a characteristic of xerophytic plants.

Wormer and Ochs (1957, 1959) confirmed and supplemented Ringoet's findings. They found a relationship between available soil moisture and transpiration, stomatal aperture, and N and P content of oil palm leaves. When the moisture content of the soil decreases, transpiration is unaffected until a critical moisture level, well above the wilting point, is reached. After that transpiration decreases and practically stops at the wilting point. The height of this level depends on the climatic factors determining transpiration. The degree of sto-

TABLE IV
Effect of Soil Moisture Content on Transpiration Coefficient of Young Oil Palms and Young Coffee Trees

	Transpiration coefficient	
Soil moisture (%)	Oil palm	Coffee (*Robusta*)
11	164	515
15	380	606
19	337	613

TABLE V
Mean Transpiration Coefficients of Various Plants at Yangambi[a]

	Mean transpiration coefficient	Range of mean for 11–19% soil moisture
Oil palm (*Elaeis guineensis*)	294	115
Upland rice (*Oryza sativa*)	413	50
Robusta coffee (*Coffea canephora*)	578	20
Palisotha thyrsiflora (wild plant)	618	10
Cacao (*Theobroma cacao*)	866	–

[a](Ringoet (1952).

matal opening appears to be a reliable measure of transpiration and may easily be estimated by an empirical method. The discovery of positive correlations between the moisture content of the soil and the N and P content of the leaves is of paramount importance for interpretation of results of leaf analysis.

Rees (1961) and Rees and Chapas (1963) provided additional evidence on the relationship between soil moisture and transpiration of nursery palms, which confirmed the results of Ringoet (1953) and Wormer and Ochs (1959).

3. *Precipitation*

The mean annual rainfall is perhaps the most misused climatic element in comparative studies on ecology of the African oil palm. Its value as a direct measure of available water is strictly limited to the site on which the palms are grown. When comparing growth and development of oil palms in different regions, more accurate estimates of the available water will be required, necessitating estimation of potential evapotranspiration and the available water stored in the soil. The mean annual precipitation in 25 of the 28 oil palm regions of the world listed by Hartley (1967) varies between 1531 mm at Sibiti (Congo) and 3634 mm at Jerangau (Malaysia). Extremes are 1217 mm at Pobé (Dahomey), 5093 mm at Barrancabermeja (Colombia), and 8430 mm at Adenau (Cameroun). A very important factor for a continuously growing and fruiting evergreen perennial such as the African oil palm is the distribution of precipitation. Prolonged, virtually rainless periods may cause gradual drying of soil in the root zone to below the permanent wilting point. Depending on the depth of the soil, its water holding capacity, and the height of the ground water table, this may affect growth and development to a greater or lesser extent. An early reaction of the oil palm to water stress is midday closure of the stomata during the hours of maximum potential evapotranspiration, e.g., around 3 PM (Hartley, 1967). This period gradually becomes longer on successive days and eventually the stomata remain closed during most of the day. Together with the semixeromorphic structure of the leaves, this provides effective protection against desiccation. A conspicuous effect of water stress for several months is failure of spear leaves to open, eventually resulting in a crown with up to six or more elongated unopened leaves. During the same time the oldest leaves continue to die at a normal or even increased rate, so that the number of functional green leaves in the crown is reduced. It is not likely, however, that this reduction of some 10% of the leaf area has more influence on photosynthesis than corresponds with the reduction in leaf area, because the average age of the functional leaves is higher. Broekmans (1957) has shown that floral abortion about five months before anthesis was related to the intensity of the dry season. This has been confirmed in Malaysia (Corley, 1973). Corley also reported that this preanthesis abortion of inflorescences may be predominantly female, resulting in an apparent change in sex ratio not resulting from the sex differentiation shortly after the initiation of the inflorescence. He also stated that moisture stress had a direct effect on sex differentiation, resulting in a higher production of male inflorescences 20 months later. He concluded that the data of Broekmans (1957) and Sparnaaij *et al.* (1963), after correction for the influence of the abortion on the apparent sex ratio, confirmed these findings. However, it appears difficult if not impossible to unravel the direct and indirect effects of moisture stress on sex ratio if moisture stress is an annually recurring phenomenon during the dry season, as is the case in West Africa.

Ng and Goh (1970) showed that fruit set, fruit/bunch and oil/bunch in Malaysia were negatively correlated with rainfall at anthesis 160 days earlier (Corley, 1973). This obviously was due to the reduction in atmospheric pollen density caused by rainfall, as shown by Hardon and Turner (1967). It seems likely that these phenomena are positively correlated with the number of rainy days. It has been suggested that drought may be one of the causes of the sudden death of adequately pollinated bunches 2–4 months after anthesis, a condition called bunch failure by Turner and Bull (1967), but no conclusive evidence is available.

It is useful to reexamine early studies on the relationship between precipitation and annual yield of oil palm in the light of the recent research. Sparnaaij *et al.* (1963) found "little or no agreement about the period of time between the occurrence of a particular climatic condition and its effect on production, or about the climatic factor or factors responsible for the major yield fluctuations." This is hardly surprising, first, because one important point was overlooked, namely, that it is not the number of months between the climatic or edaphic condition determining sex of the inflorescence and anthesis that should be compared, but rather the number of leaves produced during that period. It is obvious that the latter depends equally on water supply, temperature, soil fertility, the age and genetic origin of the palms. Research in Africa and the Far East shows this to be the case. Adult palms in Nigeria produce 22–24 leaves annually (Hartley, 1967, after Broekmans, 1957) as opposed to 25–35 leaves in Malaysia (Williams and Hsu, 1970). If the sex of an inflorescence is determined in the axil of the 18th leaf from the apex, it reaches the stage of anthesis when this leaf has reached the 66th position (Hartley, 1967, p. 177). In the meantime 48 new leaves have been initiated at the apex and the same number have unfolded. In Henry's model of development of leaves and inflorescences (Surre and Ziller, 1963, p. 21) there is no indication of the leaf number corresponding to the stage of sex differentiation in the inflorescence, but it seems likely that this is the twenty-ninth leaf (*individualisation de l'inflorescence*). When this leaf has reached the seventy-sixth position from the apex, the inflorescence reaches the stage of anthesis. In the meantime 47 new leaves have been initiated and the same number unfolded. The time required to produce 47–48 new leaves amounts to 24–26 months in Nigeria and 16–23 months in Malaysia, almost exactly the time between sex differentiation and anthesis according to Broekmans (1957) in Nigeria and Corley (1973) in Malaysia. Second, the climatic factor or factors responsible for major yield fluctuations are not identical in all oil palm regions of the world. In West Africa both solar radiation and distribution of precipitation are limiting factors. In West Malaysia occasionally occurring periods of prolonged severe drought and excessive rainfall are likely to be important.

Devuyst (1948) found at La Mé, Ivory Coast a positive correlation between annual yield and the sum of monthly rainfall up to 300 mm during 12 consecutive months 33 months earlier. His limitation of the useful rain to a maximum of 300 mm per month may seem rather arbitrary, but it should be borne in mind that Penman's and Thornthwaite's estimates of evapotranspiration were published in the same year and that Ringoet's fundamental work on transpiration of oil palm was not published before 1952. The period of 33 months between rainfall and yield suggests a period of approximately 27 months between sex differentiation and anthesis, corresponding with an annual production of about 21 leaves according to Hartley's and Henry's models. This is rather low, but not improbable in view of the occurrence of two relatively dry seasons and the rather particular palm population of the La Mé experimental station at that time (Surre and Ziller, 1963, p. 32).

Hemptinne and Ferwerda (1961) studied regression of the 12-month totals of bunch yield on the 12-month totals of precipitation in oil palm plantations in Zaire, Yaligimba situated north of the equator (2°17′N, 22°15′E, altitude 435 m) and Leverville situated south of the equator (4°15′S, 18°44′E, altitude 520 m). The 12-month totals of the precipitation varied between approximately 1500 and 2000 mm in the northern plantation, and between 1100 and 2000 mm in the southern plantation. For the northern plantation they found a negative regression of bunch yield of precipitation 31 months earlier and a positive regression on precipitation 12 months earlier. For the southern plantation they found a quadratic regression of bunch yield on precipitation 33 months earlier and no regression of bunch yield on precipitation 12 months earlier. This suggested periods of 25 and 27 months, respectively, between sex differentiation and anthesis and floral abortion some six months before anthesis in the north. Supposing again that the time between sex differentiation and anthesis equals the time required to produce 47–48 leaves, this would correspond to an annual production of 23 leaves in the northern plantation and 21 leaves in the southern plantation. These calculations are in close agreement with actual annual leaf production of palms of this genetic origin (tenera × tenera ex. Yangambi, planting 1940 and 1941) growing under the above-mentioned conditions of rainfall and temperature. Sparnaaij *et al.* (1963) misunderstood the explanation given by these investigators for these regressions. They suggested that the negative regression of yield on rainfall 31 months earlier at Yaligimba was in fact the right half of a quadratic regression curve, with an optimum between 1500 and 1600 mm, as was found for Leverville. They did not give a valid explanation for the harmful effect of excessive rain 31–33 months earlier on bunch yield, but were able to show that this was not due to insufficient solar radiation. Most of the rain at Yaligimba and Leverville falls in the afternoon and at night, so that there is no negative regression of solar radiation on rainfall. The most likely explanation in the light of our present knowledge is an increase in leaching of plant nutrients in proportion to the quantity of rainwater percolating through the soil (Olivin and Ochs, 1974) and/or an increased incidence of waterlogging of the topsoil proportional to increasing rainfall.

Ringoet (1952) estimated the water balance of an 11-year-old plantation (Table VI). Actual evapotranspiration of this plantation amounted to 1511 mm and the percolation on flat land may be as much as 439 mm. This would seem to confirm the estimation of optimum rainfall for Yaligimba by Hemptinne and Ferwerda (1961), because the climatic conditions of Yaligimba and Yangambi are very similar. It is evident that this estimate does not represent an optimum water supply for palms, because the distribution of the rainfall,

TABLE VI
Water Balance (mm) of an 11-Year-Old Oil Palm Plantation[a]

Rain	1875	Retained by vegetation	131
Dew and mist	75	Transpired by palms	400
		Transpired by cover plants	673
		Evaporated from soil	307
		Runoff and percolation	439
	1950		1950

[a]+ 150 palms/ha at Yangambi, Zaire (after Ringoet, 1952).

although quite regular, is not ideal. Short periods during which actual evapotranspiration is lower than the potential evapotranspiration may occur during the driest months.

D. Air

Only wind is considered under this heading. The composition of the air, in particular its CO_2 content, may show ecologically important differences between the oil palm regions of the world, but little is known about this. Oil palm planters make the best use of the local CO_2 content of the air in two different ways (Corley, 1973): first, by increasing leaf area index (LAI) by increasing planting density, the area of the individual leaves, and the number of leaves per palm; second, by increasing net assimilation rate by improving the water supply, mineral supply, or the genetic potential for photosynthesis.

Tropical storms may do much damage, as the speed of the wind they create may exceed 160 km/hr. Some regions are notorious for the frequency of these storms, including the islands and coastal zones of the East and South China Sea, the Bay of Bengal, the Arabian seas, the Caribbean Sea, the Gulf of Mexico, the Indian Ocean east of Madagascar, the Coral Sea, and the Pacific Ocean east of Australia and west of Mexico. Taillez and Valverde (1971) examined the effects of a cyclone that struck an oil palm plantation in the valley of the Magdalena River, Colombia, in March 1968. Although most trees survived this disaster, many had more or less leaning trunks as a consequence of roots snapping off near the bulb. Growth and bunch production after erection of the trees were closely related to the degree of inclination of the stems after the cyclone. Trees with inclinations of 0°–30° were hardly affected; those with inclinations of 30°–60° had short and sometimes deformed leaves, and did not produce any bunches during about 1 year. Thirty months after the cyclone, some of the trees with inclinations of 60°–90° still produced deformed leaves, and most produced only male inflorescences. The degree of stem inclination also was related to the age of the palms and their genetic origin. They were most susceptible to wind damage between $2\frac{1}{2}$ and $4\frac{1}{2}$ years after field planting, probably because the crown gradually reaches its final dimensions while the rooting system is still extending and brittle. The effect of genetic origin is probably due to differences in height increase. There is a positive correlation between the susceptibility to wind damage and the height of the trunk.

IV. SOILS

Semiwild palm groves and cultivated palms are found on a wide range of soils, indicating that the oil palm is not an exacting crop in this respect. This is confirmed by the fact that the natural habitats of wild oil palms, such as springs, banks of rivers and lakes, wet valleys and swamps, are too wet for dicotyledonous trees (Zeven, 1967). They grow best on fertile, deep, good structured, and well-drained soils, but are unable to compete with trees of the ombrophilous lowland forest and the tropical evergreen seasonal forest without man's intentional or unintentional protection.

1. Soil Profile

According to the Seventh Approximation of the U.S. System of Soil Classification, most oil palm soils belong to the following orders and suborders: Oxisols (Orthox), Ultisols (Udults), and Inceptisols (Aquepts). Although soil maps are available for many oil palm plantations, it is not yet possible to compare them on the common basis of an internationally used classification, such as the above-mentioned system or the FAO classification. For the present, the suitability of certain soils for cultivation of oil palm must be based on known factors limiting growth and development, such as the presence of a hardpan, poor internal and external drainage, insufficient soil depth, a high water table, poor soil structure, and low fertility.

2. Physical Properties

The texture of oil palm soils varies between loamy sand, sandy loam, sandy clay loam, sandy clay, and clay. There is perhaps a preference for soils containing 25–30% clay because of their good water holding capacity and usually good structure, but excellent plantations may be found on soils with finer or coarser textures. Modern forest clearing methods with heavy equipment tend to destroy soil structure, because they cause compaction. This may locally reduce bulk density to critical levels for root penetration and affect the growth of young palms planted on these sites.

3. Chemical Properties

Although considerable information is available on mineral reserves, clay minerals, cation exchange capacity, and exchangeable cations in soils used for cultivation of oil palms, little is known about the quantitative relationships between each of these factors and growth and development of the oil palm. Ollagnier *et al.* (1970) gave an example of fertile montmorillonite clays in Brazil, which have such poor internal drainage that yields are poor. Tinker (1963) describes chemically poor kaolinite clay giving satisfactory yields with adequate fertilizing. Tinker and Smilde (1963) estimated for Nigerian soils a release of some 20 kg of K per annum over a depth of 3 m, corresponding with 0.5 ppm from a total content of some 200–300 ppm. This covers approximately 40% of the requirements of adult palms.

a. Nutritional Deficiencies. The effects of nutritional deficiencies on growth and development are much better understood. Visual symptoms of the major element deficiencies and of some of the minor element deficiencies are known from the results of pot and field experiments. The first descriptions of K and Mg deficiency symptoms based on experimental evidence were made by Thompson (1941), of B deficiency by Ferwerda (1954), of N deficiency by Broeshart (1955), of P and S deficiency by Broeshart *et al.* (1957), and of Fe and Mn deficiency by Bull (1961b). The most thorough and complete studies were made by Broeshart *et al.* (1957) and Bull (1961a,b). Rajaratnam (1972) made comprehensive studies of external and internal symptoms of B deficiency. The visual symptoms of Cl, Cu, Zn, and Mb deficiency are still unknown, but there is evidence to suggest that Cl is an essential element for the oil palm (Ollagnier and Ochs, 1971).

Many nutritional deficiencies are known to affect growth, development, and bunch yield. A summary of these effects on young and adult plantation palms in Zaire was given by Ferwerda (1962). This summary, slightly supplemented, is reproduced in the Tables VII and VIII.

Nutritional deficiences may affect bunch production without measurable effects on growth and without causing visible leaf symptoms. Almost all nutritional deficiencies cause a decrease of the number of fruit bunches, either by changing the sex ratio in favor of male inflorescences and/or by increasing

TABLE VII
Effect of Nutritional Deficiencies on Growth, Development, and Bunch Production during the First and Second Year of Harvesting of Young Palms Six Years after Field Planting[a]

	Deficiency[b]										
	N	P	S	K	Ca	Mg	Cu	Mn	Zn	B	Mo
Leaf production	–	0	–	–	–	–	0	0	0	0	0
Length of leaves	–	0	–	–	–	–	0	0	0	–	0
Number of pinnae	–	0	–	–	–	–	0	0	0	–	0
Length of pinnae	?	?	?	–	?	0	?	?	0	–	0
Width of pinnae	?	?	?	–	?	0	?	?	0	–	0
Deficiency symptoms	+	0	+	+	0	+	0	0	0	+	0
First inflorescence	–	0	–	–	–	–	0	0	0	0	0
Male inflorescences	–	–	–	–	–	0	0	–	0	–	–
Female inflorescences	–	–	–	–	–	–	–	–	0	–	–
Fruit bunches (number)	–	–	–	–	–	–	–	–	–	–	–

[a]After Ferwerda (1962).
[b]–, reduced or retarded; 0, no effect; +, increased or accelerated; ?, effect unknown.

TABLE VIII
Effect of Nutritional Deficiencies on Bunch Number, Mean Bunch Weight, and Mortality of Adult Palms[a]

	Deficiency											
	N	P	S	K	Ca	Mg	Cu	Mn	Zn	Fe	B	Mo
Fruit bunches (number)	–	–	–	–	–	–	–	–	?	?	–	–
Mean bunch weight	0	0	?	–	?	–	0	0	?	?	–	?
Mortality	+	0	?	0	?	0	?	?	?	?	+	?

[a]–, reduced or retarded; 0, no effect; +, increased or accelerated; ?, effect unknown.

abortion of immature female inflorescences some six months before anthesis. A decrease in the number of fruit bunches does not always correspond to a decrease of mean bunch weight. Hence, reduction of the number of fruit bunches is apparently the first adaptation of the palm to adverse conditions.

Ollagnier and Ochs (1971) suggested that chlorine is an essential element for the oil palm. Their conclusion was based on a highly significant correlation between the chlorine content of leaves and bunch yield and the absence of such a correlation between the potassium content of the leaves and bunch yield in field experiments in which potassium chloride applications were used. Hew and Poon (1973) could not find confirmation for this in three manurial trials on coastal clay soils in Malaysia, although there were significant responses to fertilizing with muriate of potash and although the Cl contents of the leaves varied between 0.12 and 0.56% as compared with 0.81–1.22% for the K contents. The correlation coefficients between bunch yield and Cl content were small and insignificant. Only in one experiment was there a significant but equally low correlation of bunch yield with K content of leaves.

Van Diest and Struif Bontkes (1972) studied the influence of Cl in nutrient solutions on growth and the chemical composition of approximately one-year-old seedlings (Table IX). Addition of chlorine to the nutrient solution did not improve growth and caused a decrease in uptake of N, P, S, K, and Mg. For these reasons, it probably is too early to consider chlorine as an essential nutrient for the oil palm. Although it is tempting to interpret high correlation coefficients between bunch yield and Cl content in the leaf as indicating causal relationships, this is not necessarily the case.

Ng and Pau (1974) and Ng *et al.* (1974) presented convincing evidence that the symptoms of midcrown chlorosis of palms planted on deep, acid peat soils were due to Cu deficiency.

TABLE IX
Effect of 4-Month Exposure to Chlorine on Growth, Chemical Composition, and Organic Cation Content of Young Seedlings[a]

	%
Dry matter aerial part[b]	78
Dry matter roots[b]	84
K uptake[b]	87
Ca uptake[b]	110
Mg uptake[b]	76
Cl uptake[b]	330
NO_3 concentration[c]	140
SO_4 concentration[c]	91
H_2PO_4 uptake[b]	80
N uptake	88
Organic cations (cations–anions)	86

[a]After van Diest and Struif Bontkes (1972).
[b]Last two months.
[c]Last month.

The chemical composition of oil palm leaves is widely used for diagnosis of nutritional disorders. Between the introduction of the plant analysis technique for oil palms by Hale (1947) and Chapman and Gray (1949) and its present wide use, after having been improved by many other investigators, the confidence in the absolute value of the leaf nutrient contents as norms for deficiency or sufficiency gradually disappeared. Ferwerda (1961) stated, on the basis of Coulter's (1958) and his own investigations, that the norms for the interpretation of leaf analysis data can only be established locally by comparison with the results obtained in fertilizer experiments in the same plantation and sampled at the same time. Hartley (1967) and Bolle-Jones (1975) arrived at much the same conclusion. The latter pleaded for a more quantitative approach to uptake of plant nutrients by means of simultaneous measurements of growth and dry matter production of the analyzed plants. The uptake of a nutrient is likely to be much better related to bunch yield than its concentration in the leaf.

b. pH. The oil palm is not exacting with respect to soil. The pH of soils of good yielding plantations varies between approximately 4.0 and 8.0, but most plantations have been established on more or less acid soils (pH 4.0–6.0).

c. Salt Tolerance. Criticial salt concentrations in the soil solution may occur as a consequence of brackish groundwater or cyclic salt in coastal regions. Ferwerda and Struif Bontkes (1972) studied the influence of salt concentrations on growth of 4-leaf seedlings in sand culture. Both the nutrient solution and the additional salt were applied by continuous trickle irrigation, so that nutrient and salt concentrations in the soil solution were constant throughout the duration of each treatment. Five salt concentrations were tested by adding none, 40, 90, 140, and 190 mEq salt per liter to the nutrient solution as NaCl, half NaCl and half $CaCl_2$, or half NaCl and half $MgCl_2$. The concentration of the nutrient solution was about 10 mEq/liter, so that the range obtained was 10–50–100–150 and 200 mEq, corresponding to a conductivity of 1–5–10–15 and 20 millimhos per centimeter. Plants were exposed to the salt solutions for 2, 4, or 6 months. Those exposed to salt solutions for 2 or 6 months received a normal nutrient solution during the remainder of the time.

The time required to produce one new leaf increased from about 2 weeks to nearly 5 weeks for the highest salt concentration, but solutions with an electrical conductivity up to 10 millimhos per centimeter appeared to have little effect. The effect of pure NaCl appeared to be somewhat greater than that of the mixtures with $CaCl_2$ or $MgCl_2$ (Fig. 3).

The first symptoms of salt damage were pale green areas on the leaves, which turned yellow. Eventually the leaves became necrotic. The symptoms proceeded from the oldest to the youngest leaves. The youngest leaf was always green.

The influence on the dry matter production per plant during six months is shown in Table X. Dry matter production was inhibited by all salt concentrations, but up to electrical conductivity values of 10 millimhos per centimeter the reduction is rather limited. The duration of exposure to salt solutions was very important. If limited to two months, growth reduction was only about 10%. The

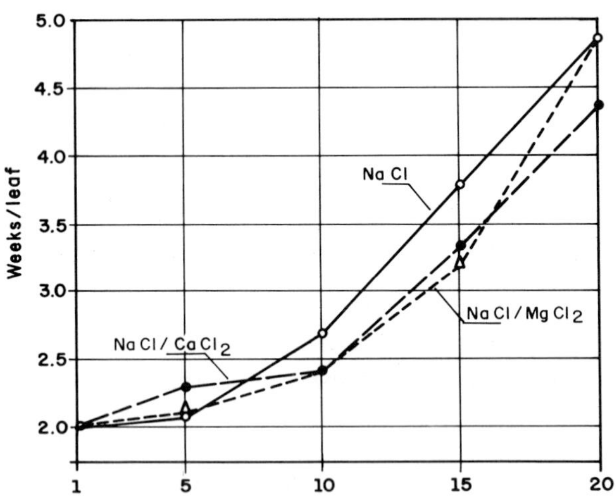

Fig. 3 Time required for the production of one new leaf and electrical conductivity (EC) of the nutrient solution in millimhos/cm (Ferwerda and Struif Bontkes, 1972).

TABLE X
Influence of Salt Concentration of the Soil Solution, Duration of the Exposure to a Salt Soil Solution, and the Kind of Salt on Dry Matter Production of 4-Leaf Seedlings Raised in Sand Culture for 6 Months[a]

		Dry weight per palm (g)	
		Aerial part	Roots
Complete nutrient solution only		125.2	17.3
Electrical conductivity	5 mmhos/cm	101.3	16.4
	10 mmhos/cm	97.0	15.6
	15 mmhos/cm	63.6	11.0
	20 mmhos/cm	33.9	6.3
Exposed for			
2 months		108.7	17.9
4 months		68.1	11.3
6 months		51.9	8.6
100% NaCl		69.0	12.0
50% NaCl + 50% $CaCl_2$		94.6	13.6
50% NaCl + 50% $MgCl_2$		79.6	12.3

[a]Ferwerda and Struif Bontkes, 1972.

composition of the salt also had much influence. Pure NaCl appeared to be more harmful than a mixture of NaCl and $MgCl_2$ and the latter more harmful than a mixture of NaCl and $CaCl_2$. From this it may be concluded that young oil palms are moderately tolerant to salts in the soil solution.

4. *Biotic Properties*

High concentrations of active tertiary and quaternary roots are found near sites with accumulations of decaying vegetable matter, such as leaves removed during pruning rounds (Tailler and Valverde, 1971). The major cause of this concentration of roots is likely to be locally increased supply of mineral nutrients. The oil palm and its cover crop provide a continuous supply of organic matter to the soil from decaying leaves and roots, and protect the soil from direct solar radiation and rain. The total annual production of leaves, stems, and fruit bunches is 20–30 tons dry matter per hectare^{-1} year^{-1} (Rees and Tinker, 1973; Corley *et al.*, 1973). This represents some 96% of the total dry matter production. The leaves, which provide 56% of this fraction (Rees and Tinker, 1973), add 11–16 tons organic matter hectare^{-1} year^{-1} to the soil. To this may be added the 10–14 tons of dry matter produced annually by the cover crop (Rombaut, 1974). Thus the primary production of an oil palm plantation amounts to 31–44 tons dry matter hectare^{-1} year^{-1} and is therefore of the same magnitude as that of ombrophilous and evergreen seasonal forest of the tropical lowland.

Even without returning the bunch refuse to the plantation and neglecting male inflorescences and dead roots, the annual supply of organic matter to the soil amounts to 21–30 tons hectare^{-1} year^{-1}. This is considerably more than required to compensate for the rate of decomposition of organic matter on similar soils under annual cropping.

V. BIOTIC FACTORS

Under this heading only the influence of higher plants will be discussed. The influence of the other biotic factors, such as lower plants and animals, is usually so complex that conclusions have very limited general validity. In addition they are often closely related to the physical environment. The most important biotic factor, man, is entirely beyond the scope of this paper, because his influence is primarily of a social, economic, or technical nature.

In oil palm plantations there is competition between neighboring palms, and between palms and the interline vegetation including cover crops, weeds, or other annual or perennial crops. The study of the competition between neighboring palms is closely related to the study of the optimal planting density. Corley *et al.* (1973) reviewed the older work on this subject, except the basic study by Marynen (1963). Ramachandran *et al.* (1973) compared wide avenue and triangular planting with densities varying from 96 to 183 palms per hectare (Corley *et al.*, 1973) and defined optimal density in three ways: (a) the highest current yield in any given year; (b) the highest cumulative yield over a given period; (c) the maximum cumulative profit over a given period. In this chapter only (a) and (b) are considered.

Prévot and Duchesne (1955) found a negative linear regression of current or cumulative yield per palm on the number of palms per hectare (Fig. 4). This may be expressed as

$$y = a - bx$$

where y is the yield per palm, x is the number of palms per hectare, and a and b are constants. It is evident that b is the amount by which the yield per palm decreased with a unit increase in density. This decrease is due both to a decrease of the mean annual number of bunches per palm and the mean annual bunch weight which was confirmed by Ramachandran *et al.* (1973). Marynen (1963) showed that this linear regression fitted just as well as two more complicated nonlinear functions describing competition within species. It is evident that this can only be true for medium densities, as the yield per palm will eventually

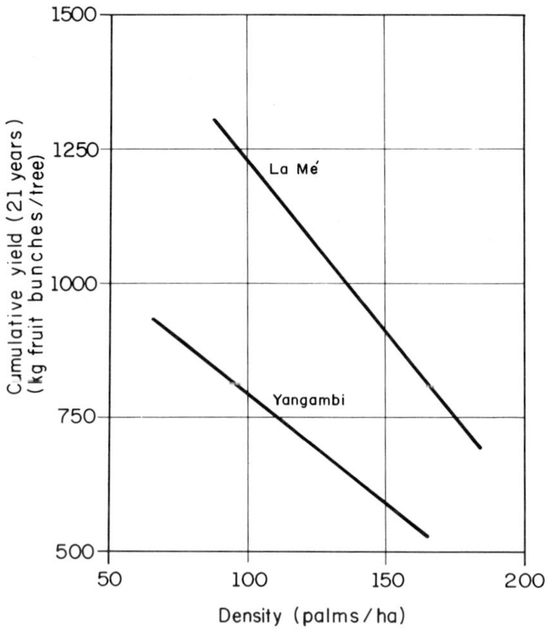

Fig. 4 Mean cumulative weight of fruit bunches per palm (21 years) and density of La Mé (Ivory Coast) and Yangambi (Zaire) (Prévot and Duchesne, 1955).

attain a maximum at very high densities and a minimum at very low ones. At medium densities the regression for yield per hectare on the number of palms is therefore a quadratic curve (Fig. 5):

$$Y = xy = ax - bx^2$$

where Y is the yield per hectare. The optimal density x_{max} is now given by

$$x_{max} = a/2b$$

As could be anticipated, the regression coefficient b (or competition factors) varies when comparing data from different planting density experiments (Fig. 5).

Corley *et al.* (1973) showed that mean area per leaf may be used as a standard for comparison of different sets of data, despite its close relationship with palm age, soil fertility, and genotype. They arrived at the important conclusion that mean leaf area can be used to estimate the current optimal density, independently of the considerable differences between actual current yields corresponding to these densities (Fig. 6). Similarly, the competition factor may be estimated from mean leaf area, independently of the actual leaf level (Fig. 7). Since leaf area reaches a maximum at 9–10 years after planting (Fig. 1) it is

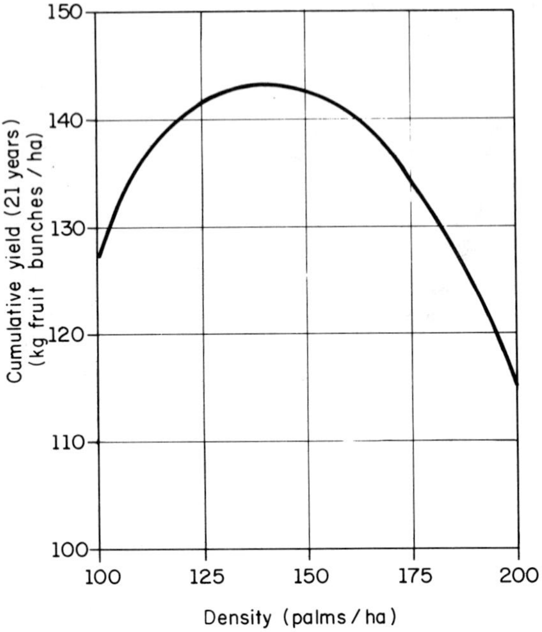

Fig. 5 Mean cumulative weight of fruit bunches per hectare (21 years) and density at La Mé, Ivory Coast (Prévot and Duchesne, 1955).

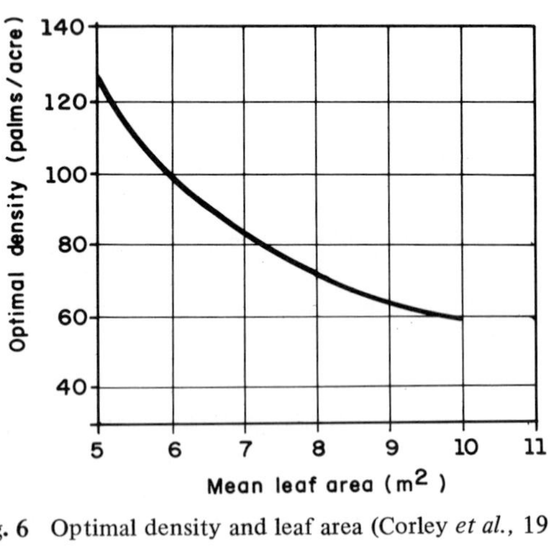

Fig. 6 Optimal density and leaf area (Corley *et al.*, 1973).

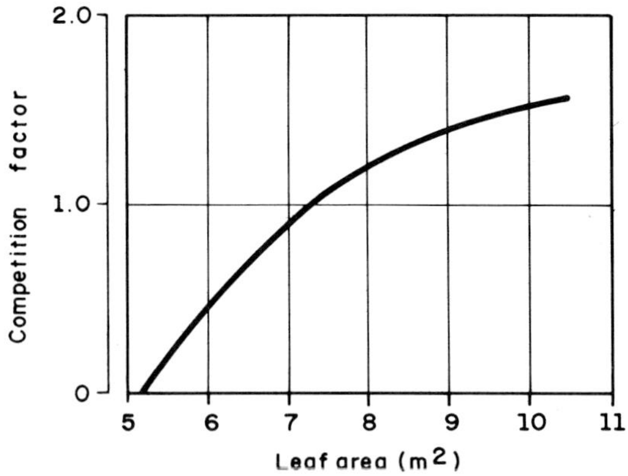

Fig. 7 Competition factor and leaf area (Corley *et al.*, 1973).

likely that the effects of density are approximately constant after this age, so that the optimal densities for 10- to 30-year-old plantations may be estimated by extrapolation.

Prévot and Duchesne (1955) presented convincing evidence that for any given density triangular spacing was superior to rectangular spacing. This was confirmed by Ramachandran *et al.* (1973) under Malaysian conditions.

VI. SUMMARY

Important contributions to the knowledge of the environmental physiology of the African oil palm, *Elaeis guineensis* Jacq., were made during the last 10 years. These would seem to suggest that the satisfactory growth, development, and bunch yield may be obtained under lower light intensities and at lower temperatures than were rather generally accepted up to the present time. The daylength is not likely to play an important role. There is much evidence to suggest that the water supply should be continuously sufficient, but not excessive, throughout the year for optimal results. Tropical storms may cause serious and long-lasting damage. The need for chlorine in mineral nutrition is questioned, but there is convincing evidence that copper deficiency may occur under field conditions. The tolerance to salts in the soil solution is moderate. The annual supply of organic matter to the soil in plantations compares favorably with that of forests of the moist tropical lowland. The mean leaf area appears to provide a good estimation for the optimal planting density. In regions with long dry seasons, the desirability of a living interline vegetation may be questioned.

REFERENCES

Anonymous. (1966). The Oil Palm in Malaya. Ministry of Agriculture and Cooperatives, Kuala Lumpur, Malaysia.

Bolle-Jones, E. W. (1975). Limitations of foliar diagnosis with special reference to the oil palm. *Trop. Agric. (Trinidad)* **52**, 11–19.

Broekmans, A. F. (1957). Growth, flowering and yield of the oil palm in Nigeria. *J. West Afr. Inst. Oil Palm Res.* **2**, 187–220.

Broeshart, H. (1955). The application of foliar analysis in oil palm cultivation. Ph.D. Thesis, Agricultural University Wageningen.

Broeshart, H., Ferwerda, J. D., and Kovachich, W. G. (1957). Mineral deficiency symptoms of the oil palm. *Plant Soil* **8**, 289–300.

Bull, R. A. (1961a). Studies on the deficiency diseases of the oil palm. 2. Macronutrient deficiency symptoms in oil palm seedlings grown in sand culture. *J. West Afr. Inst. Oil Palm Res.* **3**, 254–264.

Bull, R. A. (1961b). Studies on the deficiency diseases of the oil palm. 3. Micronutrient deficiency symptoms in oil palm seedlings grown in sand culture. *J. West Afr. Inst. Oil Palm Res.* **3**, 265–272.

Chapman, G. W., and Gray, H. M. (1949). Leaf analysis and the nutrition of the oil palm (*Elaeis guineensis* Jacq.). *Ann. Bot. (London)* [N.S.] **13**, 415–433.

Coldesa (1974). Replanting diseased oil palm areas with *Elaeis oleifera* × *E. guineenis* at "La Arenosa" estate in Colombia. *Oil Palm News* **18**, 1–6.

Corley, R. H. V. (1973). Oil palm physiology: A review. *Adv. Oil Palm Cultivation* pp. 37–51.

Corley, R. H. V., Hew, C. K., Tam, T. K., and Lo, K. K. (1973). Optimal spacing for oil palms. *Adv. Oil Palm Cultivation* pp. 52–71.

Coulter, J. K. (1958). Mineral nutrition of the oil palm in Malaya. *Malays. Agric. J.* **41**, 131–151.

Daniel, C., and de Taffin, G. (1974). Conduite des jeunes plantations de palmier à huile en zones sèches au Dahomey. *Oléagineux* **29**, 227–232.

Devuyst, A. (1948). Influence des pluies sur les rendements du palmier à huile, enregistrés à la station de La Mé de 1938 à 1946. *Oléagineux* **3**, 137–144.

de Wit, C. T. (1965). Photosynthesis of leaf canopies. *Agric. Res. Rep. (Neth.)* no. 663.

Ferwerda, J. D. (1954). Boron deficiency in oil palms in the Kasai region of the Belgian Congo. *Nature (London)* **173**, 1097.

Ferwerda, J. D. (1955). Questions relevant to replantation in oil palm cultivation. Ph.D. Thesis, Agrciultural University, Wageningen.

Ferwerda, J. D. (1961). Growth, production and leaf composition of the African oil palm as affected by nutritional deficiencies. *Am. Inst. Biol. Sci., Publ.* **8**, 149–159.

Ferwerda, J. D. (1962). Die Oelpalme. *In* "Tropische und subtropische Weltwirtschaftspflanzen" (W. Bally, ed.), Part II, pp. 309–378. Enke, Stuttgart.

Ferwerda, J. D. (1971). Oelpalme. *In* "Handbuch der Landwirtschaft und Ernaehrung in den Entwicklungslaendern" (P. von Blanckenburg and H. D. Cremer, eds.), Vol. 2, pp. 330–340. Ulmer, Stuttgart.

Ferwerda, J. D., and Struif Bontkes, T. (1972). "De zoutresistentie van de jonge oliepalm (*Elaeis guineensis* Jacq.)," Internal Research Report. Dept. of Tropical Crop Science, Agricultural University, Wageningen.

Hale, J. B. (1947). Mineral composition of leaflets in relation to chlorosis and bronzing of oil palms in West Africa. *J. Agric. Sci.* **37**, 236–244.

Hardon, J. J., and Turner, P. D. (1967). Observations on natural pollination in commercial plantings of oil palm (*Elaeis guineensis*) in Malaya. *Exp. Agric.* **3,** 105–116.

Hardon, J. J., Mokhtar Hashim, and Ooi, S. C. (1974). Oil palm breeding: A review. *Adv. Oil Palm Cultivation* pp. 23–36.

Hartley, C. W. S. (1967). "The Oil Palm." Longmans, Green, New York.

Hartley, C. W. S. (1970). Some environmental factors affecting flowering and fruiting in oil palm. *In* "Physiology of Tree Crops" (L. C. Luckwill and C. V. Cutting, eds.), pp. 169–286. Academic Press, New York.

Hemptinne, J., and Ferwerda, J. D. (1961). Influence des precipitations sur les productions du palmier a huile. *Oléagineux* **16,** 431–437.

Henry, P. (1957). Recherches sur la croissance et le développement chez *Elaeis guineensis* Jacq. et chez *Cocos nucifera* L. Comparaison avec quelques autres palmiers. Thesis, University of Paris.

Hew, C. K., and Poon, Y. C. (1973). The effects of muriate of potash and bunch ash on yield and uptake of potassium and chlorine in oil palms on coastal soils. *Adv. Oil Palm Cultivation* pp. 239–247.

Jones, L. H. (1974). Propagation of clonal oil palms by tissue culture. *Oil Palm News* **17,** 1–8.

Knecht, J. C. X., Ramachandran, R., and Narayanan, R. (1974). Some variational features of the leaf nutrient contents in oil palm sampling. *Oléagineux* **29,** 287–295.

Maillard, G., Daniel, C., and Ochs, R. (1974). Analyse des effets de la sécheresse sur le palmier à huile. *Oléagineux* **29,** 397–404.

Marynen, T. (1963). Contribution à l'étude de la densité de plantation chez les végétaux. *Publ. I.N.E.A.C., Ser. Sci.* **102,** 1–79.

Ng, S. K., and Pau, T. Y. (1974). Nutritional complexes of oil palms planted on peat in Malaysia. I. Foliar symptoms, nutrient composition and yield. *Oléagineux* **29,** 1–14.

Ng, S. K., and Pau, T. Y., Chan, E., and Cheong, S. P. (1974). Nutritional complexes of oil palms planted on peat soil in Malaysia. II. Preliminary results of copper sulphate treatments. *Oléagineux* **29,** 445–456.

Olivin, J., and Ochs, R. (1974). Le lessivage du potassium et du magnésium en colonnes de terre. Premiers résultats expérimentaux. *Oléagineux* **29,** 175–182.

Ollagnier, M., and Ochs, R. (1971). Le chlore, nouvel élément essential dans la nutrition du palmier à huile. *Oléagineux* **26,** 1–15.

Ollagnier, M., Ochs, R., and Martin, G. (1970). The manuring of oil palm in the world. *Fertilité* **36,** 3–63.

Prévot, P. (1961). Données récentes sur la physiologie du palmier à huile. *Cah. ORSTOM, Ser. Physiol. Plant. Trop. Cultiv.* **1,** 11–39.

Prévot, P., and Duchesne, J. (1955). Densités de plantation pour le palmier à huile. *Oléagineux* **10,** 117–122.

Purseglove, J. W. (1972). "Oil Palm. Tropical Crops," Vol 1, pp. 479–510. Wiley, New York.

Rajaratnam, J. A. (1972). Observations on boron-deficient oil palms (*Elaeis guineensis*). *Exp. Agric.* **8,** 339–346.

Ramachandran, P., Narayanan, R., and Knecht, J. C. X. (1973). A planting distance experiment on Dura palms. *Adv. Oil Palm Cultivation* pp. 72–89.

Rees, A. R. (1961). Midday closure of stomata in the oil palm. *J. Exp. Bot.* **12,** 129–146.

Rees, A. R., and Chapas, L. C. (1963). Water availability and consumptive use in oil palm nurseries. *J. West Afr. Inst. Oil Palm Res.* **4,** 52–65.

Rees, A. R., and Tinker, P. B. H. (1973). Dry matter production and nutrient content of plantation oil palms in Nigeria. 1. Growth and dry matter production. *Plant Soil* **19**, 19–32.

Ringoet, A. (1952). Recherches sur la transpiration et le bilan d'eau de quelques plantes tropicales. *Publ. I.N.E.A.C., Serv. Sci.* **56**, 139.

Rombau, D. (1974). Etude sur l'élevage bovin dans les palmeraise de Côte-d'Ivoire. *Oléagineux* **29**, 121–125.

Sparnaaij, L. D., Rees, A. R., and Chapas, L. C. (1963). Annual yield variation in the oil palm. *J. West Afr. Inst. Oil Palm Res.* **4**, 111–125.

Staritsky, G. (1970). Tissue culture of the oil palm (*Elaeis guineensis* Jacq.) as a tool for its vegetative propagation. *Euphytica* **19**, 288–292.

Surre, C., and Ziller, R. (1963). "Le palmier à huile." Maisonneuve & Larose, Paris.

Taillez, B., and Valverde, G. (1971). Sensibilité aux ouragans de différents types de croissements de plamier à huile. *Oléagineux* **26**, 753–762.

Thompson, H. (1941). Notes on plant diseases. *Malay. Agric. J.* **29**, 241–242.

Tinker, P. B. H. (1963). Changes occurring in the sedimentary soils of souther Nigeria after oil palm plantation establishment. *J. West Afr. Inst. Oil Palm Res.* **4**, 66–81.

Tinker, P. B. H., and Smilde, K. W. (1963). Cation relationship and magnesium deficiency in oil palm. *J. West Afr. Inst. Oil Palm Res.* **4**, 82–100.

Turner, P. D. (1974). Oil Palm pollination and pruning: Some new light on old ideas. *Planter, Kuala Lumpur* **50**, 24–31.

Turner, P. D., and Bull, R. A. (1967). "Diseases and Disorders of the Oil Palm in Malaysia." Incorporated Society of Planters, Kuala Lumpur, Malaysia.

Turner, P. D., and Gillbanks, R. A. (1974). "Oil Palm Cultivation and Management." Incorporated Society of Planters, Kuala Lumpur, Malaysia.

van Diest, A., and Struif Bontkes, T. (1972). "De invloed van de N-vorm en de toediening van Cl op de groei, de chemische samenstelling en het C-A-gehalte van de eenjarige oliepalm (*Elaeis guineensis* Jacq.)," Internal Research Report. Dept. of Soil and Fertilizers, Agricultural University, Wageningen.

Walter, H. (1973). "Vegetationszonen und Klima." Ulmer, Stuttgart.

Wessels Boer, J. G. (1965). The indigenous palms of Surinam. Ph.D. Thesis, Agricultural University, Utrecht.

Williams, C. N., and Hsu,Y. C. (1970). "Oil Palm Cultivation in Malaya." University of Malaya Press, Kuala Lumpur, Malaysia.

Wormer, T. M., and Ochs, R. (1957). Humidite du sol et comportement du palmier a huile en pepiniere. *Oléagineux* **12**, 81–89.

Wormer, T. M., and Ochs, R. (1959). Humidite du sol, ouverture des stomates et transpiration du palmier a huile et de l'arachide. *Oléagineux* **14**, 571–580.

Zeven, A. C. (1967). The semi-wild oil palm and its industry in Africa. Ph.D. Thesis, Agricultural University, Wageningen.

CHAPTER

14

Coconut Palm

D B. MURRAY

I. INTRODUCTION

Strong arguments have been advanced both for a Central American origin and a Melanesian origin for the coconut palm (*Cocos nucifera* L). Cook (1910) strongly advocating a Central American origin, stated that more species of *Cocos* were found in Central and South America than elsewhere. He also suggested that the coconut palm was distributed by the early inhabitants of South America traveling across the Pacific, a theory that led to the famous Kon Tiki expedition. Thor Heyerdahl (1952) showed that it was possible for men with only primitive boats or rafts to cross the Pacific Ocean from South America to Oceania and to carry their crops and culture with them. Whether this actually happened, however, is difficult to prove.

On the other hand, Becarri (1917) gave evidence, including the discovery of fossil nuts in New Zealand, that the coconut palm originated in Southeast Asia. He pointed out the far wider range of varieties of coconuts in the countries of the Far East as compared with the Americas and the Caribbean area. Evidence has also been advanced that while the coconut may have been on the Pacific shores of South America before Columbus' expedition, it was introduced to Puerto Rico and Brazil by the Portuguese only in the middle of the sixteenth century.

The weight of evidence supports an origin in the Old World somewhere in Melanesia. Whatever may have been the origin of the coconut palm, it is now widely distributed on the sea coasts of the tropical world (Fig. 1) (Menon and Pandalai, 1958). The extensive distribution, even where it does not appear to be associated with man's activities, has led to the widely held view that the coconut has spread on ocean currents. This view, however, is very controversial.

Although Edmonson (1941) showed that in Hawaii nuts still had the ability to germinate after floating for up to 110 days in the sea, all the Kon Tiki nuts stored below deck and exposed to seawater were ruined and only those kept dry on deck survived. Even if viable nuts are cast up on a beach, their chances of survival are very slim. Some dissemination by sea may have occurred, but the

Fig. 1 World map showing coconut producing areas.

wide distribution of the coconut palm must be due mainly to the influence of man.

II. CLIMATIC REQUIREMENTS

The Palmae are essentially tropical plants and the coconut is no exception. It is grown throughout the tropical world between latitudes 23°N and 23°S.

A. Temperature

The optimal mean annual temperature for best growth and maximum yield is considered to be about 27°C, with a diurnal range of 6°–7°C. Such conditions are usually found on tropical sea coasts where the ocean acts as a buffer against too rapid changes in diurnal temperature. With low temperatures for more than a short time, floral and fruiting abnormalities may occur. In Florida, at 25°N, palms succumb to unusually severe winters.

Temperature also determines the altitude at which coconut can grow. In Mysore State in India, plantations are found at 600–900 m above sea level; in Sri Lanka (Ceylon) they may occur up to 750 m. At Tabora in Tanzania, not far from the equator, productive plantations have been established at an elevation of 1300 m. The limit is reached at lower altitudes with increased distance from the equator, so that in Jamaica at 18°N coconuts are not grown commercially above 150m.

B. Rainfall

Even distribution of rainfall probably is the most important factor influencing coconut yield. Since the crop is produced continuously through the year and the nut takes a year to mature from pollination, ideally the tree should never undergo severe water stress. The optimum total annual rainfall lies between 1300 and 2300 mm per year, although the coconut tree will tolerate a far higher rainfall provided soil drainage is good. The monthly rainfall in a number of coconut growing areas is given in Table I. Even with high rainfall many areas have a pronounced dry season.

Uneven distribution of annual rainfall may be compensated for by special environmental conditions under which coconuts are so often grown; that is, where seepage of ground water occurs through a plantation. Where coastal beach plantings are backed by rainfed freshwater swamps and lagoons, the water level will be slightly above that of the sea so that percolation of the swamp water takes place seaward and supplies the coconuts even when rainfall is deficient for short periods. The highest yields of coconuts have been found under such specialized ecological conditions. An additional and important factor is that the

TABLE I
Monthly Rainfall in mm for Selected Coconut Growing Areas[a]

	J	F	M	A	M	J	J	A	S	O	N	D	Total
Asia													
India, Kasuragod	7	9	11	73	268	973	1.136	476	228	254	76	31	3542
Sri Lanka, Bandirippuwa	56	55	136	213	262	188	72	87	108	312	328	122	1939
Philippines, Nagacity	115	81	69	186	156	202	274	187	291	292	309	290	2452
Africa													
Ivory Coast, Port Bouet	30	50	108	160	382	570	177	35	61	207	188	113	2081
Mozambique, Porto-Belo	277	193	261	128	86	92	72	36	16	10	67	191	1429
Pacific													
New Hebrides	311	227	424	285	273	164	241	119	260	205	250	312	3071
Fiji	337	233	275	306	150	92	93	82	96	149	296	178	2287
America													
Jamaica, Port Antonio	253	88	130	247	310	372	253	167	253	434	411	574	3492
Brazil, Maceio	73	75	137	183	289	301	219	138	86	44	27	44	1616

[a]From Frémond *et al.* (1966).

freshwater carries traces of mineral nutrients which allow the coconut to flourish on almost sterile sands.

When rain-fed trees undergo a prolonged drought, the effect can persist for up to $2\frac{1}{2}$ years as shown in Fig. 2 (a) which shows the reduction in numbers of nuts picked on four occasions during each year. In addition to the reduction in nuts by drought, the amount of copra per nut is also reduced [Fig. 2 (b)] so that these two factors combined seriously reduce the total crop yield per acre.

As Patel and Anandan (1936) pointed out, the inflorescence is initiated some 16 months before the spathe opens. Severe drought at this period may kill the growing point, causing the inflorescence to abort. This will affect production of nuts up to 28–30 months later.

The amount of copra per nut also is reduced during the drought period and increases sooner as shown in Fig. 1 (b). Shepherd (1926) showed that in Trinidad there is a positive significant correlation between rainfall over a six month period and the amount of copra per nut one year later. In Malaya, rainfall in the first three months when the nuts were developing determined the size of the crop a year later (Cooke, 1953).

In Trinidad, the drop in yield under drought conditions is correlated with the original level of bearing of the tree (Murray and Lucie-Smith, 1952). Thus in a fertilizer trial which was affected by a severe shortage of rain, the trees were divided into classes on the basis of their predrought yields. The effect of the 1947 drought on the subsequent crop is shown in Table II.

C. Light Intensity

The coconut palm is a light-requiring species and does not grow well under any shade or very cloudy conditions. The etiolated appearance of trees growing under the shade of old trees is well known. In West Africa the yield of copra is related to the daily hours of sunshine during the final maturation period of the nut. With less than 2000 hr of sunlight per year or 120 per month, as measured by a Campbell-Stokes recorder, yield is reduced (Ziller, 1960).

Murray (1968) pointed out that the Campbell Stokes recorder is not a satisfactory method for determining incident light energy.

As may be seen in Table III, data obtained with the Campbell-Stokes recorder show that Ecuador receives only 35% of the hours of bright sunshine that Trinidad receives. However, in terms of Kipp Solarimeter radiation, the difference is far less pronounced, with Ecuador receiving 75% of the solar radiation received in Trinidad.

D. Wind

Coconuts grown in coastal areas are often exposed to strong sea winds. Provided adequate soil moisture is available the trees tolerate these winds well,

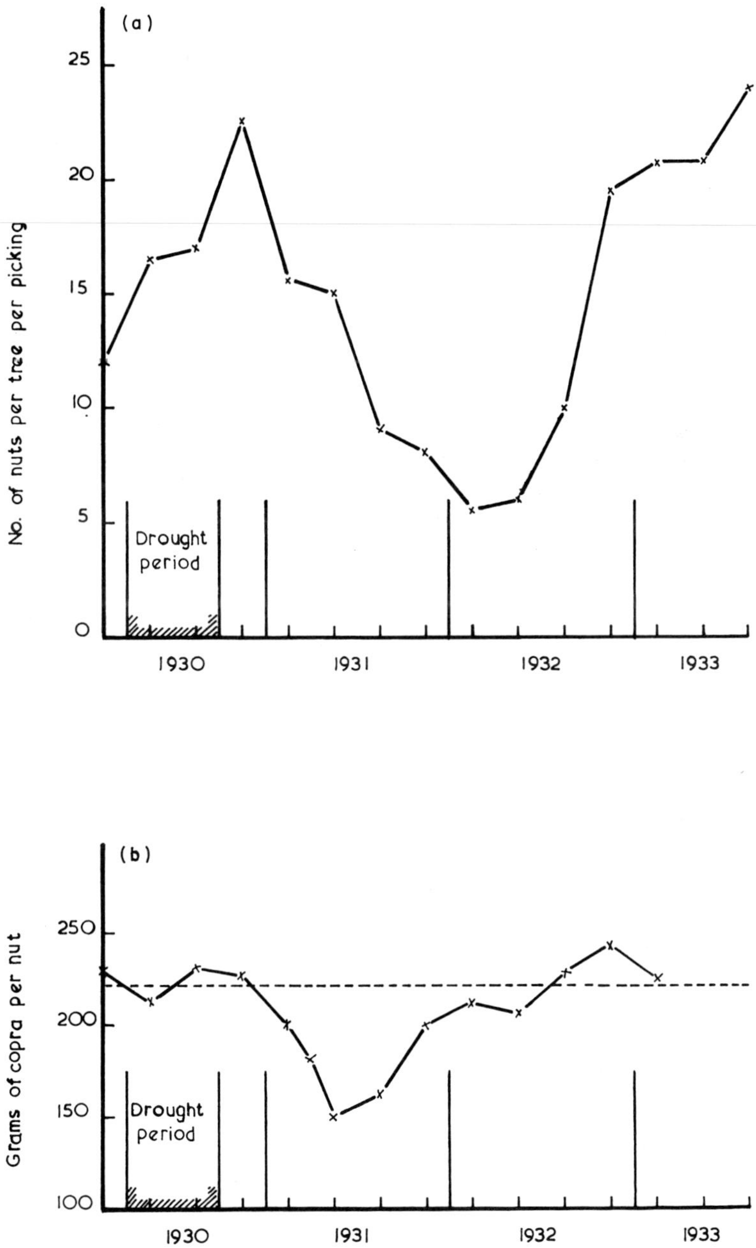

Fig. 2 The effect of drought on (a) number of nuts produced per picking, and (b) weight of copra per nut in grams.

TABLE II
Effect of Drought in 1947 on Decrease in Yield for Trees in Different Yield Classes[a]

Class (no. of nuts)	No. of trees	Average nuts/tree 1946–1947	Average nuts/tree 1947–1948	Decrease
40–60	11	53	43	10
61–80	14	73	45	28
81–100	13	90	64	26
101–120	10	108	66	42
121–140	7	129	65	64
Over 140	5	150	46	104

[a]From Murray and Lucie-Smith (1952).

although the most exposed outer row or rows may have a reduced crop. This may be associated to some extent with salt spray deposited on the leaves. Of far more importance, however, is the fact that coastally grown coconuts in some parts of the world are subjected to hurricanes or cyclones. These high velocity winds can damage the trees in two ways. Either the root system gives way and the tree falls flat or the crown may break off. In both instances the tree is destroyed; plantations in the West Indies, particularly Jamaica, have suffered severely from hurricanes in the past.

E. Nearness to the Sea

Because coconuts are so often associated with coastal regions of tropical islands, there is a common misconception that they will not grow inland.

TABLE III
Light Measurements in Trinidad and Ecuador per Month[a]

	J	F	M	A	M	J	J	A	S	O	N	D	Mean
Mean daily solar radiation (langleys) (Kipp Solarimeter)													
Trinidad	385	418	478	446	470	275	388	396	394	374	308	346	390
Ecuador	325	356	343	364	329	305	280	258	253	238	219	261	294
Hours of sunshine (Campbell-Stokes)													
Trinidad	7.6	8.2	8.0	8.1	7.8	6.7	7.2	6.9	6.5	7.0	7.0	7.1	7.3
Ecuador	2.3	4.0	3.9	4.1	4.0	3.8	2.6	2.1	1.5	0.9	0.8	1.7	2.6

[a]From Murray (1968).

Conditions near the sea may often be ideal for growing the crop, but the coconut palm will flourish hundreds of kilometers inland, provided that other conditions necessary for growth are adequate. In India good coconut plantations are found on the Mysore plateau at 240–320 km from the sea.

F. Evapotranspiration and Physiological Water Balance

A synthesis of the water balance in the soil is provided by the calculation of the evapotranspiration for a given area. Using the formula developed by Thornthwaite, Frémond *et al.* (1966) made the appropriate calculations for Port Bouet in the Ivory Coast. Table IV shows the months of the year during which evapotranspiration exceeds rainfall so that the trees undergo water stress. In other months when rainfall exceeds evapotranspiration, adequate drainage is necessary to avoid water logging of the soil.

In connection with the internal water balance of coconut trees, F. Hardy (personal communication) made the following comments:

The coconut palm possesses certain unique physiological characteristics which render it particularly susceptible to wilt in a dry environment.

(a) The size of the leaf crown is almost the same at all ages so that the amount of transpiration in an atmosphere of constant temperature, humidity, and air movement varies little at all stages of growth.

(b) The downward pull (weight) of water in the conducting elements increases with age of tree, i.e., with length of trunk or height of the tree.

(c) The frictional resistance to upward water movement in the conducting elements also increases with height (the tracheids have small diameters).

(d) The total suction force tending to abstract water from the leaf cells because of factors (a) and (b), coupled with that developed by soil colloids as the soil dries out, despite the xeromorphic features of the leaflets, causes wilting to occur at a much higher soil moisture content than that at which wilting normally occurs in other tree crops.

(e) Fluctuating water tables restrict root development in poorly drained phases of certain soil types causing coconut roots to die of asphyxiation.

(f) Wilting consequently occurs much earlier (with trees less than 20 years old) in coconut palms growing in soils having restricted root room supporting trees having small root systems than in older trees growing in soil having large root room supporting trees having large root systems.

(g) Consequently, the age and height at which coconut palms die of wilt should provide a measure of the general aridity of the environment, particularly that imposed by adverse soil conditions (unsatisfactory water relations) such as poor drainage and high water tables responsible for "physiological" drought.

TABLE IV
Evapotranspiration and Water Balance per Month at Port Bouet, Ivory Coast in mm[a]

	J	F	M	A	M	J	J	A	S	O	N	D	Total
Rainfall	30	50	108	160	382	570	177	35	61	207	188	113	2081
Potential evapotranspiration	147	153	182	178	164	132	124	107	114	138	152	155	1746
Change in soil moisture	–58				+100			–72	–28	+69	+31	–42	
Soil reserve	–	–	–	–	100	100	100	28	–	69	100	58	
Effective evapotranspiration	88	50	108	160	164	132	124	107	88	138	152	155	1466
Deficit	59	103	74	18					53				307
Excess					118	438	53				5		614

[a]Frémond *et al.* (1966).

III. SOILS

The coconut palm tolerates a very wide range of soils although the yield of nuts varies considerably on different soils. In most tropical countries the coconut palm grows on beach sands. Chemical analysis shows these sands to be almost sterile with respect to plant nutrient content; growth and cropping depend very much on topography. Provided there is higher land or freshwater swamps behind them, the slow percolation of water toward the sea will carry nutrients which will allow the trees to flourish.

On sands derived from coral, the depth of the overlying soil is of major importance to growth particularly during periods of low rainfall. The root room, that is the volume of soil capable of being physiologically exploited by the roots, is of prime importance for high yield.

A. Alluvial Soils

Some of the best coconut soils are rich river alluvial deposits, so long as internal drainage is adequate. In India and Sri Lanka, some of the highest yielding plantations are found on the loams and fine sandy transported soils of the river deltas. In the Philippines soils derived from alluvial limestone are particularly fertile.

B. Lateritic Soils

In Sri Lanka and India large areas of coconuts are found on soils derived from highly ferruginous parent material ranging from hard brick-like laterite gravels to red loam soils. The hard laterites of Sri Lanka occur in areas of high rainfall; they are highly leached. They are deficient in potash and phosphate and when shallow do not support good growth. It has been the practice in the past to use dynamite to break up the hardpan in the planting hole but this is not considered economical.

Better results are obtained on the lateritic gravels where responses to potash manuring are good. The largest area of coconuts on the west coast of India occurs on well-drained lateritic loams.

C. Volcanic Soils

Soils derived from volcanic ash and tuff are particularly fertile. The parent material weathers rapidly to release adequate supplies of minerals for the plant and internal drainage is good. Such soils are a common feature of coconut plantations in Indonesia, the Philippines, and New Guinea.

D. Clay Soils

These are the most difficult soils to manage efficiently for coconuts. Good drainage is essential and even with this they tend to be waterlogged in the wet season and to bake dry and crack in the dry season. With good management, satisfactory yields can be obtained as shown in Malaya where most coconuts are grown on clays, often with a high peat content.

E. Physical Properties

By mechanical analysis it is found that coconuts will grow on soils which are nearly 100% pure sand, through soils with an increasingly large clay fraction which may reach 70%, to peat soils in Malaya with over 80% organic matter. The most important feature already mentioned is good drainage as the coconut roots do not tolerate waterlogging for any length of time. They are also very thick and may have difficulty in penetrating compacted soils.

Together with good drainage relations, it is important to have a reasonably deep soil for the roots to exploit. Soil depth, and consequently root room, may be determined by a hardpan or a high water table and in either case growth and cropping will be poor.

F. Chemical Properties

As mentioned earlier, coconuts will grow on sands whose chemical analysis shows little or no nutrient content. The analysis is therefore of little value. It may be of use in alluvial or clay soils to demonstrate a deficiency of perhaps potash or phosphate. In a later section methods of determining the nutrient requirements of the tree are considered.

Coconuts tolerate a wide pH range, from coral-derived sands with a pH of 8 down to acid clays at pH 5, or even less. At pH 8 there is some evidence that iron is unavailable to the plant. At low pH, incompletely understood growth abnormalities occur which may be associated with toxic levels of aluminum and manganese.

IV. GROWTH PHYSIOLOGY

The general morphology of the parts of the tree has been intensively studied and described by various workers including Sampson (1923) and Menon and Pandalai (1958) and is not considered further in this chapter.

The coconut palm is a monocotyledonous tree with a single terminal bud. Hence it has a single growth pathway which makes a study of the physiological sequence leading to the production of the crop easier than in dicotyledonous trees.

A. Seed Germination

Of the ecophysiological factors influencing nut germination, the most important is the maturity of the seed nut. For planting, nuts should be fully mature, i.e., 12 months old. However, work in India and Sri Lanka has shown that there is no difference in growth or bearing if slightly immature 11-month-old nuts are used.

The time of year for harvesting is unimportant and is usually determined by the need to have the young seedlings at a proper stage of development for planting in the rainy season. In some countries the nuts are stored for a time before they are planted in the nursery, but this is unnecessary. Mature nuts can be planted as soon as they are taken from the tree.

Various methods of planting are used in different countries and nuts may be planted either horizontally or vertically with the stalk end up. Horizontal planting is recommended with the widest of the three segments uppermost. This lies over the germinating embryo and facilitates emergence of the plumule.

Germination of the normal tall variety nuts usually takes place within 11 to 12 weeks of planting. A special organ, the haustorium or apple, develops from the embryo and grows into the internal cavity of the nut. Enzymes, particularly cellulases and lipases, developed in the haustorium are responsible for breaking down the oil containing cells of the meat and converting the oil into soluble carbohydrates. These are translocated to the developing seedling and used in growth. Within 5 to 6 months after germination, the haustorium almost completely fills the cavity and persists for over a year, during which time all the reserve food in the meat is consumed. By that time the shoot has developed and photosynthesis is adequate to make the young plant independent of reserve foods in the nut.

The rate of seed germination in the nursery varies considerably. A negative correlation of 0.61 between the number of leaves produced and the time required for seed germination was obtained by Patel (1938). Correlation studies also show that palms with a number of leaves flower earlier than those with few leaves. It follows that seed nuts which sprout early give rise to palms that flower sooner and are therefore more productive than those that germinate later. This important physiological characteristic is used in selection of plants in the nursery for field planting. It is advisable to select only 50–60% of the most vigorous and advanced seedlings in the nursery for planting in the field.

B. Root Development

Since the coconut is a monocotyledonous plant it has an adventitious root system and, throughout its life, produces numerous uniformly thick roots from the base of the stem. The major roots which produce numerous secondary and tertiary roots generally are long lived; they carry physiologically adapted pneumatophores or "breathing roots" which assist in gas exchange. The major roots are thick, up to 4 mm in diameter, and this can affect their ability to penetrate compacted soils or those with a high clay content. They are well supplied with aerenchyma. Absorption of mineral nutrients and water takes place in the region just beyind the pronounced root cap since coconut roots do not have root hairs. Sampson (1923) stated that the roots carry a symbiotic fungus but little research has been done on its relationship to the tree. Old trees or those growing in waterlogged areas differentiate aerial roots from above the normal root-producing zone at the base of the stem. The capacity to produce such roots has led to the suggestion that outstanding parent trees which have become too tall may be induced to initiate roots just below the crown. The stem can then be cut below this point and the crown and new root system lowered to the ground and replanted.

C. The Stem

The true stem in tall coconuts begins to form about 5 years after planting out, and since there is only a single growing point, only one trunk can develop. Moreover since there is no cambium no secondary thickening occurs and the diameter of the tree, once laid down, does not change. Small variations in stem girth may be found as a result of unfavorable environmental factors, especially drought, when the stem is being laid down.

D. The Leaf

The terminal bud differentiates a single leaf at a time and some thirty months elapse from differentiation of a leaf to its emergence from the leaf sheath. During this long period the developing leaves are therefore subject to a range of environmental factors.

On the average a new leaf opens once a month and a mature tree has a crown with 30–40 leaves. Under favorable conditions a leaf persists on a tree for 3 to $3\frac{1}{2}$ years. On poor-bearing or ill-nourished trees, the leaf life is shorter and the number of leaves in the crown reduced.

E. The Inflorescence

The coconut palm is monoecious so that each inflorescence carries both male and female flowers. In normal tall trees flowering begins at 5–7 years; in dwarf trees it may begin at 3 years. Once flowering starts an inflorescence develops in the axil of each leaf. The first inflorescence may consist of only male flowers; female flowers also are produced in later inflorescences.

Since in tall trees the average time between production of successive leaves is about a month, the same time elapses between the opening of inflorescences. In dwarf trees, which may produce up to 18 leaves per year, inflorescences may open every 3 weeks.

The male flowers in an inflorescence are numerous, ranging from a few hundred to thousands. They open first and the pollen is disseminated by wind. The number of female flowers usually does not exceed 30–40 and seasonal differences are largely caused by variations in nutrition and water availability.

In a single inflorescence the male flowers produce pollen for about two weeks. Female flowers are not receptive until later so that cross-pollination is normal. However, pollination from the succeeding inflorescence may occur. In dwarfs, overlap of the fertile phases of the male and female flowers does occur so that self-pollination is more common.

Seasonal variation in (a) the number of female flowers per inflorescence; (b) number of mature nuts developed; and (c) the number of nuts as a percentage of the female flowers is shown in Fig. 3 (Sampson, 1923).

F. The Fruit

Up to 70% of the immature fruits are shed within a few weeks after emergence of inflorescences. Some fruit is lost because of failure of pollination but the greater loss is due to physiological reasons and may be compared with those caused by cherelle wilt in cocoa.

The fruit normally requires 12 months to reach full maturity at which time its copra content has reached a maximum. The rate of development of copra as given by Patel (1938) is shown in the following tabulation:

Maturity of nut (months)	% of copra developed
8	32
9	56
10	78
11	94
12	100

It is therefore wasteful and uneconomical to pick nuts that are not fully mature and the best practice is to allow the nuts to fall naturally and collect them from the ground.

G. Yield

The yield of the coconut palm may be expressed in terms of number of nuts, weight of copra, or weight of oil per tree or per hectare. Unless coir is manufactured the economic yield is best expressed as weight of copra per hectare.

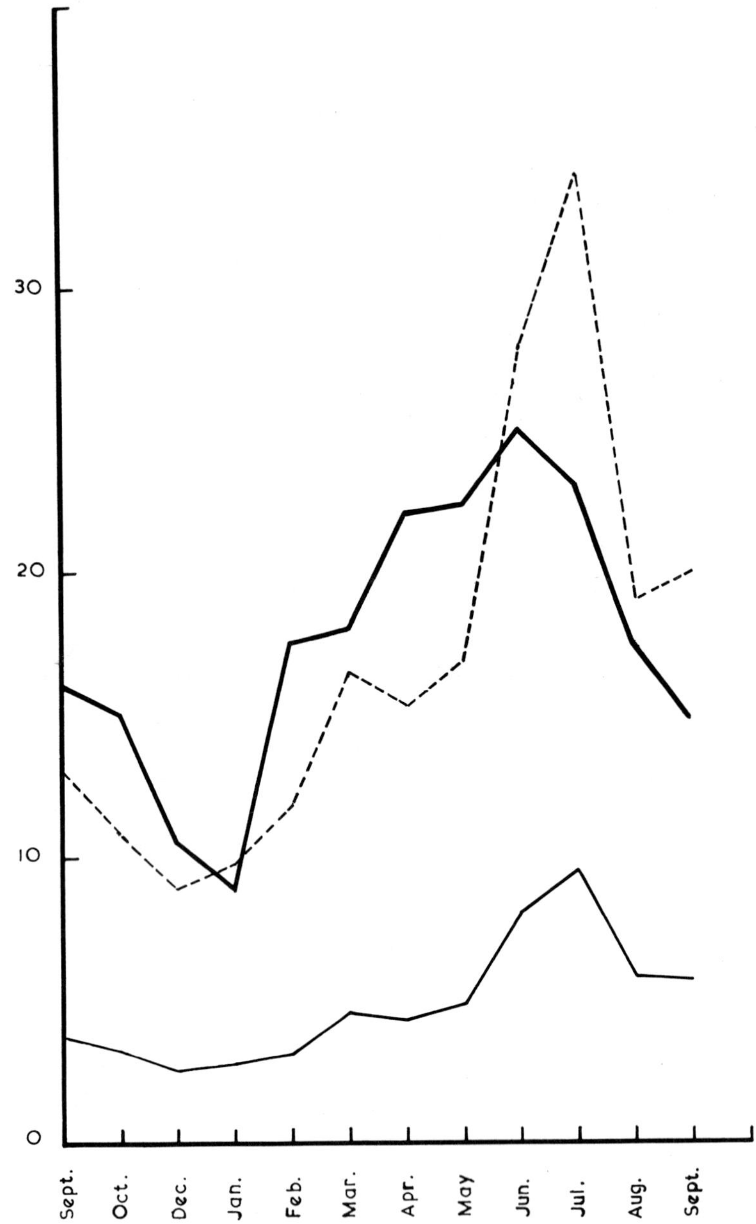

Fig. 3 Seasonal effects on (a) number of female flowers in each consecutive bunch, (b) number of ripe nuts produced, and (c) percentage of nuts to female flowers. Top curve (heavy black line): number of female flowers in each consecutive bunch; bottom curve (lighter black line): number of ripe nuts from the above; middle curve (dashed line): percentage of nuts to female flowers.

TABLE V
Correlation of Palm Characters with Yield[a]

Character	Coeff. of correlation
Total number (leaves)	0.41
Total number (opened leaves)	0.45
Number (old leaves)	0.45
Rate of production (leaves)	−0.19
Surface area of attachment (leaf to trunk)	0.27
Length (trunk)	0.42
Girth (trunk)	0.29
Number (female flowers)	0.64
Catalase activity of the tree (O_2 evolved in 3 min)	0.56

[a]From Menon and Pandalai (1958).

The factors determining yield may be classified as (1) genetic; (2) climatic; (3) edaphic (soil); (4) cultural (e.g., manuring, weed control); (5) diseases and pests.

There are obviously interactions among the above, some of which have already been discussed, i.e., the effect of water deficits in reducing crop depends on the level of bearing, a genetic factor.

Confining attention only to ecophysiological factors, Patel (1938) described the physiological stages in the early build-up of yield. Little is known, however, of factors responsible for decline of yield with age.

Monthly and annual variation in yield is determined largely by rainfall distribution. Theoretically, in a constant environment the coconut should produce exactly the same number of nuts per bunch and approximately once a month.

It is usually assumed that alternate and irregular bearing are genetically determined, but Patel (1938) suggested that irregular bearing may not be genetically determined.

Because of its pattern of growth with a single growing point it is possible to work out correlations between various vegetative characters and yield (Table V). Of particular interest is the correlation of yield with catalase activity which might be used as a method of screening young plants.

V. MINERAL NUTRITION

A considerable proportion of the mineral nutrients absorbed by the roots finds its way to the nuts. There is considerable variation in the mineral analysis of coconuts from different areas (Table VI).

TABLE VI
Ash Analysis of Ripe Nuts (Average per Nut in Grams)[a]

	Kasaragod, India (Sampson)			San Ramon, Philippines (Copeland)		
	N	P_2O_5	K_2O	N	P_2O_5	K_2O
Husk	1.25	0.18	3.05	1.609	0.017	3.915
Shell	0.45	0.015	0.17	0.660	0.459	0.947
Meat	5.13	0.77	1.35	4.683	1.740	2.475
Water	0.025	0.028	0.33	1.542	0.171	1.313
Total	6.855	0.993	4.90	8.494	2.387	8.650

[a]From Patel (1938).

Nearly half the potash found in the nut accumulates in the husk which, if not used as a source of coir, should be left to rot in the field and not burned or removed.

The quantities of nutrients removed annually from the soil by the crop also shows considerable variation (Table VII). Obviously the data of Table VII reflect the nutritional status of the soil and the size of the crop being harvested. They all demonstrate, however, the high demand for potash for the crop. Although the coconut is not usually considered to have a high phosphate demand a substantial amount is found in the crop.

A. Fertilizer Requirements

Fertilizer experiments with coconuts in the field present numerous difficulties including (1) lack of uniformity in the experimental material; (2) large area required for a statistical experiment, which increases the soil variability factor; (3) difficulties in supervising and recording field data: ideally each tree should be

TABLE VII
Annual Export of Nutrients (kg/ha)[a]

	Jacob and Coyle (1927)	Eckstein (1937)	Pillai and Davis (1963)
N	64	91	56
P_2O_5	29	40	27
K_2O	95	131	85

[a]From Frémond *et al.* (1966).

climbed and the ripe nuts harvested; (4) differential response of individual trees to manuring; (5) the long period before the effect of the fertilizers is seen; (6) seasonal effects and their interaction necessitating continuing experiments for a long time.

No attempt will be made to survey the wide field of fertilizer experiments with coconuts. A good summary has been given by Teiwes (1962). A few examples will be given to emphasize specific points.

As an illustration of differential responses to fertilizing, Murray and Lucie-Smith (1962) carried out nonstatistical trials on trees giving yields of about 50–60 nuts per tree each year, and applied nitrogen, potash, and phosphate, singly and in combination. They found a response only to nitrogen. They examined the response to nitrogen of trees in yield classes: 1–40 nuts, low yielders; 41–80 nuts, medium yielders; and over 80 nuts, the high yielders (Table VIII).

Application of nitrogen fertilizer led to a far greater increase in numbers of nuts in initially low yielding trees than in initially high yielding ones. The increase in yield in unfertilized controls was associated with general improvement in cultural practices, especially better weed control.

In contrast to the results in Trinidad, where the major response on beach sands was to nitrogen, in other countries, notably Sri Lanka and the Ivory Coast, the main mineral deficiency has been found to be potash.

Salgado (1951) carried out factorial NPK trials on two sites in Sri Lanka, one on a potash-deficient soil at Bandirippuwa and the other at Ratmalagara where the potash content of the soil was ten times as high. K_1 plots received 0.75 lb K_2O per tree and K_2 plots 1.5 lb K_2O per tree biennially. Application of potash had a much greater effect on the potash-deficient soil (Table IX).

For the Ivory Coast and Dahomey, Frémond *et al.* (1966) give the yields shown (Table X) in response to nitrogen and potash application. On potash-deficient soils, nitrogen alone will depress yield. When potash deficiency was corrected, a positive response to nitrogen was found.

TABLE VIII
Increase in Numbers of Nuts in 1947 and 1948 over Number Borne in 1946 for Different Yield Classes[a]

	1947			1948		
	Controls	Without N	With N	Controls	Without N	With N
Low yielders	7	17	30	24	20	48
Medium yielders	1	4	11	13	13	29
High yielders	–18	–10	–2	–3	12	16

[a]From Murray and Lucie-Smith (1952).

TABLE IX
Yield of Copra (lb per Acre)[a]

	Potash application		
	K_0	K_1	K_2
Bandirippuwa	891	1292	1437
Ratmalagara	1438	1466	1589

[a]From Salgado (1951).

B. Minor Elements

Not much research has been conducted on the role of minor elements in coconut nutrition but Frémond *et al.* (1966) reported a positive response to injected iron on high pH coralline soils in the Pacific, and also to manganese.

C. Leaf Analysis

Monocotyledonous trees have a great advantage over dicotyledonous trees with respect to sampling for foliar analysis. Because of the continuous process of leaf production in monocots it is possible to select the same leaf physiologically each time, and Prévot and Bachy (1962) chose the fourteenth from the youngest leaf as the standard reference leaf. Good correlation has been shown in the Ivory Coast between leaf potash and both the number of nuts per tree per year and weight of copra per nut (Table XI).

It is clear that potash increases yield of copra per acre both by increasing the number of nuts and the weight of copra in each nut.

An earlier summary of work on leaf analysis in Sri Lanka has been given by Nathanael (1958). The critical levels suggested by I.R.H.O. for the major

TABLE X
Yield in Number of Nuts per Tree with Nitrogen and Potash Manuring[a]

	Ivory Coast		Dahomey	
	N_0	N_1	N_0	N_1
K_0	44	30	36	23
K_1	72	71	60	77

[a]From Frémond *et al.* (1966).

elements in the fourteenth leaf are shown in the following tabulation:

Element	% Dry weight
N	1.8–2.0
P	0.12
K	0.8–1.0
Ca	0.5
Mg	0.3

D. Nut Water Analysis

A unique method of diagnosing potash levels in the tree was developed by Salgado (1955). The water contained in the nut carries dissolved nutrients and is a ready-made solution for analysis compared with the tedious digestion of dried leaf samples. Salgado showed in the aforementioned Bandirippuwa NPK experiment that a close correlation existed between the yield of copra per acre and the average content of potash in nut water samples (Table XII).

In Trinidad Quansah (1954) attempted to relate the other major nutrients in nut water to yield, soil, and season. No correlations with nitrate nitrogen content were found but the composition of nut water reflected not only the potash status of the soil, but also the calcium and magnesium levels.

VI. ECOPHYSIOLOGICAL DISEASES

It is not within the scope of this chapter to discuss coconut diseases that have a pathological origin. There are, however, a number of physiological diseases that

TABLE XI
Effect of Potash Manuring on Leaf K, Number of Nuts per Tree, and Copra per Nut[a]

	kg KCl/tree/year			
	0	0.5	1.0	1.5
Leaf K, % dry weight	0.16	0.38	0.53	0.77
No. of nuts per tree	62	69	74	79
Copra (gm per nut)	161	180	186	195

[a]From Ziller and Prévot (1963).

TABLE XII
Yield per Acre and Potash in Nut Water at Different Levels of Potash Fertilizing[a]

	K_0	K_1	K_2
Yield (lb/acre)	891	1292	1437
K_2O (gm/liter)	1.00	1.49	2.09

[a]From Salgado (1955).

may or may not lead to death of trees. Very often these are associated with particular soil and climatic conditions so they should more properly be called ecophysiological diseases. Among these are the following.

A. "Tapering Stem" and "Pencil Point"

These names are descriptive of the symptoms of this condition. They are manifest in a yellowing of the leaves and gradual reduction in size accompanied by a slow reduction in the diameter of the new growth at the top of the tree. Yield is greatly diminished and the tree may eventually die.

These symptoms are usually regarded to indicate slow development of deficiencies of mineral nutrients and water. They may arise from the following conditions:

(a) *Waterlogging:* Tapering stem is very common among palms planted in swampy or waterlogged areas. The development of the root system is greatly inhibited so that the entire tree undergoes physiological drought.

(b) *Hardpan:* On shallow soils over coral or ironstone pans root development is again restricted and tapering stem develops.

(c) *Infertile soils:* On infertile sandy soils receiving no fertilizer gradual mineral starvation results in tapering stem.

(d) *Senility:* This condition is often shown by very old trees.

From these descriptions it is obvious that choice of planting site is most important and is the obvious way of avoiding tapering stem.

B. Shedding of Young Nuts

During the first few months after the inflorescence appears, there is a considerable loss of the so-called "button nuts." This loss is due to physiological competition since far more female flowers are produced than can mature on the tree. The condition is analagous in many ways to cherelle wilt in cacao.

Normally only some 30% of the flowers produced reach maturity and the effect of fertilizers on increasing yield is not through decreasing fall of nuts, but rather by increasing the number of flowers produced.

C. Shedding of Immature Nuts

A more serious problem is loss by shedding of maturing nuts which have passed the button nut stage. Loss of nuts 3–5 months old is often greater at the end of a long dry spell and is sometimes accentuated when heavy rain follows a drought.

A nonphysiological shedding of nuts may occur in trees with a long, weak bunch stalk and thin leaf frond bases. The leaf may break allowing the bunch to slip downward with a jerk, thereby causing shedding of immature nuts. This emphasizes the need for selecting planting material which does not carry this undesirable character.

D. Bronze Leaf Wilt, Frond Drop, Cape St. Paul Wilt, and Kaincope Disease

These four conditions, which occur in Trinidad, Jamaica, Ghana, and Togo, respectively, are grouped because of the underlying similarity in symptoms.

According to Britton-Jones (1940), the major cause of wilt diseases is a soil factor, whether or not an organism is associated with the wilt. Although much research has been conducted on these wilts in the countries where they are found there is little evidence to contradict Britton-Jones.

In general the symptoms of bronze leaf wilt consist of yellowing of the tips of the fronds which spreads downward and becomes bronze. The fronds later wither and hang down without becoming detached from the tree. Browning of the unopened spadix occurs, followed by necrosis and desiccation. The green nuts are shed and show a blackening around the calyx end and the copra is discolored. The tree usually dies within 3–6 months after symptoms appear.

In Trinidad the disease usually occurs in mature trees (15–20 years old) and is most prevalent on heavy clay soils with impeded drainage (Bain, 1937). The theory is that on such soils the development of the root system is restricted to a shallow surface layer. In the dry season this layer dries out and soil cracks may develop so that the surface roots are damaged and killed. Absorption of water then cannot meet the transpiration requirements of the tree and wilt symptoms develop. In the land chosen for planting coconuts, great emphasis must be directed to areas where soil conditions are not such as to predispose the tree to this condition.

These wilts should not be confused with other diseases which may produce rather similar symptoms. These include red ring disease in Trinidad caused by the nematode, *Rhadinaphelencus cocophilus,* and lethal yellowing in Jamaica which is now thought to be due to a mycoplasm.

E. Lightning

Many coconut palms are lost through lightning strike. Both in Malaya and Sri Lanka lightning is a factor of the first importance in causing diseases on coconut plantations.

The sequence of events in a lightning strike is that a leader stroke develops from the lower, usually negatively charged side of a cloud. As this nears the earth it causes a local increase in the electrostatic field. When this reaches a value that causes an electric discharge from an earthed object an upward streamer develops to meet the tip of the downcoming leader. This usually occurs at an average height of 20 m or approximately the height of a mature tall coconut. The current in the upward stroke may reach 100,000 A, this resulting in intense internal heating which bursts the vascular system of a tree. When this occurs, liquids exude from lesions in the trunk. Leaf break and nut fall also occur and the tree dies. After a severe lightning strike near Singapore, 104 trees died at once or over a period of months. The condition should not be confused with bud rot due to *Phytophthora palmivora.*

REFERENCES

Bain, F. M. (1937). "Bronze Leaf Wilt of the Coconut Palm." Govt. Printing Office, Trinidad.

Beccari, O. (1917). The origin and dispersal of *Cocos nucifera. Philip. J. Sci., Sect. C* **12,** 27–34.

Britton-Jones, H. R. (1940). "Diseases of the coconut palm." Baillière, London.

Child, R. (1964). "Coconuts." Longmans, Green, New York.

Cook, O. F. (1910). History of the coconut palm in America. *Contrib. U.S. Natl. Herb.* **14,** Part 2, 271–342.

Cooke, F. C. (1953). Rainfall and Crop. *Ceylon Coconut Q.* **5,** 159.

Copeland, E. B. (1931). "The Coconut." Macmillan, New York.

Edmonson, C. H. (1941). Viability of coconut seed after floating at sea. *Occas. Pap., Bishop Mus.* **16,** No. 12, 293–304.

Frémond, Y., Ziller, R., and de la Motte, M. (1966). "Le cocotier." Maisonneuve & Larose, Paris.

Heyerdahl, T. (1952). "American Indians in the Pacific." Allen & Unwin, London.

Menon, K. P. V., and Pandalai, K. M. (1958). "The Coconut Palm." Indian Central Coconut Committee, Ernakulam, S. India.

Murray, D. B. (1968). Shade and light intensity. *Annu. Rep. Cacao Res., Imp. Coll. Trop. Agric., Trinidad, St. Augustine,* p. 47.

Murray, D. B., and Lucie-Smith, M. N. (1952). Estate fertilizer trials on coconuts in Trinidad. *Trop. Agric. (Trinidad)* **29**, 180–195.

Nathanael, W. R. N. (1958). Diagnostic approaches to problems in crop nutrition. *Ceylon Coconut Q.* **9**, 11–29.

Patel, J. S. (1938). "The Coconut–a Monograph." Govt. Press, Madras, India.

Patel, J. S., and Anandan, A. P. (1936). Rainfall and yield in the coconut. *Madras Agric. J.* **245**, 5–15.

Piggot, C. J. (1964). "Coconut Growing." Oxford Univ. Press, London and New York.

Prévot, P., and Bachy, A. (1962). Diagnostic foliare du cocotier: Influence du rang de la feuille et du development végétatif sur les teneurs en éléments. *Oléagineux* **17**, 451–458.

Quansah, S. T. (1954). A chemical investigation of mineral levels in coconut water in relation to soil, season and yield. D. T. A. Thesis, Imperial College of Tropical Agriculture, Trinidad.

Salgado, M. L. M. (1951). Soil potash and its availability in relation to potash manuring of coconuts. *Trop. Agric. (Ceylon)* **107**, 92–99.

Salgado, M. L. M. (1955). The nutrient content of nut water as a guide to the manuring of the coconut palm, a new diagnostic method. *Congr. Int. Bot., Rapp. Commun., 8th, 1954* pp. 217–238.

Sampson, H. C. (1923). "The Coconut Palm." John Bale Sons and Danielson, London.

Shepherd, C. Y. (1926). The coconut industry of Trinidad. *Trop. Agric. (Trinidad)* **3**, 186.

Teiwes, G. (1962). "Nutrition and Manuring of Coconut Palms." Verlagsges. Ackerbau, M. B. H. Hanover, Germany.

Ziller, R. (1960). Etude de quelques facteurs agissant sur la teneur en coprah des noix de coco. *Oléagineux* **15**, 73–81.

Ziller, R., and Prévot, P. (1963). Le diagnostic foliare, méthode d'étude de la nutrition minéral du cocotier. "Coconut Production, Protection and Processing." FAO, Bangkok, Thailand.

CHAPTER
15

Citrus

WALTER REUTHER

I. INTRODUCTION

During the past century, world citrus production has risen from less than 1 million to 44 million metric tons in 1973 (Food and Agriculture Organization, 1974). Of this total, about 6 million tons of fresh fruit found its way into international trade, and about 14 million tons were used for industrial purposes. Large-scale exploitation for export or industrial uses has developed almost exclusively in subtropical regions that mostly lie between 20° and 40° of latitude in both the northern and southern hemispheres (Fig. 1 and Table I).

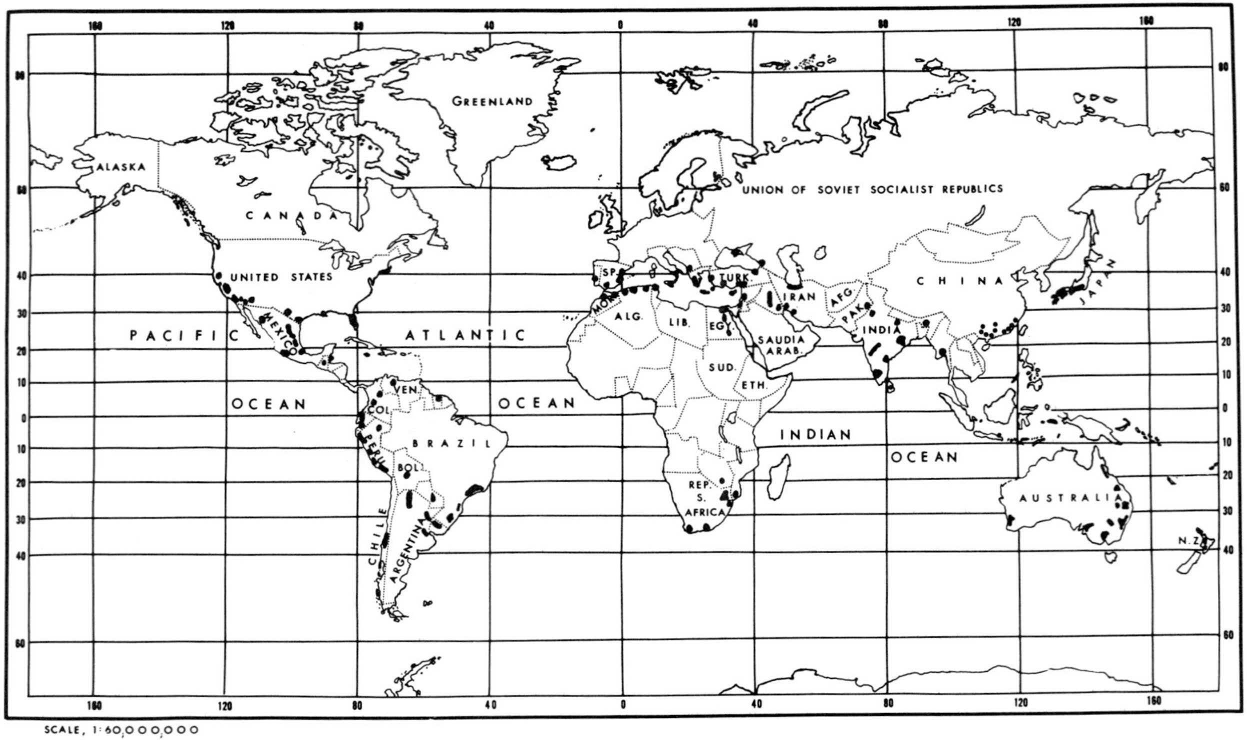

Fig. 1 Distribution of large-scale commercial citrus producing areas of the world.

TABLE I
Estimated Tropical Citrus Production, 1974[a]

Variety	Thousand of metric tons: World	Tropical	Percent tropical[b]
Oranges and mandarins	35,500	3,237	9.2
Lemons and limes	4,001	678	16.9
Grapefruit	3,709	201	5.4
Unclassified	736	470	63.8
All citrus	43,746	4,586	10.5

[a]Source: FAO Production Yearbook, 1974.
[b]Between 20°N and 20°S latitude.

Brazil, now pressing Japan for second place in world citrus production, has only about 9% of its plantings above 20°S latitude (Table II), and is probably the leading producer of tropical citrus (Table II). The amount of citrus exported as fresh fruit from tropical areas of the world is miniscule compared to subtropical areas. However, citrus is widely grown as a dooryard and orchard fruit for domestic consumption throughout most of the tropical regions of the world. Unfortunately, statistical data on citrus production in many such areas are not available, or at best are fragmentary and subject to error.

This pattern of distribution of world citrus culture has its base not only in the ecological requirements for intensive commercial exploitation of citrus, but also in such factors as communication and transportation, human health, government policy and stability, and sophistication of agriculture, which will not be considered here. If climatic and edaphic factors alone controlled world distribution

TABLE II
Distribution of Orange Planting and Production in Brazil[a]

Region	Area, 1973: Hectares (thousands)	Area, 1973: %	Production, 1973: Metric tons (millions)[b]	Production, 1973: %
Subtropical (south of 20° lat.)	416	92	3.8	91
Tropical (north of 20° lat.)	34	8	0.4	9
All	449	100	4.1	100

[a]Source: Fundação IBGE, 1975.
[b]Calculated from factor: 10^6 fruits = 170 metric tons.

of intensive citrus culture, some tropical areas would be more extensively planted.

This chapter outlines some of the basic ecological and other requirements for the culture of citrus and assess the potential of various species and varieties for intensive commercial exploitation for export as fresh fruit or as industrial products in lowland humid tropical areas.

II. ORIGIN AND NATURAL HABITAT

Most cultivated species in the genus *Citrus* appear to be indigenous to the more humid tropical and near-tropical regions of China, Southeast Asia, including the eastern areas of India and Bangladesh, and the islands of the Philippines and Indonesia. In this general area, primitive man undoubtedly played an important role in selectively disseminating and ultimately purposefully planting seed of the most attractive edible species and varieties before the dawn of history (Stebbins, 1969). The humid, tropical origin of citrus is supported by its general mesophytic nature. It has a moderately shallow and spreading rooting habit, not well adapted to withstanding long droughts. Its flat, broad evergreen leaves and thin, succulent bark of its twigs do not have specialized xerophytic features for greatly limiting transpiration or withstanding temperature extremes. The absence of a cool temperature-induced dormancy requirement for flowering and its susceptibility to damage by freezing also indicate its tropical origin.

No *Citrus* species are indigenous to the Western Hemisphere, and hence it is of interest to note the ecological conditions which favor natural spread of introduced species without cultural attention by man. After the introduction of citrus to Florida in the sixteenth century, feral groves of sour oranges, and to a lesser extent sweet oranges, Rough lemons, and West Indian limes, gradually developed in certain specialized habitats through seed spread by man or animals from cultivated sources. These "wild" groves were invariably located on comparatively rich, low, but well-drained moist lands in the vicinity of rivers or lakes known locally as "hammocks." Such habitats supported good stands of tall hardwood trees (oak, magnolia, hickory, etc.) and in many situations an admixture of sabal palms and pines. The topsoil was porous, acid, high in organic matter and fertility, and often overlaid a calcareous subsoil. In such natural groves, varying from 1 to 100 or more hectares, citrus trees tended to form a dense undergrowth under the shading canopy provided by the foliage of the taller native hardwoods and palms (Webber, 1943). The main attributes of this hammock environment favoring feral citrus growth appear to be a reasonably uniform year-round supply of moisture in the first meter of topsoil, but without prolonged periods of waterlogging, and a protective canopy of forest trees. Such a protective leaf canopy not only reduces the transpirational demand for soil moisture, but also affords significant protection against moderate frosts. Observations in Florida, Central and South America, the Caribbean islands, and Southeast Asia indicate that *Citrus* species must have substantially the environment of a tropical rain forest to successfully invade native plant communities. The key factors for survival of seedlings appear to be a fairly well-drained and fertile topsoil with a fairly uniform year-round moisture supply, and freedom from extremes of temperature, high wind velocity, and fluctuations in water table. Of course, neither the native habitats nor adopted ones are infallible indications of optimal conditions for commercial exploitation of crop plants but merely indicate the basic requirements for survival.

Before World War I there were many commercial citrus orchards, especially in the Indian River section of the Atlantic coast of Florida, which were planted under the shade of native stands of sabal palms. Under these conditions fruit quality was excellent, but yields were low and hence these shaded groves were rapidly transformed as soon as it became clear that citrus, like coffee, can attain its maximum yield potential only with full sunlight.

III. CLIMATE

Tables III and IV present some climatic parameters in selected citrus regions in the American lowland humid tropics as compared with some potential citrus regions in the Amazon basin. Unfortunately, from such data it is possible to predict only in broad general terms the probable climatic adaptation of citrus variaties and species to a prospective new region. Much remains to be learned about complex interactions of a crop such as citrus with diverse seasonal climatic regimes in order to predict yield potential, season or seasons of maturity, and market quality of fruit, particularly in the tropics.

Perhaps one of the most important elements of climate from the standpoint of growth, production, and ultimate market quality of the fruit is the seasonal energy or heat flux regime, which controls the temperature regimes of plant tissues during various stages of development. Tissue temperature, among other factors, influences plant moisture stress (water deficit). Thus it is often difficult to differentiate between direct effects of tissue temperature on metabolism, growth, and development from indirect effects because of complex interactions with moisture stress and related factors (Elfving *et al.,* 1972). The ambient temperatures experienced by various tissues or organs of the tree vary with time and with location and orientation on the tree, and are influenced by such factors, among others, as air temperature, wind velocity, atmospheric humidity, solar radiation, shading, cloud cover or fog, attitude in relation to sun, sky, and soil surface, and reflectance characteristics of the soil or ground cover surface (Reuther, 1973). These fixed and dynamic factors interact to govern the energy

TABLE III

Temperature Parameters in Citrus Growing Regions of the American Low Altitude Humid Tropics Compared with Potential Regions in the Amazon Basin

Monthly means (°C)	Month											
	1	2	3	4	5	6	7	8	9	10	11	12
Puerto Cortez (Cortez), Honduras 15 3/4°N Lat., Alt. 15 m												
Maximum	27.5	28.6	29.7	30.9	31.5	31.7	31.4	31.7	31.7	30.0	28.4	28.0
Minimum	19.5	19.1	21.2	21.2	22.2	22.8	23.0	23.5	23.0	21.8	20.9	20.1
San Pedro Sula (Cortez), Honduras 15 1/2°N Lat., Alt. >50												
Maximum	28.5	30.4	32.4	33.8	34.0	33.7	32.4	33.2	33.0	30.7	29.4	27.9
Minimum	17.9	17.8	19.5	20.0	21.7	22.1	21.8	21.7	22.0	20.9	19.9	18.9
Port of Spain, Trinidad 10 3/4°N Lat., Alt. 7 m												
Maximum	29.4	30.0	30.6	31.1	31.7	30.6	30.6	30.6	31.1	•31.1	30.6	30.0
Minimum	19.4	19.4	19.4	20.6	21.1	21.7	21.1	21.7	21.7	21.7	21.1	20.6
Aracataca (Magdalena), Colombia 10 3/4°N Lat., Alt. 33 m												
Maximum	33.3	33.2	32.8	34.6	33.8	33.0	33.7	33.5	33.9	33.7	33.0	34.0
Minimum	21.4	21.7	22.3	23.2	22.9	22.7	22.4	22.3	22.4	22.2	22.1	21.7

Salvador (Bahia), Brazil 13°S Lat., Alt. 47 m												
Maximum	30.0	30.0	30.0	30.6	30.6	31.7	32.2	32.2	31.7	31.1	30.6	30.6
Minimum	19.4	19.4	20.0	21.1	22.2	23.3	22.8	22.8	22.8	22.8	21.7	20.6
Cruz das Almas (Bahia), Brazil 12 1/2°S Lat., Alt. 295 m												
Maximum	31.2	31.0	31.7	29.2	27.1	26.2	25.7	26.6	26.8	29.1	30.1	31.2
Minimum	21.4	21.3	21.7	21.3	20.2	19.1	18.2	18.4	18.4	20.5	20.7	21.2
Cañete (Lima), Peru 13°S Lat., Alt. 111 m												
Maximum	28.4	29.3	29.5	28.1	24.7	21.4	20.2	20.0	21.0	22.4	23.9	25.6
Minimum	19.9	20.3	20.0	18.2	16.1	15.2	14.9	14.4	14.8	15.5	16.4	18.0
Manaus (Amazonas), Brazil 3°S Lat., Alt. 48 m												
Maximum	30.0	29.9	30.0	29.9	30.7	31.1	31.6	32.7	33.1	32.7	32.0	31.7
Minimum	23.3	23.2	23.3	23.3	23.6	23.4	23.2	23.5	23.9	24.1	24.0	23.7
Altamira (Pará), Brazil 3 1/4°S. Lat., Alt. 80 m												
Maximum	30.3	32.3	30.2	30.1	30.3	31.7	30.7	31.5	31.7	31.9	31.4	31.2
Minimum	21.2	21.9	20.4	21.4	21.4	20.9	20.3	20.7	21.0	21.3	21.3	21.3
Belém (Pará), Brazil 1 1/3°S Lat., Alt. 24 m												
Maximum	31.0	30.4	30.3	30.8	31.4	31.8	31.7	32.0	31.9	32.0	32.2	31.8
Minimum	22.6	22.7	22.8	23.0	22.9	22.5	22.2	22.1	22.0	22.0	22.1	22.4

TABLE IV

Precipitation, Relative Humidity, and Insolation Parameters in Citrus Growing Regions of the American Low Altitude Humid Tropics Compared with Potential Regions in the Amazon Basin[a]

Values (mm, %, and hour)	Month 1	2	3	4	5	6	7	8	9	10	11	12	Annual total
San Pedro Sula (Cortez), Honduras 15 1/2°N Lat., Alt. >50 m													
Ppt	72	52	51	36	92	159	154	117	191	179	144	120	1642
RH	79	75	72	70	73	76	79	76	79	83	83	85	
Port of Spain, Trinidad 10 3/4°N Lat., Alt. 7 m													
Ppt	58	30	36	33	71	163	198	193	175	142	165	120	1384
RH	92	92	92	91	90	91	93	93	94	93	93	94	
Aracataca (Magdalena), Colombia 10 3/4°N Lat., Alt. 33 m													
Ppt	2	8	9	52	216	178	125	182	275	344	239	31	1661
RH	72	77	67	74	82	82	83	84	83	84	84	81	
Insol.	265	240	273	230	224	204	241	213	212	222	206	270	2790
Salvador (Bahia), Brazil 13°S Lat., Alt. 47 m													
Ppt	66	135	155	285	274	239	183	122	84	102	114	142	1900
RH	88	89	89	89	88	87	86	85	86	87	88	88	

Cruz das Almas (Bahia), Brazil 12 1/2°S Lat., Alt. 295 m													
Ppt	55	64	97	116	137	128	130	96	40	49	103	85	1100
Cañete (Lima), Perú 13°S Lat., Alt. 111 m													
Ppt	0.8	1.8	0.8	0.7	5.0	9.2	1.5	6.7	5.8	1.8	2.5	1.1	38
RH	85	83	83	82	86	88	89	89	89	88	86	85	
Insol.	195	178	198	203	145	69	52	54	62	100	135	172	1563
Manaus (Amazonas), Brazil 3°S Lat., Alt. 48 m													
Ppt	276	277	301	287	193	98	61	41	62	112	165	228	2101
RH	88	88	88	88	86	83	80	77	78	79	82	85	
Insol.	120	112	111	118	168	208	245	256	225	206	177	153	2097
Altamira (Pará), Brazil 3 1/4°S Lat., Alt. 80 m													
Ppt	216	275	346	278	176	73	49	22	30	44	65	106	1680
Belém (Pará), Brazil 1 1/2°S Lat., Alt. 24 m													
Ppt	318	407	436	382	261	165	161	116	120	105	90	197	2761
RH	89	91	91	90	87	84	83	83	84	83	82	85	
Insol.	157	112	102	132	196	239	268	267	235	247	221	213	2389

[a]Sources: Tables of Temperature, Relative Humidity and Precipitation for the World, 1958. Meteorological Office Air Ministry, London and climate summaries from official publications of Peru, Colombia, and Brazil. Methods and times of reporting relative humidity vary among countries.

balance in tissues. It is not possible to derive a useful index of this balance from conventional, generally available weather data, such as those in Tables III and IV.

At one extreme Cañete in almost rainless coastal Peru, although near the equator in a low altitude region of high atmospheric humidity, has a more subtropical than tropical climate (Tables III and IV) from the standpoint of citrus behavior. The cooling influence of the Humbolt current and the very high incidence of cloudiness and fog combine to produce a mild climate in which irrigated citrus grows less rapidly than in the hotter location in Table III.

At Cañete, varieties of the climate-sensitive navel orange group develop good juice quality and acceptable rind color, and store well on the tree, but require 8 or 9 months between anthesis and market maturity. In this type of climate, it seems probable that the effects of cool, humid ocean breezes and the high incidence of cloudiness and fog combine to moderate mean temperature as well as to dampen both temperature and moisture stress oscillations in tissues, producing a favorable effect on ultimate market quality of oranges. Since high yields are obtained in the best citrus orchards in this region, it appears that any light starvation effect on net photosynthesis is of minor importance.

At the other extreme is the hot, humid climate of the Caribbean coast of Honduras represented by Puerto Cortez (coastal) and San Pedro Sula (about 25 km inland). Here citrus grows very rapidly and, when irrigated, fruits the year around. In this truly humid tropical climate most orange varieties mature in 6 or 7 months from anthesis and are of poor export market quality. On the other hand, grapefruit may develop good export market quality, when well managed. The orchard represented in Tables VI and VII has a climate intermediate between that of Puerto Cortez and San Pedro Sula. In this region, although rainfall is high during the wet season, cloudiness is mostly transitory and daily hours of sunshine are high, probably similar to Aracataca, Colombia or Belem, Brazil (Table IV). The climates represented by Puerto Cortez, San Pedro Sula, Port of Spain, Aracataca, Manaus, Altamira, and Belem are all rather typical of a very common type found in lowland coastal plains and valleys of tropical American with a monsoon-type rainfall distribution. Most of the Amazon basin has some variation of this year-round humid, hot, and sunny climate.

The climates of Salvador and Cruz das Almas in Bahia, Brazil represent an intermediate type between the atypical humid, mild, and cloudy climate of Cañete and the typical humid, hot and sunny climate of most of the Caribbean and Amazon basins, even though their latitude is only about 13°S. Monthly mean minima in the coolest months at Salvador are around 3°C cooler, and at Cruz das Almas around 4°C cooler, than Belem, for example. The year-round monthly mean maxima of Salvador and Belem are not greatly different, but the maxima of Cruz das Almas, about 40 km inland from Baia de Todos os Santos and 295 m elevation, are 4.5°–6.5°C lower in the four coolest months. Unfortunately no data are available to compare the cloudiness of these locations. The performance of citrus at Cruz das Almas (Passos and Sobrinho, 1970) suggests that moderate cloudiness as well as mild temperatures during the April–July period of highest rainfall, among other factors, contribute to the good eating quality of navel oranges (Passos *et al.*, 1971). At Cruz das Almas, oranges usually have a major bloom during the September–November period, resulting in mature fruit in the April–June period.

Navel oranges grown in the year-round humid, hot sunny climates such as San Pedro Sula, Honduras, Port of Spain, Trinidad, or Aracataca, Colombia (Tables III and IV) do not develop acceptable eating quality, and most orange varieties produce fair to poor internal quality. Thus the climate of lowland humid tropical areas like the Amazon basin is such that its adaptation to intensive citrus culture is likely to be limited to a few varieties discussed in a subsequent section.

IV. TREE GROWTH AND REPRODUCTION

In tropical regions near the equator at elevations of 1000 m or lower, both day and night temperatures are within the range favorable for very rapid growth of citrus during 12 months of the year. In a subtropical climate like Riverside, California, diurnal ranges favorable for very rapid growth are in only five or six of the warmest months. As a result, with well distributed rainfall or supplemental irrigation in monsoon-type climates, citrus trees grow about twice as rapidly in lowlands near the equator as in comparable conditions 25° or 30° from the equator (Table V).

TABLE V
Comparative Size of Valencia Orange Trees on Routh Lemon Stock in Some Tropical and Subtropical Areas[a]

Location	Tree age (years)	Number of trees measured	Mean trunk cross-section (cm^2)	Mean tree height (cm)	Mean canopy diameter (cm)
Cali, Colombia 3½°N	5½	20	416	576	554
Buga, Colombia 4°N	5	25	203	411	493
Clermont, Florida	5	324	101	–	–
28½°N	11	324	222	347	470
Monte Verde,	7	97	151	–	–
Florida 28½°N	13	96	289	403	475

[a]From Reuther (1973).

Observations and very limited reliable records suggest that commercial yields of citrus orchards in tropical regions are commonly one-half or less of those obtained in subtropical regions. Whether climate in lowland tropical regions is always a primary limiting factor to yield is an open question. Locally developed and adapted production technology systems for citrus culture are for the most part lacking in tropical areas. Also, with the exception of a few varieties there was little selection of varieties specifically adapted for tropical conditions. The major commercial orange, mandarin, and lemon varieties were all selected in subtropical areas for adaptation to subtropical climates. Hence, it has not been possible so far to evaluate the yield potential for a variety such as Valencia orange, for example, in an area such as the Amazon River valley under optimal cultural conditions providing favorable soil moisture, fertility, rootstocks, and control of pests and diseases.

During extensive travels in tropical regions, I have observed that citrus yields and vigor are commonly much lower than in most subtropical areas. The grapefruit yields of Trinidad, Jamaica, and Belize provide typical examples of production levels common in the humid tropics. The small grapefruit industry of Trinidad ($10\frac{1}{2}°$N latitude; rainfall from 1500 to 2500 mm, with a 3–4 months dry period), with a little over 1300 ha of grapefruit-bearing orchards, produced crops averaging 12.8 metric tons/ha in 1956, of which only 19% was exported (Burke, 1967; Trinidad and Tobago, 1959). This suggests that the better orchards produced about 25 tons/ha and the poorer about $6\frac{1}{2}$ tons/ha. In Jamaica, the 1960 production of 3314 ha of grapefruit averaged 10.7 tons/ha (Food and Agriculture Organization, 1970). In 1963, Belize averaged 24.7 tons/ha on 445 ha of grapefruit. In contrast, the 1956 production of about 45,000 ha of grapefruit in subtropical Florida (25°N to 29°N latitude; rainfall from 1250 to 1500 mm with a dry spring) averaged 34.5 metric tons/ha, and in 1967 averaged 50.5 tons/ha from 35,200 ha (Shuler *et al.*, 1970). In Florida, per hectare production of oranges averages around 60% that of grapefruit.

I have encountered good vigor and heavy production in only a few exceptional citrus orchards in the tropics. Table VI gives examples of such orange orchards growing near the equator at around 1000 m of elevation in the Cauca valley of Colombia with about 1000 mm of annual rainfall, and a bimodal distribution so that a three month dry period occurs in July–September, and another six week dry period in January–February. These yields were obtained *without* supplemental irrigation and *without* a well-developed crop protection technology, but on exceptionally deep, well-drained, and fertile soils. Table VII shows the exceptionally high yields that can be obtained with grapefruit in a lowland humid tropical climate with very favorable soils and excellent management.

Yields and sizes of 5-year-old experimental trees in Table V commonly were

TABLE VI
Comparison of Yields of Oranges in Tropical Colombia and Subtropical Florida[a]

Crop year	Mean yield (metric tons per hectare)			
	Colombia overall	Florida		
		Lower 25%	Overall	Upper 25%
1965	32.7	9.7	24.3	42.8
1966	33.2	10.8	24.4	58.6
1967	32.8	11.9	30.7	50.7
1931–69	–	–	27.3	–

[a]Sources: *Colombia*—adapted from Rios-Castaño *et al.* (1968). Each value is the pooled mean of 16 "Valle" navel and 16 "Indian River" orange trees in a rootstock experiment located at the Centro National de Investigaciones Agropecuarias Palmira, Palmira, Colombia at an altitude of 1006 m and $3\frac{1}{2}$°N latitude. The trees, planted on Rough lemon rootstock in 1960, were spaced 7 × 7 m and were over 4 m high by 1965. *Florida*—adapted from Brooke (1970) and Savage (1967). These yields are from 50 to 60 representative commercial orchards over 10 years of age in major citrus growing areas. The early, midseason, and late varieties represented are pooled; most are on Rough lemon rootstock but some are on Sour orange, Cleopatra mandarin, or Sweet orange.

equal to 10-year-old Florida trees. Similarly, 7-year-old grapefruit trees in Table VII equaled 14- or 15-year-old Florida trees in size and productivity.

The foregoing yield and growth observations and data indicate neither the high year-round temperatures nor the lack of a cool-winter short-day dormancy period per se limit citrus yields in the humid tropics. Rather, they point to the lack of a locally adapted, well-developed production technology system as the primary limiting factor. In addition, they suggest that yield potential may be even greater in some tropical regions than in the established subtropical zones.

V. GROWTH AND FRUITING IN RELATION TO CLIMATE

Specialists have long recognized that there are large differences among citrus growing regions, or climatic zones within regions, in growth and yield of trees and in physical and chemical characteristics of the fruits. Also, there are marked differences in the time lapse between anthesis and market maturity and in the period that marketable fruit may be stored on the tree. Indeed, these factors,

TABLE VII
Yields[a] of an Exceptionally Well-Managed Marsh Grapefruit Orchard in the Humid Tropics

Year	Yield (metric tons per hectare)	Percent exported
1972	94.3	71.5
1973	92.2	67.5
1974	79.5	71.0

[a]Calculated from total production of more than 100 ha of orchard on sour orange rootstock ranging in age from 7 to more than 50 years. Trees located on deep, well-drained soil on a Caribbean coastal plain having around 2600 mm annual rainfall fairly well distributed. Fertility is controlled by leaf and soil analysis, and pests are controlled by a well-developed program which includes 6 or 7 pesticide sprays annually.

among others, determine the adaptability of a specific variety for commercial exploitation in a particular citrus growing region.

Only in recent years have there been studies with citrus aimed at disentangling climatic influences from cultural and edaphic factors. The limited data so far obtained indicate that within fairly broad parameters of adaptable soil and reasonably good cultural practices, climate is the most important component of the climate–soil–culture complex causing differences in growth, yield, and fruit qualities among commercial citrus producing areas. Some influences of major climatic components are discussed in the following sections.

A. Rainfall, Humidity, and Wind

It is not feasible here to discuss the influence of rainfall or humidity on citrus responses per se because it is difficult to isolate them as discrete aspects of the environmental complex. Rainfall and relative humidity are interrelated, and both moderate energy flux, which in turn influences tissue temperature and the vital processes within tissues. The amount and distribution of annual precipitation have a major direct effect on soil moisture, which can be manipulated artificially by irrigation and drainage.

In spite of the above limitations, it is possible to make some gross estimates of the parameters of evapotranspirational loss of water from citrus orchards in relation to climate. Available data in the United States suggest that the annual use of water by evapotranspiration in a well-watered mature citrus orchard ranges somewhere between about 750 mm in a cool coastal subtropical climate like Santa Barbara, California and 1250 mm in the hot semiarid subtropical climate of Weslaco, Texas (Newman, 1968). Of course, in commercial practice considerably more water is required to compensate for losses from runoff, deep percolation, and other factors influencing uniformity of distribution of rainfall and irrigation and their efficiency. Thus, it is suggested that somewhere between 1000 and 2000 mm of rainfall annually are required to maintain good soil moisture conditions for citrus culture throughout the year. The lower value would be for a region with a rainfall distribution well correlated with seasonal temperatures and daylength in a cool, foggy subtropical climate, and on a deep, retentive, but well-aerated soil. The higher value would be required on a deep, well-drained and retentive soil in a hot lowland tropical climate like the Amazon valley, but with fairly uniform distribution of precipitation throughout the year.

In lowland tropical areas of very high rainfall, such as in southeastern Thailand, with around 3500 mm of precipitation fairly well-distributed, and high humidity, commercial citrus culture is marginal. Without benefit of a well-developed pest control system in a very wet climate, fungus diseases become rampant, and older trees accumulate a heavy burden of epiphytic and semiparasitic plants which compete for light and favor pests and diseases (Reuther, 1973).

A humid climate, when compared to an arid one, tends to produce fruit with smoother, thinner skins and trees with a more open growth habit (Reuther, 1973). But its most significant effect is on the ability of fungus diseases to parasitize foliage, flowers, fruit, branches, and bark, and thus disease control is likely to be one of the most difficult production problems in the humid tropics.

The amount, velocity, and seasonal distribution of winds are of some importance in determining the adaptability of a region to citrus culture. Hot winds tend to cause excessive evapotranspirational losses and often injury or death of exposed leaves and twigs through dehydration. Cold winds also cause similar damage in the winter of some subtropical regions. In addition, excessive wind, especially when the fruits are young, causes scarring and consequent loss of market value of fruits, and stimulates abscission of leaves and fruit. Planted windbreaks in the form of perimeter rows of tall-growing trees offer protection against excessive wind damage (Platt, 1973). Fortunately excessive windiness is not as common a problem in tropical as in subtropical regions. Gentle air movement or breezes with velocities up to 15 or 20 km per hour are often beneficial and promote drying of tissues, thus reducing fungus diseases as indicated above.

B. Temperature and Moisture Stress

1. Vegetative Growth

Most citrus species make little or no shoot elongation at constant temperatures in the neighborhood of 12° to 13°C (Reuther, 1973). Above this temperature, growth rate increases progressively until a maximum is achieved somewhere above 30°C, but growth virtually ceases at constant temperatures above 40°C. However, experimental difficulties tend to entangle effects of temperature per se, and its indirect, and often overshadowing effects on internal moisture stress and related factors. Thus, in the tropics, high moisture stress in tissues probably limits growth more often than temperature. A further difficulty is that, in the orchard, root systems normally are exposed to different temperature parameters than the tops. Unfavorable temperatures of either the root system or the top probably limit growth about as effectively as if both portions were subjected simultaneously to an unfavorable temperature (Reuther, 1973).

2. *Flowering*

In subtropical areas during the winter months mean air and soil temperature fall below about 15°C for several months (Figs. 2 and 3) during the winter. This causes growth to cease, and the plant to become dormant for three or four months. This dormancy induces greater tolerance to frost damage, and causes changes within the buds which induce flowering when warmer temperatures in the spring induce resumption of vegetative growth. Thus, in a typical subtropical climate, with cool winters, citrus tends to bloom profusely in the spring and set

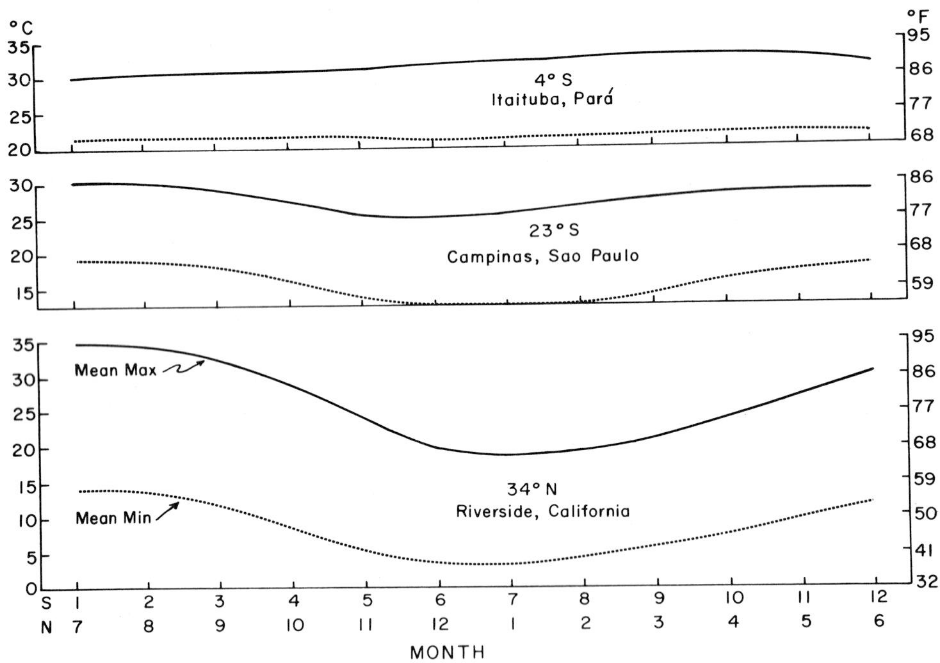

Fig. 2 Comparison of monthly mean maximum and minimum temperature regimes in the Amazon valley with those of major subtropical citrus growing regions in Brazil and the United States.

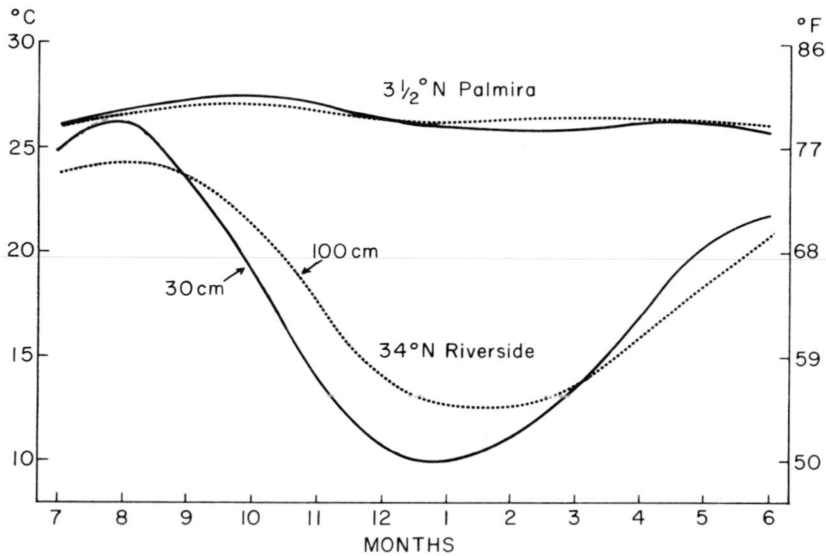

Fig. 3 Comparison of soil temperatures (monthly means) under grass sod in tropical Palmira, Colombia (1008 m elevation) and subtropical Riverside, California (250 m elevation). (From Reuther, 1973.)

only one major crop of fruit which matures in late fall, or winter, or even the following spring, depending on variety and seasonal temperature regime.

However, a period of low temperature-induced dormancy is not essential for flower induction. Dormancy induced by periods of low soil moisture supply is about equally effective. In a tropical climate with no cool periods and ample soil moisture throughout the year, either by natural rainfall distribution or through supplemental irrigation, most citrus species produce some bloom and mature fruit in every month of the year. However, with periods of less than ample soil moisture, flushes of bloom and vegetative growth tend to follow periods of drought of six weeks or more in duration, with a lag of about four weeks after resumption of effective rainfall or application of supplemental irrigation. Without irrigation in some tropical regions such as Cauca valley of Colombia, with about 1000 mm of rainfall annually in a bimodal pattern of distribution, oranges tend to bloom after the "long" and "short" drought periods, and the two major harvests occur about seven months after the dry periods end (Torres-M. and Rios-Castaño, 1968).

Table VIII shows the seasonal distribution of harvest (and by inference, flowering) in a large, unirrigated but highly productive (see Table VII) grapefruit orchard in a coastal Caribbean location in the humid tropics. As in most of the

TABLE VIII
Seasonal Distribution of Flowering and Harvest in a Marsh Grapefruit Orchard[a] in a Humid Tropical Climate

Four week harvest periods ending	Metric tons harvested	Percent of annual total	Estimated mean date of anthesis
First 16 weeks, 1973			
Jan. 27	1110	8.5	July 13, 1972
Feb. 24	1808	13.9	Aug. 10, 1972
March 24	2054	15.8	Sept. 10, 1972
April 21	1556	12.0	Nov. 5, 1972
Totals (16 wk)	6528	50.2	
Middle 20 weeks, 1973			
May 19	376	2.9	Nov. 5, 1972
June 16	0	0.0	–
July 14	17	0.1	Dec. 41, 1972
Aug. 11[b]	0	0.0	–
Sept. 8[b]	465	3.6	Feb. 22, 1973
Totals (20 wk)	858	6.6	
Final 16 weeks, 1973			
Oct. 6[b]	1586	12.2	March 23, 1973
Nov. 3	2562	19.7	April 21, 1973
Dec. 1	1273	9.8	May 18, 1973
Dec. 29	200	1.5	June 16, 1973
Total (16 wk)	5621	43.2	
Totals (annual)	13007	100.0	

[a]Described in Table III.
[b]World supplies scarce.

Caribbean, the mean annual rainfall in this orchard is highly variable, but averages around 2600 mm. The monthly distribution is also highly variable, but about one year in two, little or no severe moisture stress occurs during the course of the year. When droughts occur, they are likely to be in the drier April through July period, when multiannual monthly precipitation means range between 90 and 130 mm (data not presented). Temperatures in this region are less variable than rainfall, and are intermediate between Puerto Cortez and San Pedro Sula (Table III). Thus, temperatures are favorable for very rapid growth in all twelve months of the year. Humidity is very high throughout the year in this coastal plain, but most rain normally falls in almost daily showers during the wettest season. Prolonged periods of cloudy, rainy weather do occur, but are not the rule. Thus, flowers, foliage, and fruit normally do not remain wet for more than a few hours at a time.

Under the above conditions, some flowering and fruit set occur in all twelve months of the year, but in the 1972–1973 season, most occurred in the four month period July through November, 1973, from which mature fruits were harvested approximately six months after anthesis in the January through April period of 1973 (Table VIII). Another substantial bloom period occurred during the March through May period of 1973, from which mature fruits were harvested about six months later, during the September through November period of 1973.

Only minor blooms occurred during June and the November through February "winter" months (Table VIII). Although practically no fruits were harvested six months later from mid May through mid August, some fruits matured during this time, but were deemed to be too few to warrant the costly pest control program required to produce fruits suitable for export, so they remained unsprayed and unharvested.

The seasonal pattern of bloom and harvest indicated in Table VIII would vary appreciably from season to season. The major factors determining bloom distribution in the 1972–1973 season, when only minor moisture stress periods occurred, are difficult to isolate. Probably the major factor was the growth and fruiting cycles of branches established the previous season when some dry periods did occur. Observations in the tropics indicate that without periods of dormancy induced by moisture stress or cool temperatures to synchronize the flowering of a major portion of shoots, each shoot or branch independently goes through a variable period of vegetative growth after bearing fruit before it again flowers. The result is a protracted and sporadic flowering habit as shown in Table VIII.

The above considerations indicate that flower induction in citrus is not fundamentally dependent on low temperature, short days, or dormancy per se. Evidently there are cyclic internal factors involved with the past physiological and fruiting history which will ultimately cause flowering of a particular branch in the absence of conditions favoring dormancy. Dormancy tends to cause a major portion of the shoots to flower at one time, but in the absence of dormancy, individual shoots or branches tend to flower sporadically during the year.

3. Fruit Set

As with vegetative growth, it is difficult to delineate clearly temperature and moisture stress effects per se on fruit set, except at the extremes. Satisfactory set of fruit may be observed in a very wide range of thermal conditions in many diverse citrus growing regions, including the humid tropics. In subtropical regions, citrus trees normally produce a large number of blossoms, along with a heavy "flush" of vegetative growth, in the spring each year, but only a very small percent of these "set" and ultimately develop into mature fruits. The final crop reaching maturity depends not so much on the initial number of blossoms as on the severity of attrition of flowers and developing fruitlets, particularly in the 2-

to 4-month period following anthesis. This attrition may be influenced in a few self-unfruitful varieties by interplanting with suitable varieties for cross-pollination (Frost and Soost, 1968). However, with most varieties, factors within the tree (food reserves, mineral nutrition, hormonal balance, etc.) as well as environmental factors such as temperature, wind, humidity and soil moisture supply, and, of course, pests and diseases play the major role in determining the rate of attrition of developing flowers and fruitlets. After the fruit reaches a diameter of 15–20 mm, the rate of abscission drops dramatically, and is said to be "set."

During the period from anthesis until the fruit attains about 15–20 mm in diameter, abscission may be greatly stimulated by high temperature and thus seriously reduce set. Temperatures of 40°C and above, particularly in combination with low humidity and wind, cause poor set from time to time in the hotter, drier regions of California, especially in the sensitive navel group of varieties. High temperatures should not often limit set in the Amazon valley, since available weather data indicate that temperatures of 40°C or above occur only rarely in most humid tropical areas.

4. Fruit Growth and Maturation

A comparison of the rate of fruit growth and maturation in four diverse climatic situations is shown in Fig. 4. Seasonal temperatures at Palmira, in the Cauca valley of Colombia, are similar to Itaituba, Pará (Fig. 2), but average about 3°C cooler at night, the annual mean about 2.5°C lower, and rainfall and humidity are also somewhat lower. Likewise seasonal soil temperature regimes in much of the Amazon valley should approximate those in Palmira (Fig. 3), but perhaps average 1° or 2°C higher. With irrigation, growth and maturation of Valencia oranges in the Amazon valley should approximate the rates obtained in Palmira and the Caribbean lowland tropics (Table VIII). Thus, most citrus varieties should reach market maturity 6–7 months after anthesis. This compares with 8–13 months required in subtropical climates,e.g., California and Florida.

In subtropical climates there is a spread of several months in the rate of ripening of early orange varieties, e.g., Hamlin, and late varieties, e.g., Valencia. For example, in the climate of central Florida, Hamlin oranges commonly reach market maturity in November or December, 8–10 months after blooming in February or March. Valencia oranges under comparable conditions reach market maturity in February or March of the following year, or about 11–12 months after anthesis. In a hot, humid tropical climate this difference in maturation rate between early or later varieties probably would be reduced to about two weeks instead of two or three months. In fact, because of the sporadic blooming tendency in the tropics, it is frequently difficult to distinguish differences in rate of maturation between early and late varieties without tagging fruitlets to log the approximate date of anthesis.

Just as the fruit matures more rapidly under the influence of continuous warm air and soil temperatures in the tropics (Figs. 2 and 3), it also becomes senescent and unsuitable for the market much more rapidly. Thus, I would predict that in the Amazon valley it would not be possible to store citrus fruit on the tree much beyond three to four weeks after market maturity has been reached. In contrast, in subtropical cool-winter climates, citrus may be stored on the tree from three to five months after harvest maturity is reached, depending on the local climatic conditions (Fig. 4).

5. *Fruit Quality*

It is difficult to make a generalized comparison of the physical aspects of fruit quality produced in tropical and subtropical climates. In the humid tropics, fruit size tends to be large, rinds thin and smooth, and juice content high as compared to arid or semiarid subtropical regions. Similarly, fruit in the humid

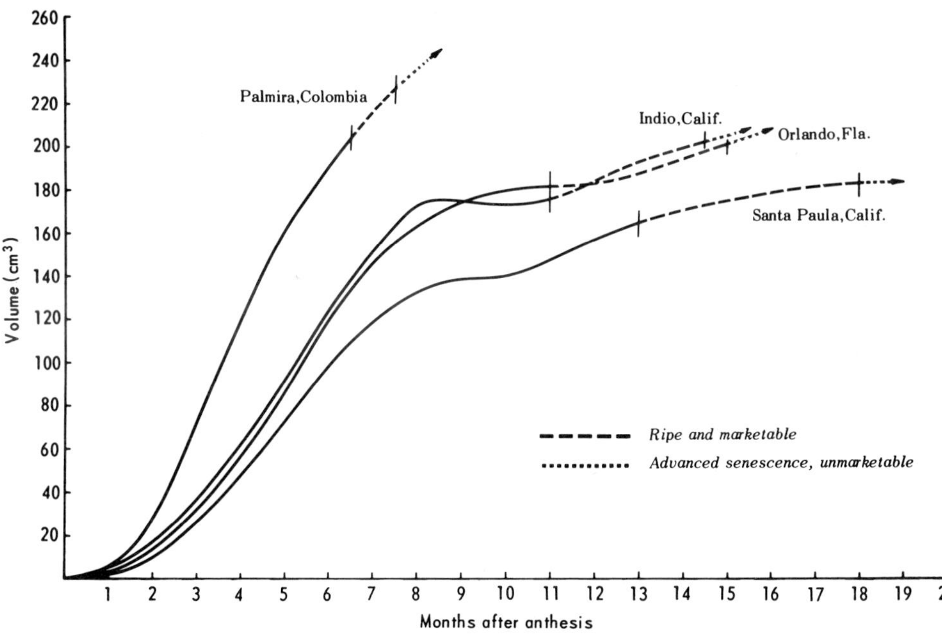

Fig. 4 Schematic comparison of typical "Valencia" orange growth curves in a tropical (Palmira) with three subtropical climatic situations. All have adequate or nearly adequate soil moisture throughout the year from irrigation natural rainfall, or both. In addition, comparative periods of immaturity, market maturity, and advanced senescence are approximated. (From Reuther, 1973).

tropics tends to have lower total soluble solids and acid concentration in the juice than fruit from subtropical climates. The data in Fig. 5 illustrate the pronounced effect of prevailing temperature on the rate of decrease of acid in juice as the Valencia orange matures. The acid concentration in all orange, grapefruit, pummelo, and mandarin varieties behaves similarly in response to temperature. However, acid citrus fruits do not exhibit this effect. Acid concentrations in lemons or limes may be just as high in tropical as subtropical regions (Reuther and Rios-Castaño, 1969).

The data presented in Fig. 6 show the acid data of Fig. 5 as a ratio to total soluble solids. Taste panel studies have shown that acceptable eating quality of oranges (market maturity) is reached when the solids to acid ratio is about 9 to 1 or above. However, for reasons I do not understand, Brazilians—as the people of Southeast Asia and other regions—appear to prefer less acid in their citrus fruit than do most North Americans and temperate zone Europeans.

Perhaps the most striking effect of tropical temperature regimes on fruit quality is on rind color (Fig. 7). In subtropical climates, the cool nights during fall and winter when most varieties are rapidly gaining in juice and total soluble solids content cause changes in the rind so that an acceleration of the breakdown of chlorophyll and the synthesis of carotenoid pigments occur (Erickson, 1968). Chilling temperatures of 15°C and lower are associated with this change, which seems, in turn, to be related to ethylene produced as a result of a chilling injury effect (Cooper *et al.*, 1969). In the tropics, lack of chilling temperatures is associated with a very slow breakdown of chlorophyll and synthesis of caro-

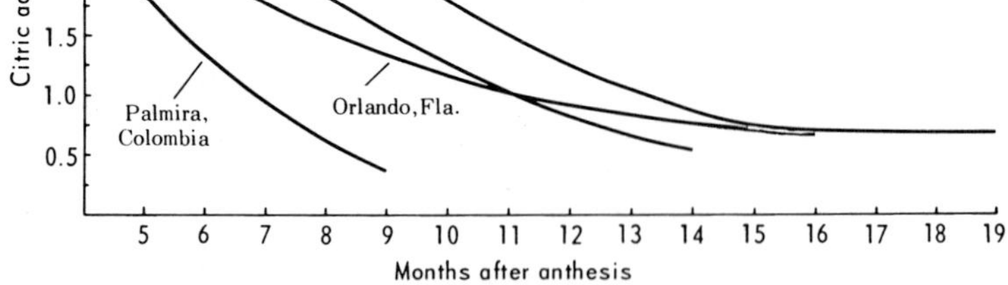

Fig. 5 Schematic comparison of seasonal trends in percent acid in juice of "Valencia" oranges in a tropical (Palmira), a cool tropical (Medellín), and two subtropical climates. (From Reuther, 1973.)

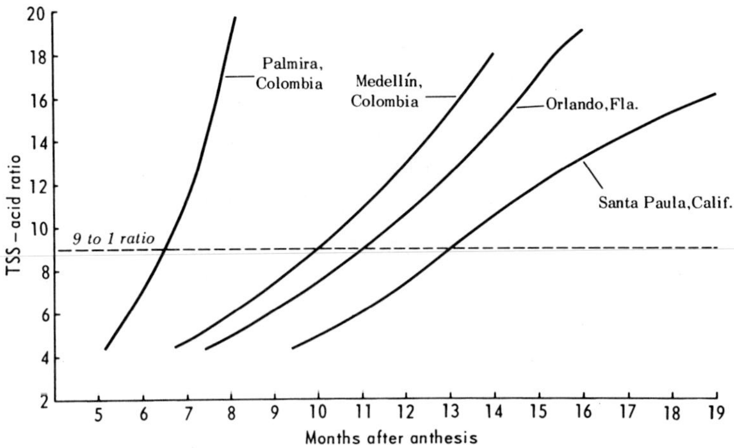

Fig. 6 Schematic comparison of seasonal trends in the ratio of total soluble solids to acid in juice of "Valencia" orange in four widely different climatic situations. The approximate dates of anthesis were: Palmira, Nov. 1; Medellín, Jan. 20; Orlando, March 15; Santa Paula, May 1. (From Reuther, 1973.)

tinoids in the rind, so the fruit may be pale green, or at best pale yellow when it reaches market maturity. This precludes acceptance of tropical oranges and mandarins in most of the great markets of the world, because the consumer is used to the bright orange color of oranges and mandarins produced in subtropical climates. However, grapefruit, pummelos, and limes produced in the tropics have quite attractive rind colors readily accepted in European and North American markets.

Although treatment with ethylene known as "degreening" is used in Florida and elsewhere to improve rind color of some early varieties (Grierson and Newhall, 1960), such treatment would not be sufficiently effective to transform the rind color of tropical oranges and mandarins to a hue competitive with subtropical fruits. However, research now in progress shows promise of developing bioregulators that greatly stimulate the synthesis of carotenoid pigments in the rind, even if applied after the fruit has been harvested (Yokoyama *et al.*, 1976).

One of the red carotenoid pigments, lycopene, is synthesized in the peel and flesh of some grapefruit and pummelo varieties, as well as in such common fruits as the tomato. A lycopene pigmented variety such as "Redblush" grapefruit becomes very highly pigmented when grown in hot tropical climates, but develops only a faint pink tinge in the cooler subtropical climates. On the other hand, the so-called blood orange varieties, which synthesize the crimson to

Fig. 7 Temperature and rind color of Valencia oranges. Upper: Green fruits from a simulated tropical climate (30°–34°C day and 20°–24°C night) in a controlled temperature greenhouse. Lower: Bright orange fruits from ambient temperatures at Riverside, California (Fig. 2). The upper fruit are from trees placed in hot greenhouse Oct. 8, 1969. Both upper and lower fruits were harvested and photographed Feb. 2, 1970.

purple flavonoid pigment anthocyanin, develop the "blood" color only in the cooler subtropical climates, and are bloodless when grown in the tropics (Reuther, 1973).

VI. SOIL

Citrus is adapted to a wide range of soil types provided they are reasonably deep and well aerated. In this respect, its requirements are similar to most tree fruits and other perennial woody plants like coffee. Texture and level of natural fertility are of less importance than depth to the water table, hardpan, gravel, or rock. With proper adjustment of fertilizer and nutritional spray programs, citrus can be grown on soils ranging from calcareous to quite acid. Low inherent fertility may be overcome by proper fertilization, but poor physical conditions for deep rooting cannot be corrected as easily or economically.

About two-thirds of Florida's 350,000 ha of citrus are planted on undulating to hilly, deep, well-drained fine sand too low in moisture holding capacity and fertility for profitable production of most other crops. On such soils, citrus sends down an extensive root system enabling it to utilize moisture in the top 3 or 4 m of soil. Modern fertilizer and nutritional spray programs supply good nutrition so that yields as high as anywhere in the world are achieved. Much of the remaining one-third of Florida's citrus is planted on flat terrain on soils which, in their virgin state, were imperfectly drained. By construction of extensive systems of drainage works, and planting on raised beds, the water table is controlled at a level near 1 m by pumping water out in the wet and in during the dry season. Investments to make citrus culture feasible on these types of soils may be very high.

VII. MINERAL NUTRITION

Most of the common citrus species have nitrogen requirements rather similar to most other tree fruits. On most soils, vigor and yield are stimulated sharply by nitrogen fertilization up to about 50–75 kg per hectare, and then more slowly up to about 100–125 kg per hectare (Embleton *et al.*, 1973b). Liberal nitrogen fertilization is an almost universal requirement for intensive citriculture.

In contrast, only citrus orchards on very phosphorous-deficient soils benefit from phosphorous fertilization (Embelton *et al.*, 1973b). Citrus appears to be exceptionally efficient at accumulating phosphorus from low P soils, perhaps because of endomycorrhizae in the roots (Menge *et al.*, 1975).

Citrus orchards respond to potassium fertilization primarily on acid sandy soils low in exchangeable bases, on soils high in calcium, and on soils having a high K fixing capacity.

Magnesium deficiency symptoms are common in many citrus growing regions due primarily to low native magnesium content of the soil, and to high Ca-K saturation of the exchange complex in relation to magnesium. Also, heavy K fertilization frequently induces a deficiency of magnesium (Embleton *et al.*, 1973b).

The calcium requirement of citrus is high in relation to many crop plants, perhaps because of sensitivity to secondary toxic effects associated with low pH on low base saturated soils. Citrus grows well on soils of pH 5.5 to about 8.0. Citrus is grown extensively on highly calcareous soils, but not without certain problems (Embleton *et al.*, 1973b).

Most citrus species are relatively poor foragers for micronutrients, especially the heavy metals. Deficiencies of zinc, manganese, and copper once were serious yield-limiting factors in commercial orchards around the world, but now are commonly controlled efficiently and cheaply by foliage sprays, and in some situations, by soil applications. Boron deficiency is common in many regions, especially on acid soils in regions of high rainfall.

The level of mineral nutrients in the tree may influence the market quality of fruit (Embleton *et al.*, 1973a). Thus, manipulation of fertilizer doses and timing, within limits, can provide a means for improving market quality of fruit, expecially on tropical soils of low native fertility.

In common with most other tree fruits, citrus species have a low tolerance to a high content of soluble salt or boron in the soil (Richards, 1954).

It seems likely that successful intensive citriculture on the dominant deep, well-drained, acid, infertile, red and yellow Latosols (Oxisols) of the Amazon basin (Bennema, 1975; Dos Santos, 1972; Falesi, 1972) will require the development of efficient means of supplying all nutrient elements. The two elements that may provide special difficulties are Ca and P (Blue, 1974), and perhaps novel methods will have to be developed to supply these elements economically. Fortunately tissue analysis techniques have been well developed for citrus (Embleton *et al.*, 1973a), and should prove an invaluable research tool in solving such problems. Unfortunately, little is known about the relative tolerance of citrus to the high exchangeable aluminum common in Oxisols. Similarly, little is known of the implications of the high year-round temperatures in tropical soils on nutrient uptake by citrus roots.

VIII. IRRIGATION

Although under comparable conditions, less water is used by mature citrus trees per hectare than by cotton, for example, estimates for citrus in the subtropical United States indicate an evapotranspirational use of 120–160 mm during the hottest months (Van Bavel *et al.*, 1967). Similar levels of evapotranspiration will probably occur with citrus in the humid tropics during dry periods with good soil conditions and efficient irrigation. Thus, an estimate of the combined rainfall and supplemental irrigation need for a mature citrus orchard in the hottest and driest months would be around 200 mm, taking into account runoff and other losses, and inefficiencies in distribution. Of course, this would vary with depth and moisture holding characteristics of the soil, and other factors. Typically, in the driest areas of the United States, two or three irrigations per month are applied in midsummer, ranging between 50 and 100 mm of water per application.

Rainfall data in the Amazon valley (Bastos, 1972) show that annual precipitation ranges between 1071 mm (Barra do Corda, Maranhãs) and 3654 mm

(Taracua, Amazonas). In common with most American lowland tropical regions (Table IV), precipitation in most of the Amazon valley has a monsoon pattern of distribution, with dry periods lasting from 2 to 6 or 7 months, especially in areas with less than 2500 mm. In all such areas irrigation will be necessary for intensive exploitation of citrus. In the drier areas on poorer soils, irrigation will be required for tree survival.

I predict that citrus will do best in regions with 2500 mm or less of annual rainfall. With very high rainfall, free water remains too long on tissue surfaces, and fungus diseases become major limiting factors.

In a region such as Altamira (Table IV), with 6 months of rainfall below 100 mm, it would be possible to manipulate irrigation to influence time of bloom (and harvest). Irrigation could be withheld about 6 to 8 weeks during the beginning of the dry period or long enough so that the trees become dormant, and flower induction takes place—perhaps by August 1 in a normal year. Then irrigation could be applied, causing a major bloom in September, and fruit ready for harvest the following February or March. Another possibility would be to irrigate during the normally dry months of June, July, and August, then withhold irrigation until about November 15. A major bloom would then be induced in December, which would produce fruit ripening about June or July. Fruit ripening during this period, if stressed somewhat for moisture in April, May, and June, would have a higher sugar content than fruit which ripens during periods of ample soil moisture (Reuther, 1973). Thus, manipulation of irrigation has the potential in the tropics for regulating season of maturity and with favorable timing, improving fruit quality.

IX. VARIETIES

Most, but not all, of the important cultivated varieties of citrus have been selected in subtropical regions on the basis of their adaptation to subtropical climates. Thus, most cultivars produce fruit of better market appeal in an environment that is more subtropical than tropical. Hence, the list of citrus varieties well adapted to a lowland tropical climate is short and poorly documented.

Among orange varieties, only Valencia stands out as being fairly well adapted among the well-known, widely grown commercial varieties. The navel group, though of Brazilian origin, would likely become too large, coarse, and insipid in the lowland hot humid tropics, although it develops fair to good quality in a slightly cooler climate at 1000 m elevation in the Cauca valley of Colombia, and at 300 m in the Cruz das Almas area of Bahia, Brazil. Most other orange varieties become too low in acid before the sugar content reaches an acceptable level for good eating quality in most hot, humid lowland tropical areas. Although the Valencia variety would not be suitable for export as a fresh fruit when grown in most lowland tropical areas because of poor rind color and only fair eating quality, it probably could be grown to produce industrial juice concentrates for export.

Among citrus varieties that should be well enough adapted to the climate of the Amazon valley to be exportable are grapefruit, West Indian lime, and the Persian (Tahiti) or Bearss lime. Among the unknown hazards of importance is the virulence of virus diseases like tristeza in the Amazon valley, and the prevalence and efficiency of its vectors. Both grapefruit and West Indian limes are known to be severely affected by some strains of this virus complex in certain regions, regardless of the rootstock used.

Among varieties of grapefruit to try is Marsh, the most widely grown commercial sort. It is good for both fresh fruit and industrialization. The pigmented sports of Marsh such as Redblush, can also be grown but are less adapted to industrial use than the white-fleshed Marsh. The Duncan variety, containing many seeds, is suitable only for industrial uses such as juice concentrates and canned juice or segments. Both the West Indian and Persian limes should be suitable for export fresh and for industrialization.

Lemon varieties are poorly adapted to the hot, humid tropics, being especially susceptible to fungus diseases of the bark, and having a tendency to produce large, coarse fruit.

Most of the widely grown mandarin and mandarin hybrid varieties are not well adapted to hot tropical climates because of low acid, coarse texture, and poor rind color. However, the varieties Willowleaf, Ponkan, Ortanique, Minneola, Temple, Fortune, Murcott, Kara, Encore, and Kinnow (see Hodgson, 1967) should be tried for local use. It probably would not be worthwhile to try satsuma types in the Amazon basin.

A group of citrus fruit known as pummelos, resembling large grapefruit, are examples of citrus varieties selected especially for their adaptation to tropical conditions. This highly prized citrus fruit is grown extensively as a commercial market fruit in Thailand and other sections of Southeast Asia. Select tropical varieties, when grown under subtropical conditions, are of poor quality. Such varieties as Kao Pan, Kao Phuang, and Thong Dee from Thailand should be tried as a garden and local market fruit in the Amazon basin, which has a range of climatic conditions similar to those found in Thailand.

Because of diseases and other pests, it would not be meaningful to speculate on the best rootstock varieties for the Amazon valley. Ultimately, this will have to be determined by local experimentation in the major ecological zones. It is clear, however, that tristeza-susceptible stocks such as sour orange should be avoided. There will undoubtedly be great differences among rootstocks in their adaptation to the acid oxisols, but no data are available so far.

X. RESEARCH

Although it is possible to predict in a general way the behavior of a crop such as citrus in a new ecological situation, our knowledge of the precise growth and fruiting response of citrus to the complexities of climate, especially the thermal aspects, is rather limited. The soils of the Amazon basin are largely an unknown in relation to culture and rootstocks. Similarly, we are unable to give a very precise appraisal of how the relation of pests and diseases to citrus as a host will be affected in new ecological situations. It is common experience to find that most, but not all, pest control measures developed in subtropical areas are not effective in the tropics, or at least require substantial modification to be useful. In addition, new regions usually turn up new pests or diseases which require research in order to develop effective control measures.

In view of the above and other considerations, I cannot stress too strongly the importance of establishing a thorough program of research to assess the potential of citrus, or any other crop, for commercial exploitation in the Amazon valley. If large-scale commercial exploitation of citrus is attempted without first assessing its adaptability and developing systems of cultural technology suited to the various ecological situations, it will surely fail. The Amazon basin is among the largest remaining undeveloped area on this planet. The potential for producing food and fiber for the burgeoning population of Brazil and the world can only be gauged by substantial investments in agricultural research. Its efficient exploitation will have to be based on a system of locally developed and adapted production technology. Although Brazil has some of the best citrus specialists in the world, they will need new information obtained from experimentation in the Amazon valley to develop a sound system of production technology for the various ecological regions.

In my view, any plan to develop the agricultural potential of the lowland humid tropics should emphasize research on tropical tree fruits and nuts. Among these, citrus should occupy an important place. Despite its tropical origins, almost all varieties now grown extensively around the world were selected in subtropical regions for their adaptation to a cool winter climate. The time is long overdue to undertake, somewhere in the lowland humid tropics, a long-range variety improvement program aimed at discovering the genetic potential for citrus as a tropical crop. For example, grapefruit, which probably originated as a chance hybrid seedling in the Caribbean tropics, suggests that an array of completely new varieties, of hybrid or other origin, might be developed suitable for large-scale exploitation in tropical regions.

Only a well-conceived, long-range, and well-supported program of research can indicate the economic and horticultural potential of citrus culture in the Amazon valley and similar humid tropical areas. However, my present knowledge and experience suggest that there are some reasons for optimism that citrus will ultimately prove to be of economic significance in the development of such areas.

REFERENCES

Bastos, T. X. (1972). O estado atual dos connecimentos das condicões climáticas da Amazônia Brazileira. *Bol. Tec. Inst. Pesquis. Exp. Agropecu. Norte* **54**, Part 2, 68–122.

Bennema, J. (1975). Soil resources of the tropics with special reference to the well drained soils of the Brazilian forest region. *Ecophysiol. Trop. Crops* **1**, 1–47.

Blue, W. G. (1974). Management of Ultisols and Oxisols. *Proc. Soil Crop Sci. Soc. Fla.* **33**, 126–132.

Brooke, D. L. (1970). Orange production costs and return in Florida season 1968–69 with comparisons. *Univ. Fla., Coop. Ext. Serv., Econ. Ser.* **70-3**, 1–19.

Burke, J. H. (1967). The commercial citrus regions of the world. *In* "The Citrus Industry" (W. Reuther, L. D. Batchelor, and H. J. Webber, eds.), Vol. 1, pp. 40–189. Div. Agric. Sci., University of California, Berkeley.

Cooper, W. C., Rasmussen, G. K., and Waldon, E. S. (1969). Ethylene evolution stimulated by chilling in citrus and persea sp. *Plant Physiol.* **44**, 1194–1196.

Dos Santos, W. H. P. (1972). O estado atual dos connecimentos sôbra os solos da Amazônia Brazileira. *Bol. Tec. Inst. Pesquis. Exp. Agrupecu. Norte* **54**, Part 1, 9–67.

Elfving, D. C., Kaufmann, M. R., and Hall, A. E. (1972). Interpreting leaf water potential measurements with a model of the soil-plant-atmosphere continuum. *Physiol. Plant.* **27**, 161–168.

Embleton, T. W., Jones, W. W., Labanauskas, C. K., and Reuther, W. (1973a). Leaf analysis as a diagnostic tool and guide to fertilization. *In* "The Citrus Industry" (W. Reuther, L. D. Batchelor, and H. J. Webber, eds.), Vol. 3, pp. 183–210. Div. Agric. Sci., University of California, Berkeley.

Embleton, T. W., Reitz, H. J., and Jones, W. W. (1973b). Citrus fertilization. *In* "The Citrus Industry" (W. Reuther, L. D. Batchelor, and H. J. Webber, eds.), Vol. 3, pp. 122–182. Div. Agric. Sci., University of California, Berkeley.

Erickson, L. C. (1968). The general physiology of citrus. *In* "The Citrus Industry" (W. Reuther, L. D. Batchelor, and H. J. Webber, eds.), Vol. 2, pp. 86–126. Div. Agric. Sci., University of California, Berkeley.

Falesi, I. C. (1972). Solos da rodovia transamazônica. *Bol. Tec. Inst. Pesquis, Exp. Agrupecu. Norte* **55**, 196 pp.

Food and Agriculture Organization. (1970). "Report on 1960 Census of Agriculture." FAO, Rome.

Food and Agriculture Organization, Production Yearbook, 1974. FAO, Rome.

Frost, H. B., and Soost, R. K. (1968). Seed reproduction: Development of gametes and embryos. *In* "The Citrus Industry" (W. Reuther, L. D. Batchelor, and H. J. Webber, eds.), Vol. 2, pp. 290–234. Div. Agric. Sci., University of California, Berkeley.

Fundação Instituto Brazileiro de Geografia e Estatistica, Ministerio do Planejamento e Coordenação Geral, Anuário Estatistico do Brazil, 1975, Rio de Janeiro.

Grierson, W., and Newhall, W. F. (1960). Degreening of Florida citrus fruits. *Fla., Agric. Exp. Stn., Bull.* **620.**

Hodgson, R. W. (1967). Horticultural varieties of citrus. *In* "The Citrus Industry" (W.

Reuther, L. D. Batchelor, and H. J. Webber, eds.), Vol. 3, pp. 431–591. Div. Agric. Sic., University of California, Berkeley.

Menge, J. A., Gerdeman, J. W., and Lambright H. (1975). Mycorrhizal fungi and *Citrus*. *Citrus Ind. Mag.* **56,** 16–18.

Newman, J. E. (1968). Estimating water needs of citrus orchards. *Citrus Ind. Mag.* **49,** 19–24.

Passos, O. S., and Sobrinho, A. P. DaC. (1970). "Cultura dos citros no estado da Bahia." IPEAL/SUDENE, Cruz das Almas, Bahia, Brazil.

Passos, O. S., Rodrigues, E. M., and Caldas, R. C. (1971). "Citrus Fruit Maturity in the State of Bahia, Brazil." Instituto de Pesquisas Agropecuárias do Leste, Cruz das Almas, Bahia, Brazil.

Platt, R. G. (1973). Planning and planting the orchard. *In* "The Citrus Industry" (W. Reuther, L. D. Batchelor, and H. J. Webber, eds.), Vol. 3, pp. 48–81. Div. Agric. Sci., University of California, Berkeley.

Reuther, W. (1973). Climate and citrus behavior. *In* "The Citrus Industry" (W. Reuther, L. D. Batchelor, and H. J. Webber, eds.), Vol. 3, pp. 280–377. Div. Agric. Sci., University of California, Berkeley.

Reuther, W., and Rios-Castaño, D. (1969). Comparison of growth, maturation and composition of citrus fruits in subtropical California and tropical Colombia. *Proc. Int. Citrus Symp., 1st, 1968* Vol. 1, pp. 277–300.

Richards, L. A., ed. (1954). "Diagnosis and Improvement of Saline and Alkali soils," Agric. Handg. No. 80. U. S. Department of Agriculture, Washington, D.C.

Rios-Castaño, D., Torres-M., R., and Camacho-B., S. (1968). Ensayo de patrones para las naranjas 'Indian River' y 'Valle Washington' en Colombia. *Agric. Trop.* **24,** 491–496.

Savage, Z. (1967). 35 years of orange costs and return in Florida. *Univ. Fla., Agric. Ext. Serv. Econ. Ser.* **67-6,** 1–23.

Shuler, P. E., Todd, J. W., and Scarborough, E. F. (1970). "Florida Agricultural Statistics. Citrus Summary 1970." Florida Department of Agriculture, Tallahassee.

Stebbins, G. L. (1969). The effect of asexual reproduction on higher plant genera, with special reference to citrus. *Proc. Int. Citrus Symp., 1st, 1968* Vol. 1, pp. 455–458.

Torres-M., R., and Rios-Castaño., D. (1968). Crecimiento y maduratión de frutos citricos en el Valle de Cauca. *Agric. Trop.* **24,** 502–517.

Trinidad and Tobago. (1959). "Administrative Report of the Agriculture Department." Government Printing Office, Trinidad.

Van Bavel, C. H. M., Newman, J. E., and Hilgeman, R. H. (1967). Climate and estimated water use by an orange orchard. *Agric. Meteorol.* **4,** 27–37.

Webber, H. J. (1943). Plant characteristics and climatology. *In* "The Citrus Industry" (W. Reuther, H. J. Webber and L. D. Batchelor, eds.), Vol. 1. Univ. of Calfornia Press, Berkeley.

Yokoyama, H., Hsu, W., Poling, S. M., Hayman, E., and DeBendict, C. (1976). "Enhancement of Color and Provitamin A Quality of Citrus Fruit," Citrus Res. Conf., 1976, Mimeo Summ. U. S. Dept. of Agriculture, Fruit and Vegetable Chemistry Laboratory, Pasadena, California.

CHAPTER

16

Banana

EGBERT A. TAI

I. INTRODUCTION

The statement that "bananas are the most important of the tropical fruits" (Purseglove, 1972) cannot be successfully challenged. It is generally known that in world production of fruits bananas rank after grapes.

The cultivated banana is believed to have originated in Southeast Asia–India, Indochina, Malaysia, and the Philippines–and probably became domesticated before historical times. The naturally occurring propagation by vegetative means made possible the perpetuation of seedless types, and selections from these were almost certainly the progenitors of present-day cultivars. Simmonds (1959) cites Reynolds (1951) as claiming that the earliest references to banana cultivation occur in Indian writings of 500–600 BC and long after, circa AD 200, mention of the fruit first appeared in Chinese literature.

The likely course of spread appears to be through Malaysia to the western Pacific area at a very early stage; subsequent distribution in the Pacific islands took place as the result of movement by the Polynesians. There is speculation on how bananas reached Africa; although there is no record of the crop in East Africa prior to 1300 AD (Greenway, 1944) the opinion is held (Baker and Simmonds, 1951; Thomas, 1955) that the introduction was effected long before, possibly by the Indonesians who settled in Madagascar prior to the days of the Arab traders. [The Arab traders are credited by Reynolds (1951) with taking bananas to the coast of East Africa during the first millenium AD.] From East Africa the spread across the continent to the west coast was simple, and the crop was well established there by the time Europeans colonized the area.

The Portuguese took the banana to the Canary Islands at the beginning of the fifteenth century (Reynolds, 1927) and later to the New World. In 1516 Friar Tomas de Berlanga brought bananas to Santo Domingo, and since then many introductions have been made in that region. The possibility exists that bananas were known in the Americas in pre-Columbian times, but sufficient evidence has not been found to confirm this.

Export trading in bananas started with small shipments from the Caribbean to the United States of America early in the ninteenth century; these consisted of bunches, probably Gros Michel, transported in unrefrigerated space on schooners. Later came the steamships, extension of the trade to Europe, and the use of the first refrigerated banana boats to take fruit from Jamaica to Bristol in 1901 (Simmonds, 1959). Today the greater part of world banana trade is still based in the Americas, but large amounts of the fruit are also produced and sold in foreign markets by Asian and African countries, Australia, and islands of the Pacific. In addition, it is to be noted that Cavendish types have largely replaced the Gros Michel in commerce.

II. WORLD PRODUCTION AND TRADE

Table I giving approximate areas, production, and export of banana from some countries in which the crop is of major importance was compiled from figures supplied in the Commonwealth Secretariat publication, FRUIT 1972.

TABLE I
Approximate Areas, Production, and Export of Banana

Country	Production ha[a]	Production tons[a]	Export tons[a]
Americas			
Brazil	273	9266	204
Colombia	728	6048	257
Costa Rica	25	1018	867
Dominican Republic	24	435	400
Ecuador	179	4330	1364
French Antilles	19	311	229
Honduras	98	3571	828
Jamaica	38	142	136
Mexico	77	1000	n.a.
Nicaragua	11	81	n.a.
Panama	37	929	600
Venezuela	115	1249	15
Windward Is.	30	192	134
Europe			
Canary Is.	11	416	391
Africa			
Burundi	127	1278	n.a.
Cameroun	52	1188	50
Egypt	3	68	–
Equatorial Africa	61	711	n.a.
Ethiopia	17	n.a.	–
Ivory Coast	187	1252	142
Kenya	121	n.a.	–
Madagascar	10	132	12
Somalia	11	110	100
South Africa	19	n.a.	–
Tanzania	189	n.a.	–
Uganda	464	n.a.	–
Zaire	263	1776	–
Asia			
India	229	3155	8
Malaysia	19	322	21
Pakistan	49	797	–
Philippines	217	755	n.a.
South Vietnam	18	164	n.a.
Taiwan	47	458	n.a.
Thailand	198	1334	n.a.
Australia	9	131	–

[a]Measured in thousands.

Although the information should not be regarded as truly up-to-date and precise, it provides an idea of the coverage of the banana trade. It is noteworthy that several of the largest producers are not exporters; domestic consumption may even require importation. Figure 1 indicates the main regions of production and the flow of trade.

III. BOTANICAL CLASSIFICATION

The practice of assigning species status to groups of the edible bananas and plantations, e.g., *Musa sapientum, M. paradisiaca, M. Cavendishii,* is no longer favored and has been superseded by a genome nomenclature for cultivars in recognition of their derivation from two wild species: *Musa acuminata* and *Musa balbisiana.* Simmonds and Shepherd (1955) devised a method of indicating the relative contributions of these two species to the constitution of any cultivar and applying the derived information to arriving at a suitable designation. Fifteen characteristics were chosen for pointing out differences between *M. acuminata* and *M. balbisiana* and showing how their contributions can be identified; ploidy was then taken into consideration. By this procedure naturally occurring edible bananas fall into six groups: two diploid, AA and AB; three triploid, AAA, AAB, ABB; and one tetrapoloid, ABBB.

The best known bananas of commerce belong to the AAA group which includes the Gros Michel and Cavendish; other groups of some importance are AAB, most prominent of which are Mysore, Silk, and the Plantains, and AAB which includes Bluggoe and Pisang Awak. Each of the cultivars mentioned has numerous other names and it is recognized that "it will be many years before an authoritative list of names and identities can be compiled" (Simmonds, 1959). To add to the confusion, bananas are highly subject to somatic mutation which may introduce different degrees of variation in nearly any plant characteristic, resulting in the need for additional names.

Of some interest are representatives of other naturally occurring groups: Sucrier AA, Lady's Finger AB, and Tiparot ABBB. In the extensive breeding work initiated in 1922 at the Imperial College of Tropical Agriculture and carried out in Jamaica and Honduras since then, AAAA bananas were produced of which IC2 and Bodles Altafort (Osborne, 1962) are examples, and recently attention is being directed to production of AAA triploids.

IV. GEOGRAPHICAL DISTRIBUTION

Bananas are essentially plants of the humid tropics and do not thrive in areas where the temperature falls below 15°C or where annual rainfall is less than

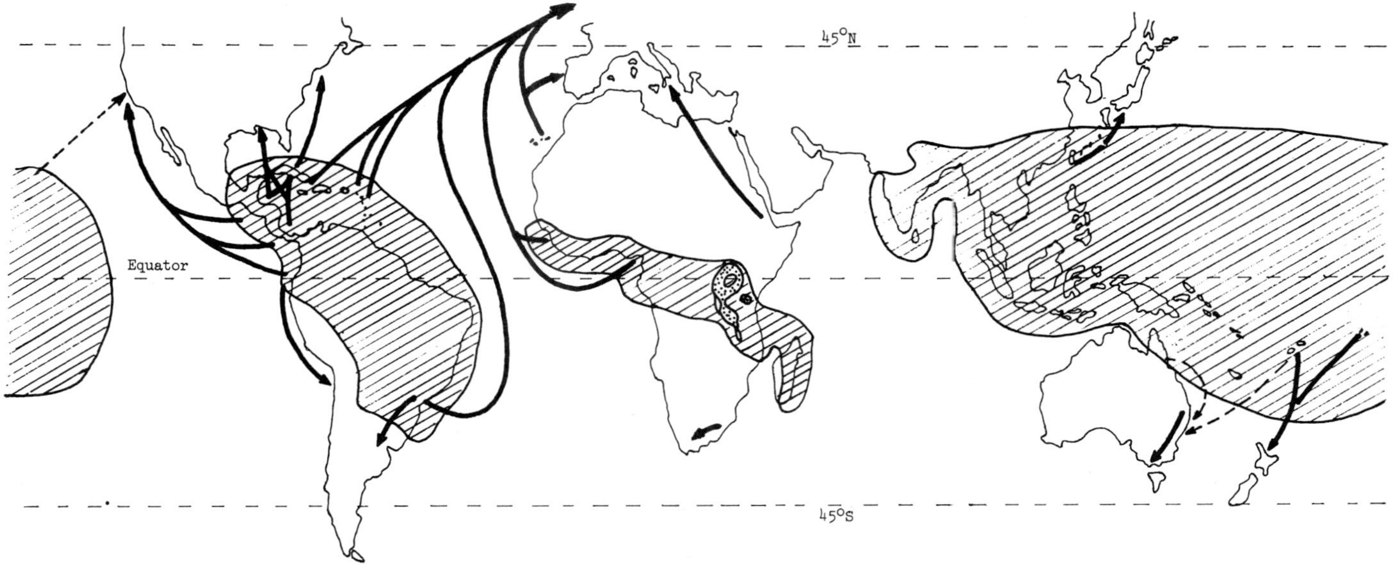

Fig. 1 Banana trade routes (from Simmonds, 1959). Diagonally ruled areas indicate areas of production.

2000 mm. Suitable conditions for banana occur within the region lying between 30°N and 30°S of the equator and very little banana growing is done outside of this region; Israel in the Northern Hemisphere and New South Wales, Australia in the Southern Hemisphere are the limits of commercial banana production. Figure 2 shows the world distribution of the banana growing areas.

The *balbisiana* genome is associated with greater drought hardiness and resistance to disease than occur in *M. acuminata.* It is not surprising, therefore, that hybrids of AB, AAB, and ABB constitution are better adapted to monsoon areas with marked dry seasons; these bananas show wide variability and are grown chiefly for local consumption as in India and Uganda. The AA and AAA bananas are produced mainly in areas where rainfall is equally distributed throughout the year or where water for irrigation is readily available; the greater uniformity of the AAA types makes them particularly suitable for large scale production for export as in Central America, the Caribbean islands, and northern South America.

The Gros Michel cultivar was, up to about fifteen years ago, the most important of all the bananas. Its susceptibility to the Panama disease wilt has been responsible for reducing the acreage to almost negligible proportions in the majority of banana-growing countries today; only where new land can be cleared for planting is it cultivated as in parts of Ecuador, Colombia, and Panama. The hope is sustained that the highly desirable quality of the Gros Michel fruit will be preserved in the commercial hybrids expected to result from the breeding programs now in progress by using it or its dwarf mutant, Highgate or Cocos, as the female parent.

Nearly every place where banana crop is commercially important, clones of the Cavendish subgroup immune or resistant to Panana disease are grown. Selection among these has yielded several cultivars of value and promise. The Dwarf Cavendish, under a multiplicity of names, is the most widespread, especially in the cooler regions of the "banana zone"; it is the basis of trade in Brazil, Canary Islands, Israel, Somalia, and South Africa, and of significance in Australia and the French Antilles. The Giant Cavendish is less common but is gaining ground in several areas; it achieved greatest prominence in the French Antilles and Australia, but its cultivation on an increasing scale has also been noted in the Caribbean and Central American areas.

The Robusta is grown in the Caribbean islands for export, and is believed to be on the increase in several other banana areas such as Australia, Brazil, and islands of the Pacific; it is frequently confused with the cultivar named "Lacatan" in Jamaica but is claimed by Simmonds (1959) to be "Pisang masak hijau." This cultivar is of significant importance only in Jamaica where it replaced Gros Michel when revival of banana production was given attention after World War II.

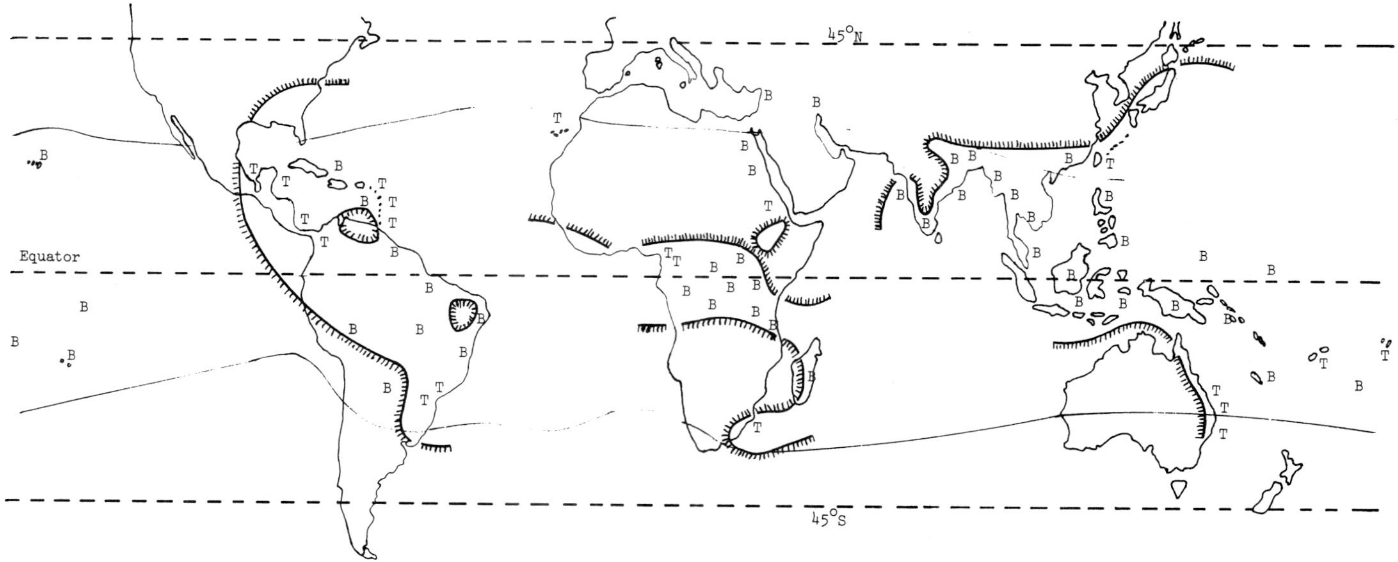

Fig. 2 Distribution of banana crops (from Simmonds, 1959). B, bananas grown for local consumption; T, bananas grown for export. ——, 15°C isotherm; wetter/drier (hatched line), 2000-mm isohyet.

A clone that came to fore about 1953 is the Valery, selected in Honduras by the United Fruit Company. It spread in Central America and Jamaica initially, but is now grown also in other West Indian islands and in northern South America.

V. ENVIRONMENTAL REQUIREMENTS

Ideal conditions for banana growth are a mean atmospheric temperature of 27°C, monthly rainfall of 100 mm, exposure to full sunlight, protection from wind, deep soil of high fertility, free drainage, and a pH of 6.5–7.0. In very few, if any, instances can all these conditions be satisfied, and commercial production is practiced where these conditions vary greatly.

A. Temperature

The growth of banana plants is greatly reduced at temperatures lower than 16°C, and at that temperature leaf emergence comes to a complete stop (Champion, 1963) in at least one commercial cultivar, the Poyo, which is of major importance. If low temperature is prolonged, the expanded leaves may develop symptoms similar to those due to water shortage; they lose turgor, become yellow, and die prematurely. The dwarf Cavendish is capable of withstanding lower temperatures than other commercial cultivars but is killed by the slightest frost; this fact limits the extent to which this crop may be grown in subtropical areas such as Israel, South Africa, and Australia.

Reduction in vegetative growth due to cold is accompanied by an extension of the period from planting to appearance of the first inflorescences. In areas of best production, the peak of the "plantilla" harvest may be reached in a year, whereas a much longer time may be required elsewhere; Martin-Prével (1974) reports an average of 18 months, in South Africa, for instance. Even in the tropics, variation in prevailing temperatures resulting from differences in altitude causes significant differences in the planting–shooting period. In Jamaica, Lumsden (1960) found that the period increased by approximately one month for every 100 m rise in elevation above sea level, and in Martinique it was observed at the Institut des Fruits et Agrûmes Coloniaux (Montagut, 1961) that a difference of 70 m in altitude was reflected in the lengthening of the planting–shooting period by 46 days.

A condition known as "choke" occurs frequently with dwarf banana cultivars as one effect of low temperatures: elongation of the internodes of the true stem and also of the leaf sheaths making up the pseudostem is reduced so that vertical separation of the laminae may be absent at the "throat" and the crown takes on the appearance of a rosette; the bunch is not properly shot and the individual hands may be malformed. In the Canary Islands (Holmes, 1930) and Guinea (Champion, 1951) "choke" is well known in the Dwarf Cavendish.

Continuing cold weather affects the rate of fruit production, the time taken to reach maturity, and ultimate yield; in addition a "chill" effect on fruit is often observed in the field as a result of even short cold spells. Banana latex coagulates in the pericarp of the fruit at 12°C, imparting a dull brown pigmentation (Slocum, 1933; Puvis, 1945) in subepidermal streaks which detract considerably from the appearance of the fruit when it ripens.

Other effects of chilling may not be immediately apparent, but are nonetheless serious; these include uneven softening of the pulp in ripening and predisposition toward development of storage rots and other blemishes. In banana-consuming countries the variation between "summer" fruit and "winter" fruit is fully recognized.

B. Light

There is no evidence of photoperiodic response in bananas but the length of the vegetative cycle is extended at low light intensity (Murray, 1961), although new leaves will emerge and unfurl in complete darkness (Skutch, 1931). In the shade, the pseudostems grow to a greater height than in open sunlight and, when interplanted with tall-growing coconut palms in Jamaica Lacatan, bananas have been reported to require up to two months longer to mature fruit than when planted in the open.

The photosynthetic activity of banana leaves was found by Brun (1961) to increase rapidly at illuminations between 2000 and 10,000 lux and more slowly between 10,000 and 30,000 lux when measurements were conducted on the abaxial surfaces where stomata are more abundant. It is doubtful whether full use is actually made of the high light intensities present in Egypt and Israel where up to 77,000 lux have been recorded (Ticho, 1960). In fact, "sunburn" of fruit can result from exposure to high light intensity, especially when this is accompanied by high temperatures, as is evidenced by fruit harvested from plants on which the foliar canopy has been reduced by sigatoka leaf disease, low temperature, or other causes.

C. Air Movement

Because of its large saillike leaves the banana plant is highly susceptible to wind damage; the lamina is split into strips by even gentle air movement and it is rare to encounter entire banana leaves. Mild tearing of the laminae normally is not serious, but severe shredding caused by winds of up to 20–30 km an hour prevailing during the trade wind season in the Caribbean regions can, together with accompanying distortion of the crown, reduce yield of marketable fruit.

The use of effective windbreaks can minimize losses, and plants of the vigorous "wild" banana, *Musa balbisiana,* are satisfactory for this purpose. Windbreaks also mitigate the desiccating action of steady air movement over the leaves and help to maintain the turgor desirable for assimilatory processes.

Winds of a velocity greater than 40 km per hour frequently cause breakage of the pseudostems and sometimes uproot single shoots or even complete mats, depending on the particular cultivar, the age, and state of health of the plants; young plants tend to break, whereas older ones, especially those with hanging bunches, become uprooted, and any condition leading to a poorly developed root system (attack by burrowing nematode is an example) will also be uprooted.

Breakage of leaves due to high winds is not very common among the commercial triploid banana cultivars; the AAAA tetraploids produced in the existing breading programs are, however, susceptible to leaf petiole breakage as the result of even light winds. Simmonds (1962) attributed this to the fact that "tetraploids in relation to triploids have a high regression of leaf weight on area and a low regression of strength on weight."

Windstorms, which occur frequently in the tropics, are destructive to banana populations, and the "hurricane season" is usually dreaded in banana producing areas. Blow-down constitutes the greatest single source of loss to banana plantations; for instance, Simmonds (1959) estimated the loss in Cameroun at a mean of over 20% per annum based on reported (Borel and Pélegrin, 1951) blowdowns of 55, 69, and 53% of samples of 1000 plants in the three years 1949 to 1951. There is no answer to the problem of windstorm damage to banana as no windbreaks are effective against cyclones or hurricanes.

D. Rainfall and Humidity

It is generally accepted that the water needs of bananas are ideally supplied by 100 mm of evenly distributed rainfall each month and that serious shortage is experienced when there is less than 50 mm in any month. Nowhere, however, do perfect rainfall conditions exist, and irrigation is required at certain times of year to obtain optimum growth and production in most areas. This is by no means surprising when the very large transpiring surface presented by each banana plant is taken into consideration. According to Champion (1963), at the usual density of 900 mats/ha, Gros Michel bananas actively transpire from a surface of 40,500 m^2 and Dwarf Cavendish at 2500/ha transpire from 37,500 m^2. If each plant produces a new leaf each week there is no difficulty in appreciating that the water needs of a banana area must be constantly high. The studies of Morello (1954) in Brazil and Shmueli (1953) in Israel indicate that water loss by banana leaves in full sun amounts to 40–50 $mg/dm^2/min$ when the stomata are wide open.

When atmospheric humidity is high and the rate of transpiration is matched by the rate of absorption of water, the two sides of each fully expanded leaf lamina lie in the same plane and the stomata remain open; in dry air or under even slight water stress the half-laminae move through 90° at their junction with the midrib, as if on hinges provided by the pulvinar bands, and lie in parallel planes reducing the exposure of the lower surfaces while at the same time the stomata close. This action tends to aid in conservation of water in the tissues and in preventing or delaying the consequences of severe deficit; it occurs on each day of sunny weather in the life of the banana plant.

Extended periods of heavy rainfall may cause saturation of the soil to considerable depths and so give rise to drainage problems as banana plants are intolerant of "wet feet"; waterlogged conditions for 24 hours result in dying of roots and yellowing of leaves. Rain also creates high atmospheric humidity which may lead to more intense and greater spread of destructive leaf diseases such as sigatoka.

Acute drought produces wilting and collapse of leaves followed by desiccation and breaking of the pseudostem above ground; the roots also dry out and die, but usually the corm retains moisture and has the capacity to regenerate shoots long after the original pseudostem has disappeared. Dry weather which is not sufficiently severe to cause death of the shoots induces "choke" in cultivars of the Cavendish subgroup; in some instances the plants take on the appearance of *Ravenala* (Portères, 1950). "Choke" may occur in the purely vegetative development, involving only the crown of leaves, and then is known as "engorgement foliare" by French workers, or it may also involve the emerging inflorescence, when it is termed "engorgement floral." In the tall-growing cultivars of banana the inflorescence may fail to reach the throat and burst through the side of the pseudostem on the way up. Freeman (1938a,b) mentions that, apart from being necessary for proper elongation of the inflorescence-bearing organ, water in the form of a thin film is required as a lubricant to facilitate its passage through the closely appressed leaf sheaths of the pseudostem.

Dry weather always adversely affects fruit development, and even when the "shooting" occurs under favorable conditions faulty filling out as a result of subsequent periods of water stress is sometimes evident in harvested bunches; all of the fruit or part, consisting at times of single hands, may show the effects. In Fig. 3 some effects of dry weather on banana development are illustrated.

E. Soil

Bananas are grown commercially on a wide range of soils on varied topography; good aeration is common to most productive sites. The natural internal drainage is often supplemented with the removal of excess water by artificial devices to make available as great a volume of soil as possible for occupation by the banana roots. Measures to ensure adequate fertility and moisture are also necessary.

Fig. 3 Effect of dry period on development of banana bunch. A, Poorly filled fingers on distal hands; B, poorly filled fingers on otherwise satisfactorily developed hand.

Root development is best on soils of crumby structure as are found in the alluvia of Central America. These vary in texture from heavy clays to stony sands but all possess abundant pore spaces. In the compact clays and fine impermeable sands of Trinidad and eastern Venezuela banana roots do not survive for long; overall growth is poor and conditions favor attack by moko disease

Soil reaction does not in itself appear to have any significant effect on banana growth, and vigorous, healthy, productive plants occur on acid, neutral, or basic soils of pH 4.5 to 8.0 in Jamaica. It is known, however, that at low pH the effects of sigatoka disease are more severe (Leach, 1943). Outside the pH 6.0 to 7.0 range the absorption of some mineral nutrients is affected: on acid soils banana plants frequently suffer from phosphorus deficiency while on soils of basic reaction potash nutrition is often limiting.

The mineral composition of soils for the growing of bananas is not regarded as of equal importance to the physical characteristics mentioned. Adequate nutrition for satisfactory growth can usually be ensured by addition of the appropriate fertilizers. The use of the major elements was determined by Baillon *et al.* (1933) for Dwarf Cavendish banana grown in the Canary Islands and by Martin-Prével (1962) for the same cultivar in Guinea; there is evidence that the greatest demand is for potash so that the potash status of the soil merits attention if bananas are to be cultivated. Accumulation of salt in the soil above a concentration of 500 ppm is toxic to banana (Jacks, 1936); such accumulations sometimes occur in irrigated lands or in areas reclaimed from saline swamps.

VI. PESTS AND DISEASES

Bananas are subject to attack by many natural enemies which cause damage to aerial and underground organs. Insects, nematodes, fungi, bacteria, and viruses restrict development of the crop in different areas and at times are sufficiently serious to call for extensive coordinated campaigns for effective control. A very useful annotated list of recorded pests is given by Simmonds (1959) and the diseases are extensively treated by Wardlaw (1961).

A. Insects

The most widespread banana pest is the weevil borer, *Cosmopolites sordidus*, believed to have originated in Southeast Asia, and now almost universally associated with banana growing. Both larvae and adults feed on tissues of the corm and weaken the plants mechanically by their tunneling as well as physiologically by reducing the store of assimilated material. Detailed studies of the life history of the insect have been made in several areas (Froggatt, 1925; Leonard, 1931; Cuillé, 1950) and recommendations given for control by trapping, use of insecticides, or predators and parasites (Vilardebo, 1953; Leach, 1959). Good weed control and crop sanitation have proved effective in keeping down losses due to this pest. Other beetle borers attack banana corms–*Phyllophaga* and *Metamasius* species in the Caribbean area, *Oryctes* sp. in West Africa and the Pacific islands–but do not attain the level of importance of *Cosmopolites.*

Insects which affect the leaves, petioles, and pseudostem include locusts, grasshoppers, caterpillars of several lepidopterous species, scales, and aphids. These are not of general economic importance although in individual instances severe losses may be incurred. The aphis, *Pentalonia nigrinervosa,* has been identified as the vector of "bunchy top," the most serious virus disease of banana (Magee, 1927; Hardy, 1941).

Disfigurement of banana fruit is caused by several insects; although the damage seldom extends to the pulp the value may be reduced to practically nothing. Banana thrips, *Scirtothrips signipennis,* is serious in Queensland, Australia, where it is responsible for losses due to "banana rust" (Caldwell, 1938; Smith, 1947). This and other species of thrips are known in the Caribbean and the Americas—*Paleucothrips musae* and *Systenothrips latens* in Ecuador (Hood, 1956), *Frankliniella* sp. in St. Lucia and St. Vincent (Fennah, 1947)—but do not appear to have a serious impact on banana production. They occur mainly in conditions favorable to the development of "choke." Scarring of fruit is caused by the scab moth, *Nacoleia octasema,* a beetle, *Colaspis hypochlora,* and slugs and snails in wet locations or during long spells of rainy weather.

B. Nematodes

A strain of the burrowing nematode *Radopholus similis* is a destructive pest of banana; it causes the blackhead toppling disease (Leach, 1958) and has been the subject of study by Loos and Loos (1960), Holdeman (1960), Vilardebo (1960), Edmunds (1968), and others. Because of its specificity to banana, effective control is feasible through the use of clean seed (Loos, 1961), and satisfactory procedures have been evolved for this purpose. Other nematode species attack banana; *Meloidogyne* spp. produces galls on the roots while *Helicotylenchus multicinctus* kills the rootlets, adversely affecting absorption of water and nutrients, especially on light-textured soils. Treatment with nematocides and elimination of alternate hosts from the vicinity of banana plants have proved effective methods of control when carefully applied.

C. Fungal Diseases

Banana wilt caused by *Fusarium oxysporum* f. *cubense* has had more effect on banana production than any other single factor. It is basically responsible for the trend away from the susceptible Gros Michel cultivar to others of the resistant Cavendish subgroup and also for the initiation of banana breeding programs of considerable magnitude in the West Indies and Central America. There is no known effective method of controlling this disease which has spread to all areas of commercial banana production despite elaborate precautions in some instances. The relative importance of Panama disease is now less than it was a decade ago although it is still desirable to take measures to exclude it from "new" banana plantings such as occur when virgin forest is cleared for this purpose. Literature on Panama disease is extensive as is information on the range of relevant research; very useful descriptions are provided by Wardlaw (1961) and Stover (1962).

Sigatoka (*Mycosphaerella musae*) and black leaf streak (*M. fijiensis*) are the most virulent leaf diseases. The former occurs throughout banana-growing areas but the latter is at present unknown in the Americas. Economic control measures for sigatoka disease by aerial spraying have evolved from considerable research (Leach, 1946; Calpouzos, 1955; Guyot, 1954) and are widely adopted, but these are not fully effective against black leaf streak which has wiped out much of the banana formerly grown in the Fiji Islands. Acid soils and humid atmosphere with inadequate air drainage favor the development of these two diseases.

Several other leafspots caused by fungi sometimes appear on banana leaves; they are seldom of significance as only plants weakened by other causes are usually attacked. Among the best known are cordana leafspot (*Cordana musae*), black spot (*Deightoniella torulosa*), and leaf speckle (*Chloridium musae*), which are readily controlled by copper sprays.

On the fruit the tips of the fingers are subject to cigar-end rot attributed to *Verticillium theobromae* on Dwarf Cavendish in the West Indies and *Trachysphaera fructigena* on Gros Michel in Cameroun (Brun and Champion, 1955). Infections of the skin which results in fruit rots during storage and ripening after harvest sometimes occur in the field. The fungi *Piricularia grisea* and *Gloeosporium musarum* are most frequently responsible.

D. Bacterial Diseases

A wilt with symptoms similar to those of Panama disease is caused by certain strains of *Pseudomonas solanacearum*. The organism may gain access through the roots especially on ill-drained land or any injured area of the aerial part of the banana plant. Losses from the disease are of considerable proportion in Central America, but application of satisfactory preventive measures (Buddenhagen and Sequeira, 1958; Hildreth, 1961) can obviate these. A rhizome rot is caused by *Erwinia carotovora* (Loos, 1962) in wet locations.

E. Viral Diseases

Bunchy, top occurs widely in Asia, Australia, and the Pacific, but has not been recorded in the Americas. It is spread by the use of infected planting material and is believed to have been first observed in Fiji. The aphid, *Pentalonia nigrinervosa,* is the only known vector of the disease (Magee, 1927, 1940b). Other virus infections of banana are known; of very wide distribution is infectious chlorosis or heart rot (Reinking, 1950; Wardlaw, 1948) caused by a strain of cucumber mosaic virus and transmitted mainly by *Macrosiphon solanifolii* or *Asphis gossypii*; and Rayadilla, described by Cardenosa (1952) in the Cauca valley of Columbia and Roxana (Wardlaw, 1961) in Costa Rica, is also worthy of mention.

VII. MANAGEMENT PRACTICES

It is unlikely that perfect conditions for banana production exist in any location. However, knowledge of the requirements, preferences, and tolerances of the banana plant can be applied to devise ways and means of modifying existing conditions to make them as ideal as possible.

A. Establishment

Selection of the site largely determines the management practices to follow. Effective procedures for ensuring continued good growth of the plants and satisfactory yields will depend on the nature of the microclimate, soil type, exposure, and topography. Deep well-drained soil of crumby structure, availability of reasonably abundant water from rainfall or other sources, existence of good air drainage, and shelter from strong wind are important considerations.

Preparation of the land for planting involves preliminary tillage to accommodate the "seed" which is best obtained from a source free of pests or disease. The "seed" may be set in individual holes or furrows before being covered with soil; in either event the need for including soil conservation measures, such as basing all operations on contours when the land slopes, must not be overlooked. Greater freedom, and more work, exists where new land is being cleared in forests for cultivating bananas.

B. Regulation of Soil Moisture

The water content of banana tissue is high, and since it must come from the soil, provision of a regular supply in adequate amounts is necessary. Excess at any time is undesirable and should be drained away; shortages at other times occur even in wet areas and then, unless water is added by one of the approved methods of irrigation, the plants may suffer and be unproductive. In some instances it is possible to store water drained away in times of surplus for use in irrigation in times of drought.

C. Maintenance of Soil Fertility

The nutrient needs of bananas cannot be supplied indefinitely by any soil; use of fertilizers and manures becomes necessary to replace the mineral elements removed at harvest. There should never be a need to replenish basic reserves once initial fertility is built up. In this regard, determinations of the nutrient status by chemical analysis of soil and the banana plant leaves should be helpful.

In addition to the major nutrients, N, P, and K, other elements are essential for good growth of bananas, and deficiency symptoms appear when they are missing. Descriptions of symptoms and measures for correcting the deficiencies have been given by Cooil and Shoji (1953), Moity (1954), Murray (1959), and Jordine (1962).

D. Crop Protection

Reduction of the need for banana plants to compete for essential water and nutrients is achieved by weed suppression. Use of mechanical means for this purpose has largely given way to the application of herbicides. Protection against natural enemies such as pests and disease organisms can be achieved through field sanitation and other culture measures or directly with pesticides.

It is sometimes advisable to provide protection from natural phenomena such as spells of unfavorable weather. Windbreaks offering shelter from the drying effects of even gently prevailing winds are of great benefit to banana plants, and under special circumstances artificial heating may be employed against cold, as is done in Queensland (Trim, 1948), by burning sawdust in the fields.

REFERENCES

Baillon, A. F., Holmes, E., and Lewis, A. H. (1933). The composition of, and nutrient uptake by, the banana plant with special reference to the Canaries. *Trop. Agric. (Trinidad)* **10,** 139–144.

Baker, R. E. D., and Simmonds, N. W. (1951). Bananas in East Africa. I. *Emp. J. Exp. Agric.* **19,** 284–290.

Borel, E., and Pélegrin, P. (1951). La culture du bananier au Cameroun. *Fruits Outre Mer* **6,** 421–427.

Brun, J., and Champion, J. (1955). La culture bananière au Cameroun brittanique. *Fruits Outre Mer* **10,** 163–170.

Brun, W. A. (1961). Photosynthesis and transpiration from upper and lower surfaces of intact banana leaves. *Plant Physiol.* **36,** 399–405.

Buddenhagen, I. W., and Sequeira, L. (1958). Desinfectants and tool disinfection for banana plantations of Costa Rica, and their relationship to bacterial wilt of bananas. *Phytopathology* **50,** 660–664.

Buddenhagen, I. W., and Sequeira, L. (1958). Desinfectants and tool desinfection for prevention of spread of bacterial wilt of bananas. *Plant Dis. Rep.* **42,** 1399.
Butler, A. F. (1960). Fertilizer experiments with the Gros Michel banana. *Trop. Agric. (Trinidad)* **37,** 31–50.
Caldwell, N. E. H. (1938). The control of banana rust thrips. *Queensl. Agric. J.* **50,** 114–163, 295–316, 422–449, and 576–584.
Calpouzas, T. (1955). "Studies on the Sigatoka Disease of Bananas and its Fungus Pathogen." Atkins Gardo Res. Lab., Cuba.
Cardenosa, R. (1952). "El genero Musa en Colombia." Publ. Tech. Estac. Agric. Exp., Palmira.
Champion, J. (1951). Aperçus sur la culture du bananier Nain a Guinée française. *Fruits Outre Mer* **6,** 466–472; **7,** 9–12.
Champion, J. (1963). "Le Bananier." Maisonneuve Larose, Paris.
Cooil, B. J., and Shoji, K. (1953). Studies reduce banana chlorosis. *Hawaii Farm Sci.* **1,** 1.
Cuillé, J. (1950). Recherches sur le charançon du bananier *Cosmopolites sordidus* **Germ.** *I.F.A.C., Ser. Tech.* **4.**
Edmunds, J. E. (1968). Nematodes associated with bananas in the Windward Islands. *Trop. Agric. (Trinidad)* **45,** 119–124.
Fennah, R. G. (1947). "Insect Pests of Food Crops in the Lesser Antilles." Grenada & Antigua, B.W.I.
Freeman, H. J. (1938a). Banana growing in Queensland. *Queensl. Agric. J.* **49,** 44–57, 121–125, 238–241, 327–331, 437–442, and 636–643.
Freeman, H. L. (1938b). Banana growing in Queensland. *Queensl. Agric. J.* **50,** 34–3.
Froggatt, J. L. (1925). The banana weevil borer. *Queensl. Agric. J.* **24,** 558–593.
Greenway, P. J. (1944). Origins of some East African food plants. *East Afr. Agric. J.* **10,** 34–39 and 115–119.
Guyot, H. (1954). La lutte contre *Cercospora musae* dans les bananeraies de Guadeloupe. *Fruits Outre Mer* **9,** 293–296.
Hardy, G. H. (1941). Aphididae in Australia. II. Subtribe Pentaloniina. *Proc. R. Soc. Queensl.* **52,** 36–40.
Hasselo, H. N. (1961). Premature yellowing of Lacatan banana. *Trop. Agric. (Trinidad)* **38,** 1.
Hildreth, R. C. (1961). Certified banana seeds. *Trop. Agric. (Trinidad)* **39,** 103–107.
Holdeman, Q. L. (1960). Nematodes that attack bananas. *Res. Newsl. U.F. Co.* **7,** 12–19.
Holmes, E. (1930). The Canary banana; banana cultivation in the Canaries. *Trop. Agric. (Trinidad)* **7,** 320–325.
Hood, J. D. (1956). Two new Thripidae (Thysanoptera) from banana. *Proc. Biol. Soc. Wash.* **69,** 215–218.
Jacks, G. V. (1936). Tropical soils in relation to tropical crops. *Imp. Bur. Soil Sci., Tech. Commun.* **34.**
Jordine, C. G. (1962). Metal deficiencies in bananas. *Nature (London)* **194,** 1160.
Leach, R. (1943). Soil conditions affecting production of perithecia in Banana Leaf Spot Disease. *Nature (London)* **151,** 199.
Leach, R. (1946). "Banana Leaf Spot (*Mycosphaerella musicola*) on the Gros Michel Variety in Jamaica," Bull. No. 46. Govt. Printer, Jamaica.
Leach, R. (1958). Blackhead toppling disease of banana. *Nature (London)* **81,** 204.
Leach, R. (1959). Annual Report, 1957/58. Banana Board Res. Dept., Jamaica.
Leonard, M. D. (1931). A bibliography of the banana root weevil *Cosmopolites sordidus. J. Dep. Agric. P. R.* **15,** 147–176.

Loos, C. A. (1961). Eradication of the burrowing nematode *Radopholus similis* from bananas. *Plant Dis. Rep.* **45**, 457–461.

Loos, C. A. (1962). Factors affecting incidence of rhizome-rot caused by *Erwinia carotovora* of Gros Michel and Lacatan bananas. *Phytopathology* **52**, 2.

Loos, C. A., and Loos, B. (1960). The Black-head Disease of bananas (*Musa acuminata*). *Proc. Helminthol. Soc. Wash.* **27**, 189–193.

Lumsden, V. R. (1960). Some observations on the effect of altitude on time taken to shoot by Lacatan bananas in Jamaica. *Reun. Int. Banan. FAO & CCTA, Abidjan, 1st.*

Magee, C. J. P. (1927). Investigation on the Bunchy Top Disease of bananas. *Aust., Commonw., Counc. Sci. Ind. Res., Bull.* **30**, 64.

Magee, C. J. P. (1940a). Transmission of infectious chlorosis or heart rot of the banana and its relationship to cucumber mosaic. *J. Aust. Inst. Agric. Sci.* **6**, 44–47.

Magee, C. J. P. (1940b). Transmission studies on the Banana Bunchy Top virus. *J. Aust. Inst. Agric. Sci.* **6**, 109–110.

Martin-Prével, P. (1962). Les éléments minéraux dans le bananier et dans son régime. *Fruits Outre Mer* **17**, 123–128.

Martin-Prével, P. (1974). Notes sur la production et de la recherche bananières en Republique sud-africaine. *Fruits Outre Mer* **29**, 637–644.

Martin-Prével, P., and Tisseau, R. (1961). Les éléments minéraux dans le régime de bananes. *I.F.A.C. Doc. Int. R.A.*

Moity, M. (1954). La carence en zinc sur le bananier. *Fruits Outre Mer* **9**, 354.

Montagut, P. (1961). Etude de la croissance du bananier, comparison de deux essais. *I.F.A.C. Doc. Int. R.A.* p. 60.

Morello, J. (1954). Transpiracion y balance de agua de la bananera en las condiciones de la ciudad de Sao Paulo. *Univ. Sao Paulo, Faci Filos., Cienc. Let., Bol., Bot.* **10**, 27–97.

Murray, D. P. (1959). Deficiency symptoms of the major elements in the banana. *Trop. Agric. (Trinidad)* **36**, 100–107.

Murray, D. P. (1961). Shade and fertilizer relations in the banana. *Trop. Agric. (Trinidad)* **38**, 2.

Osborne, R. E. (1962). Bodles Altafort. A new banana for Jamaica. *Banan. Board Res. Dep., Occas. Bull.* **3**.

Portères, R. (1950). La maladie physiologique de l'éventail foliare et de l'engorgement du stipe des bananiers dans l'Ouest africain. *Fruits Outre Mer* **5**, 6.

Purseglove, J. W. (1972). "Tropical Crops," Vol. 2. Longmans, Green, New York.

Puvis, J. (1945). Contribution à l'étude de la pigmentation des bananes en Guinée française. *Fruits Outre Mer* **1**, 4.

Reinking, O. A. (1950). Banana virus studies and observations made at Tawau, Semporna, Mostyn and Sebatik Island, British North Borneo; suggested studies for investigations to clarify the abaca Bunchy Top relationship with the banana virus complex. *Plant Dis. Rep.* **34**, 66–69.

Reynolds, P. K. (1927). "The Banana." Houghton, Boston, Massachusetts.

Reynolds, P. K. (1951). Earliest evidence of banana culture. *J. Am. Orient. Soc., Suppl.* **12**, 28.

Shmueli, E. (1953). Irrigation studies in the Jordan valley. I. Physiological activity in relation to soil moisture. *Bull. Res. Counc. Isr.* **3**, 228–247.

Shmueli, E. (1960). Chilling and frost damage in banana leaves. *Bull. Res. Counc. Isr., Sect. D* **8**, 225.

Simmonds, N. W. (1959). "Bananas." Longmans, Green, New York.

Simmonds, N. W. (1962). "The Evolution of the Bananas." Longmans, Green, New York.

Simmonds, N. W., and Shepherd, K. (1955). The taxonomy and origins of the cultivated bananas. *J. Linn. Soc. Lond., Bot.* **55**, 302–312.

Skutch, A. F. (1931). Some reactions of the banana to pressure, gravity and darkness. *Plant Physiol.* **6**, 73.

Slocum, A. F. (1933). Effect of low temperature upon cell structure. *U. F. Co. Res. Dep. Bull.* **48.**

Smith, W. A. (1947). Banana Rust Thrips control. *Queensl. Agric. J.* **40**, 508–524; **41**, 5–29.

Stover, R. H. (1962). Fusarial wilt (Panama Disease) of bananas and other *Musa* species. *Commonw. Mycol. Inst. Phytopathol., Pap.* **4.**

Thomas, A. S. (1955). The coming of the banana to Uganda. *Uganda J.* **19**, 211.

Ticho, R. J. (1960). The banana industry in Israel. *Rapp., Reun. Int. Banan. FAO & CCTA, Abidjan, 1st.*

Trim, L. G. (1948). Prevention of frost damage in Lady Finger bananas. *Queensl. Agric. J.* **66**, 155–157.

Vilardebo, A. (1953). Etude de l'action de l'H.C.H. sur le bananier. *Fruits Outre Mer* **8**, 490–496.

Villardebo, A. (1960). Premiers essais de lutte contre les nematodes du bananier en Guinée Française. *Proc. Int. Congr. Crop Prot., 4th, 1957* pp. 595–598.

Wardlaw, C. W. (1948). Infectious Chlorosis of bananas in the Cameroons. *Nature (London)* **162**, 894.

Wardlaw, D. W. (1961). "Banana Diseases Including Plantains and Abaca." Longmans, Green, New York.

CHAPTER

17

Cashew

M. C. NAMBIAR

I. INTRODUCTION

The economic yield of any crop is the ultimate result of a series of biosynthetic processes, which in turn are influenced by an array of environmental factors. Any attempt to improve the crop must, therefore, take into consideration these factors and their interactions. The relative impact of various environmental factors on the overall growth and productivity of a plant is dependent to a great extent on its genetic background as well as on its morphological and physiological characteristics.

Substrate constituents (soil), conducting fluid (water), differential reactions to variations in photo- and thermoperiods and the resultant changes in climatic parameters are the important ecophysiological factors that influence crop production. Apart from these ecological factors which determine the physiological limit of crop production, the biotic environment can limit crop distribution even in areas where climatic and edaphic factors are favorable for growth. Social environment also will have a great influence in determining whether a particular crop is actually grown in an area which has potential crop producing ability.

Cashew (*Anacardium occidentale* L.) is found growing in many areas of the tropics either as a naturalized or cultivated plant. In recent years the crop has attained great economic importance in India, East African countries, and Brazil. In addition to the highly nutritious kernels, the cashew nut shell oil forms an important raw material for industrial purposes. The enlarged peduncle of the fruit known as cashew apple is used for various purposes, including alcoholic beverages. Despite this great economic importance of the crop, studies on the role of the ecophysiological factors on growth and production have been very limited. Isolated attempts have been made to study the influence of these factors, mainly on flowering. The effects of ecophysiological factors on vegetative and reproductive growth of the cashew plant, in relation to its geographical distribution, are reviewed briefly.

The French, Portuguese, and Dutch seafarers described cashew from Brazil in the sixteenth century, and the first illustrative description of cashew was given by Thevet in 1558 (Johnson, 1973). The fact that cashew is not mentioned in Spanish chronicles covering the exploration of the Caribbean islands probably shows that it was introduced after the arrival of the Spaniards. Cashew was introduced to the Malabar coast of India in the sixteenth century by the Spaniards, and probably served as a dispersal point for other centers in India and Southeast Asia (da Costa, 1578; van Linschoten, 1598). The popular explanation that cashew was introduced to the coastal areas of India to check soil erosion was not acceptable to Johnston (1973), chiefly because soil erosion was not known until the twentieth century. He believed that the Spaniards who were aware of the uses of cashew in medicines, foods, and beverages visualized the potential importance of this crop to India. From India it was carried eastward to Amboina in Indonesia (Rumphius, 1962); dispersal of the species to Southeast Asia appears to have been carried out by birds, bats, monkeys, and human agents (Burkill, 1935; Johnson, 1973). The earliest record of cashew growing in Africa is that of de Loureiro (1790), although Johnson (1973) surmised that it was introduced to that continent by the Portuguese much earlier.

Cashew is believed to have originated in southeastern Brazil. "Acaju" is the name given to cashew by the native Tupi Indians of Brazil. According to Johnson (1973) this became "Caju" in Portuguese, cashew in English, etc. Most of the names used in India and Southeast Asia are derived from the Portuguese "Caju" for cashew. This also serves as a solid piece of evidence that the cashew originated in Brazil.

The cashew growing regions of the world have been separated into five climatic types (Johnson, 1973) based on "Koppen" climatic types. The largest areas among these, the "A" type, include the east coast of India, Indonesia, East Africa (Tanzania and Mozambique), Northwest Malagasy, West Africa, Caribbean islands, and Central America. The cashew areas are concentrated mostly in the coastal regions in Malaysia (Ridley, 1930), Sri Lanka (Paul and Canagarathnam, 1937), India (Rao, 1958a), Mozambique (Ascenso, 1970), and Venezuela (Araque, 1968), although cashew is found in inland regions in limited areas up to elevations of 1000–1200 m (Martins, 1965; Argles, 1969). In India, cashew is grown mainly in the coastal regions of southern India and to a limited extent in West Bengal and Assam.

Argles (1969) discussed the wide range of ecological factors under which cashew had become naturalized. In areas where cashew is found rainfall varies from 500 mm in Angola, Colombia, and Northern Venezuela, to 3500 mm on the west coast of India (Naik, 1949; Argles, 1969; Johnson, 1973). Though cashew trees are drought resistant under conditions of limited rainfall, closely spaced trees with a closed canopy may undergo severe moisture stress compared to widely spaced and clean-weeded trees.

Cashew is generally grown in sandy, lateritic, and rocky soils considered to be unsuitable for most other crops, with a preference for loams and sandy loams (Argles, 1969; Johnson, 1973). Free drainage and absence of brackish conditions are essential for cashew; it grows poorly in heavy clay (Morton, 1961; Argles, 1969).

The cashew plant is extremely sensitive to frost and excessively hot, dry weather, as found in northern parts of India (Argles, 1969). However, it is being cultivated up to 27°N (southern Florida) and 28°S latitude (South Africa) (Joubert and Thomas, 1965).

II. BOTANICAL FEATURES

Cashew belongs to the family Anacardiaceae, which includes many economically important tropical and subtropical trees and shrubs characterized by resinous bark and alternate leaves. *Anacardium occidentale* L., the cultivated cashew, is a low spreading evergreen tree reaching a height of 10–15 m. The tree is usually branched near the base, giving a spreading and procumbent appearance. It has a large tap root and an extensive network of lateral and sinker roots. The leaves are alternate, simple, smooth, leathery, obovate, rounded, and notched at the apex, tapering at the base, and crowded at the ends of the branches.

The inflorescence is a terminal panicle more or less conical in shape, with secondary branches appearing in racemes. The ultimate branchlets of the panicles, however, are cymes, the terminal buds opening earlier than lateral buds. The secondary and tertiary branches are subtended by bracts. The average length of panicle varies from 14 to 21 cm and the number of flowers per panicle varies from 200 to 1600 (Damodaran *et al.,* 1966). According to Dasarathi (1958), cymose branching of the panicle continues until the sixth or even the seventh order.

Cashew flowers are small, regular, pentamerous, white or light green at the time of opening, later turning to pink. Flowers are either bisexual or male, and both occur intermixed in the same inflorescence. The structure of the male and bisexual flowers is similar except that the pistil is either rudimentary or absent in the male. Rarely, female and neutral flowers have also been reported.

The androecium consists of one fully developed stamen and 7–9 staminodes (Damodaran *et al.,* 1966). The filament of the well-developed anther in the hermaphrodite flower is only half as long as that of the staminate flower, and the level of its anther is much lower than that of the level of stigma (Damodaran *et al.,* 1966). According to Northwood (1966), the male flowers usually have one large exerted stamen and 5–9 small inserted ones; most of the small stamens also produce pollen. In hermaphrodite flowers, the pistil normally is longer than the large stamen, but occasionally it is shorter or about the same size.

The ovary is superior, reniform, and monocarpellate. The cashew fruit is a kidney shaped nut 3–5 cm long and 2.0–3.5 cm wide, weighing about 3–20 gm.

III. GROWTH IN RELATION TO ECOLOGICAL FACTORS

Harmony between climatic and vegetative rhythms of crops is very rarely obtained. During certain "critical" stages of development, the plants either make more definite or exacting demands on environmental factors or are more easily affected adversely by these factors at suboptimal or supraoptimal levels. This is particularly so with perennial crops such as cashew in which distinct growth patterns exist during the prebearing or juvenile stage and the bearing stage. Also, in spite of the capacity of cashew plants to tolerate a wide range of climatic and edaphic conditions, plant responses to variations in climatic and other environmental factors differ markedly during specific critical growth stages such as germination, vegetative growth, flowering, and fruit set, and development.

This section briefly summarizes growth characteristics of cashew as a prelude to a discussion of the functioning of the plant system in relation to environmental factors.

Marked variations among progenies of individual trees occur in shape, size, and weight of nuts, and shape, size, and color of apple. This is to be expected in a heterozygous, cross-pollinated, and seed-propagated population. Maturity and selection of seeds, storage and planting methods, and their effects on germination and yield potential have been investigated. Studies in India showed that medium-sized nuts germinated better than heavier or lighter nuts (Rao and

Hassan, 1956; Shetty and Bhatkal, 1965; Rao, 1974). Argles (1969) recommended consideration of specific gravity in selection of nuts. Seeds with higher specific gravity germinate more rapidly, show greater viability, and produce more vigorous seedlings. These produce a higher percentage of inflorescences in the first season of flowering (Turner, 1956). Northwood (1967) observed that seedlings from the higher density group grew better and had higher yields during the first three years of harvest, although the differences between these and low-density groups disappeared by the fourth year. Studies in India have indicated that there is no appreciable difference in germination of nuts collected at different periods of the season (Rao and Rao, 1957). The observation that the nut attains its full growth and maximum specific gravity by the fortieth day after pollination when the apple just begins to increase in size (Dasarathi, 1958) suggests that the nut is physiologically mature and capable of producing a normal plant when the apple has hardly started to develop.

Seeds stored in airtight containers retain their viability up to 11–12 months and still give 95–98% germination (Rao *et al.*, 1957). Rocchetti and Panerai (1970) found that seeds can be stored under dry airtight conditions up to 2 years without affecting viability.

Depth and position of sowing the seeds were also studied in India (Aiyadurai, 1966). Better germination and straight stemmed seedlings were obtained when seeds were sown slanted at an angle of 45° at a depth of 5–8 cm with the hilum upward. As depth of planting increases, germination is delayed as well as decreased (Rao *et al.*, 1957; Rao, 1974).

Rocchetti and Panerai (1970) demonstrated that temperature greatly affected germination of cashew nuts. Their studies with temperatures ranging from 10° to 40°C showed that the highest percentage of germination occurred at 35°C. The percentage of germination declined appreciably at 30° and 25°C and was highest at 20° and 40°C. The speed of germination was also very rapid at 35°C. These observations suggest that there is a critical temperature range for obtaining maximum percentage and rate of germination of seeds, and different germination percentages reported from the center may be due to the sharp fluctuations in temperature (Argles, 1969).

Various methods of vegetative propagation of cashew such as grafting, layering, and use of cuttings are possible. Air layering was attempted early in India and Tanzania and has been successful under the humid conditions of the west coast of India (Rao, 1958b). Seasonal variations in the percentage of successful establishment of layers under different climatic conditions in India were reported by Aiyadurai (1966) and Rao (1958b). Maximum success in air layering in Ullal, India was obtained during the hot weather period when the trees were in full flush. But in Tanzania the rainy season was best for air layering when the trees were not in flower (Northwood, 1964). However, in both cases best rooting was obtained when the trees were in the peak period of the vegetative growth phase. This important observation led Argles (1969) to suggest that the vegetative growth phase, and not rainfall or presence or absence of flowers and fruits, determines the success of air layering. Observations in India (Aiyadurai, 1966) that nonflowering shoots had significantly higher rooting percentages than flowering shoots also tend to confirm this hypothesis. The available information also indicates that vegetative propagation is most successful when temperature ranges are reasonably high and moisture stress is not severe. At cashew research stations, Ullal and Vengurla (India), trees showing complete failure of root formation in air layers have been identified (D. G. Bandekar, personal communication, 1972). Tree to tree variation in root-forming ability also has been reported in Tanzania (Northwood, 1964). While the variation between trees could be due to fluctuations in climatic factors such as temperature and moisture, differences maintained consistently throughout the year suggest inherent variation in root-forming ability. Inarching was found feasible, with 40–75% success in India. The best results with approach grafting were achieved when the plants were growing, and the least success when the trees were dormant or just beginning growth flushes (Rao and Rao, 1957). Side grafting is also successful under conditions in India. Sixty to seventy percent success in side grafting was obtained at Vengurla on the west coast of India during September–October. The results were almost similar at Bhubaneswar on the east coast. Side grafting has been successful with plants of up to 3–4 years old. Veneer grafting was equally successful at Vengurla during July–August on younger seedlings.

Shield budding has been the most efficient and economic method of grafting in the Philippines, Malaysia (Argles, 1969), and Mozambique (Albergaria, 1968). Although this method has been found to be successful in India also, intensive studies for economic exploitation of this method were initiated only very recently. The best results were obtained at the cashew research stations at Vengurla and Vridhachalam (India) during August–October.

Thus it appears that only air layering and approach grafting have been subjected to any detailed investigation. Approach grafting is highly desirable and has many advantages over air layering, since it gives a rooting system comparable to that in seedlings, at the same time maintaining genetic purity of the mother tree through the scion. However, approach grafting is more expensive and cumbersome than air layering. With all its limitations, air layering is still regarded as the most satisfactory method of vegetative propagation of cashew.

Side grafting, veneer grafting (Fig. 1), and budding are three potentially economic methods of propagation. The fact that they can be used in the field makes them all the more important. Economic exploitation of these methods, therefore, deserves highest priority. Whatever the efficacy of the method of propagation, an efficient method of clonal propagation useful on a commercial scale is yet to be standardized. Large scale orchards are even being raised from elite seeds alone.

Fig. 1 Veneer grafts of cashew.

Difficulties experienced in transplanting seedlings have been attributed to the inability of the tap root to recover from injury (Argles, 1969). More difficult is the problem of establishing approach-grafted or layered plants after they are separated from the parent trees (Alvim, 1973). Investigations in Jamaica (Tai and Topper, 1947) and India (Hassan and Rao, 1957) showed that transplanting losses could be reduced by lifting seedlings when seeds have just germinated or by cutting one-half of two-thirds of the top of the older plant, but not by severing the tap root several weeks in advance of transplanting or by halving the leaves. Tap root dominance may not play such an important part in cashew as generally assumed (Argles, 1969). Tsakiris and Northwood (1967) showed that tap root dominance was relatively short lived. Argles (1969) interpreted the above data in terms of fluctuation in top–root ratio (ratio of the transpiring top to absorbing root) and explained how heading back, but not tap root severance, resulted in higher survival. The layered plants also do not endure transplanting well. The usual practice is to harden them for a few days after separation from the parent trees. Observations in India show that precuring the layered twig by removing all its leaves about a week before the separation fo the twig and transplanting immediately after separation from the parent tree improves plant establishment. Precuring probably helps to activate the growing bud. The observation conforms to the view expressed by Argles (1969) that the top–root ratio is at its lowest point and the trees are in the best condition to survive transplanting when the terminal buds are just starting to swell preparatory to making a new growth flush.

IV. REPRODUCTIVE GROWTH

In a bearing cashew tree, two or three peak periods of growth are usually observed, even though under favorable conditions of soil moisture and other environmental factors, as in parts of Tanzania, stray shoot growth may occur almost every month (Argles, 1969). The pattern of growth of a bearing tree, best characterized as one of indeterminate flushing, consists of a generative flower flush and a vegetative flush. The vegetative flush, consisting of lateral shoots, always develops soon after the main fruit crop has matured. Flowering is terminal and is universally preceded by the vegetative flush. Rao (1956) recognized two major flowering flushes in cashew, one in November and the other in May. Dasarathi (1958) observed two main flushes in December and May and another minor flush in September resulting in off-season flowering in some trees. Galang and Lazo (1936) also recognized three corresponding flushes in November, May, and July in the Philippines.

Correlation between growth and fruiting has been long known (Naik, 1949; Rao and Crane, 1956; Mallik, 1957). Fruiting in any crop is the ultimate end product of a long complicated series of physiological and growth features preceding flowering. In both mango and cashew, Galang and Lazo (1936) stressed the significance of growth features for an understanding of bearing tendencies. They found that productivity was associated with leaf area and internode length. Dasarathi (1958) reported that trees with excessive vegetative growth and long internodes bore less than those with slow or medium vegetative growth, and concluded that high yield was associated with medium internode and moderate vegetative growth (Rao and Crane, 1956). Dasarathi (1958) attempted to trace the growth of shoots from the vegetative phase in two consecutive flowering seasons and correlated the rate of growth with productivity. Cashew flowers in February on current shoots which commence growing in December under east coast conditions in India. Normally a flower shoot produces two lateral shoots in June–July after fruiting, and they flower in February after the growth in December. The minor flushes of September result in off-season flowering in November. After growth, the shoots produce up to four laterals and about 30% of these flower in the main season. Thus, in cashew some twigs support two flowering periods in a year without any rest, whereas in mango, a shoot rarely bears more than one crop annually. Given proper care, cashew always is a regular bearer unlike mango, which is either biennial or irregular in bearing habit (Naik and Mohan Rao, 1943).

Dasarathi (1958) recognized two types of branching in cashew, intensive and extensive (Figs. 2 and 3). The intensive shoot grows to a length of about 25–30 cm and terminates in a panicle. Simultaneously, three to eight laterals arise within 10–15 cm of the apex. Some of these lateral shoots may also terminate in panicles in the same flowering season. These flowering laterals act as leaders in the following season, repeating the growth process described above, giving a well-covered bushy appearance to the tree. The extensive shoot grows to a length of 20–30 cm and rests. Further growth is a bud sprouting 5–8 cm below the apex. This process of growth continues 2–3 years without any flowering. This pattern of growth produces a spreading tree. Both types of branching are observed in all trees, but one type dominates a tree. High yielding trees have more than 60% intensive branches whereas low yielders have less than 20%. The intensive type of branching has a distinct advantage in that the number of laterals produced and percentage of laterals that flower are greater in the extensive type. In the intensive type of branching up to 75% of the shoots flower but in the extensive type only 12% flower.

The desirability of having erect growth in cashew has been emphasized by Morton (1970). Trees exhibiting the usual uncontrolled sprawling growth have nothing but a tangled mass of base and dead branches. This is very similar to the extensive type of branching described by Dasarathi (1958).

If growth of cashew is unchecked it develops into a low spreading plant. Regular pruning is used in Tanzania, Mozambique, Venezuela, and Colombia to induce plants to grow into acceptable shapes (Mutter and Bigger, 1962; Goldson, 1966; Northwood, 1966; Albergaria, 1966; Araque, 1968). Rao (1958a) and Dasarathi (1958), however, do not agree with this view and found that, after pruning, the tree loses potential flowering branches with consequent adverse effects on yield.

Cashew is extremely sensitive to light and produces more foliage, flowers, and fruit on exposed than on shaded bunches (Morton, 1970). Close planting may result in overlapping of branches of adjacent trees; shading decreases stem diameter and in extreme cases results in dead twigs. According to Argles (1969),

Fig. 2 Intensive branching of cashew.

closely planted trees may undergo moisture stress whereas widely spaced trees flourish with an annual rainfall of 75–90 cm. In the light of the limited information available, it is difficult to determine whether productivity is actually influenced by moisture stress or by shading.

Flowering in cashew is profuse. About 85–90% of the shoots of a bearing tree flower every year (Dasarathi, 1958; Rao, 1956). The number of flowers per panicle may vary from 200 to 1600 and the proportion of staminate flowers from 75 to 90% (Damodaran *et al.*, 1966; Rao and Hassan, 1957). According to Dasarathi (1958), the maximum number of perfect flowers per panicle is found in the middle of the flowering season, with a gradual reduction toward the

Fig. 3 Extensive branching of cashew.

beginning and end of the season. Northwood (1966) reported a ratio of 3.7:1.0 male to hermaphrodite flowers in Tanzania. The above observations indicate that the proportion of male flowers to bisexual flowers varies considerably depending on season, location, and even tree characteristics. The flower opens in Tanzania between 8:30 AM and 2:00 PM, with a peak opening between 10:00 AM and 12:00 noon (Northwood, 1966). Under Bapatla conditions (east coast of India) anthesis starts as early as 1:00 AM. The peak is attained between 8:00 AM and 12:00 noon (Dasarathi, 1958). The stigma is receptive as soon as the flower is open, and receptivity lasts for about 2 days. Anthesis starts 1–5 hr after the opening of the flowers. Cloudy weather retarded the opening of flowers on the west as well as east coasts of India (Rao, 1956; Dasarathi, 1958). Temperature was considered to be more important than humidity in controlling flower opening (Dasarathi, 1958).

As mentioned, cashew flowering is delayed as one moves from the equator and to high altitudes. Conceivably the delay in flowering is due to low temperature. Despite the absence of sufficient experimental evidence, low temperature probably retards bud break in cashew.

Another interesting observation made by Dasarathi (1958) is that "the period of minor flush generally coincides with the interval between the monsoons. During monsoon the plants remain dormant till December when intensive growth starts again to terminate in flowering." In other cashew growing areas in India, if the temperature range is favorable, intensive growth starts immediately after the heavy rains. The bud remains dormant during the rainy season and becomes active after the rains. Alvim (1960) demonstrated that moisture stress was a prerequisite for bud break and that dormancy could be prolonged indefinitely provided the plants were irrigated frequently. High plant water potential (low water stress) is necessary for the opening of flowers, but it has no effect if the plants were not previously submitted to moisture stress. The response of cashew is slightly different from the "hydroperiodic" response reported for other plants (Alvim, 1960). The growth flush and flowering in cashew start after the rains when there is a gradual decline in relative humidity. Bright sunshine hastens the process. A favorable temperature range, low relative humidity, and bright sunshine probably provide the optimal conditions required for reduction in growth inhibiting factors, resulting in bud growth. Water potential of the plant never appears to be a problem in cashew for further development of the bud.

A. Pollination

The extent of self-pollination in cashew has not been determined with accuracy. The style is longer than the filament and consequently the stigma is at a higher level than the anthers, showing that cross-pollination is favored. The occurrence of strong scented flowers and sticky pollen not only support this view but also emphasize the importance of insects over wind as pollinating agents (Rao, 1956; Dasarathi, 1958; Northwood, 1966). In addition, Northwood (1966) reported that when the inflorescences were bagged, they failed to produce nuts in the absence of hand pollination. Artificial pollination increases fruit set (Northwood, 1966; Rao and Hassan, 1957). Under natural conditions, pollination is not very efficient and only 3% set was observed on the west coast of India (Rao, 1956). Under east coast conditions, Dasarathi (1958) noted an average fruit set of 6–12%. By hand pollination he increased fruit set to a maximum of 80%, with an average of 40%. Failure of pollination was, therefore, one of the reasons for low fruit set in cashew under conditions in India. On the other hand, Northwood (1966) observed that in Tanzania flower production, pollination, and fruit set were efficient, and those factors did not limit the yield. A close ratio between style and stamen was helpful in obtaining good set in mango (Naik and Mohan Rao, 1943) and cashew (Rao, 1956). Northwood (1966) reported that the pistil is normally longer than the large stamen, but occasionally it is shorter or about the same size. The fact that sometimes dehiscence of anthers starts even before flowers open and the stigma is receptive (Dasarathi, 1958) is of great significance. If such a pattern of development occurs in flowers having stigmas equal or nearly equal in size to styles (Northwood, 1966), it might help in self-pollination. It would also help explain why pollination is not a limiting factor for fruit set in Tanzania.

The time span from fruit set to maturity is about 60 days. The true fruit or nut completes its growth in about 40 days after the apple just starts increasing in size, and another 10–20 days are required to reach maturity. However, at high temperatures, the number of days from pollination to fruit set decreases.

B. Flowering and Latitude

Variations in the flowering season of cashew in different countries are related to latitude (Fig. 4). The flowering season is from June to November in Tanzania, with a peak in August–September (Northwood, 1966). The peak flowering period is August–September in Brazil (Popenoe, 1924), October in Mozambique, and March in the Philippines (Galang and Lazo, 1936). Seasonal variations in flowering also occur in different parts of India. At Anakkayam and Mangalore on the west coast, 13°N of the equator, the peak flowering is in early January and the peak harvest in early April (Rao, 1956). At Bapatla, 17°N latitude on the east coast, peak flowering occurs between January 15 and February 15 and the crop is harvested by the end of April (Dasarathi, 1958), whereas further north in Berhampur harvest is prolonged until the end of May. There is a gradual delay in flowering as one moves north or south from the equator. It is interesting to note that the crop is ready for harvest in the summer both above and below the equator. Dasarathi (1958) also observed a strong influence of temperature on growth, flowering, and fruiting of cashew. The isochrones drawn from the data collected under the All India Phenological Scheme on mango, neem, tamarind, and babul on various photophases show similar variation in the flowering periods as noted by Randhawa (1957). He stated that for each degree of change in latitude northward there was a shift of approximately four days in each photophase. The relationship between latitude and flowering probably reflects the effect of latitude on temperature. Away from the equator the minimum temperature in the winter is low for a longer time. The minimum temperature main-

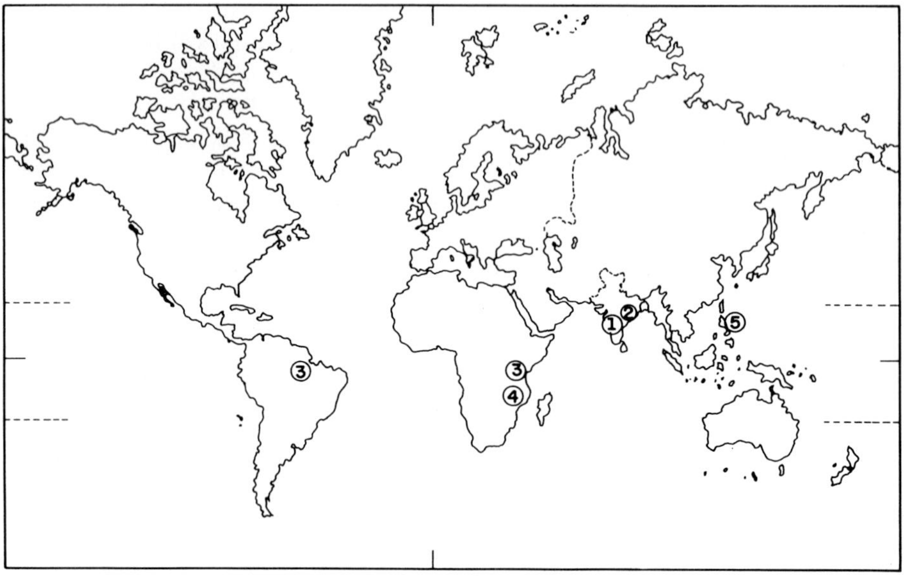

Fig. 4 Peak flowering periods of cashew in different parts of the world. (1) India–west coast (January); (2) India–east coast (January–February); (3) Tanzania and Brazil (August–September); (4) Mozambique (October); (5) Philippines (March).

tained during this period probably is well below the critical temperature required for bud break. The observation that cashew trees at higher altitudes, irrespective of latitude, show almost the same tendency for late flowering and fruiting tends to support the above views. It is well known that night temperature, humidity, and daylength affect floral initiation in both annuals and perennials.

V. CROP IMPROVEMENT

The aims of crop improvement programs are (a) to build plants of desirable genetic potential and ideal plant type and (b) to create the necessary environment for growth in order to exploit the genetic potential. The history of geographical distribution shows that variability in the cultivated cashew has been very low. The original introduction probably consisted of nuts from a few trees and thus had a very limited genetic base from which all the presently cultivated cashews in India evolved. This may be true for most other cashew growing areas except those of South America. Hence, there is an urgent need to increase genetic variability of the breeding materials in cashew by introducing additional collections from its original home. The results of such introduction could be the direct utilization of new varieties, parent materials in hybrids, or genetic stocks resistant to disease and pests. It is well established that even if introductions are made from similar ecophysiological regions, their direct adoption in new countries is rather limited because of failure to compete with local germplasm. But the aim of introduced germplasm of cashew should be its use in a hybridization program to evolve locally adapted, high yielding, and disease/pest resistant types.

The success of any breeding program depends on the selection procedure adopted. The characteristics selected should be present in high frequency and should have high heritability. The heritability of desirable characteristics, such as medium length of internode, intensive branching, intensive flowering, high proportion of bisexual flowers, short flowering phase, nut size, and shelling percentage, and correlation of these and other morphological characteristics with yield must be estimated. The possibilities of both vegetative and seed propagation of cashew are of considerable advantage in estimating the additive and nonadditive components of heritability and interaction of genotype with environment. These genetic parameters could be estimated rapidly in cashew as compared to plants propagated by seeds alone. Correlating seedling characteristics with subsequent yield will be an added advantage in selection. After establishing the selection criteria along the above lines, a short-term approach would be to make a complete survey of the individual plantations within and outside the country to identify the best hybrids. Rapid multiplication of these natural hybrids should make it possible to replace the existing plants with superior genotypes.

Large scale genotypes can be synthesized through hybridization of new introductions from the center of origin of these collections. Wherever possible, it would be desirable to develop inbred lines, either from the original introductions or from synthesized genotypes, for their specific combining ability. Inbred lines with good combining ability could be used to synthesize desirable hybrids and to exploit hybrid vigor. Clonal propagation of cashew is an advantage since heterosis can be fixed and the need for continuous seed production for fresh planting can be obviated.

Cashew is a unique crop, wherein the advantages of both seed and vegetative propagations can be exploited. While production of hybrid strains with favorable gene interaction can be maintained through vegetative reproduction, composite or synthetic varieties could easily be developed with seeds. For development of synthetics, different genotypes can be tested for general combining ability and good combiners can be used to form the base population. The parental lines of the synthetics can be maintained through vegetative propagation and, wherever deterioration sets in the original strain, it can be reestablished through a fresh cycle. Mass pedigree or recurrent selection can be used to improve the synthetics.

Yield is a complex characteristic influenced by physiological and biochemical processes in plants, morphological architecture of plants and its interaction with environmental factors, and resistance to diseases and pests. Most plant breeding programs follow two approaches to develop varieties with high yield potential which Donald (1968) termed "selection for yield" and "defect elimination." In addition, a new approach involves breeding of model plants or ideotypes with a view to tailoring new varieties of cashew for different environments.

Comparatively little work has been done in forest and other perennial cultivated plants on the influence of canopy structure and leaf geometry on growth and productivity because of the difficulty in obtaining phytometric measurements. Critical comparison of photosynthesis in cashew plants with different canopies is desirable to determine the optimum plant canopy for high yield. The current practice of allowing unlimited canopy growth will have to be modified accordingly.

The available evidence indicates that cashew ideotypes should have intensive branching with medium internodes, more flowering laterals, a short flowering phase, high percentage of bisexual flowers, high fruit set, large nuts with high shelling percentage, and kernels with high protein content. Cashew breeders should endeavor to synthesize such varieties with high adaptability to varying ecophysiological conditions.

REFERENCES

Aiyadurai, S. G. (1966). "A Review of Research on Spices and Cashewnut in India." Indian Counc. Agric. Res., New Delhi.

Albergaria, M. H. S. (1966). O problema da producão do caju. *Gaz. Agric.* **18,** 106–108

Albergaia, M. H. S. (1968). Enxertia nos cajueiros. *Gaz. Agric.* **20,** 172–174.

Alvim, P. de T. (1960). Physiology of growth and flowering in coffee. *Coffee (Turrialba)* **2,** 57–62.

Alvim, P. de T. (1973). Factors affecting flowering in coffee. *J. Plantation Crops* **1,** 37–43.

Araque, R. (1968). El merey. *Ser. Cult. Ven.* **15,** 1–20.

Argles, G. K. (1969). *Anacardium occidentale* (Cashew): Ecology and botany in relation to propagation. *Pap., FAO Conf. Prop. Trop. Sub-trop. Fruits,* pp. 1–24.

Ascenso, J. C. (1970). Patrimonio fruteiro de Moçambique. *Abric. Mocamb.* **4,** 221–226.

Billings, W. D. (1952). The environmental complex in relation to plant growth and distribution. *Q. Rev. Biol.* **27,** 251–265.

Billings, W. D. (1957). Physiological ecology. *Annu. Rev. Plant Physiol.* **8,** 375–392.

Burkill, I. H. (1935). "A Dictionary of the Economic Products of the Malay Penninsula," Vol. I. Crown Agents for the Colonies, London.

da Costa, C. (1578). "Tractado de las drogas y medicinas de las Indias Orientales" (Lisbon, 1964).

Damodaran, V. K., Abraham, J., and Alexander, K. M. (1966). The morphology and biology of the cashew flower *Anacardium occidentale* L. II. Anthesis, dehiscence, receptivity of stigma, pollination, fruit set and fruit development. *Agric. Res. J. Kerala* **4,** 78–84.

Dasarathi, T. B. (1958). A study of the blossom biology and growth features of cashew-nut–*Anacardium occidentale* Linn. M.Sc. Thesis, Andhra University, Waltair (unpublished).

de Loureiro, J. (1793). "*Flora Cochinchinensis.*" Berlin.

Donald, C. M. (1968). The breeding of crop ideotypes. *Euphytica* **17,** 385–403.

Galang, F. G., and Lazo, F. D. (1936). Fruiting as related to vegetative growth in cashew, *Anacardium occidentale* L. *Philipp. J. Agric.* **7,** 21–33.

Goldson, J. R. (1966). "Cashews." Train. Cent. Improv. Hort. Tech., FAO, Nairobi.

Hassan, M. V., and Rao, Y. N. M. (1957). Studies on the transplantation of seedlings of cashew (*Anacardium occidentale* Linn.). *Indian J. Agric. Sci.* **27,** 177–184.

Johnson, D. (1973). The botany, origin and spread of the cashew, *Anacardium occidentale* L. *J. Plantation Crops* **1,** 1–7.

Joubert, A. J., and Thomas, D. des S. (1965). The cashew nut. *Farming S. Afr.* **40,** 6–7.

Mallik, P. C. (1957). Studies on floral biology of *Momordica charantia. Indian J. Hortic.* **14,** 42–47.

Martins, J. D. de B. (1965). O cajueiro. *Gaz. Agric. Ang.* **10,** 32–36.

Morton, J. F. (1961). The cashew's brighter future. *Econ. Bot.* **15,** 57–58.

Morton, J. F. (1970). Prospects for cashew advancement in Colombia and Venezuela. *Proc. Fla. State Hortic. Soc.* **83,** 405–410.

Mutter, N. E. S., and Bigger, M. (1962). Cashew. *Tanganyika, Min. Agric. Bull.* No. II.

Naik, K. C. (1949). "South Indian Fruits and Their Culture." P. Varadachary & Co., Madras.

Naik, K. C., and Mohan Rao, M. (1943). Studies in blossom biology and pollination in mangoes. *Indian J. Hortic.* **1,** 107–119.

Northwood, P. J. (1964). Vegetative propagation of cashew (*Anacardium occidentale* L.) by the air-layering method. *East Afr. Agric. For. J.* **30,** 35–37.

Northwood, P. J. (1966). Some observations on flowering and fruit setting in the cashew, *Anacardium occidentale* L. *Trop. Agric. (Trinidad)* **43,** 35–42.

Northwood, P. J. (1967). The effect of specific gravity of seed and the growth and yield of cashew (*Anacardium occidentale* L.). *East Afr. Agric. For. J.* **33,** 159–162.

Odum, E. P. (1954). "Fundamentals of Ecology," 2nd ed. Saunders, Philadelphia, Pennsylvania.

Paul, W. R. C., and Canagarathnam, C. (1937). The production of cashew nuts in Mannar. *Trop. Agric. (Trinidad)* **88,** 6–11.

Popenoe, W. (1924). Economic fruit-bearing plants of Ecuador. *Contrib. U.S. Natl. Herb.* **24,** Part 5.

Randhawa, M. S. (1957). "Flowering Trees in India." Indian Counc. Agric. Res., New Delhi.

Rao, S. N., and Crane, J. C. (1956). Relationship of leaf area and internodal length to fruiting habit in Fig. *Indian J. Hortic.* **12,** 57–63.

Rao, V. N. M. (1956). Multiply better-yielding cashewnut. *Indian Farming* **6,** 15–17.

Rao, V. N. M. (1958a). "Cashewnut Cultivation in India," Farm Bull. No. 9. Indian Counc. Agric. Res., New Delhi.

Rao, V. N. M. (1958b). Studies on the vegetative propagation of cashew (*Anacardium occidentale* L.). I. Air-layering by cincturing and etiolation of shoots. *Indian J. Agric. Sci.* **28,** 181–197.

Rao, V. N. M. (1974). "Crop Improvement of Cashew," mimeo., pp. 128–134. Report of the All India Summer Institute on Improvement and Management of Plantation Crops.

Rao, V. N. M., and Hassan, M. V. (1956). A note on the variations in seed characters of cashew. (*Anacardium occidentale* L.). *Indian J. Agric. Sci.* **26**, 211–216.

Rao, V. N. M., and Hassan, M. V. (1957). Preliminary studies on the floral biology of cashew (*Anacardium occidentale* L.). *Indian J. Agric. Sci.* **27**, 277–288.

Rao, V. N. M., and Rao, I. K. S. (1957). Studies on the vegetative propagation of cashew (*Anacardium occidentale* L.) approach grafting (inarching) with and without the aid of plastic film wrappers. *Indian J. Agric. Sci.* **27**, 267–275.

Rao, V. N. M., Rao, I. K. S., and Hassan, M. V. (1957). Studies on seed viability in cashew. *Indian J. Agric. Sci.* **27**, 289–294.

Ridley, H. N. (1930). "The Dispersal of Plants Throughout the World." L. Reve & Co. Ltd., Ashford, Kent, U.K.

Rocchetti, G., and Panerai, L. (1970). Ulteriori indagini sulla germinazione delle castagne di anacardio. *Rast* **64**, 151–160.

Rumphius, G. E. (1962). "*Herbarium amboinense.*" Uitgevery en Drukkerij Hollandia N.V., Baarn.

Shetty, K. T., and Bhatkal, R. (1965). Cashew. Note on raising seedlings and planting. *Lal Baugh* **10**, 34–35.

Tai, E. A., and Topper, B. E., (1947). Transplanting cashew seedlings. *J. Jam. Agric. Soc.* **51**, 116–118.

Tsakiris, A., and Northwood, P. J. (1967). Cashew nut production in southern Tanzania. IV. The root system of the cashew nut tree. *East Afr. Agric. For. J.* **33**, 83–87.

Turner, D. J. (1956). Some observations on the germination and grading of cashew nuts. *East Afr. Agric. For. J.* **22**, 35–39.

van Linschoten, J. H. (1598). "The Voyage of John H. van Linschoten to the East Indies" (London, 1885).

CHAPTER

18

Mango

L. B. SINGH

I. INTRODUCTION

Mango (*Mangifera indica* L.), which has been grown in India for the last 6000 years (Hill, 1952) is an important commercial crop of several countries, particularly those of Southeast Asia: the Philippines, Indonesia, Java, Thailand, Burma, Malaya, and Ceylon. It is also quite popular in Egypt, Southeast Africa, South Africa, Hawaii, and the islands of the West Indies (Fig. 1). Considerable effort have been made to develop mango as a potential market crop in Brazil, the United States (Florida), Israel, tropical Australia, and Mexico. In India, mango occupies an area of 2.2 million acres out of a total acreage of 3.16 million acres of orchards.

Mukherjee (1972) pointed out the local origin on the mango from wild *M. indica,* without intervention of other species. The mango spread eastward out of

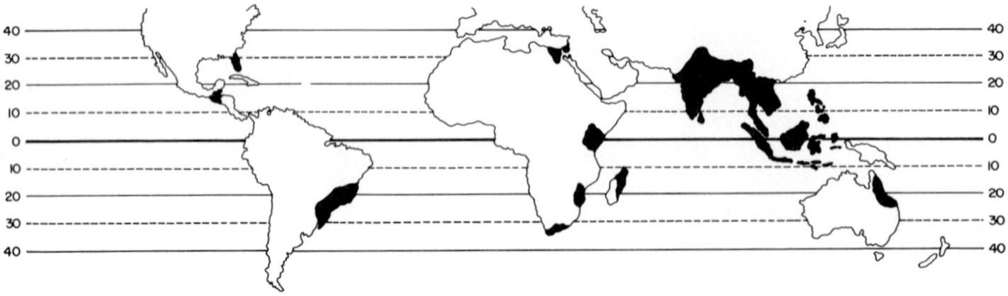

Fig. 1 Mango regions of the world.

India perhaps as early as 400–500 BC in the hands of Buddhist monks but did not become diffused through the islands and into the Pacific until much later. Mukherjee (1972) believes that it was moved westward as late as the sixteenth century by Portuguese travellers from Goa by way of East Africa to West Africa and the New World. Mangoes were first grown in the West Indies in the eighteenth century and reached Florida in 1833. The Portuguese were also the first to introduce mango in Brazil by the beginning of eighteenth century (Pope, 1929).

The biggest obstacle to the spread of mango in both hemispheres, except in India, has been the short-lived character of its seed, which was formerly the only agent known for its propagation and dissemination. With recent advances in vegetative propagation and means of transport, grafted varieties of mango are being introduced to all countries found suitable for cultivation of the species.

II. GROWTH CHARACTERISTICS

Mango belongs to the dicotyledonous family Anacardiaceae and is said to have descended from a single species, *Mangifera indica* L. Elongation of the long tap root seems to continue until it reaches the water table, and only a few anchoring branches are developed at that time. Mango roots reaching a depth of 18 feet have been recorded (Stephens, 1949). In Bihar (India) it has been reported that a 60-year-old Frazli tree had anchor roots which descended nearly 4 m and then became horizontal, but the finer roots were confined to the top 50 cm, with the majority in the top 25 cm on deeper and lighter soils. After the elongation phase, the surface roots begin to develop, and form a dense network just below ground level. At Saharanpur (India) a thin branch root from a 30-year-old seedling mango tree extended for 25 feet on one side of the tree.

The aerial part of the tree develops as an evergreen, erect, and branched structure, and the foliage forms a beautiful canopy. The top of the tree is usually oval, elongated, or dome shaped. The dark gray bark is thick, rough, with numerous small fissures, and foliating scales. Mango trees grown from seeds are mostly erect and robust, while grafted trees generally are dwarfed with spreading branches. Sometimes mango trees are very vigorous; a specimen with a spread of 120 ft and a girth of 20 ft has been reported in Brazil (Popenoe, 1920). Probably the largest mango tree, whose main trunk has been damaged, is near Chandigarh (India). Its circumference is 28 ft and it spreads 150 × 120 ft. Some varieties such as Machhali and Brindabani can also be trained as creepers.

Branches of mango are very numerous, the lower ones spreading horizontally to a great extent, the upper ones gradually ascending until they become nearly erect in the center. The branchlets are rather thick and robust, with alternating groups of long and short internodes. They are terete and glabrous yellowish green when young.

The leaves are simple, alternate, irregularly placed along the branchlets, sometimes remote and at other times crowded, long-petioled, oblong ovate to oblong lanceolate, base acute to cuneate, narrowed, apex acute to acuminate, entire often with wavy margins, and glabrous on both surfaces. The length and width vary from 5 to 15 inches and 1 to 5 inches, respectively, depending on variety and culture. The secondary veins are quite prominent, numerous, pinnately nerved, and distinctly reticulate veined.

New shoots arise mostly as laterals from axillary buds around the stump of the previous year's fruited twig. Such growth either remains unextended or continues to make extension growth in subsequent periods, depending on the variety. Terminal growth is always in the form of an extension of shoots already produced. Axillary adventitious buds may also produce new shoots. Mango shoots grow in periodic flushes with frequency of flushing mainly regulated by climatic factors. Flushing may also be a varietal characteristic. In India 2–5 flushes are reported in different climatic regions. Under conditions in southern India, where summers and winters are milder than in northern India, shoot growth occurs more or less continuously rather than in distinct flushes.

The mango is commonly subject to alternate or irregular bearing. Vegetative and floral growth are closely interrelated. Inflorescences are generally borne on shoots nearly one year old, and the shoots which bear panicles of flowers do not ordinarily produce vegetative growth until after harvest. However, if fruit set fails or fruits fall early, lateral shoots may be produced early enough to flower the following spring.

III. EFFECT OF ENVIRONMENT ON GROWTH AND YIELD

Mango tolerates a wide range of climate, and is particularly well adapted to tropical and subtropical climates. Temperature and rainfall within a clearly defined range are essential.

A. Temperature

The minimum endurance temperature lies between 34° and 36°F. In Florida young mango plants up to 5 years old were killed at a minimum temperature of 26.5°F (Popenoe, 1920). At Saharanpur (India), a temperature of 31° and 32°F killed 1-year-old plants whereas only the growing tips of 4-year-old plants were killed (Singh and Singh, 1955). Apart from the direct damage, frost injury even in adult trees is a deterrent to subsequent growth (Oppenheimer, 1947; Singh and Singh, 1955). Susceptibility of mango to frost varies also with the variety and the age and condition of the tree. Some varieties can withstand as much as 22°F frosts without damage, while others are severely injured by 27° or 28°F. The total amount of heat during the growing season is also very important and may be the limiting factor at high elevations. Oppenheimer (1947) suggested that about 1000 heat units were the minimum requirement for mangoes of high quality.

High temperature is not as injurious as low temperature but if accompanied by low humidity and high winds, it affects the trees adversely. In such conditions, due to high transpiration, there are quick fluctuations between day and night temperatures and the normal balance of the tree is upset. The optimum growth temperature for mango has been reported to be 75° to 80°F (Woodrow, 1910). Temperature affects directly the time of flowering in mango and the ripening period of particular varieties depends to some extent on flowering time. Since the flowering time of mango differs from country to country, and even from one locality to another in the same country, its ripening time also differs accordingly.

B. Rainfall, Clouds, and Dew

In India, mango grows in places where annual rainfall is only 10 inches, but it also grows in regions with a maximum rainfall of 100 inches (Singh, 1960). In areas of heavy rainfall the rate of vegetative growth is high but occurs at the expense of fruiting, and crop yield is low (Harris, 1901). In Rio de Janeiro and moist localities such as Castleton (Jamaica), mango does not set fruits, whereas in comparatively drier localities it yields a commercial crop (Popenoe, 1920). Similarly, in the humid tracts of Bengal, Assam, Kerala, and in southern–eastern Madras state, it grows well, but does not fruit equally well. In the Philippines also, commercial mango cultivation is not possible in areas of heavy rainfall. Rain at the time of flowering washes away the pollen. The mango flower is very delicate and easily injured in moist weather. During rainy days pollinating insects remain inactive and effective pollination cannot occur. After continued moist weather, heavy attacks of mango hoppers and certain fungi cause heavy shedding of flowers and fruits, and often result in crop failure. Clouds affect the crop by intercepting the radiation of heat from the soil and thus to a certain extent cause drop of flowers and fruitlets (Woodrow, 1910).

C. Altitude and Wind Velocity

For each 400 ft of altitude and each degree of latitude south or north of the tropics, flowering is delayed by four days (Hopkins, 1938). High velocity winds cause rapid evaporation of water from the soil and reduce the moisture which is necessary for optimum plant growth. Storms not only tear off limbs and branches but sometimes even uproot fairly large trees.

D. Water Relations

Climatic conditions may influence the time of the year when permanent wilting percentage of soil is reached. It was reached several weeks later in a cool season with occasional showers, than in a warm clear one (Veihmeyer and Hendrickson, 1950). There may be places where because of humidity, temperature, and rainfall conditions, loss of moisture through transpiration and evaporation may be so small throughout the year that the soil never reaches the permanent wilting percentage or does so for only a few days. This explains why the mango does not need irrigation in many parts of India.

Withholding water for two to three months preceding flowering induces vigorous flowering (Singh, 1960). Excess soil moisture after October may even retard formation of blossoms (Gandhi, 1955) or may cause them to be malformed (Watts, 1891). The former may be due to stimulation of vegetative growth during October–November, which interferes with flowering in February (Hayes, 1953). Fruit drop as well as size and quality of fruits of mango are believed to be influenced by irrigation at the appropriate time. During fruit development there is a need for irrigation of very poor soils in order to reduce fruit drop and hasten development. Further irrigation from the time the young fruit is set to the time the fruit begins to ripen greatly improves its size and quality. In addition, fruit drop is reduced, and large juicy fruits are favored by regular irrigation during April and May (Singh, 1960).

IV. HORMONAL CONTROL OF GROWTH

Inhibition of lateral bud growth in mango is due to an endogenous growth inhibitor and not to auxins. Decapitation resulted in the reduction of inhibition level and outgrowth of lateral buds. Growth substances, both promoters and inhibitors, are translocated from fruits to shoots, especially in the early stages (Singh and Singh, 1972).

A naturally occurring growth substance in shoots of mango exhibited biological properties similar to those of auxins (Elias *et al.*, 1972). It exhibited the same R_F value as IAA but chromogenic tests did not confirm its identity with IAA. An inhibitor extracted from mango shoots had many characteristics of abscisic acid and was present in large quantities in developing mango fruits.

Reece *et al.* (1946) concluded that flower-inducing hormones are transmitted from leaves to axillary buds of Haden mango. Singh and Singh (1956) hypothesized that hormonal substances were responsible for changing a leaf bud into a fruit bud. Reece *et al.* (1949) and Singh (1958) observed that the flower-inducing hormones can move proximally and distally at the same time and that leaves of young branches produce a substance which has an inhibitory effect on the flowering hormone. Sen (1962) and Holdsworth (1963) reported similar observations. Sen and Chaudhary (1969) studied changes in hormone levels in shoots of the biennial bearing variety Langra and perpetual bearing variety Baramasi of mango. They did not find differences in the trend of promotive and inhibitory substances by the wheat coleoptile straight growth test. However, there was greater inhibition in Langra than in Baramas, in December. An inhibitor was found in mango shoots by Chacko (1968), but its role in initiation of flowering and flower bud formation has yet to be assessed.

REFERENCES

Chacko, E. K. (1968). Studies on the physiology of flowering and fruit growth in mango (*Mangifera indica* L.). Ph.D. Thesis, Division of Horticulture, I.A.R.I. (referred from R. N. Singh, "Problem of Biennial Bearing in Mango").

Chacko, E. K., Singh, R. N., and Kachru, R. B. (1972). Studies on the physiology of flowering and fruit growth in mango (*Mangifera indica* L.). VII. Naturally occurring auxins and inhibitors in the shoots of flowering (on) and vegetative (off) mango trees. *Indian J. Hortic.* **29**, 115–125.

Gandhi, S. R. (1955). "The Mango in India," Farm. Bull. No. 6. Indian Counc. Agric. Res., New Delhi.

Harris, W. (1901). The mango. *Bull. Bot. Dep. Jam.* **8**, 161.

Hayes, W. B. (1953). "Fruit Growing in India." Allahabad Kitabistan.

Hill, A. F. (1952). "Economic Botany." McGraw-Hill, New York.

Holdsworth, M. (1963). Intermittent growth of mango trees. *J. West Afr. Sci. Assoc.* **7**, 163–171.

Hopkins, A. D. (1938). Bioclimatics. A science of life and climate relations. U.S. Dep. Agric., Publ. **280.**

Mukherjee, S. K. (1972). Origin of mango (*Mangifera indica* L.). *Econ. Bot.* **26**, 260–266.

Oppenheimer, C. (1947). The acclimatization of new tropical and sub-tropical fruit trees in Palestine. *Bull., Agric. Exp. Stn., Rehovot* **44**, I.

Pope, W. T. (1929). Mango culture in Hawaii. *Hawaii, Agric. Exp. Stn., Bull.* **58.**

Popenoe, W. (1920). "Manual of Tropical and Sub-tropical Fruits." Macmillan, New York.

Reece, P. C., Furr, J. R., and Cooper, W. C. (1946). The inhibitory effect of terminal bud on flower formation in the axillary buds of the Haden Mango (*Mangifera indica* L.). *Am. J. Bot.* **33**, 209.

Reece, P. C., Furr, J. R., and Cooper. W. C. (1949). Further studies of floral induction in the Haden mango (*M. indica* L.). *Am. J. Bot.* **36**, 209.

Sen, P. K. (1962). Flowering and bearing habit with special reference to the mango (*Mangifera indica* L.). *Indian J. Plant Physiol.* **5**, 264–271.

Sen, P. K., and Chaudhary, J. M. (1969). Studies of growth substances in the shoots of two varieties of mangoes–Langra (biennial) and Baramasi (perpetual) in relation to the growth and fruit bud formation. *Proc. Indian Sci. Congr.*

Singh, A. B., and Singh, R. N. (1956). Floral induction in axillary buds of mango shoots. *Proc. Am. Soc. Hortic. Sci.* **68**, 265.

Singh, L. B. (1958). Deblossoming in relation to biennial bearing in mango (*Mangifera indica* L.). *Hortic. Adv.* **3**, 50–75.

Singh, L. B. (1960). "The Mango." Leonard Hill, London.

Singh, L. B., and Singh, R. N. (1955). Frost injury to tropical and sub-tropical fruits of Saharanpur. *Indian J. Hortic.* **12**, 108.

Singh, R., and Singh, R. N. (1972). Lateral bud growth in *Mangifera indica* L. in relation to auxin and inhibitor content of shoots and fruits. *Symp. Mango & Mango Cult., Acta Hortic.* **24**, pp. 175–184.

Stephens, S. E. (1949). The mango (in Queensland). *Queensl. Agric. J.* **68**, 71, 146, and 208.

Veihmeyer, F. J., and Hendrickson, A. H. (1950). Response of fruit trees and vines to soil moisture. *Proc. Am. Soc. Hortic. Sci.* **55**, 11.

Watts, G. (1891). "A Dictionary of the Economic Products of India." Allen, London.

Woodrow, G. M. (1910). "Gardening in the Tropics." Alexander Gardener, London.

Index

D

E

T

U

V

W

X

Y

Z

A 7
B 8
C 9
D 0
E 1
F 2
G 3
H 4
I 5
J 6